I0073299

Mobile Working Machines

Mobile Working Machines

BY MARCUS GEIMER

Warrendale, Pennsylvania, USA

400 Commonwealth Drive
Warrendale, PA 15096-0001 USA
E-mail: CustomerService@sae.org
Phone: 877-606-7323 (inside USA and Canada)
724-776-4970 (outside USA)
FAX 724-776-0790

Library of Congress Catalog Number 2020936341
http://dx.doi.org/10.4271/9780768099546

ISBN-Print 978-0-7680-9432-9
ISBN-PDF 978-0-7680-9954-6

To purchase bulk quantities, please contact: SAE Customer Service

E-mail: CustomerService@sae.org
Phone: 877-606-7323 (inside USA and Canada)
724-776-4970 (outside USA)
Fax: 724-776-0790

Visit the SAE International Bookstore at books.sae.org

Chief Growth Officer
Frank Menchaca

Publisher
Sherry Dickinson Nigam

Director of Content Management
Kelli Zilko

Production and Manufacturing Associate
Erin Mendicino

contents

CHAPTER 3

Power Train 61

Swen Bosch, Prof. Danilo Engelmann, Dr.-Ing. Gerhard Geerling, Edwin Heemskerk, Dr.-Ing. Torsten Kohmäscher, Fabrizio Panizzolo, Dr.-Ing. Christian Pohlandt, and Prof. Dr.-Ing. Heinrich Steinhart

CHAPTER 4

Machine Control Concepts 211

Detlef Hawlitzek

used symbols

Symbols

a/A	Ampere	current
A	mm^2	area
A	kg	mass of trailer
B	-	differential blocking factor
b	mm	side gear contact face width
b_e	g/kWh	specific fuel consumption of a combustion engine
D	m	torque converter fluid annulus output diameter
D	kN	D-value
d_e	mm	side gear outer pitch diameter
f	Hz	frequency
f	mm	point of horizontal shift
f	mm	flange distance to which are attached the wheels
F	N	force
F_G	N	static tire load
F_U	N	wheel peripheral force
F_Z	N	net traction force
g	m/s^2	gravitational acceleration (9.81 m/s^2)
G	kg	maximum permissible tractor weight
H_u	kJ/kg	lower heating value
i	-	gear ratio
i_0	-	Standard ratio in planetary gear set
K	$\frac{rpm}{(Nm)^{\frac{1}{2}}}$	capacity factor (TC)
L	Henry	inductance
l	mm	length
l_0	mm	length of a rolling wheel without slip
$\dot{m}$	m^3/s	mass flow
M	Nm	momentum
n	rpm	rotational speed
n_T	rpm	rotational speed (TC)
n_P	rpm	rotational speed (TC)

o	mm	wheel (or rim) offset
p	N/mm^2	pressure
P	W	power
Q	L/min	volume flow
R	mm	average brake lining radius
R	Ohm	Ohmic resistance
r_0	mm	average side gear pitch radius
r_{dyn}	mm	dynamic wheel radius
R_{EXT}	mm	outer radius of brake or clutch lining
R_{INT}	mm	inner radius of brake or clutch lining
R_m	mm	mean clutch lining radius
s	-	number of sliding surfaces
s	-	slip
s	mm	spool position
S	-	slip (at induction machines)
T	Nm	torque
t	s	time
t	mm	wheel tread
T_{in}	Nm	input torque
T_{LEFT}	Nm	left differential output torque
T_{out}	Nm	output torque
T_P	s	period time
T_{RIGHT}	Nm	right differential output torque
T_{SW}	s	switching time
T_T	Nm	wheel torque
T_v	Nm	braking torque generated by limited slip differential system
v/V	Volt	voltage
v	m/s	speed
V, V_g	m^3	positive displacement of hydrostatic unit (pump or motor)
v_{DC}	Volt	DC-link voltage
v_L	Volt	load voltage
$v_{vehicle}$	m/s [mph]	vehicle speed
z	-	gear tooth number
z	-	Tooth count of mechanical gear

Greek Characters

μ	-	labeling of the phase in AC grids ($\mu = 1, 2, 3$)
μ	-	transmission conversion ratio
μ	-	wet brake lining slipping coefficient = 0.08 (for paper lining)
w	-	regenerative/not regenerative power flow direction sign
α	-	labeling of the normal axis in the αβ-frame
α_0	°	straight bevel gears pressure angle
α_D	-	flow coefficient
α_{Hybrid}	-	degree of hybridization
β	-	labeling of the orthogonal axis in the αβ-frame
γ	-	rotor angle
δ	°	differential bevel gears working angle
ϵ	$Nm \cdot \frac{s^2}{m^5}$	torque converter characteristics comparison function
σ	N/mm^2	stress
τ	-	transmission ratio
φ	-	phase angle
ψ	-	magnetic flux
ω	rad/s	angular speed
η	-	efficiency
η_{Diff}	-	differential efficiency value
κ	-	net traction ratio
ρ	-	rolling resistance ratio
ρ	kg/m^3	density
Δ	-	indication for delta value - e.g., pressure difference Δp

Indices

A	sun gear in planetary gear set
B	ring gear in planetary gear set
C	lock-up point (TC)
C	planet carrier in planetary gear set
d	labeling of the direct axis in the dq-frame
$Diff$	differential mechanism
e	engine

eff	effective value (in difference to the theoretical value)
F	direct drive point (TC)
GH	ground horizontal
GV	ground vertical
hm	hydro-mechanical
in	input
losses	transmission losses
M	design point at max efficiency (TC)
M	motor
max	maximum value
out	output
P	pump
PG	planetary gear set
Pl	planet gear in planetary gear set
PM	prime Mover
q	labeling of the quadrature axis in the dq-frame
R	reactor (TC)
R	rotor
ref	reference
S	losses (volumetric and hydro-mechanical in hydrostatic unit)
S	stall point (TC)
S	stator
Stall	torque converter stall condition
T	turbine (TC)
th	theoretical value (in difference to the effective value)
v	vehicle
vol	volumetric
w	efficiency index to indicate power flow

Abbreviations

µC	Microcontroller
A	Ampere
AC	Alternating Current
ACK	Acknowledgment
ACV	Alternating Check Valves

ADC	Analog-Digital Converter
ADT	All Terrain Dump Truck
AEM	Association of Equipment Manufacturers
ANSI	American National Standards Institute
AOC	Ammonia Oxidation Catalyst
ASABE	American Society of Agricultural and Biological Engineers
ASAE	American Society of Agricultural Engineers (Predecessor of ASABE until 2005)
AUX-N	Auxiliary Control
BIM	Building Information Modeling
BPD	Break Pressure Defeat
CAD	Computer-Aided Design
CAN	Controller Area Network
CB	Central Bypass
CC	Closed Center
CCF	Common Cause Failure
ccm	cubic centimeter(s)
cf.	compare (from Latin: confer)
CiA	Control in Automation
CMP	Compare Value
CNG	Compressed Natural Gas
CNH	Case New Holland
COG	Center of Gravity
CPR	Constant Pressure Rail
CRC	Cyclic Redundancy Check
CSMA/CA	Carrier Sense Multiple Access with Collision Avoidance
CTR	Counter
CUNA	Commissione Tecnica di Unificazione nell'Autoveicolo (EN: Technical Commission for Standardization in the Motor Vehicle)
CVT	Continuous Variable Transmission
DC	Diagnostic Coverage
DC	Direct Current
DC	Displacement Control
DEF	Diesel Exhaust Fluid
DEL	Delimiter
DEM	Discrete Element Method
DIN	Deutsche Industrie Norm (EN: German Industry Standard)
DLC	Data Length Code

DLG	Deutsche Landwirtschafts Gesellschaft (EN: German Agricultural Society)
DOC	Diesel Oxidation Catalyst
DOF	Degrees of Freedom
DP	Dynamic Programming
DPF	Diesel Particle Filter
DSP	Digital Signal Processor
EAF	Agricultural Industry Electronics Foundation
EC	European Commission
ECU	Electronic Control Unit
EDC	Electronic Displacement Control
EDS	Electronic Data Sheet
EGR	Exhaust Gas Recirculation
EHA	Electro Hydraulic Actuation
EHR	Electrohydraulic Hitch Control
EMC	Electromagnetic Compatibility
EMI	Equipment Manufacturers Institute
EOF	End of Frame
EPA	Environmental Protection Agency
EPC	Electronic Positive Control
EU	European Union
FAO	The Food and Agriculture Organization of the United Nations
FEA	Finite Element Analysis
FOC	Field Oriented Control
FPGA	Field Programmable Gate Array
FSM	Functional Safety Management
FWD	Forward
GND	Ground
GPS	Global Positioning System
GSM	Global System for Mobile Communication
hm	hydro-mechanic
HMI	Human-Machine-Interface
HMT	Hydro-Mechanical Transmission
HMU	Hand Metering Unit
HPRV	High Pressure Relief Valve
HRC	Highly Regulated Countries
HST	Hydrostatic Transmission
I/O	Input/Output

IC	Input Coupled (Hydro-Mechanical Transmission)
IC	internal combustion (short form of ICE: internal combustion engine)
ID	Identifier
ID	Intermediate Drop (transmission)
IDE	Identifier Extension
IGBT	Insulated-Gate Bipolar Transistor
IH	International Harvester
IM	Induction Machine
IODD	Input Output Device Description
IoT	Internet of Things
IP	Ingress Protection
ISO	International Organisation for Standardization
ISTR	Indirect Self-Tuning Regulator
ITM	Intermission
KBA	Kraftfahrtbundesamt (EN: Federal Motor Transport Authority)
KP	Kiss Point
L	Load
LD	Long Drop (transmission)
Le	Leakage, external
Li	Leakage, internal
LLC	Load Limit Control
LMS	Least Mean Square
LPG	Liquid Propane Gas
LR	Power Control (limiter)
LRC	Low Regulated Countries
LS	Limited Slip Differential
LS	Load-Sensing
LU	Lock-Up feature (TC)
MBS	Multi Body Simulation
MCV	Main Control Valve
MES	Manufacturing Execution System
MF	Massey-Ferguson
MOSFET	Metal-Oxide-Semiconductor Field-Effect Transistor
MPN	Multi Pressure Network
MTTF	Mean Time to Failure
MU	Motor Unit
MWM	Mobile Working Machine
NAFTA	North American Free Trade Agreement

NC	Negative Control
NEMA	National Electrical Manufacturers Association
NFC	Near Field Communication
NFPE	Non-Feedback Proportional Electric
NO-Spin	Locking differential mechanism
NR	Non Reactive
NRTC	Non-Road Transient Cycle
OC	Open Center
OC	Output Coupled (Hydro-Mechanical Transmission)
OEM	Original Equipment Manufacturer
OSI	Open System Interconnection
PA	Pilot Actuation
PC	Parallel Channel
PC	Positive Control
PCA	Pump-Controlled Actuation
PCO	Pressure Cut-Off
PCOR	Pressure Compensation Over-Ride
PDU	Protocol Data Unit
PFD	Probability of Failures on Demand
PFH	Probability of Failures per Hour
PG	Planetary Gear
PGN	Parameter Group Number
PL	Performance Level
PL	Pressure Limiter
PLC	Programmable Logic Controller
PM	Prime Mover
PMD	Photonic Mixer Device
PMSM	Permanent Magnet Synchronous Machine
PRV	Pressure Relief Valve
PST	Power Split Transmission
PTO	Power Take-Off
PU	Pump Unit
PWM	Pulse Width Modulation
RAR	Regulation to the Administrative Requirements
REF	Reference
Reg	Regeneration
RFID	Radio-Frequency Identification
RMS	Root Mean Square

ROPS	Roll Over Protective Structure
RTC	Rough Terrain Cranes
RTR	Remote Transmission Request
SA	Source Address
SAE	Society of Automotive Engineers
SAHR	Spring Applied Hydraulically Released
SbW	Steer-by-Wire
SCR	Selective Catalytic Reduction
SCU	Steering Control Unit
SD	Short Drop (transmission)
SIL	Safety Integrity Level
SM	Synchronous Machine
SOF	Start of Frame
SOF	Shift On Fly
S-P	Series-Parallel
SP	Spring
SPN	Suspect Parameter Number
SRR	Substitute Remote Request
StVO	Strassenverkehrsordnung (EN: Road Traffic Regulations)
StVZO	Strassenverkehrszulassungsordnung (EN: Road Traffic Licensing Regulations)
SV	Shuttle Valve
TBH	Telescopic Boom Handler
TC	Technical Committee
TC	Torque Converter
TC-BAS	Task Controller Basic
TC-GEO	Task Controller Geo-based
TCO	Total Costs of Ownership
TC-SC	Task Controller Section Control
TCU	Transmission Control Unit
TECU	Basic Tractor ECU
TIM	Tractor Implement Management
TMR	Tractor Mother Regulation
TÜV	Technischer Überwachungsverein (Technical Supervisory Association)
UDDC	Urban Dynamometer Driving Schedule
UK	United Kingdom
UT	Universal Terminal
V	Volt
VBO	Virtual Bleed Off

VCA	Valve-Controlled Actuation
VDMA	Verband Deutscher Maschinen- und Anlagenbau (German Mechanical and System Engineering Association)
vol	volumetric
WD	Wheel Drive, i.e., 2-wheel drive (2WD) or 4-wheel drive (4WD)

preface

In 2005, I started my professorship at the University of Karlsruhe, which is renamed today as Karlsruhe Institute of Technology (KIT). The institute was established as an endowed chair for Mobile Working Machines and supported by twelve industrial partners. The most often question I was asked was "What is a Mobile Working Machine?" Because there was no common agreed definition in the community available, I tried several definitions and identified the machines' characteristics. These are

- The machines have a certain task of doing a working process.
- They are mobile.
- They have a significant energy share in their working functions.

The definition presented in this book is based on these experiences: "Mobile working machines have a certain task of doing a working process and they are mobile…"

The industrial partners' expectations from the institute were clearly defined: they want to establish a research institute to strengthen their market position. The mentioned machines should be as productive, efficient and of high quality as possible. All these machines in the field of agriculture, forestry, construction, logistics, the municipal sector, and other special applications work in different applications. But many technologies placed in the machines are the same, similar, or comparable.

For that reason, I strongly feel that different branches can learn from each other. The book is structured accordingly and starts with a description of the machines' structure and possible construction concepts. Following the energy flow and the forces guided through the machine, drive technologies are described in the next step.

The importance of electronic controls in the machines is continuously increasing. Control architectures are shown and the CAN-BUS system is described in detail.

Considering today's established technologies, where will this lead to in the future? The innovation chapter tries to answer this question by showing promising technologies and research trends. As references, the most relevant conferences and journals for research and development trends are presented.

The book concludes with typical examples of mobile working machines. The examples are chosen to show the wide variety, beginning with the tractor as a universal machine and ending with the wheel loader representing a specialized machine.

In summary, this book gives a wide and deep view of the technologies used in mobile working machines. It addresses as well the new engineers to get involved in this topic as engineers who want to widen their knowledge.

I would like to thank all the authors who have put a lot of effort into their chapters as well as their sometimes long intensive discussions. My special thanks also to the SAE Team who have made this book possible.

Research and development are very dynamic; therefore, suggestions for further improvements to the book are very welcome.

Karlsruhe, 2020
Marcus Geimer
Institute of Mobile Machines (Mobima) at Karlsruhe Institute of Technology (KIT)

foreword I

Dear reader,

When Professor Marcus Geimer called and asked me to write a foreword for this book, I felt both very honored and excited to have an opportunity to look back at a success story.

As with so many things in life, something new was born out of necessity. In this case, the predicament for the manufacturers and suppliers of mobile working machines was the closure of a renowned chair that for decades has had a worldwide influence on research and teaching in its field. We, at the time a core group of Verband Deutscher Maschinen- und Anlagenbau (VDMA) member companies from the agricultural engineering, construction machinery, and materials handling branches of power transmission and fluid power technology, saw this as an opportunity to take the future into our own hands. The end of this journey was the creation of the endowment chair for mobile working machines in Karlsruhe under Professor Marcus Geimer.

Then CEOs Manfred Witte of Bosch Rexroth and Georg Härter of ZF were key drivers of their time, both were also chairmen of the Power Transmission and Fluid Power Engineering Associations, of which I was the managing director.

Today's "Förderverein Mobile Maschinen" (Association for the Promotion of Mobile Working Machines) grew quickly. The industry was convinced that a center for research and teaching in the field of mobile working machines was needed. Cross-sectional technologies, be it the powertrain or the man-machine interface, required a more system-oriented approach in research and teaching.

Professor Marcus Geimer answered the call to Karlsruhe. The Association, of which I had the privilege of being the first managing director, took it upon itself to provide more than just a chair for the community. We created the basis for the financing of research projects, initiated a renowned conference on hybrid drives for mobile working machines, and also provided a media home for the industry by founding the German "Mobile Maschinen" (Mobile Working Machines) trade journal. In its 15 years of existence, the chair has expanded education and generated many important research projects – the KIT has become a beacon of research and teaching.

The chair makes an important contribution to addressing the cross-sectional technologies of mobile working machines in an international context of industry and science.

The VDMA estimates the global mobile working machine market volume at €220 billion, i.e. $250 billion. However, beyond the financial element, the products, construction machinery, vehicles for intralogistics, etc. are core elements of goods and materials handling – they are used to build our world, feed us, and transport goods and are central to industrial life.

And who would have thought 15 years ago that mining machines or other mobile working machines could also be used for the mining of data in the machine, its working processes, and its environment?

The designers, developers, and engineers of mobile working machines also move digital worlds. Furthermore, they are aware of their impact on the environment, whether in terms of the carbon footprint and sustainability, minimizing noise or soil compaction, or enabling food production for a growing population. Mobile working machines are used to meet many challenges.

This book will surely also help you, dear reader, to do your part even better, as it invites you to get to know mobile working machines and make your own contribution to the further development of this exciting world.

And for this, we owe a debt of gratitude to the innovators and authors!

Hartmut Rauen

Deputy Executive Director of the German Engineering Association
(VDMA – Verband Deutscher Maschinen- und Anlagenbau)

foreword II

The megatrends of a steadily growing world population and rapid acceleration of urbanization around the globe are the key drivers for the demand for sufficient, high-quality, safe, and sustainable food supply and reliable, effective, and integrated infrastructure supporting our needs from energy, communications, and housing to roads and bridges. Mobile off-road working machines are essential to the productive outcomes across these industrial sectors. Mobile off-road working machines are broadly understood to be nonstationary equipment designed to perform productive and specific work tasks benefitting from machine mobility.

The book addresses mainly mobile working machines designed, developed, and deployed in agricultural farming, forestry, supply chain logistics, and construction worksites.

Subsets of these machines mentioned in this book, including agricultural tractors, have well-defined ANSI/ASAE or ISO engineering standards.

The primary objective of the book is to introduce the inclined reader to technical and technological advancements occurring in the mobile machinery market, specifically focusing on the abovementioned industry segments.

As of particular value is the fact that in the book these technical and technological developments are put into a historical perspective spanning several decades of industrial and engineering progress in agricultural arming and infrastructure construction, as well as adjacent industries like forestry, logistics, or the municipal sector.

The book is a thorough reference that addresses historically significant and recent technology innovations. It even includes recent breakthrough technology innovations, like electrification, automation, and autonomy, that will revolutionize the off-road industry in the years ahead. The authors describe with an appropriate level of detail the underlying sensor needs, embedded software architecture requirements, and operation/user experience expectations. And there is recognition that the needs for off-road machines parallels other industries, such as the automotive industry, but includes the unique requirements for harsh off-road environments.

The book is organized into seven dedicated chapters, which are well-illustrated and inclusive of detailed and in-depth information. The editorial style of the materials is easy to read, and provides a compelling frame and structure for understanding the key principles of mobile working machines. Also, the chapters of the book have leveraged industry-wide recognized experts in their technical field with a deep and comprehensive understanding of the subject matter.

I'm suggesting this book as a "must read" for anyone who is interested in studying engineering in academia or is working as a professional in the off-road industry. You will learn and expand your insights and knowledge of amazingly complex machines, their functional subcomponents, and, last but not least, their invaluable contributions to productivity improvements in farming, construction, and our daily lives.

CONGRATULATIONS and THANK YOU to Prof. Dr.-Ing. Marcus Geimer and all contributing co-authors for publishing this interesting and comprehensive book. The authors have drawn from the high-quality historical teaching materials blended with new content that exquisitely fills many existing knowledge gaps in the broad, increasingly complex, off-road equipment industry.

Prof. Dr.-Ing. habil.
Klaus G. Hoehn

about the editor

Prof. Dr.-Ing. Marcus Geimer (born 1966) is working at the Institute of Mobile Machines (Mobima), Karlsruhe Institute of Technology KIT, Karlsruhe (D), as a full professor. The research focus of the institute is on drives, traction and function drives, and controls, such as predictive control and machine learning methods, for these machines. For the research, modern methods, such as simulation methods, are developed and used.

He studied at the RWTH Aachen University (D) and received his Dr.-Ing. degree in 1995 from the Institute of Hydraulics and Pneumatics (IHP), now the Institute of Fluid Power Systems and Drives (IFAS). After his education, he went into different industrial enterprises and was responsible for the development and construction of hydraulic systems and components. In 2005, he was appointed to the former University of Karlsruhe (TH), today KIT.

CHAPTER 1

Introduction

Prof. Dr. Ludger Frerichs, Prof. Dr.-Ing. Marcus Geimer, and Prof. Kalevi Huhtala

The present book gives an overview of the design and the techniques used in mobile working machines. Although there is a large variety of these machines, the used basic technologies are similar. Before going into detail, one seemingly easy but important question has to be answered.

What Is a Mobile Working Machine and How Is It Defined?

Going back to the roots, we have to consider that Mechanical Engineering is "the branch of engineering dealing with the design, construction and use of machines" [1] and a machine therein is "an apparatus using mechanical power and having several parts, each with a definite function and together performing a particular task" [2]. Machines can be subdivided into stationary and mobile machines, whereby vehicles represent one big group of the mobile machines.

Looking on legislation directives [3], "machines" are characterized, as follows:

> *... an assembly, fitted with or intended to be fitted with a drive system other than directly applied human or animal effort, consisting of linked parts or components, at least one of which moves, and which are joined together for a specific application,*
>
> ...

> *an assembly referred to in the first indent, missing only the components to connect it on site or to sources of energy and motion,*
>
> *... [3]*

Taking combustion engines, electric batteries, and other primary energy provisions into account, machines like excavators or forklift trucks fulfill the requirement of this "machine" definition. Nevertheless, how can they be delimited from stationary machines? Emission regulation laws can probably help, because they are made for vehicles:

> *non-road mobile machinery shall mean any mobile machine, transportable industrial equipment or vehicle with or without body work, not intended for the use of passenger- or goods-transport on the road, in which an internal combustion engine [...] is installed, [4]*

The United States Environmental Protection Agency has defined regulations for emissions from nonroad vehicles and engines. Examples for mobile working machines can be found in the Regulations for Emissions from Heavy Equipment with Compression-Ignition (Diesel) Engines:

> *This page provides regulations for nonroad compression-ignition (diesel) engines that are used in machines that perform a wide range of important jobs. These include excavators and other construction equipment, farm tractors and other agricultural equipment ... [5]*

So, what are the common characteristics of mobile working machines? It is their mobility and their work task! Thus, mobile working machines are defined here as follows:

> ***Mobile working machines have a certain task of doing a working process and they are mobile. Therefore they use a drive technology with a traction drive and they have a work function with significant energy shares in both, in mobility and in work function.***

Typical machines that are included in this definition are excavators, combine harvesters, forklift trucks, and waste-collecting vehicles. We can also find vehicles that can provide attachments with power like a tractor with a baler or a universal vehicle with an attachment for cleaning a street. According to EU legislation directive [3], some attachments fulfill the requirements of a "machine" as defined there and thus can be seen as a mobile working machine as well. The combination of an attachment and a vehicle that provides the attachment with power is a mobile working machine, according to the definition above.

Machines that are also included are cable-electric excavators, tamping machines, or snowcats. A full list of mobile working machines would be very long, but these examples show the far-reaching significance. On the other hand, trucks do not belong to mobile working machines because they typically do not have a significant energy share in a work function.

All these mobile working machines have in common the combination of work functions, their drives and a built-in or external traction drive, a chassis, and an operator's working place. **Figure 1.1** shows the general structure of the machines.

For the mentioned machines, combine harvester, excavator, forklift truck, and tractor-baler combination example structures are shown in **Figure 1.2**. The four main

FIGURE 1.1 Division of a mobile working machine.

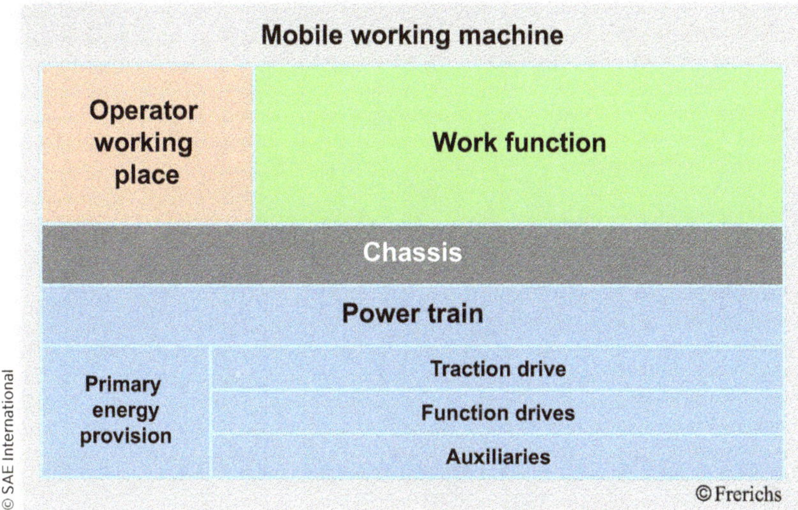

parts, that is, operator's working place, process technology, drive train, and chassis, are colored.

The **operator's working place** is arranged on the machine in such a way that the driver can see the essential work functions well and can look in the direction of travel. The working place is equipped with operating and monitoring devices for driving and working. The workplace is usually protected by its own surrounding frame structure or by the machine structure.

Future worksites will be characterized by different levels of automation of mobile working machines. How these machines are working individually and how a fleet of these machines cooperates will be developed in the future. Those machines can be controlled manually, remote operated, or autonomously. Control of those machines consists of static or dynamic mapping, path planning, obstacle observation, and collision avoidance. The perceptions of environmental stimuli actions have to be composed by information of sensors that detect the condition of the machine. In the autonomous machines and in machines where an operator assistance system is used in the future, the situational awareness plays a key role.

The **chassis**, as this term is used in this book, consists of a frame, means of locomotion, a steering system, and attachments. The frame is the central load-bearing assembly of the chassis. It can consist of one solid body or several parts connected by joints. Connected to the frame are the means of locomotion such as wheel axles or crawler tracks. These devices transfer all horizontal and vertical forces to the ground via the fixed, hinged, or suspended connections. The steering system actuates parts of the locomotion system or parts of the frame. All working functions, the drive system, and the operator's workplace, and also all basic components such as battery, tank, and toolbox, are mounted directly or indirectly on the frame.

To work with a mobile working machine, the **power train** requires a primary energy provision. Internal combustion engines with tank and fuel as energy storage, but also battery-electric driven machines are widely used. Hydrogen or methanol fuel-cell systems

FIGURE 1.2 Structure of four typical machines.

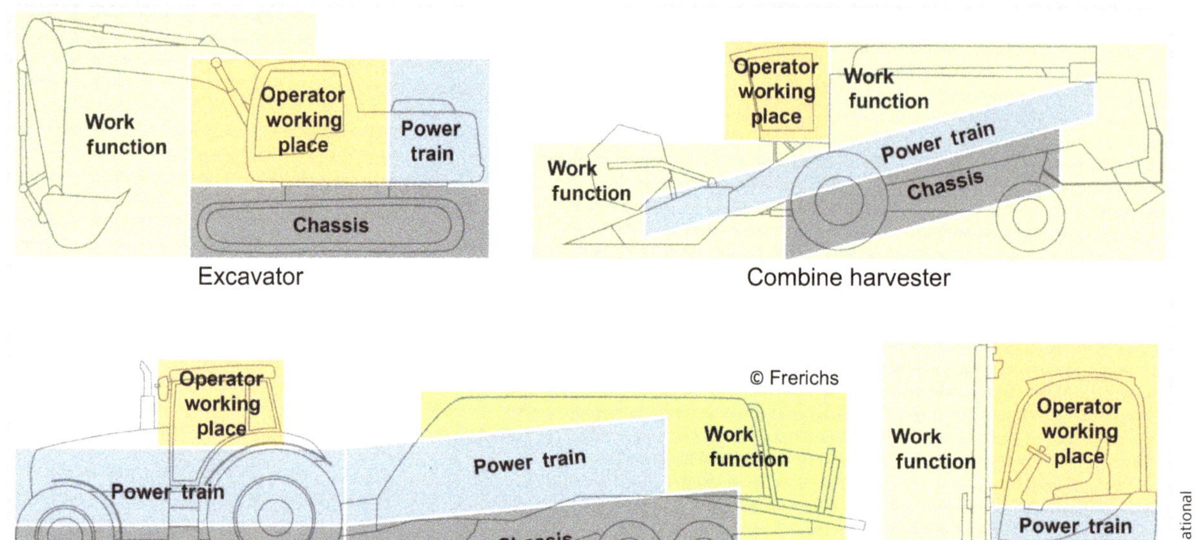

are already available for some machines. Since the variety of these machines and the respective requirements are very different, more or less all mechanical, electrical, and hydraulic drive principles for traction drive, function drives, and auxiliaries can be found.

The **work function** determines typically the name but also the technique principle and construction of the machine. Work functions differ from highly complex process technologies for instance in combine harvesters up to relatively simple lift function in forklift trucks. To realize these work functions, the respective parts and drives possibly have to be distributed over the entire machine or can be concentrated locally, see Figure 1.2.

The complexity and variety of mobile machines are already evident from the previous descriptions and the markets are just as different as the machines. In order to meet the respective market requirements, the products are categorized. A technology-oriented approach is the division into technology levels. *Renius* defines in "Fundamentals of Tractor Design" five levels, from low powered and quite simple machines (level 1) up to fully equipped, very high-end ones (level 5) [6]. More marketing-oriented is the division into product classes such as "Premium," "Value," and "Basic." The technical equipment also plays an important role here, but aspects such as quality, value, and the selling price are more important. This classification is not made globally but according to local understanding and requirements. Thus, the same product in one world region can already be perceived as a premium product, but in other regions only as a value product. The manufacturers use such categorizations to offer a far-reaching product portfolio, preferably based as product platforms and modular concepts.

The requirements for mobile machines are as diverse as the products themselves. An often used clustering of requirements looks like the following:

- *Functionality, productivity, and efficiency*
- *Manufacturer and customer economics*

- *Easy maintenance*
- *Reliable availability*
- *Safety and security*
- *Compliance with standards and legislation*

In the following chapters, the machine and module-specific requirements are taken up in detail.

The structure of the book is close to the structure of the mobile working machines. Chapter 2 will therefore start with the construction of the machines. The requirements and influences will be discussed; typical structures, like block or frame chassis, be explained; and design advices be given. The locomotion system consisting of axles, wheels, or crawler tracks will be presented in detail. In this regard, the tire-soil interaction is of great importance. The chapter will close with a view on different steering systems, which enable the machines to move in their defined space.

The disciplines and technologies of traction and function drives for mobile working machines are introduced in Chapter 3. **Figure 1.3** shows a system view up from the entire machine down to the components. The descriptions follow the power flow through the machine: from the introduction of different energy and power sources, going forward to the various disciplines of power transmission, and summing up in the power utilization at the wheels or the work function. The basics and characteristics of typical conversion systems will be explained as well as the subsystems and components. The fundamentals of mechanical, hydraulic, and electric energy conversions are introduced and definitions are aligned before going into the detailed technical implementations.

The control of a mobile working machine on a higher technology level is today mostly electronically realized and can be done with a centralized or a decentralized control. The structure of controls and the used technologies are presented in Chapter 4. For the controls of components, like the transmission control or the control of the combustion engine, a reference is made to the relevant literature. Looking on research and future machines, the control is expected to play a key role in automation and autonomous operation. New controls in the field of artificial intelligence are discussed and can influence the machines significantly.

New, innovative machine concepts are discussed in Chapter 5. They are divided into evolutionary and revolutionary concepts. Examples are displacement-controlled hydraulic systems, hybrid systems, digital hydraulics, and combinations of the traction and function drives. Because this field is very wide, most promising technologies are presented as well as main research activities. With the references to the conferences and journals, the reader can follow future trends.

After the description of the technologies used in mobile working machines, selected examples of machines are shown in Chapter 6. The examples are chosen as a typical representative in the field of agriculture (tractor), construction (wheel loader), and intralogistics (forklift truck). In addition to the chapters earlier, specialized technologies used on these machines are as well described as the structure and task of these machines. Important standards are as well included as legislation aspects.

A short summary of the book and its aim is given in Chapter 7.

FIGURE 1.3 Vertical division of an example of a mobile working machine.

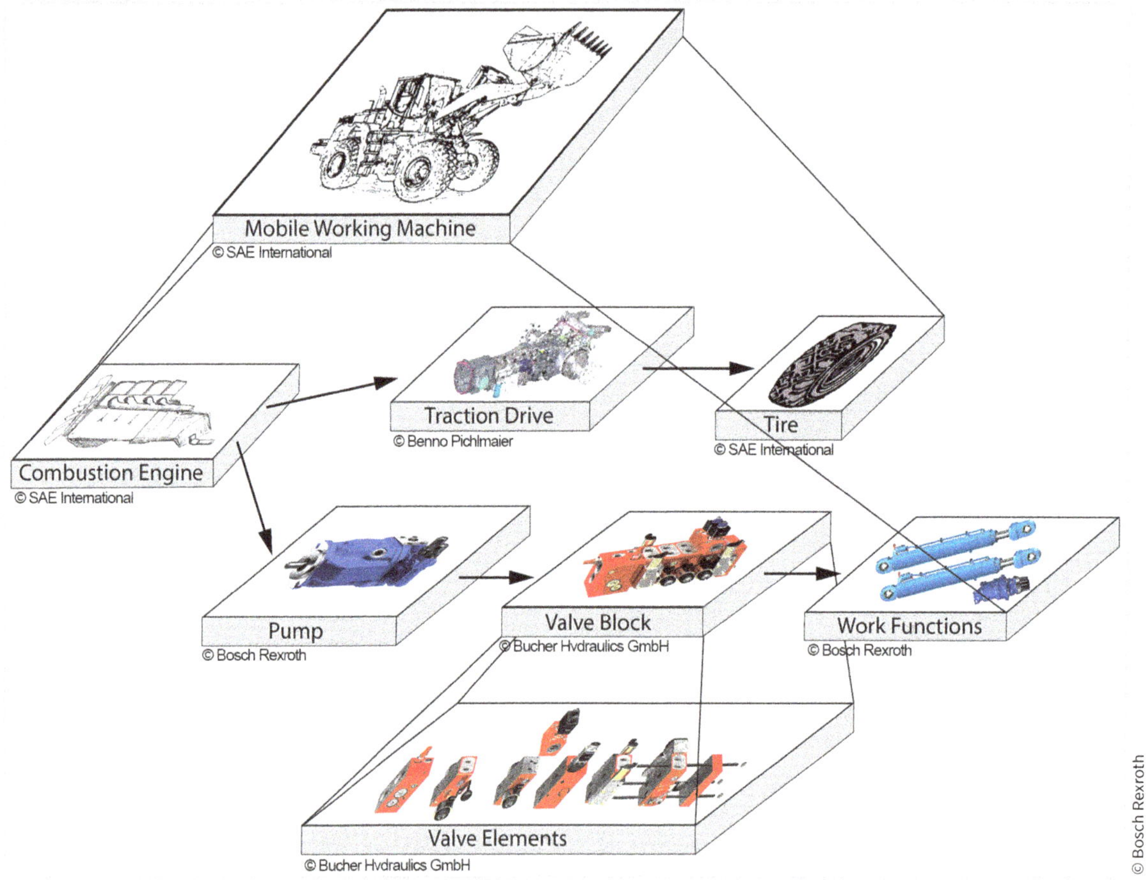

References

1. "English Oxford Living Dictionaries," https://en.oxforddictionaries.com/definition/mechanical_engineering, last view January 26, 2018.
2. "English Oxford Living Dictionaries," https://en.oxforddictionaries.com/definition/machine, last view January 26, 2018.
3. Directive 2006/42/EC of the European Parliament and of the Council of 17 May 2006 on Machinery, and Amending Directive 95/16/EC (Recast).
4. Directive 97/68/EC of the European Parliament and of the Council of 16 December 1997 on the Approximation of the Laws of the Member States Relating to Measures against the Emission of Gaseous and Particulate Pollutants from Internal Combustion Engines To Be Installed in Non-Road Mobile Machinery.
5. "Regulation for Emission from Heavy Equipment with Compression-Ignition (Diesel) Engines," United States Environmental Protections Agency (EPA), https://www.epa.gov/regulations-emissions-vehicles-and-engines/regulations-emissions-heavy-equipment-compression, last view March 13, 2018.
6. Renius, K.Th., *Fundamentals of Tractor Design* (Cham: Springer Nature Switzerland, 2019).

CHAPTER 2

Chassis

Prof. Dr. Ludger Frerichs, Prof. Dr.-Ing. Marcus Geimer, Markus de la Motte, Fabrizio Panizzolo, and Thomas Pippes

This chapter will address the general structure of mobile working machines and their main systems. Therefore, it will start with the construction of mobile working machines and then take a look at different subsystems. The subsystems follow the forces on the machines which can be, as an example, guided through the process tools and the means of locomotion, that are the axles (Section 2.2) and the tires (Section 2.3), into the ground. The last section (2.4) looks at the steering systems used in mobile working machines.

2.1 Construction of Mobile Working Machines (By Ludger Frerichs)

Since the diversity of mobile work to be done is enormous, the variance of mobile working machines is very high. The work function but also the operational environment, and consequently, the process technology and the procedure of machine use, largely determines the construction of the machines. Having the basic structures in mind, Figures 1.1 and 1.2, concerning the construction in all machines, the chassis is of central importance, because it brings all modules together. That is why this book and this chapter start with chassis requirements and show the design variances and details.

The chassis of mobile working machines combines mainly the frame and the means of locomotion including the steering. This short definition is required because the term chassis is used differently in some industries. Sometimes only the frame is included in the definition or it includes almost the entire vehicle including traction drives and operator workplace but without body [1, 2].

2.1.1 Requirements and Influences

The chassis design's primary goal is being able to do the intended work task in the best possible way. However, this task not only consists of the work function to be performed but also results from the economical use of a machine in the entire process chain. In practice, this has led to totally different concepts, specialized machines with decidedly restricted tasks, and to universal concepts with far-reaching flexibility. Chassis concepts must be understood and developed from this systemic perspective.

Figure 2.1 shows a selection of the machine concepts that makes the different chassis concepts obvious. A motor grader can be considered a representative of a specialized machine on the construction site. An excavator is specialized as well, but due to interchangeable equipment, a certain flexibility is given. Universal machines are carrier vehicles in municipal use with the shown attachments for mowing, sweeping, or snow pushing. Such machines with a wide range of different equipment are to find in all sectors from industry to airport to agriculture, and so on. A similar flexibility is offered by tractors and their attached machines and implements. A tractor has its own chassis, which is coupled with the chassis of the working machine. The tractor chassis is

FIGURE 2.1 Machine concepts determine the chassis design. (I) Self-propelled mobile working machine. Motor grader as an example. (II) Universal carrier vehicle with various equipment. Municipal vehicle with (a) mower, (b) road sweeper, (c) snow blade, and (d) hopper. (III) Tractor differently coupled with equipment: (a) trailed drawbar trailer, (b) semi-mounted baler, and (c) mounted cultivator.

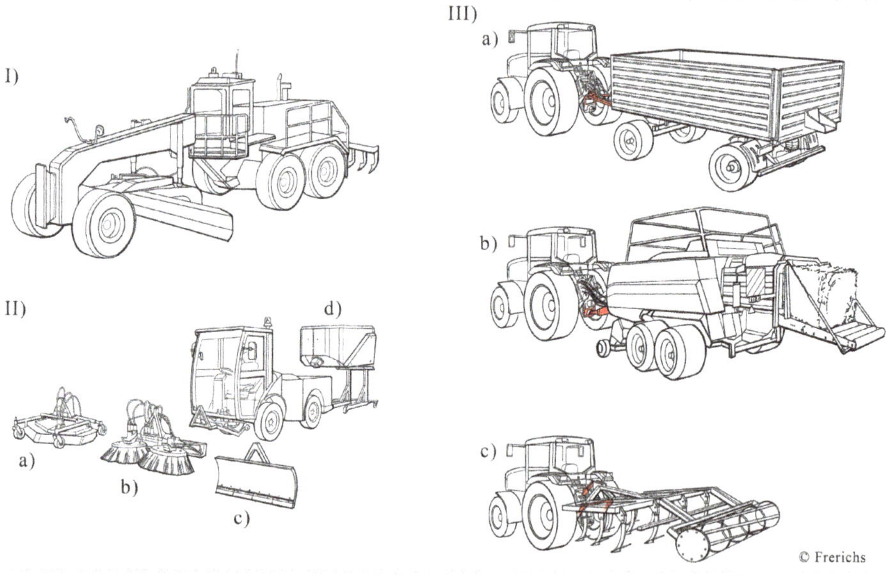

determined by the general functions of pulling, driving, and carrying. By contrast, the chassis of the coupled working machine is determined by the work function and by the coupling with the tractor. This type of coupling is differentiated according to trailed, semi-mounted, and mounted machines. Self-propelled machines are often developed from such coupled machines. Decisive reasons are the better, more uncompromising realizability of the work function and often also the limited performance of the tractors or universal vehicles. If the annual capacity utilization and market size are sufficient, self-propelled machines for users and manufacturers are economical.

The requirement headlines have already been presented in Chapter 1. Following this structure, the essential criteria to be considered for designing a chassis can be set. Off course, the requirements described below must be specified for the individual kind of mobile working machines.

First and foremost to be mentioned are the requirements for **functionality**, **productivity**, and **efficiency**. The chassis itself may be part of the work function or it must carry the assemblies of the work function and support the forces occurring. The design is determined by the needed space for the functional elements like process devices. Depending on the task, sufficient ground clearance may also have to be provided. The power train significantly influences the chassis design, today this is typically determined by the needs of combustion engines. For instance, an electric power train (based on batteries or fuel cells) will enable and need innovative machine and chassis designs with positive effects on the mentioned requirements above.

However, chassis concepts, in particular in conjunction with the drive units and the operator working place, must enable sufficient noise and vibration isolation. Good visibility to the work function and driving lane are necessary for high productivity and efficiency of the operation. On the other hand, future driverless machines may work without large-volume operator cabins. This development can be used for a further optimization of the chassis design. Thus, future-oriented efforts must be made to make use of the new possibilities offered by electrification and automation.

Wheels and axles, crawler tracks, walking mechanism, so the means of locomotion and steering, must fulfill the vehicle-specific required moving function. In this regard, a challenging requirement for the chassis design often is the balanced combination of a large-volume function unit with suitable means of locomotion and steering. However, the maneuverability and thus the usability of the machine essentially depend on this topic. In many applications, the travel movement of the machine is the feed motion of the work function. For this reason, the requirements for a high functionality (e.g., infinitely variable speed adjustment) are coupled with ride comfort and safety issues. That means, depending on the machine and application, the task may be to realize high speeds on road, but under working conditions off-road also to support the massive forces of the machine and to maintain the tipping stability. Therefore, the chassis may have to be prepared for the integration of quasi-static and dynamic stability control systems.

In other applications, good traction properties are required. An effective transmission of forces between the means of locomotion and the soil surface is needed. The machine weight and additional weights, the weight distribution as well as external forces have to be utilized in a proper way. By choosing suitable chassis concepts and using smart adjustability options, a high traction efficiency can be achieved. Since mobile machines often operate on ground surfaces with limited load capacities, low loads and low contact pressures assure the machine stability or underground protection. Especially in agriculture, soil protection is important and not at least, earlier and longer opportunities to drive on the fields lead to improved plant production and higher productivity.

Universal machines require space and interfaces for mounting the work functions. If third parties are allowed to mount work aggregates, they must adhere to general standards. In addition, an installation guideline is required. The more standardized and generous the mounting space is defined, the more flexible the basic machine can be utilized and the work function can be designed. Therefore, the chassis requires a suitable load capacity as well as good driving behavior. The chassis must carry and transfer the machine, implement and function loads horizontally, vertically, quasi-static, and dynamic. As previously mentioned, it is important to pay attention to an optimal load distribution to the ground contact and possibly, to allow targeted ballasting.

Despite the complexity of these many functional requirements, a well-designed chassis is characterized by simplicity. A mobile working machine should be easy to use; otherwise, the installed functionality, productivity, and efficiency cannot be utilized.

Simplicity is also an essential requirement in the category **manufacturer and customer economics**. It supports the demand for a good manufacturability of the chassis. The use of existing production facilities as well as the entirety of the company-specific decision-making factors of the "make or buy" must be considered. Essential for the manufacturer's profitability are primarily the production costs and the marketable quantities, but also the development and distribution costs. The quantities and the production costs, however, influence each other. Therefore, an essential chassis requirement is a quantity-matched design. Here, special attention should be paid to the joining technology. The cost-effectiveness of so-called tool-falling parts but also the complexity costs have to be calculated individually.

In order to master the complexity of the product and variant diversity, a modular design must be sought for chassis and component design. The essential requirement is the clear definition and compliance with standardized and fixed interfaces between the modules. Such modules have to be differentiated according to criteria such as option catalogues or product categories. However, modular concepts typically reach their limits when the goal of global solutions leads to a technology and cost mismatch. In these cases, series or standalone platforms adapted to the requirements of the respective markets are to be preferred.

The profitability of a mobile machine from the customer's perspective depends on the entire cost of the operation, the TCO (total cost of ownership). Of course, the price and the resale value of the machine are important. Thus, the quality and durability of the chassis play an important role. In addition, the chassis design must support the productivity and efficiency of machine use. These include the modularity and flexibility requirements outlined above to reach the versatility of use.

The variable costs are essential to the TCO. With regard to the chassis, the wear and maintenance costs must be particularly minimized. In terms of **ease of maintenance**, this means that access to the maintenance points must always be ensured and it must be possible to replace wearing parts without much effort, especially because mobile machines have to be maintained at all times of the year, often under adverse conditions and not in the workshop. Fulfilling these requirements is crucial not only because of the costs, but also for the satisfaction of the customer. This includes also the frequency of maintenance and the tuning of the maintenance cycles of the various assemblies.

Mobile working machines are integrated in process chains with several interacting stakeholders. In addition, the use is often time-critical, the weather depending work in agriculture or the penalties for delayed completion of construction work are illustrative examples. The **reliable availability** of the machines is therefore an important requirement for the machine and all subassemblies. The functional and operational safety must be guaranteed under all on- and off-road conditions. For the chassis, the abovementioned functional requirements are repeated but also the manufacturing quality and the ease of maintenance.

Special attention must be paid to the topics **safety and security**. Mobile machines have a hazard potential by the motion of the machine and by operating with the work functions. In most countries, regulations stipulate that a product may only be placed in the market if it is designed in such a way that the safety and health of users and third parties are not jeopardized when it is used. Due to the diversity of the machines, the potential hazards must always be assessed systematically, individually, and in detail. The spectrum in relation to the chassis ranges from the attachment of handles and steps up to the functional safety of, for example, automatic steering systems. There is a large number of directives and regulations specific to each country, industry, and product. Essential is the documentation of the assessment and any measures taken. The risk assessment is an essential part of the development process.

Security also plays an important role due to the often quite high values but also due to the described hazard potential of mobile machines. The unauthorized operation or theft must be prevented. Mechanical and electronic access and locking systems, possibly in conjunction with a localization, must also be taken into consideration for the chassis design. With increasing connectivity, measures must also be taken against external manipulation.

Compliance with standards and legislation: In development and operation of mobile machines, a large number of regulations in the form of standards, guidelines, directives, and laws must be observed. In addition, these requirements to be considered differ internationally in some cases quite clearly. This ranges, for example, from recommendations of good technical practice, general regulations (road traffic), company, and ISO standards, to very demanding regulations with regard to emissions, energy efficiency, or even product liability [5, 3, 4, 6]. Because the chassis is central to the essential functions of the mobile working machine, stringent compliance with standards and regulations is highly recommended.

2.1.2 Basic Structures and Frame Concepts

The basic chassis structures of mobile working machines are derived essentially from four aspects. The first decisive factor is the distinction between **special and universal machines** (Figure 2.1). The essential constructive shape receives the chassis by the choice of the **frame concept** (second) and the **means of locomotion** (third). This determines or at least influences the fourth aspect, the concept of **steering** [7].

TABLE 2.1 Frame concepts and characteristics.

Basic design	Combinations	Characteristics	Example
Block construction		Housings of engine, clutch, transmission, rear axle, and support bracket for the front axle are directly connected to each other; supplemental support frame possible	Tractor, backhoe loader
	Half-frame	Half-frame is firmly connected with transmission housing of the transaxle assembly (axle and transmission); engine has no load-carrying function	Tractor
	Three-quarter frame	Three-quarter frame is firmly connected with axle housing of the transaxle; engine and transmission have no load-carrying function	Tractor, forklift truck
Frame construction **Ladder frame** **Space frame** **Box frame**	All frame combinations	Frame takes all assemblies and aggregates directly or indirectly as well as all resulting forces and torques. Combinations of ladder, space, and box frame structures possible, connection firmly or articulated	Universal vehicle, tractor, excavator, dozer, wheel loader, grader, skidder, div. harvester, road sweeper, trailer
	Combinations of frame and self-supporting structures	Combination of frame and self-supporting structure, which carries all assemblies and aggregates and takes all loads	Scraper, square baler
Self-supporting construction		Structural elements and housings of work functions take the load-bearing function; no explicit loads bearing frame	Combine harvester, slurry tank

The conceptual and constructive chassis design must be carried out in the systemic entirety considering these four aspects in parallel. However, the complexity requires the individual description of the possible solutions. Therefore, the different frame concepts are presented first and in the next chapters, details on the means of locomotion and steering will be explained.

The frame concept defines the carrying system of a machine in conjunction with the support points for the individual function modules and the functionally necessary spaces [8]. The frame may be a dedicated assembly but also functionally integrated in certain assemblies. It may consist of a solid body, typically composed of individual parts, or of several individual assemblies, which are connected by limber joints.

Table 2.1 shows the scheme of the usual frame concepts as well as the characteristics and examples. The overarching structure is derived from the common conventions of various machines and industries, from tractors, various mobile machines but also from trucks and cars. There are three basic designs, the **block**, the **frame**, and the **self-supporting construction**. In the historical development process, these basic designs have emerged from both disruptive and evolutionary developments. Every design has its machine-related authorization. In addition, due to the variety of functions and requirements to be met, combinations of the designs have developed, which, in turn, have become characteristic of certain machine types.

Block construction, also called block chassis, comes from the history of tractor development. For the first precursors, all modules were assembled on a frame. However, in 1917, Henry Ford revolutionized the tractor development with his "Fordson" and introduced the affordable block chassis tractor in conjunction with his mass production concept [7]. Block construction is characterized in that the housings of the engine, clutch, transmission, rear axle, and support bracket for the front axle are directly connected to each other to one carrying unit. A supplemental support frame can be used for partial support, for example, for the front loader attachment. The component housings are designed as load-bearing and function-integrated cast designs. This design is characterized by the manageable complexity, low number of parts, and good accessibility. An extremely important functional advantage arises for the steering angle of the steering axle, which is not limited by a frame. However, since they are designed to be load-bearing units, the engine and transmission blocks, as cast elements, must be larger in size and heavier than functionally required. This design prevents the direct use of large-series engines, for example, from the commercial vehicle sector, which are designed as built-in engines for the frame installation.

Block construction is globally the predominant design for tractors even though the combination with frame elements has been increasing in the last decades. In the construction equipment sector, the block chassis concept can be found in some designs for backhoe loader, which derives from the tractor.

In the transition between the basic designs, block and frame construction, typical **combinations** are widely used. For these designs, the terms half-frame and three-quarter frame have become common. Figure 2.2 shows the basic differences. In the case of the **half-frame** construction, a transaxle assembly together with a firmly to the transmission connected half-frame forms the carrying unit. In the transaxle assembly, the load-bearing cast housings of transmission and axle are firmly flanged.

Together with attached or incorporated auxiliary drives, hydraulic pumps as well as various mechanical coupling devices, transaxles are composed as highly integrated preassembled modules. Connected with the half-frame, these modules carry all machine components. The engine and the front axle are mounted on or in the half-frame, whereby the engine housing must have no supporting function and greater flexibility in the engine selection is given.

The half-frame can be shaped very differently. Holistic cast structures can be found as well as welded and bolted structures made of longitudinal members (L- or U-profiles or thick-walled

FIGURE 2.2 Transition from block to frame construction: (a) block, (b) half-frame, (c) three-quarter frame, and (d) (full) frame construction.

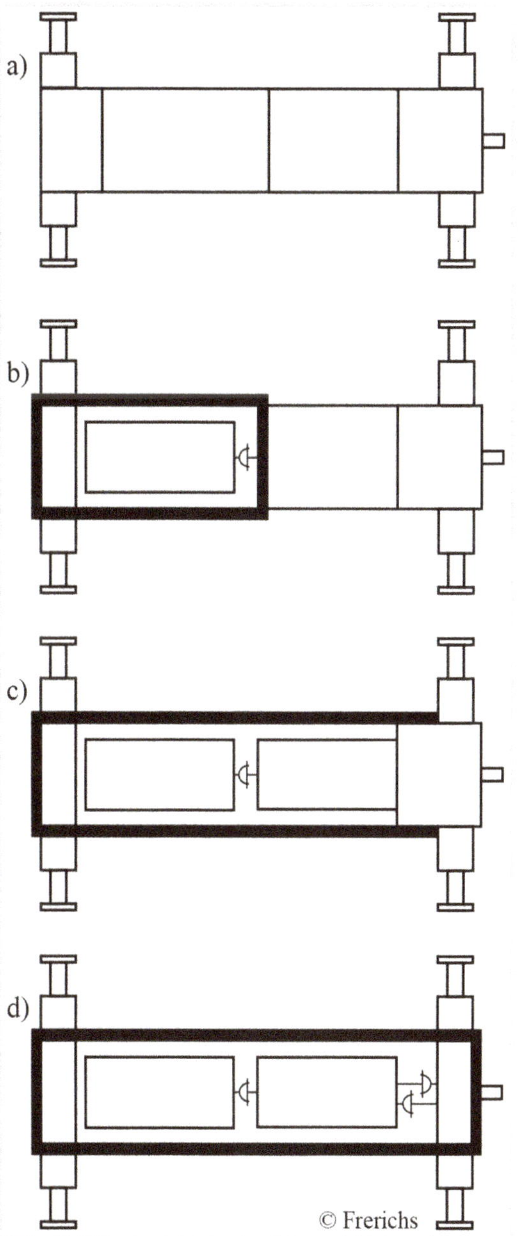

sheet metal) with cross-members. The engine is elastically and damped mounted between the lateral frame members. This reduces both transfer of loads on the engine and transfer of noise and vibrations in the vehicle. Power is transmitted from the engine to the transmission via a flexible element such as cardan joint or shaft.

A major disadvantage of such a frame construction is the limitation of maneuverability due to the limited steering angle of the wheels by the frame elements. This is especially true for machines that are often used in confined spaces and for those that are equipped with large wheels because of the traction and load capacity. For large tractors, this restriction is often unacceptable. A strongly integrating variant almost indistinguishable from the block construction is a frame below the engine. In order not to restrict the ground clearance, this frame can be used as an oil pan of the engine [10]. The transition to block construction is fluent at this point, especially when larger parts of the engine housing have to be load bearing, and in that regard, specialized. However, the advantages such as elastic engine mounting and free engine selection become less. But it remains with the advantages that the half-frame compared to the block design offers more possibilities, especially through easier-to-integrate functions. Examples are the suspension of a pendulum axle or the replacement of such axle by independent wheel suspensions.

In the **three-quarter-frame** design, the frame is attached to the axle. The load-bearing unit thus consists of the frame and the axle. Due to this direct connection to the axle, the engine and the transmission do not have to perform supporting functions. In this type of construction, the transmission is connected to the rear axle in the design of a transaxle. The connection to the engine can be carried out rigidly or elastically. Engine and transmission are elastically connected to the frame to compensate relative movement. This results in the advantages that now both engine and transmission housing can be made lighter and the choice of options can be made more flexible. This concept supports the modular design and the optional offer of different engines and transmissions.

The three-quarter-frame construction can be found in tractors as well as in other mobile machines. For example, in forklifts with manual or torque converter transmissions: the three-quarter frame with the rigidly connected counterweight in the back combined with the transaxle in the front forms the holistic frame structure. In this example, the driven front axle is bolted integrally to the frame. The load-bearing support points for the mast often are directly located on the casted axle, while the tilting cylinders are supported on the frame, see Section 6.3.

The (full) **frame construction** is the predominant design with regard to the totality of all mobile working machines. It offers maximum flexibility in terms of work and driving functions. The frame takes all assemblies and aggregates directly or indirectly as well as all resulting forces and torques. It fulfills the complete task of a carrying structure with all supporting functions and coupling possibilities. Conceptually, different properties can thus be generated, for example, a flexible machine designed for comfort and high speeds or else a rigid structural system dimensioned for high loads. It is possible to install or integrate large-series components such as engines and chassis components or even special modules designed for the application flexibly. Although frame construction seems to be more expensive than block construction, the later presented self-supporting construction outweighs the advantages from a systemic point of view because of the additional necessary frame components. It leads to lower individual component costs and supports the modular machine design as well as the production of small quantities with the extensive variants required by the market. For tractors, the frame construction method can be found in small garden tractors, in system tractors (e.g., mid-cab tractors) and heavy-duty four wheels drive (4WD) and track tractors. But especially in mobile working machines, decisive advantages result from the possible integration and combination of work functions, which can also simplify the accessibility and thus maintenance.

Frame structures can be divided into **ladder frames**, **space frames**, **box frames**, and various **combinations** thereof, see Figure 2.3. The frame can be a dedicated, central assembly or individually composed based on the targeted functions. A classic **ladder frame** is characterized by two longitudinal members with a number of cross-members. The ladder frame itself is self-supporting and often the central assembly in mobile working machines. The shape and dimension of the side and cross-members are determined primarily by the loads to be absorbed and the desired stiffness. Longitudinal members are often made of standard profiles (rectangular and round tube profiles or open U, L, and Z profiles); for cross-members, additional profiles are used. The concrete

FIGURE 2.3 Exemplary frame design of mobile working machines: (a) ladder frame of a forage harvester, (b) space frame with distributed bar structure of a pulled carrot harvester, (c) box frame of an excavator undercarriage with crawler carrier, (d) front box frame articulated combined with the rear ladder frame of a wheel loader, (e) rear ladder frame articulated connected with a central box frame of a motor grader, and (f) self-supporting construction of a combine harvester.

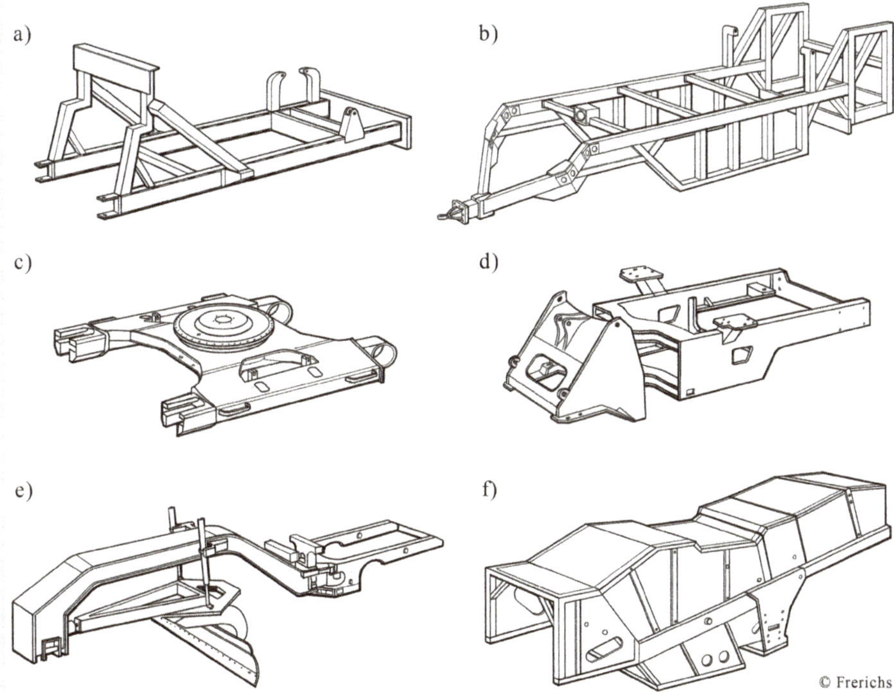

design depends on the individual function and task of the mobile working machine. Therefore, in addition to the standard profiles, thick-walled sheet metal and specially shaped profiles or partial reinforcements are also used, which allow the adaptation of the section moduli to the load curve. In addition, functions or functional elements are often integrated, such as hydraulic tanks or support elements for work functions.

In some off-road vehicles, ladder frames are made soft torsional to reduce the suspension travel of axles or wheels. Especially in case of universal vehicles, the ladder frame represents the defined interface to the mounted body or work functions, just like a truck. Depending on the body, they must be connected via a stiffening or torsion-compensating subframe. However, because in the mobile working machines, most of the aggregates to be assembled are inherent stiff or must be kept stiff, for example, an engine block or a rotating aggregate, the frames are typically designed with high bending and torsion stiffness. Nevertheless, this is a veritable design challenge in detail, since such machines are often unsprung and, because of the off-road use, large forces are acting. For this reason, additional subframes or special fasteners are often used. Thus, the design of stiffness has to be given great attention already in the early design phase and the early and consistent use of modern design methods is strongly recommended, see Section 2.1.3.

Figure 2.3a shows the ladder frame of a forage harvester with a massive rear cross-member to mount counterweights. The longitudinal members consist of high rectangular reinforced profiles. In the front area, the mounting points for the chopping drum and the bolted front axle as well as the frame extensions for carrying further process aggregates and the operator workplace are placed. In the back, the rear axle support bracket and the fasteners for the engine are shown.

Space frames can be composed of single bars to a symmetric space truss structure. These are today rarely found in chassis of self-propelled mobile machines but sometimes in tractor-coupled machines. Typically, such frames are designed in individual non-regular distributed bar structures, which are adapted to the functionally best position of the working aggregates, Figure 2.3b. Main target is the design of a lightweight structure, which takes all acting forces. For the bars and struts, typically tubes, sometimes open profiles are used; moreover, the combination with surface or contour elements for a reinforcement or for an integration of work functions is common. However, such integration of work function elements, if those take loads, is the transition to the combination of frame and self-supporting structures.

Box frames are predominantly made of surface elements, and, if necessary, cast elements or profiles, which are assembled into machine-specific more or less closed assemblies. Due to their function and strength, these can assume irregular contours or be symmetrical, centrally arranged frames. Typical for a box frame is the creation of very rigid assemblies and high load-bearing structures. The undercarriage of crawler excavators, as shown for example in Figure 2.3c, is designed as a box frame.

The transition and combination of box frames with space frame structures is as fluent as the combination with ladder frame structures. Such connections can be done in firmly connected or articulated manner. Figure 2.3d and e show the design combination of two typical representatives. Wheel loaders are offered as machines with all-wheel steering and articulated steering. Figure 2.3d shows the articulated frame, which consists of two assemblies. The front part with the connection points for the mast is designed as a very torsion-resistant box frame. The front axle is firmly connected to the frame. The connection to the rear frame part can be designed as a swiveling (see Figure 2.3d) or a swiveling and pendulum joint. Depending on this, the rear axle is connected to the rear frame part fixed or pendulum. The rear frame, which carries the engine, cooling and further assemblies, is designed as a ladder frame with longitudinal members made of solid sheet metal, a rear cross-member as a counterweight, and further cross-members for stiffening.

Figure 2.3e shows the frame of a motor grader. The rear ladder frame, carrying rear axle, engine, and operator workplace, is articulated connected to a central box frame. At the box frame, which is designed as a square tube, the pendulum suspended front axle is arranged. About midway between the front and rear axles the working equipment (the blade) is located.

The **self-supporting construction** is common in the automotive sector with the monocoque or unibody construction of cars or the integral space frame construction of buses [1, 11]. Even mobile machines are built in such a construction type in the composition of surface elements and bars. Body and frame merge into a holistic structure by the integration of working aggregates or components of work functions in the supporting structures. Accordingly, these machines do not have an explicit all loads bearing frame and are characterized by that. The self-supporting construction thus comes close to the definition of the block construction of tractors. But even some combinations of dedicated frame structures with integrated work functions are hardly distinguishable from a self-supporting structure.

As a self-propelled mobile working machine, the combine harvester is a typical representative of the self-supporting construction method, Figure 2.3f. The bolted housing modules of the threshing, separating, and cleaning functions, which are made of sheet metal reinforced by profile elements, together with the grain tank and the cross-member for the rear axle suspension, form an integral, rigid support structure.

The presented three frame concepts block construction, frame construction, and self-supporting construction with the different features and possibilities come along with the variety of mobile working machines and their applications. Because of their universality, tractors and carrier vehicles have different requirements than self-propelled or coupled working machines designed for a special scope of tasks. Regularly new application-related variants and combinations are created from the basic structures shown. This applies with a view to the upcoming technologies in particular. With the increased use of electric drive technology, it will also be possible to come up with completely new chassis and frame concepts for mobile working machines. Just as the introduction of the internal combustion engine has originated the block construction of tractors, the degree of freedom gained through its omission creates completely new design options.

The same applies to the development towards more autonomous machines. Two fundamentally different concepts are being pursued. Existing machine concepts may be highly automated while maintaining the operator workplace, in order to keep the variety of possible applications. Thus, the machine can operate driverless or the operator stays on the machine and the operator can still perform partial tasks manually or he can monitor the automated use of his machine or even several machines. However, as diverse examples show, autonomous and therefore conceptually driverless systems can do without a driver's working place. This results in completely new disruptive possibilities for machine and chassis design, not least in combination with new drive systems.

2.1.3 Design Principles

Mobile working machines are characterized by their task-related individual solutions and technical details. However, for the development of all these machines, the generally applicable engineering methods and design principles can and must be applied [12, 17, 13, 14, 15, 16]. At this point, some aspects are emphasized that are special for the development of mobile working machines in particular.

The development of mobile machines takes place systemically according to the requirements in the functional chain of **work function-machine-procedure**. For the conceptual design of the machines, the aspects from this three (work function, machine, procedure) have equal importance. For the subsequent construction, however, there is an almost hierarchical dependence on this functional chain. That means the chassis and frame construction are subordinated to the work function and must positively support the use of the machine. An example of this precedence of the work function is the self-supporting construction of combine harvester shown in Figure 2.3. Here everything is subordinated to the functionally complex harvesting process. A second example, the special frame construction of electric counterbalance forklift trucks, see Section 6.3, allows the basic function of a forklift, the handling of loads, but at the same time it supports the rapid lateral battery change. While in old concepts the battery had to be taken out laboriously with a crane, today's frame concepts with the lateral changing device support the productive use of those machines.

For the economic success of the manufacturer as well as for the user of mobile working machines, a successful **modular design** and machine-specific **platform** is crucial. Bonvoisin et al. gives a literature overview on the general thematic [18]. The entire construction, including the frame, must support the modular design. On the manufacturer side, this is the combination of variants at defined interfaces in order to be able to represent the customer- or brand-specific design economically, even for small quantities. On the customer side, the modularity supports the optional variety and enables the required task-specific equipment. An excellent example of this is the excavator, where various chassis or undercarriage modules are combined with different modules of the upper structure, boom, and work equipment. A successful and not necessarily visible for the customer modularity extends over series, production locations, or even brands. It may make sense, depending on the product range and on the global access of the individual company to the different markets, to create more than only one modular platform. Finally, only with a modular design, the complexity in development and production as well as in sales and service can be mastered successfully.

Sometimes the methodological approach for the development of mobile working machine varies in details significantly. This is mainly due to the diversity of the work function to be developed as well as the available development resources and competences. A functionally complex machine requires a dedicated predevelopment phase to realize, for example, a process sequence like cutting, conveying, separating, and so on, under all operating conditions reliably and with the expected quality. In contrast to realize comparatively simple work functions such as lifting, lowering, and tilting, a comprehensive functional development can take several years to consider all the diverse operating conditions. This is especially true in regards to the construction of complex process technology, the respective interaction of the machine with the process material such as soil, gravel, plants, and debris, and also with biological and chemical substances requiring a wide range of know-how. Comprehensive functional development takes more time especially for the development of machinery such as agricultural machinery, where real practical testing is limited to short phases in plant production.

With regard to the differences of resources and competences, it should be noted that besides a small number of manufacturers with very high turnover, the mobile machinery sectors are predominantly made up of medium-sized family businesses. Even if these companies stand out because of their high innovative strength, the absolute development resources, and therefore also the breadth (disciplines) or the depth (training) of the competences, are very different within the companies. Development expenditures of approximately 2%–6% of revenue are usual. Overarching generalizations cannot be deduced, but from experience: in the area of agricultural machinery about 4%–6% of revenue is reinvested in R&D and, in the areas of intralogistics and construction machinery, more than 2%–4% of revenue is reinvested in R&D.

Against this background, the innovation and product development processes do not differ in principle but are specific to the company in terms of the development methodology. In general, a computer-aided development is always combined with the extensive practical use of machines. The development takes place with company on-board and in addition with externally acquired resources and competencies. For the design sector, the use of 3D-CAD tools has become standard. This also applies to the use of the FEA (finite element analysis) tools, although it becomes clear on closer inspection that often calculation tools for FEA or MBS (multi-body simulation) are still more used for "post-calculation", especially when problems occur, than for "pre-calculation" and for development support. However, it is a trend that even in small and medium-sized companies, modern design methods and thus the interaction of proactive and accompanying load determination, simulation, and testing are finding more and more application.

In the end, of course, the cost and benefit of development methods and digital models (up to comprehensive digital twins) have to be examined in terms of their efficiency. However, the efficiency gain is much more than a reduction in effort or timesaving (capacity, number of physical prototypes, reworking). Especially with limited resources, the gained capacity can be translated into a higher innovation power or through a better understanding into a higher solution quality, not to mention legislative requirements and guidelines. Thus, not only in the event of damage, evidence is required that products have been developed according to the state of the art, and thus using common methods and tools.

The available modeling and simulation tools for chassis development are numerous. The calculation and the combination of possibilities of the programs are constantly increasing, for example, the dynamic interaction of load and stress. Another and methodically innovative approach based on FEA is topology optimization. In the early development phase, based on the acting loads and the fixed attachment points, in a given installation space an optimal structure is developed by filling with material or with

FIGURE 2.4 Topology optimization for a forklift frame from available space to structural design.

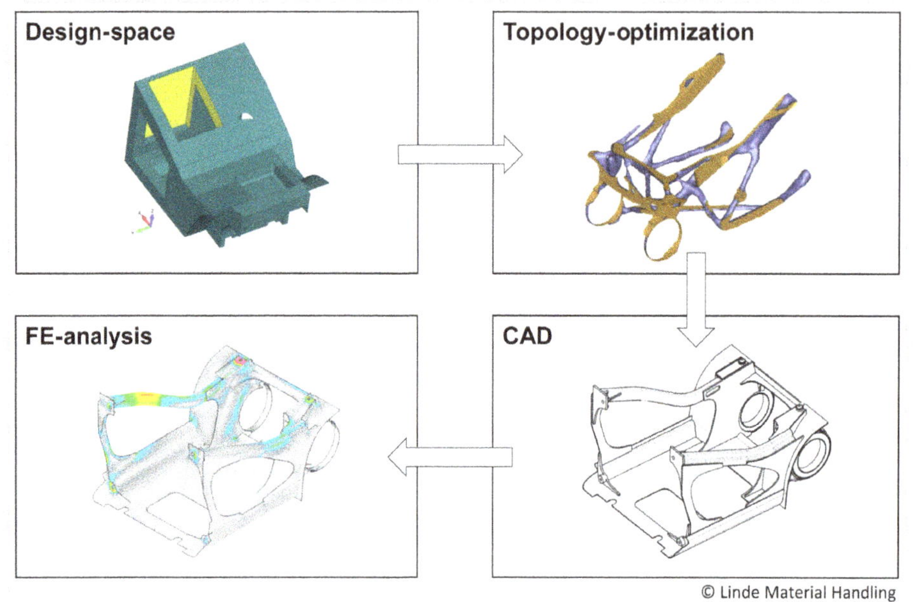

molded elements [19]. The optimization is carried out according to predetermined criteria (minimum mass, best rigidity, lowest stresses). This results in a design proposal, like one shown in Figure 2.4 for the frame of a counterbalance truck, which is in the next step designed in an implementable form by using CAD and FEA tools.

The biggest challenge for dimensioning lies in accessing realistic application profiles and loads. Against the background of the large number of mobile working machines multiplied by the product variants and the extremely different load situations, individual experience and know-how both still play a very important role in the design departments today. However, with the telematics systems that are now available, systematic recording of usage and load increasingly takes place in many companies. These data have more and more input into the methodical development process. In addition, prerequisites are being created in order to simulate the machines and the usage of the machines in their process chains holistically with new modeling approaches and to derive load variables therefrom [20, 21, 22].

A still relatively young method, the discrete element method (DEM), increasingly opens up the possibility of deriving the load specification from the simulation of the work function. For example, the dynamic filling process of a wheel loader bucket and the load effect on the work equipment and on the frame can be determined in a simulation [23, 24]. The DEM is continuously being further developed in science and increasingly used by manufacturers to determine the loads and stresses up to wear and tear effects at the early stage of development and consistently at the model level [25]. Thus, the frontloading in the development process is supported by the independence from machines and test benches as well as from season and weather influences.

For the early validation of the created models and to accelerate the development, the model quality is examined in combination of simulation runs, test bench, or machine tests. This methodology is called "X in the Loop" [26, 27], where X stands for Model, Software, or Hardware. In terms of frame development, this means that frame models on the computer, the real frames, or even complete machines on corresponding test benches are loaded not only with data from field trials but also with input from dynamic load models. Therefore, the methodical, computer-aided chassis development cannot be done without physical prototypes and concrete test bench and practice examinations. The validation of the models and simulations with the variance of real operational data is mandatory.

The **constructive design** of the load-bearing structure has to be oriented at the main three technical aspects: **functional design**, **production-oriented design**, and **load-oriented design** [8]. Of course, this is always accompanied by the need to reach the economical optimum for the manufacturer as well as for the customer. For the chosen basic frame design, the designer performs the **functional design** in such a way that the work function meets the expected quality and reliability. The term "quality" is to be understood very broadly; it refers to the achieved results of the work functions as well as the interaction with the operator and the environment. High productivity with low effort and low losses has to be achieved. The general requirements have been described in Section 2.1.1; few concrete additions shall be made here.

The basic design and the constructions in detail must support, as mentioned before, the optimal layout of the work functions. For a machine with connected process sequences and flow of material, the construction must support a continuous material flow with smooth transitions from one sequence to the next. Thus, the design of the frame may also need to support the integration of the function. This can be done in different ways as shown with a ladder frame for a forage harvester or for the self-supporting design of the combine harvester in Figure 2.3.

For discontinuous material-handling-oriented work functions, such as in a forklift, wheel loader, or excavator, particular requirements for the operator (like visibility) and environment (length, maneuverability, stability) must be considered for the chassis design. In Chapter 6, some explanations are given to such machines. Taking the example of a motor grader, it becomes clear by looking at the used central tube frame how the frame design has been influenced by the necessary visibility of the function, Figure 2.1. Even if the blade is automatically guided by sensors and control systems, the operator still has to monitor the process.

Special attention requires the design of the interfaces of the machines. They represent functional coupling points as well as force introduction points for the load-bearing structure. Such coupling points are required to be able to use optional and different working aggregates on the same frame. In such cases, the statements made above regarding the modular design from the manufacturer point of view apply. Coupling points have another meaning if they are to be suitable for the flexible change of implements, working units or parts of work functions on the operator side. Here, a distinction must be made between standardized interfaces and individual, machine-specific interfaces. For the tractor as a universal power unit, the mechanical coupling points (e.g., 3-point-linkage, towing and ball coupling) are standardized. In various ISO standards, the shape, dimensions, and clearances are categorized according to load ranges and the calculation instructions are given. The concrete components of the coupling points are mainly sourced from specialized suppliers. Section 6.1 introduces some coupling points and standards for tractors.

Machine-specific non-standardized coupling points are particularly created for universal machines and system tractors, which are equipped by the manufacturers, third parties, or by the operators flexibly with various bodies. The design and layout have to be based on the expected and approved loads. For bodies that are frequently changed by the operator, the coupling system, including the implement-side supports, must be designed for rapid replacement. The acceptance of a universal machine is judged on the exchange system. In some universal vehicles or trucks and applications, the bodies are bolted firmly and directly to the frame. For this, the frames are manufacturer-specific provided with hole patterns. For example, in commercial vehicles, the mounting area above the frames are to be defined as free areas and are to be kept free consistently in further developments [2]. Company-owned bodies or those of external companies must comply with all requirements. Extensive bodywork guidelines are created for this purpose.

Some questions surrounding **production-oriented design** were already highlighted when presenting the general requirements at the beginning of this chapter. In production-oriented design, the expected quantities as well as the available production facilities of the manufacturer and the suppliers play a decisive role. Independent of chosen production strategies, finally an economical production with high product quality is expected.

Due to the high product variance, modern production strategies pursue the goals of producing parts and assemblies in larger quantities in the first steps of production and then build different machines with varying configuration in single piece numbers on the assembly line. This concept underlines the importance of modular design and combines different variants and customer options at defined interfaces. For frame production, this leads to larger parts quantities and therefore, more advantageous special profiles, tool-falling parts, and cast designs can be used.

Clamping and screw connections are the usual way of connecting assemblies to the frame. By contrast, the load-bearing frame structures of the machines themselves are predominantly welded. One major cause lies in the complexity and integral design of most of the frames. The targeted design and selection of special parts, such as tailored blanks or lightweight parts, makes it possible to combine optimum shape and strength or specific properties of functional components easily in terms of welding technology [9, 28]. For complex frame constructions, the cost targets are typically achieved by robotic production and for simple structures, the targets are often realized by external production.

The qualitative definition of the supporting structures and components takes place with the **load-oriented design**. These structures must be designed in such a way that the occurring loads can be absorbed without damage and with the least possible amount of material. Based on static basic structures, the real system is designed in the kind of a synthesis taking all requirements into account. The best possible utilization of the structural elements is sought, considering the known basic frame designs as well as the functional and production aspects.

When designing a mobile working machine, optimizing the weight is an essential requirement. Therefore, the principles for the construction of light structures have to be applied. The challenges lie in the observance of permissible axle loads for road travel and for ground pressure, and in the compliance with relevant statutory requirements. In addition, the frame's structural strength and the manufacturing costs play an important role. Some basic lightweight design principles are [1, 29]:

- The degree of utilization, the ratio of the occurring component tension to the permissible tension, should be as close as possible to "1."
- Avoidance of tension peaks by elimination of stiffness jumps and achievement of a uniform undisturbed force or tension flow.
- Utilization of the material used by concentration in areas of high stress and cross-sectional adaptation to the stress.
- The force input is to be distributed over large component areas.
- Off-center force input points should be avoided.
- Adapted design of the components to the stress.

The transition from the qualitative design to the concrete dimensioning should take place via the calculations of the operational stability and according to defined safety concepts. The design and verification of a machine structure requires a systematic approach in permanent matching the steps measuring, checking, and calculating. Prerequisites for this are methodical, computational, and experience skills; equipment for recording application profiles and measurement data as well as access to test facilities are also needed. The principle methodology for the calculations of operational stability is shown in Figure 2.5. Using the damage accumulation hypothesis, from the application

FIGURE 2.5 Basic procedure for the operational stability design of parts.

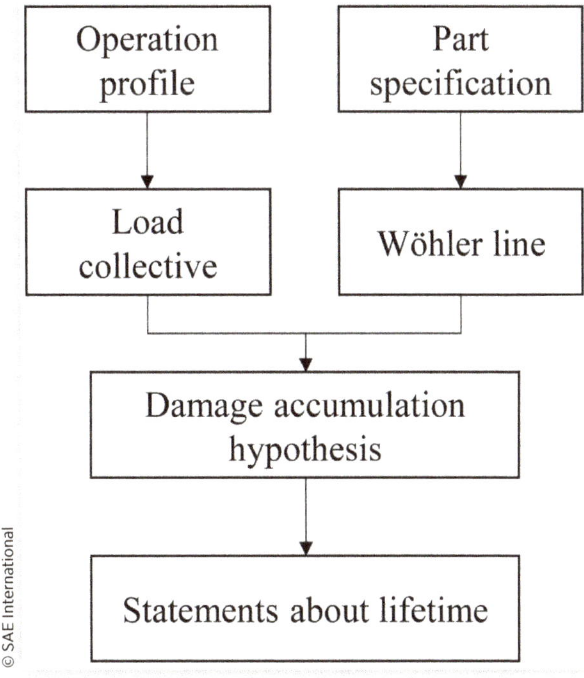

profile and the associated load spectra compared with the component Wöhler line, the damage and lifetime statements can be compared [8, 30, 31, 32]. The methodological procedure and the results have to be documented.

In the future, load-bearing structures in mobile machines are also likely to be produced from alternative or composite materials [1, 29]. This is already state of the art for some cladding and functional parts. Additive manufacturing processes will also play a role. From both research and the automotive and aviation sectors, there are promising approaches to create load-bearing structures. Efficient multi-material manufacturing processes are being developed by integrating manufacturing technologies. These make it possible to produce hybrid structures with metal, carbon fiber, and plastic components in one operation [33]. On this basis, completely new function-integrated chassis and machine concepts can be realized.

2.2 Axles (By Fabrizio Panizzolo)

When considering the entire drivetrain from engine to the wheels, the axle is the last fundamental mechanical assembly on the chain providing torque amplification and corresponding reduction of revolution to the wheels, connected to axle hubs, with the scope to fulfill assigned vehicle task. The main functions to be covered and executed by an axle are the following:

1. Structurally support the current dynamic distribution of vehicle weight between front and rear axles either left or right side, as well to let the current dynamic torque to be properly distributed between front and rear axles either left or right vehicle side.
2. Drive the wheels of those axles adopted for this purpose and properly distribute torque and revolutions between left and right hubs and connected tires.
3. Allow for directional control of those steering axles integrating steering mechanisms.
4. Decelerate and control the vehicle speed utilizing those axles integrating dedicated braking devices.
5. Interface with vehicle chassis, when either rigidly or pivoting connected and interface as well with the chassis when the suspension function will be either integrated on the axle structure or when dedicated spring suspensions will be interposed between axle structure and vehicle chassis.

Available groups of axles can be split between either rigid or steering driving axles and rigid or steering not driving axles. Driving rigid and steering axles are largely utilized on different mobile working machines while not driving and steering axles are instead mostly utilized for two wheel drive tractors (in the front) and for some self-propelled vehicles like municipal vehicles (in the front) and combine harvester as well fork-lift truck (on the rear side), while not driving rigid axles are largely utilized instead for trailers. There are few specialized companies producing for globally requested different families of driving steering and rigid axles tailored for different applications like:

A. 4WD tractors equipped with engine power between 35 and 500 horsepower (front steering and driving axle).
B. Industrial fork-lift trucks with lifting capacity from 1.5 to 45 tons (front rigid driving axle).
C. Construction and forestry machinery like wheeled loaders in the range of 8–30 tons and even higher empty vehicle weight and wheeled excavators having an empty vehicle weight ranging between 6 and 24 tons (and in some cases also above 30 tons).
D. Telescopic boom handlers capable of lifting capacity between 3 and 20 tons.
E. Rough terrain cranes covering lifting capacity between 20 and 80 tons or more.
F. Compactors with a vehicle weight between 5 and 25 tons.
G. Special axles for multipurpose self-propelled machines.

2.2.1 Axle Architecture

Taking into account design requirements needed to fulfill different applications task, tailored axle architectures have been developed over past years and are available on the market. What distinguish the different axle architectures is the position along axle drive chain of braking devices and axle frame interface.

2.2.1.1 Inboard Wet Disc Brakes: Inboard brakes are largely adopted for most applications except for wheeled excavators where there is a specific strong requirement to keep the vehicle blocked during excavation without generating discomfort for the driver because of disturbing presence of amplified oscillations arriving to driver seat inside the cab. Those oscillations are strictly related to the cumulative sum of all different mechanical plays resulting from the contribution of each different braking mechanical component interposed in the reaction chain between chassis (frame) and vehicle wheels. Inboard wet disc brakes represent the best compromise between cost and performances for those general applications and vehicle use where the lubricating and brake cooling oil temperature [even when traveling at high speed up to 40–50 km/h (25–31 mph)] will stay below reasonable and allowable limit of 80°C–85°C (176°F–185°F). For those axles where it will be additionally required the presence of both combined parking and emergency brake functions (as also normally allowed by vehicle regulations), there is the possibility to choose dedicated and modular design for axle housings suitable to integrate different brake functions. Those sandwich mounted and clamped together housings will integrate an independent service brake actuation piston and at the same time as well a second independent piston suitable to release, when pressurized, both combined parking and emergency brake functions. Figure 2.6 presents the design of an inboard wet disc brake made possible by alternating discs with linings (paper friction material in normal cases) with counter-discs to generate a multi-disc package. The brake is actuated by the

FIGURE 2.6 Inboard wet disc service and SAHR brake.

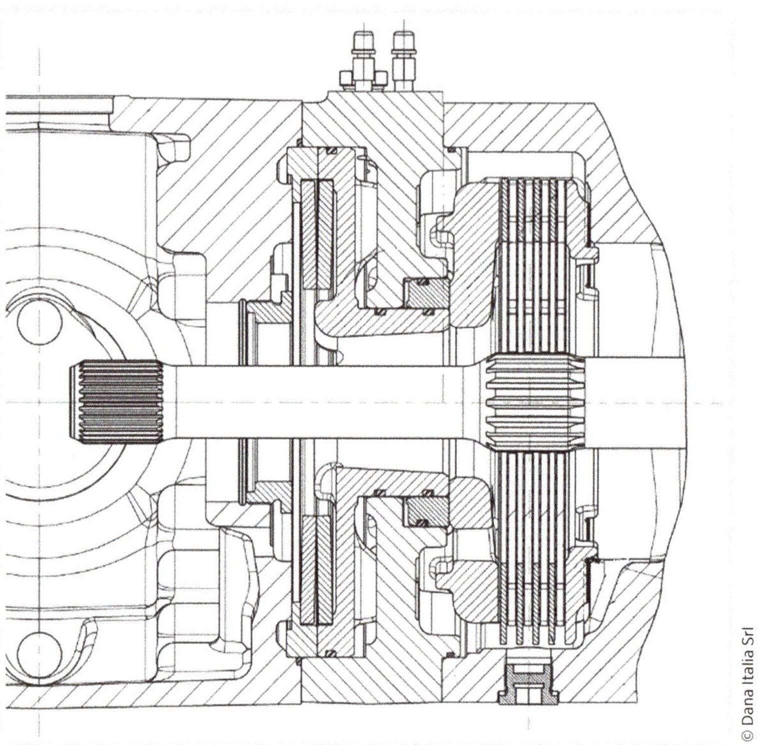

piston feed by volume and pressure generated by the braking pump and the consent to the brakes actuation is provided by braking pedal pushed by the operator in the cabin. The braking valve directly connected with the braking pedal will connect the supply pump with the brake's actuating pistons inside the axle. Interesting to note also how a spring applied hydraulically released (SAHR) emergency and parking brake functions can be integrated utilizing, when independently actuated, the same service brakes disc packages. Once the vehicle will be switched on, pressure available from vehicle circuit will act against spring opening the parking brake, allowing than to the vehicle to move immediately after.

2.2.1.2 **Wheel Speed Axle Brakes:** Outboard positioned wet disc brakes inside axle hub directly and mechanically connected to the wheels are called "wheel speed axle brakes."

For heavy-duty applications, like professional mid-to-high range wheel loaders, the severity of continuous cycling followed by braking will result in high-power dissipation and increased cooling oil temperature above allowable limits. For those heavy-duty application the shift from inboard to "wheel speed brake" axle configuration will be the most convenient design to be adopted from performances point of view. This architecture will represent, for braking function, the most expensive alternative. As shown in Figure 2.7, it is possible to recognize how a wheel speed wet disc brake has been integrated within a steering knuckle of an axle hub. The amount of friction discs and

FIGURE 2.7 Wheel speed axle brake.

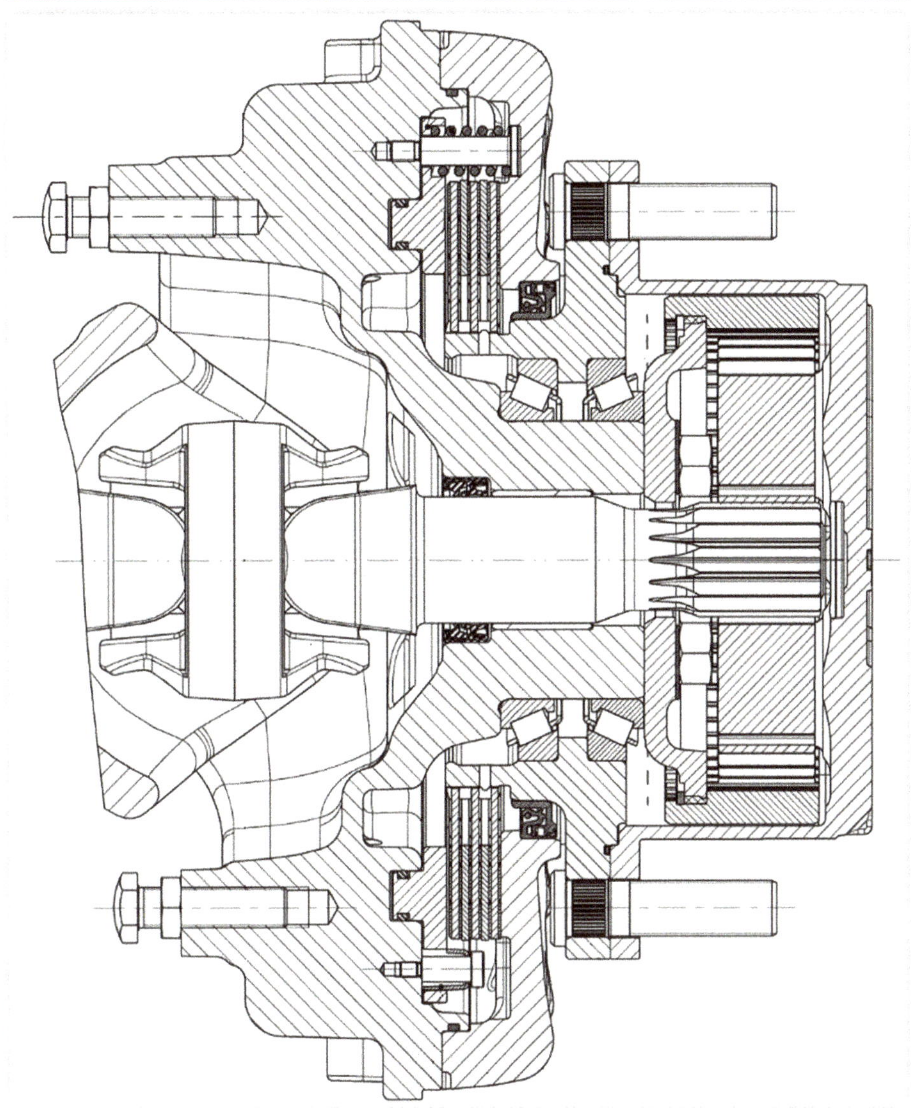

counter-discs is very low but for wheel speed brakes the maximum braking torque to be assumed for component dimensioning is equal to wheel's required maximum braking torque. For such reason to achieve required braking performances, high diameter and expensive discs will have to be selected.

2.2.1.3 **Outboard Positioned Wet Disc Brakes But before Hub Planetary Reduction:** Outboard wet disc brakes are mostly utilized on tractor front driving and steering axles where front-assisted braking function will be mainly utilized for pulling heavy trailers and tractors traveling at high speed. The cost for this brake design architecture will be between the cheapest inboard wet disc brakes design and the more

FIGURE 2.8 Outboard wet disc brake before hub planetary reduction.

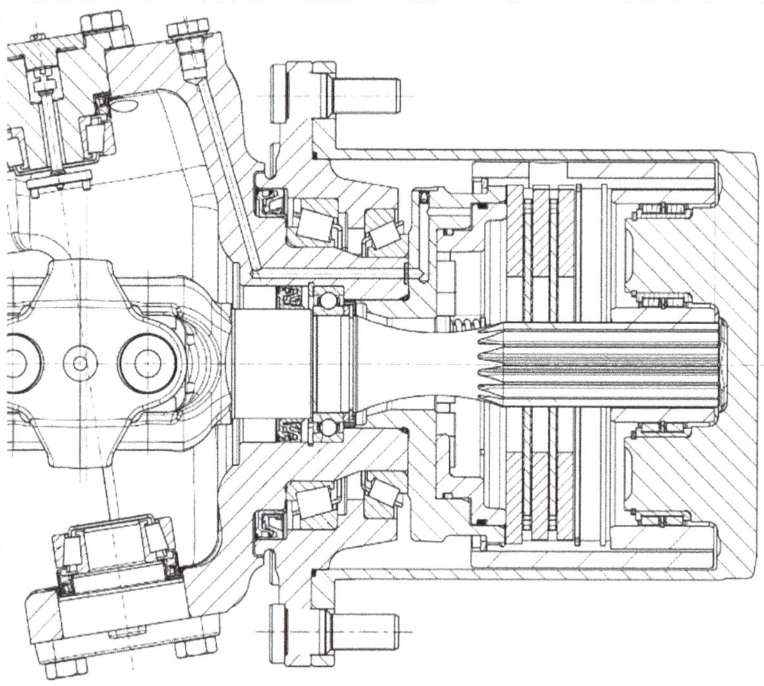

expensive wheel speed brakes design due to step-by-step involved complexity and size of braking components (see Figure 2.8).

2.2.1.4 **Suspended Axle and Frame Interface Example:** As anticipated above, one other axle architecture variant is represented by the integration or not of suspension function. There are few different design solutions adopted by different tractor manufacturers. Most successfully adopted technology foresees the adoption of a hydro-pneumatic and hydraulically controlled suspension systems. Depending on the size of the tractor, a more sophisticated damping suspension function can be offered as a way to optimize suspension behavior during different working conditions, for example, the presence of heavy implements attached to the rear 3 points hitch of the tractor, heavily influencing the dynamic front and rear tires weight distribution and having a huge impact on vehicle and driver comfort, when travelling in off road as well on road conditions. For reference, see Figure 2.9 as well Figures 2.10 and 2.11 where a complete high pin pivot mounted front driving and suspended axle is shown. This axle belongs to a complete family of suspended axles developed to equip a tractor brand. It is advantageous to have an easily installed and simplified interface connection. This is possible while integrating the suspension linkages and hydraulic suspension cylinder within the axle structure, as well as by fixing the entire suspension control valve block and sensors to the axle structure beam, leaving few plugs and fittings to be connected, respectively, with the wiring and pressure supply system, during tractor assembly. In the Figure 2.9, the presence of the axle steering system is shown by the presence of a double-effect double-rod cylinder fixed in the middle of the axle beam followed by two tie-rods linked,

FIGURE 2.9 High pin pivot mounted suspended axle assembly.

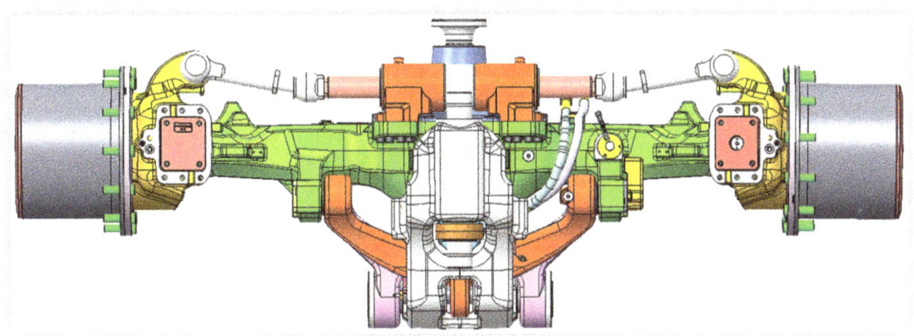

FIGURE 2.10 High pin pivot mounted suspension system.

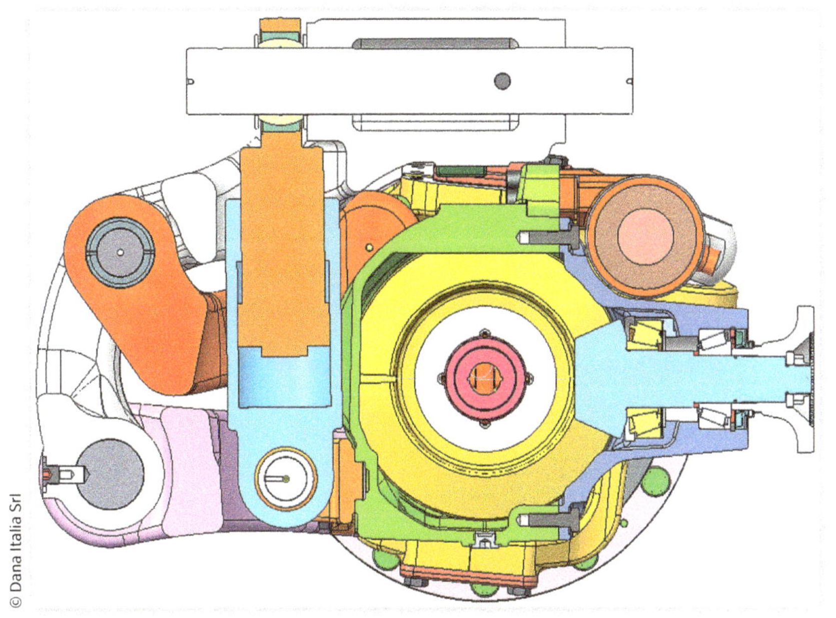

FIGURE 2.11 High pin pivot mounted suspended axle steering linkages.

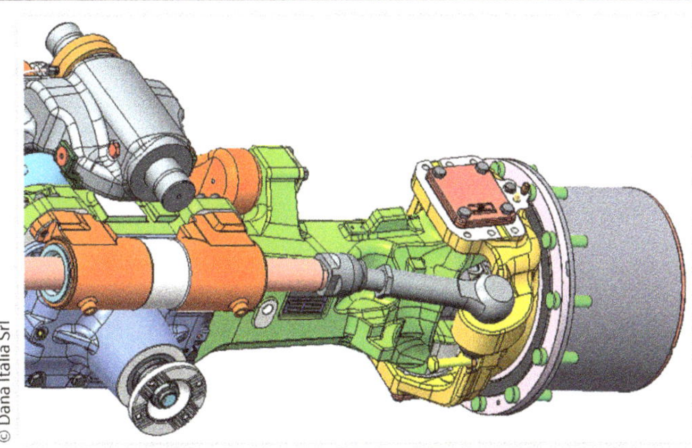

respectively, via two ball joints to the two cylinder rods extremities while the other two 90° ball joints are rigidly connected to the two steering arms integrated on castings of axle steering knuckles.

2.2.2 Fundamental Axle Components

Considering a complete axle gear train from the input flange to the wheel hub, it is important to mention some fundamental axle components including the bevel gear set, open and lock differential mechanisms, and hub reductions. Please refer to Chapter 3, where is described in detail the principle of planetary hub reduction units.

A bevel gear set (pinion and crown wheel) that depending from max needed reduction ratios can be split in two families, respectively: spiral bevel set suitable to cover with robust design ratios between 1 and 4.0 max and hypoid bevel set utilized to cover higher reduction ratios up to 51/7 = 7.28. It is important to note that a hypoid bevel set is utilized with the purpose to achieve the highest axle reduction ratio possible by the combination of hypoid bevel set and wheel planetary hub reduction ratios. Unfortunately, due to the asymmetric strength/resistance capacity of hypoid bevel set, it is mandatory necessary to use at the best and leverage on hypoid bevel set presence only when the vehicle will have to face heavy loads just in one single travelling direction: the forward one.

2.2.2.1 Open Differential Mechanism and Limits: The differential mechanism consists of a differential case/carrier driven by bevel crown gear. Inside the differential case, there is a spider rigidly connected (rotating with the case) and on spider pins are idling rotating differential components called satellites. All satellites (2 or 4) are, by matching teeth, engaged with side differential gears (called also sun differential gears). The two side gears are respectively left and right rigidly connected to corresponding left and right axle half-shafts. As anticipated the main function of the differential is to allow a proper torque and revolution distribution/repartition between left and right wheel when approaching a corner in the presence of a common and constant surface condition for both left and right wheel side.

It could happen that, on some extreme soil condition, one tire could act on "normal" soil surface, while the opposite tire could face mud or, even worst, ice. Under this condition the first tire could transfer to the soil, by friction and without slipping, only the maximum amount of torque (and related force) allowed by the internal efficiency of an open differential mechanism. The opposite tire, acting on muddy surface, will transfer to the soil, by friction, a relatively low force. In the worst condition, acting the opposite wheel on the ice, a zero force will be transferred by friction. If this side presents an icy condition, the friction coefficient will be approximately near to zero; in worse icy condition, the tire will start to slip at a rotational speed that is double of current differential case speed divided by interposed hub reduction; the opposite tire will provide instead his small traction force contribution but unfortunately in stand still condition. In this extreme condition, traction force contribution provided by the sliding wheel will be near zero. The direct consequence of what reported above will be the severe risk for the vehicle to remain blocked (loose of traction for example for a tractor while plowing, reducing accordingly vehicle productivity and loose for example for a military vehicle the possibility to escape from an incumbent attack a cause of missing mobility). For this reason, different differential locking devices have been developed and depending from the different vehicle needs/requirements more or less sophisticated design concepts have been selected. Clearly most sophisticated design solutions have been chosen for those applications where the differential locking function is a must and where for securing the presence of this function the market will be prone to accept to pay for such complex and expensive solution.

2.2.2.2 **100% Differential Lock Mechanism:** Taking under consideration the most effective design among differential locking devices, it is interesting to review a particular 100% differential lock solution adopted on heavy-duty vehicles like high-power tractors, heavy-duty wheel loaders, and military vehicles. Please see Figure 2.12 where a wet brake actuated differential lock is presented. The brake is hydraulically actuated, and it is normally integrated within the differential case. Very significant is to realize the advantage that this solution is offering: the brake can be engaged under load (while slipping) and while moving once the driver fill that is going to face severe working soil condition and potential loose of traction on one axle side (tire). This represents the best performing differential lock mechanism but also the most expensive solution. The consent to the actuation of locking mechanism is given by driver in the cabin while pusching a dedicated differential lock button.

FIGURE 2.12 100% mechanical diff-lock.

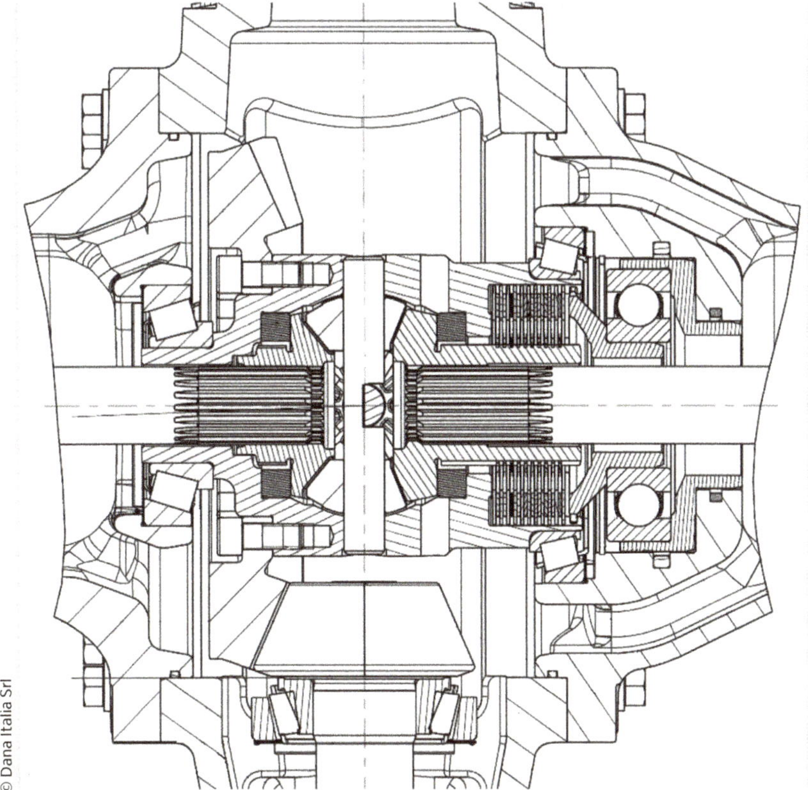

2.2.2.3 Limited Slip Differential Lock Mechanism: There are applications where a compromising, but cheaper solution could be adopted. This is the case well accepted by the majority of manufacturers of small- and medium-size mobile working machines. Those manufacturers that do not need an expensive 100% differential lock but instead for those will suffice to have an automatic self-actuated limited slip (LS) differential allowing to achieve a difference of transferred torque between left and right wheels and vice versa of a percentage up to 45% if compared to the 100% input torque available at differential carrier, for those manufacturers the LS will represent the best compromise between functionality and system costs. In other words, integrating within a differential case a LS self-actuating differential locking system, it will be possible to transmit on one wheel the 27.5% of the differential carrier input torque amplified at wheel by planetary reduction ratio factor while on the wheel located in the opposite position on the axle the transmitted torque will arrive to a percentage of torque with an order of magnitude of 72.5% of differential carrier input torque amplified at wheel by planetary reduction ratio factor. All this is possible thanks to the introduction of a dedicated wet brake package acting on both differential side gears that are connected as well to the axle half-shafts. Axial braking clamping loads are, for this particular mechanism, generated by differential straight bevel gear's teeth interacting inside differential case. LS differential presented here is belonging to the family of load dependent self-locking differential [34, 35].

$$T_V = BT = \frac{\mu \cdot R \cdot s \cdot tan\alpha_0 \cdot sin\,\delta}{r_0} \cdot T \tag{2.1}$$

where

T_V is the braking torque generated by LS differential system [Nm]

T is the differential carrier input torque [Nm]

B is the differential blocking factor

μ is the wet brake lining slipping coefficient = 0.08 (for paper lining)

s is the number of sliding surfaces of each brake package located on both differential side gears

α_0 is the straight bevel gears pressure angle [°]

δ is the differential bevel gears pitch angle [°]

R is the average brake lining radius calculated according to the following formula [mm]

$$R = \frac{2}{3}\frac{R_{EXT}^3 - R_{INT}^3}{R_{EXT}^2 - R_{INT}^2} \tag{2.2}$$

where

R_{EXT} is the outer radius of brake lining [mm]

R_{INT} is the inner radius of brake lining [mm]

r_0 is the average side gears pitch radius [mm] calculated according to the following formula:

$$r_0 = \frac{1}{2} \cdot (d_e - b \cdot \sin\delta) \tag{2.3}$$

where

d_e is the side gear outer pitch diameter [mm]

b is the side gear contact face width [mm]

For detail, see Figure 2.13.

If we consider that LS differential blocking factor B can also be rewritten as max percentage difference of torque between two differential output sides when compared to differential carrier input torque, we arrive to the following equation:

$$B = \frac{T_{RIGHT} - T_{LEFT}}{T_{RIGHT} + T_{LEFT}} \tag{2.4}$$

FIGURE 2.13 Extract from differential gears dimension sheet.

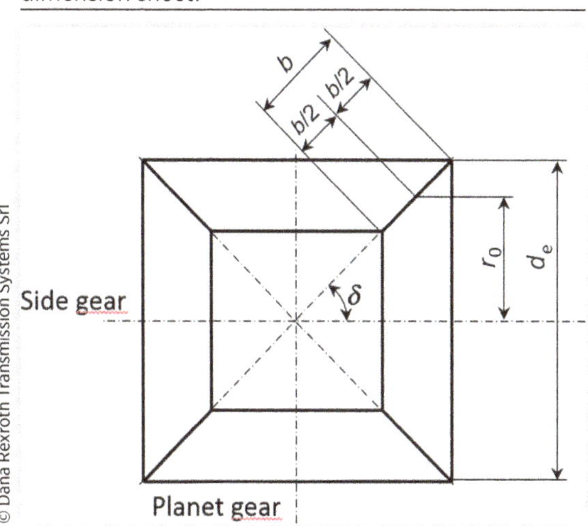

From this, it will be quite easy derivate, after few passages, that the LS differential internal efficiency can be written as

$$\eta_{Diff} = \frac{1 - B}{1 + B} \tag{2.5}$$

For considered configuration of differential lock, having differential blocking factor value of $B = 0.45$, the differential efficiency value is $\eta_{Diff} = 0.38$. This gives evidence how a low value of internal differential efficiency, for selected differential locking mechanism, is contributing to higher locking effect, if compared to a full open differential mechanism, see Figure 2.14.

In Figure 2.15, it is possible to recognize how the differential side braking discs have been integrated inside differential case. The lining material utilized for wet brake for the highest percentage of utilization is obtained from vegetal fibers and the different lining materials available on the market are belonging to the so-called paper-based friction materials. In the past, sintered lining material enhanced by the presence of the molybdenum was utilized and, despite the highest achievable LS performances

FIGURE 2.14 Torque behavior (approaching a corner) of a LS load-dependent automatic differential lock designed with differential blocking factor $B = 0.45$.

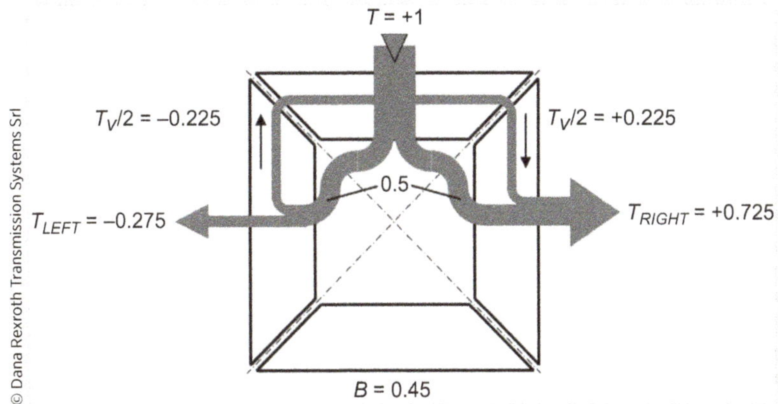

FIGURE 2.15 LS diff-lock.

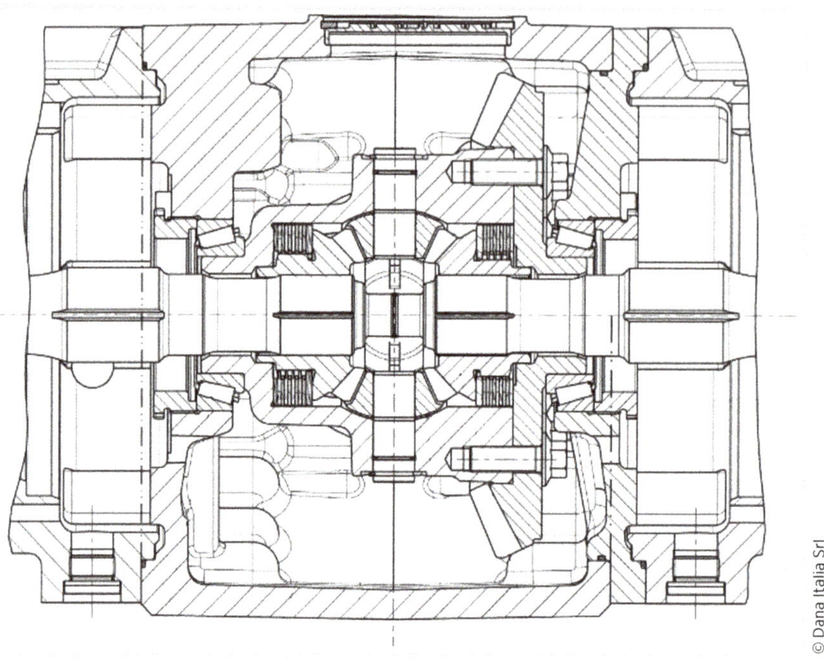

(and despite the advantage of limited packaging space), this friction material was too much sensible to stick-slip-related phenomenon and it was prone to noise generation. For high demanding LS performances for severe working conditions that could easily result in overheating of the brake linings are available on the market some friction materials based on mixture of paper (vegetal fibers) and Kevlar as well more sophisticated carbon fiber-based linings utilized for very extreme conditions.

2.2.2.4 **Other Differential Lock Mechanisms:** There are many other self-locking differential mechanisms available on the market. For particular applications it is interesting to mention the No-Spin mechanism still currently utilized on heavy-duty wheel loaders, on scrapers and graders. This solution was utilized in the past also on tractors and some mobile working machines. When approaching a corner, the outer wheel will be forced to rotate at a speed higher than the differential carrier revolutions divided by interposed hub reduction ratio of an amount proportional to the current outer wheel corner radius, while the inner wheel will be rotating at a corresponding lower revolutions that for this mechanism is equal to the differential carrier revolutions divided by hub reduction ratio. Inside No-Spin mechanism there is a cam that while approaching a corner will disengage the outer wheel half-shaft and corresponding connected axle side from internal left or right face teeth driving component directly connected with the differential carrier (acting as a spider). As a consequence of said outer wheel disconnection, the vehicle will be in the condition to approach under load a corner while having the total amount of propulsion torque and corresponding tractive effort passing through a single half-shaft driving the inner corner wheel. This unsymmetrical behavior of

No-Spin (vs. longitudinal axis of the vehicle) will have to be taken in consideration during dimensioning of the half-shafts and axle hub gears that for such reasons will have to be designed in such a way to withstand the 100% of the axle-driving output torque. This kind of consideration is anyhow valid also for wet brake actuated 100% differential lock system adopted for heavy-duty application in construction machineries and mid-to-high tractor power segment as well. It is important to note anyhow that for this 100% differential lock configuration breaking transitory torque peak, during differential lock brake engagement, will be less impacting than No-Spin differential lock because mitigated and modulated by the presence of the brake slipping under load.

2.3 Soil-Tire Interaction (By Marcus Geimer)

The means of locomotion is characterized by the soil-tire interface and the forces applied through the traction drive to the ground. This subchapter will focus on tires as the most common used technology on the machines. Typical alternatives, like crawler tracks, or railway tracks, as they are used in tamping machines, are not taken into account here. More information about crawler tracks can be found in Refs. [36, 37] and about railway tracks in Ref. [38].

The soil-tire contact can vary in a wide range at mobile working machines. Typical contacts in the field of agriculture can be characterized from a good drivable ground with medium dry conditions to very wet conditions. Extremely wet conditions can be reached in the field of forestry when harvesting trees in musty ground, as an example. Nevertheless, the contact here can be seen as frictional engagement. Best soil-tire contacts with frictional engagement can be achieved on asphalt with a net traction ratio[1] of about 0.9-1.0, or even a bit above 1.0. On the other hand, construction machines can drive on massive rock structures. Here, the contact will become similar to positive engagement and the net traction ratio can be above 1.0.

The soil-tire contact can be calculated with different simulation models. The models can be classified in empirical and physical ones. **Figure 2.16** compares the different simulation models according to accuracy, effort, calculation time, and number of specific tests. The empiric models are further subdivided into models based on experiments and derivations from measured curves. The effort and accuracy increase in the models based on experiments. On the other hand, tests that are more specific are necessary with the models derived from measured curves [39, 40].

Examples for empirical models are the so-called Magic Formula from Pacejka [41], the TM Easy from Hirschberg [44], or the Hohenheimer Reifenmodell [40, 42]. With the Magic Formula, slip curves can easily be defined [41]:

$$\kappa = D \cdot \sin\left\{ C \cdot \arctan\left[B \cdot s - E \cdot \left(B \cdot s - \arctan(B \cdot s) \right) \right] \right\} \tag{2.6}$$

[1] The "net traction ratio" will be defined later in the chapter.

FIGURE 2.16 Division of simulation models. Adapted from Pacejka [39] and Ferhadbegovic [40].

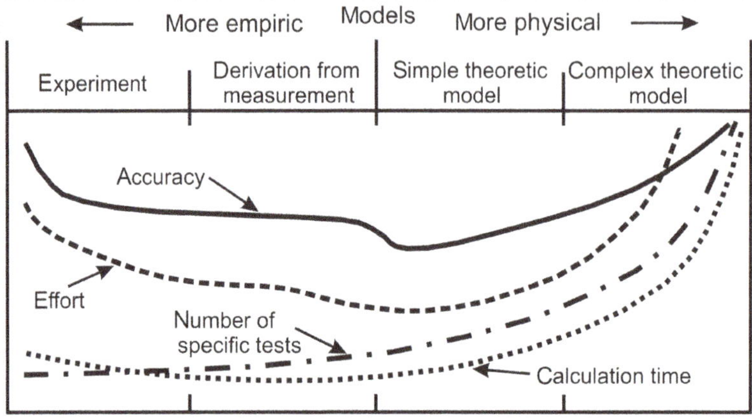

Adapted/modifed from Pacejka, H.B.: Tyre factors and vehicle handling. International Journal of Vehicle Design 1 (1979) H. 1, p. 1-23 and Ferhadbegovic, B.: Entwicklung und Applikation eines instationären Reifenmodells zur Fahrdynamiksimulation von Ackerschleppern (Development and application of a transient tire model for driving dynamics simulation of agricultural tractors), dissertation at the Technical University of Munich (D), 2008.

FIGURE 2.17 Slip curve calculated with the "Magic Formula." Adapted from Pacejka [39].

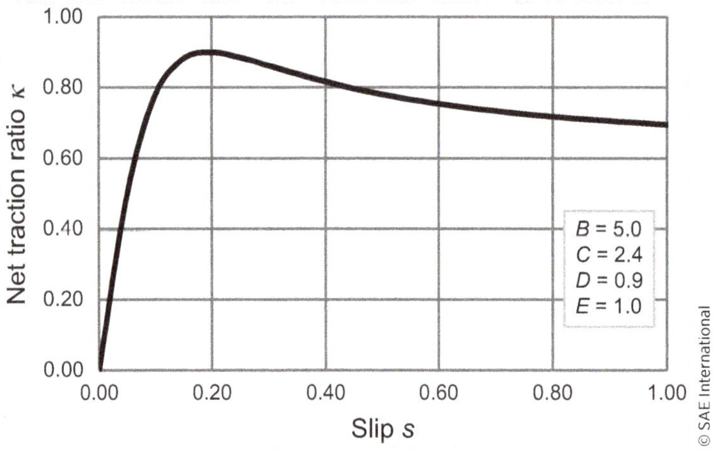

Figure 2.17 shows a slip curve created with the Magic Formula that runs through the origin of the coordinate system and where the net traction ratio reaches its maximum at 0.9, that is, $D = 0.9$. In the formula, B represents a stiffness factor, C represents a form factor, and E represents a deflection factor. The slip curve can be modified with these parameters.

The physical models are based on simple or complex theoretical models, see Figure 2.16. They usually describe the behavior of the tire with spring-damper elements. The accuracy especially for the complex models like finite-element analysis or discrete-element method is increasing as well as the calculation time. Therefore, more easy spring-damper models are developed and can be purchased commercially, like the well-known

FTire [43]. The dynamic of the tire is modeled with a coupling of a spring and a damper between belt and hub or/and in between the belt elements. A more realistic adaption to the soil profile can be reached by 2D models. These models are especially suitable for driving comfort [45].

The brush model [46, 47] assumes that the material of the tire can be approximated as small brush elements. The models rely on the assumption that the slip is generated by the deformation of the rubber between the tire and the ground. The model is easy to parameterize but not as flexible as the Magic Formula.

In addition to the simulation models shown, the forces acting on a tire can also be explained on the basis of the resulting forces. The variables shown below are based on American Society of Agricultural and Biological Engineers (ASABE) standards [49] and the works of Bock [50], Söhne [51], and Steinkampf [52]. In the stationary case, as shown in Figure 2.18, the resulting horizontal and vertical forces acting on the tire are equal:

$$\text{Wheel torque:} \quad T_T = F_Z \cdot r_{dyn} + F_G \cdot f \tag{2.7}$$

FIGURE 2.18 Forces on a tire. Adapted from Renius [48].

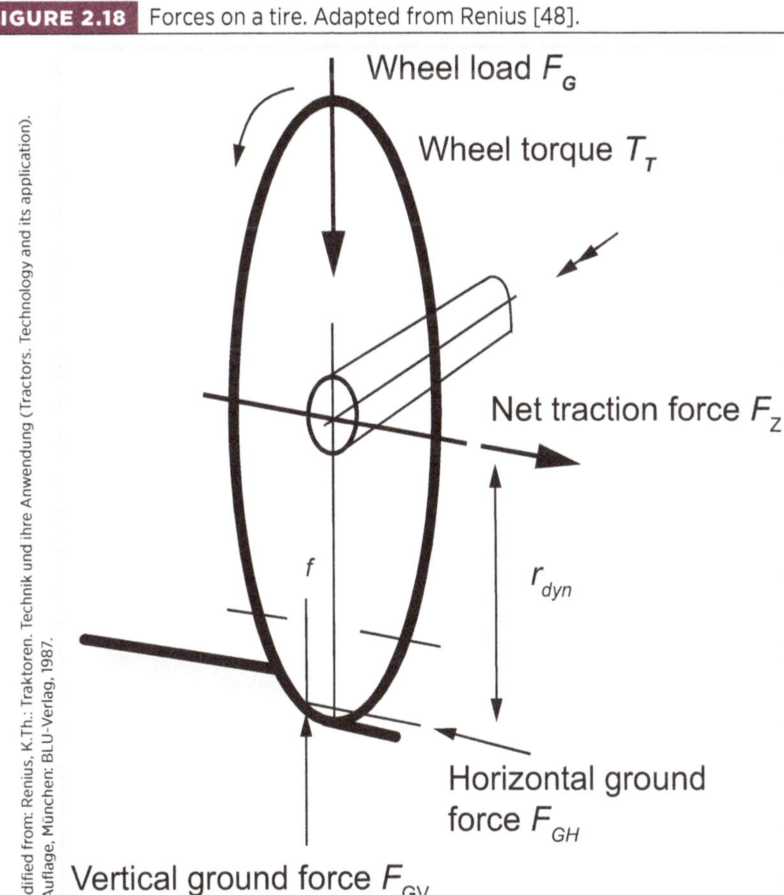

Modified from: Renius, K.Th.: Traktoren. Technik und ihre Anwendung (Tractors. Technology and its application). 2. Auflage, München: BLU-Verlag, 1987.

$$\text{Net traction force: } F_Z = F_{GH} \tag{2.8}$$

$$\text{Static wheel load: } F_G = F_{GV} \tag{2.9}$$

In terms of technical mechanics, the forces shown are the forces cut free on the tire. The forces resulting between the tire and the ground are summarized in one point. The torque of the driven wheel T_T is balanced with the forces between tire and ground, the horizontal ground force F_{GH}, and the vertical ground force F_{GV}, see Figure 2.18. The ground forces on the other hand are balanced with the net traction force F_Z and the static wheel load F_G. According to ASABE, the horizontal traction force is named "net traction force" to distinguish it from the gross traction [49]. The gross traction has two parts, the net traction and the motion resistance of the traction device. Because motion resistance and gross traction are acting in one point, the two parts are summarized here.

For practical use, the efficiency of the tire-soil contact is of great importance. It is calculated taking the power from the traction drive as the input power into account as well as the vehicle speed and pulling force, see also tractive efficiency in ASABE's *General Terminology for Traction of Agricultural Traction and Transport Devices and Vehicles* [49]:

$$\eta = \frac{P_{out}}{P_{in}} = \frac{F_z \cdot v_{vehicle}}{T_T \cdot 2\pi \cdot n_T} = \frac{\kappa}{\kappa + \rho} \cdot (1 - s) \tag{2.10}$$

The velocity of the vehicle $v_{vehicle}$ and the speed of the wheel n_T can be both measured. According to ASABE [49] as well as Söhne [51] and Steinkampf [52], wheel specific values are defined:

$$\text{Net traction ratio: } \kappa = \frac{F_Z}{F_G} \tag{2.11}$$

$$\text{Wheel peripheral force: } F_U = \frac{T_T}{r_{dyn}} = F_Z + F_G \cdot \frac{f}{r_{dyn}} \tag{2.12}$$

$$\text{Rolling resistance ratio: } \rho = \frac{f}{r_{dyn}} \tag{2.13}$$

The net traction ratio is an important value to characterize the tire-soil contact. It is transferable to different loads and characteristic for the contact. A characteristic curve has already been shown in Figure 2.17. The wheel peripheral force is in balance with the wheel torque and has two parts because the resulting point of the acting forces must not necessary lay in a vertical line below the wheel hub. Therefore, the dynamic wheel radius r_{dyn} and the point of horizontal shift f have to be taken into account. The rolling resistance ratio can be calculated with the mentioned values [48].

The vehicle speed $v_{vehicle}$ can be measured or calculated in dependence of the slip s and the covered distance l. The covered distance is the distance a vehicle moves when the wheel is turned once; therefore, the vehicle speed can be calculated as follows:

$$v_{vehicle} = l \cdot n_T \tag{2.14}$$

The lower picture of Figure 2.19 shows the covered distance l of a wheel with a pulling load, which creates a slip between tire and ground. On top of the figure, the distance l_0 of a slip-free movement of a tire on the ground is shown. In case of positive pulling loads l will be lower than l_0 and the slip s is positive; in case of breaking loads, the slip will become negative because $l > l_0$:

$$s = \frac{l_0 - l}{l_0} \tag{2.15}$$

FIGURE 2.19 Slip of a tire, distances at one turn of a wheel.

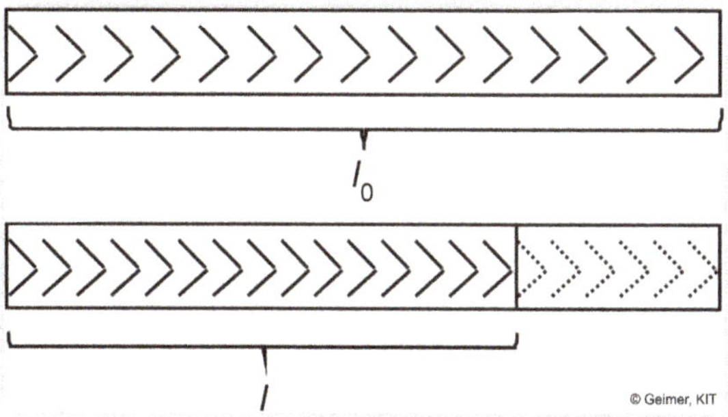

Taking these equations into account, the following dependency between the vehicle velocity and wheel speed can be found:

$$v_{vehicle} = (1 - s) \cdot 2\pi \cdot r_{dyn} \cdot n_T \tag{2.16}$$

In all these formulas, the dynamic wheel radius r_{dyn} is needed. It is defined in dependence of the length l_0 of a slip-free wheel:

$$l_0 = 2 \cdot \pi \cdot r_{dyn} \tag{2.17}$$

Looking at the ideal conditions shown in Figure 2.20, the slip can easily be defined in dependence of the theoretical distance covered. When neither the wheel nor the ground is deformable, r_{dyn} is equal to the outer diameter of the wheel and therefore just dependent on the geometric size.

Taking the deformations of the wheel and the ground into account, the dynamic wheel radius cannot be found easily. Therefore, a definition of zero slip has to be made, but today no common definition can be found. ASABE [49] defines, as an example, four possible conditions for zero slip.

Taking a nondeformable surface as ASABE's first condition for self-propelled conditions and the third for towed conditions into account, it is obvious that the dynamic radius varies between these two conditions: At the first condition, the self-propelled wheel has to overcome the rolling resistance by the driving torque. Here, no traction force is generated. On the other hand, a towed condition, like mentioned as third condition, will create a traction force. Two different points are therefore defined in a slip diagram with the two mentioned conditions.

FIGURE 2.20 Movement between tire and ground.

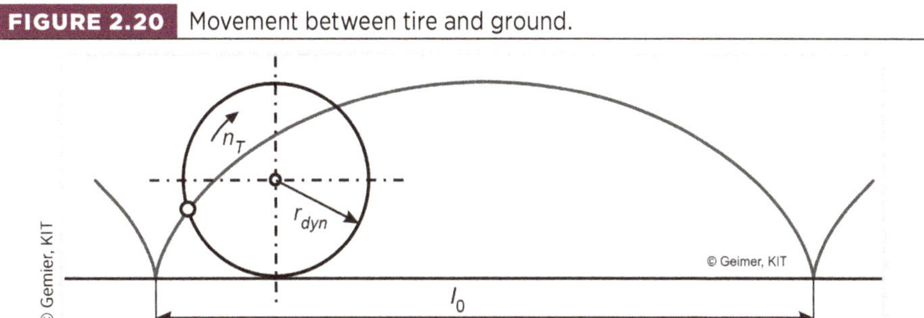

A more general description, when slip occurs, can be found in Bock [50]:

> *Slip is the relative "loss of distance" due to the sliding of the wheel on the ground, as well as soil and tire deformation.*

Looking more into detail, three definitions of slip are today in use:

1. The slip is zero if the horizontal pulling force is zero, in line with ASABE's *General Terminology for Traction of Agricultural Traction and Transport Devices and Vehicles* [49], first condition.

   Here, the torque on the wheel has to overcome the rolling resistance.
2. The slip is zero in between a pulling free and a torque free wheel [50, 51, 52]:

   Here, the assumption is made that a driven wheel without a pulling force has a small positive slip and a pulled wheel without a torque has a small negative slip. This definition is usually used for a real tire on a deformable ground.
3. If the torque on the wheel overcomes the rolling resistance, the slip is zero [53]:

   The torque on the wheel is equal with the inner rolling resistance of the tire deformation. An outer rolling resistance from the ground deformation is representing a pulling force.

The last definition is a theoretical one, equal to the first one on a non-deformable ground and difficult to measure in reality on a deformable ground [54]. An assumption, which is also in line with the car research, takes the type of contact into account [52]:

> *there is no slip if the tangential acting force between tire and ground is zero.*

This assumption seems to be plausible for frictional engagement because there cannot be any movement between two bodies if there are no moving forces, Newton's Law. Taking the example of Figure 2.20, there will be no slip if there is no pulling force on the tire.

For the discussion of the dynamic wheel radius, different situations are presented and discussed here. First, it is assumed that the ground cannot be deformed but the wheel is deformed, see **Figure 2.21**. This is the typical case if a car or machine is driving on a road.

FIGURE 2.21 Forces on a wheel at zero slip (left) and a slip curve (right).

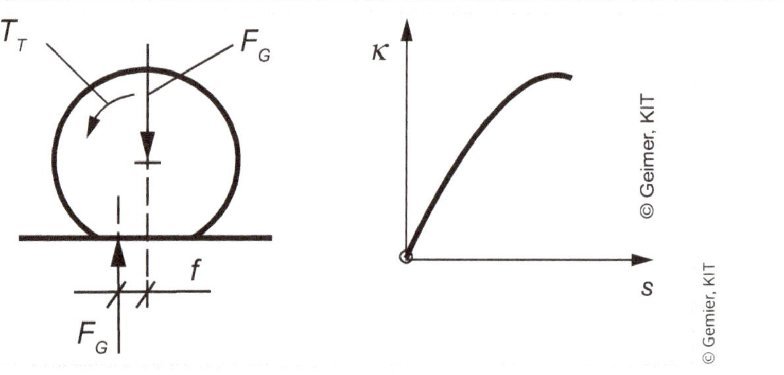

Because the ground is flat, no tangential forces between tire and ground are acting, when the pulling force F_Z is zero. For this case, the forces are shown in the left side of Figure 2.21. The resulting force between tire and ground is vertical because the pulling force is zero. The torque T_T on the wheel has to overcome the rolling resistance; thus, the resulting force between ground and tire has a defined distance f from a vertical line through the wheel hub. A traction-slip curve is shown on the right side of Figure 2.21 and the described point ($s = 0$; $\kappa = 0$) is marked with a circle in this picture. This is in line with the first definition of ASABE's *General Terminology for Traction of Agricultural Traction and Transport Devices and Vehicles* [49] and the third one of Grecenko [53]. In measured slip curves, it has been shown that r_{dyn} lies between the radius of the undeformed tire and the distance between the axle and the ground.

Taking a deformed wheel and a deformed ground as the second case now into account, the shape and force for zero slip according to the assumption of Steinkampf [52] cannot be predicted. According to the second definition of Bock [50], Söhne [51], and Steinkampf [52], a first measuring (a) can be done with a pulled wheel. The torque on the wheel is zero and therefore the reacting force between tire and ground is running through the wheel hub. It is assumed that this will result in a small breaking slip.

A second measuring (b) can be done with zero pulling force. The horizontal forces are therefore zero and the driving torque will result in a small traction slip. This procedure is also in line with the second and fourth definition of zero slip from ASABE [49]. According to the second definition of Bock [50], Söhne [51], and Steinkampf [52], l_0 and r_{dyn} are calculated as the average of the distance of a pulled machine without load and distance of a tractor that is driven by itself without a pulling force:

$$l_0 = \frac{l_{(a)} + l_{(b)}}{2} = 2 \cdot \pi \cdot r_{dyn} \tag{2.18}$$

At these ground conditions, a pulled machine (a) has a small negative slip and the driven-by-itself machine (b) has a small positive slip. The two situations and these points in the slip curve are shown and marked in **Figure 2.22**. With this definition, zero slip can be measured easily.

With the abovementioned method, it is possible to create slip curves for tires and to calculate the behavior of the soil-tire contact. In the field of agriculture, a maximum traction force is reached at about 20% slip on a dry surface. Therefore, in practical

FIGURE 2.22 Definition of slip on a deformed wheel and deformed ground.

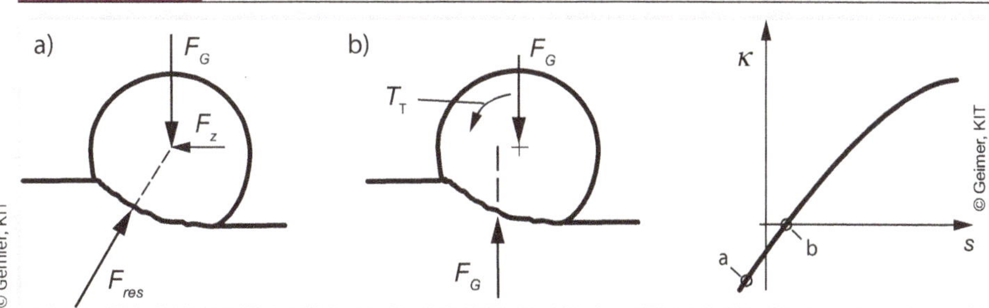

operation a slip near to the maximum efficiency at about 10% is suggested. Nevertheless, the slip is much dependent on the tire pressure, and for a comparison of different published slip curves, the used method and values have to be taken into account.

2.4 Steering Systems (By Markus de la Motte and Thomas Pippes)

There are three functional principles for steering systems at mobile working machines: Ackermann steering, articulated steering, and skid steering, see Figure 2.23.

While skid steering systems use different tire speeds to create a steering movement, Ackermann steering or articulated steering systems are operated typically by a steering actuator, for example, hydraulic cylinder, in order to create a steering force. The Ackermann steering is for instance the standard steering for tractors and a great variety of mobile machines like excavators and telehandlers with one or two steered axles. Articulated steering, the front frame is twisted in relation to the rear frame, is widely used in wheel loaders and similar mobile working machines.

Due to the demand of high steering forces at mobile working machines, mechanical steerings (steering force transmitted purely mechanical) have been replaced by hydrostatic steering systems in the 1960s and 1970s. Furthermore, the replacement of the rigid mechanical connection between steering wheel and steering axle by hydraulic hoses allows a great construction flexibility for the chassis design. Nowadays, electrohydraulic steering or steer-by-wire systems are introduced in mobile machines to improve driving comfort or driving safety and to realize functions like autonomous driving or variable steering ratio.

According to ISO 10998, the steering command can be transmitted purely mechanical, purely hydraulic, pure electric, or in a combination as a hybrid steering transmission. In addition to that, steering systems can be differentiated as passive, semi-active, or active systems as shown in Figure 2.24. Compared to passive systems, the steering command in active steering systems can be provided by an electronic control unit.

FIGURE 2.23 Functional principles for steering systems.

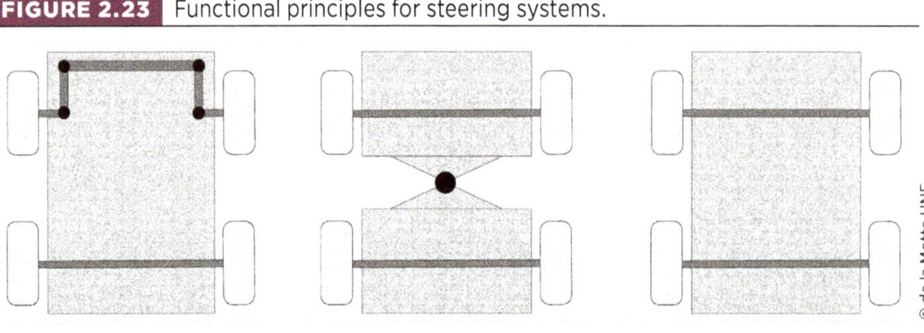

FIGURE 2.24 Steering systems for mobile machines according to Bosch Rexroth Projekthandbuch Hydrostatische Lenksysteme (Project Handbook hydrostatic steering systems) R961009559 [56].

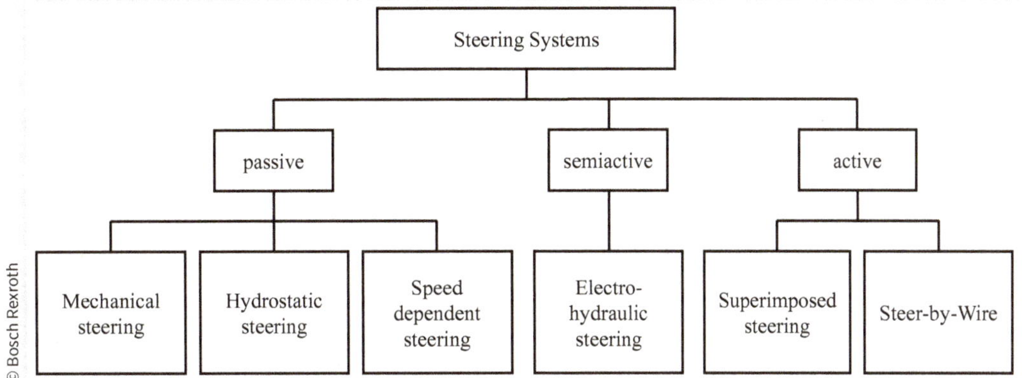

2.4.1 Design and Functions of Hydraulically Assisted Steering Systems

Hydraulically assisted steering systems belong to the passive steering systems. They are amplifying the driver's steering effort by a hydrostatic steering gear.

Figure 2.25 shows the structure of an assisted steering. This steering system consists of hydraulic oil reservoir (1), pipes and hoses (5), hydraulic pump (2), steering cylinder (6), steering control unit (SCU) (4), and steering column with steering wheel (3).

Most of the required input energy for steering is provided by the hydraulics. The effort of the driver for turning the steering wheel is only a very small part of the complete steering input energy.

The principal structure of a hydraulically assisted open center (OC) steering system is shown in the block diagram Figure 2.26.

While Figure 2.26 describes the energy supply of one pump only to the steering system (OC with dedicated oil supply), Figure 2.27 shows that the pump charges a steering system and other consumer with lower priority (low-ranking part-system). A hydraulically

FIGURE 2.25 Example of hydraulically assisted steering system.

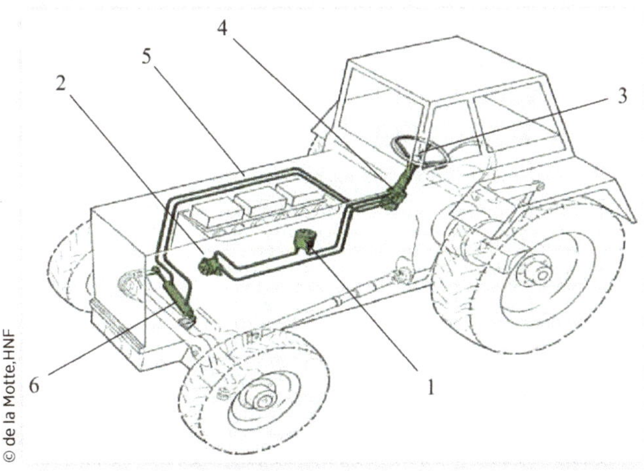

FIGURE 2.26 Hydraulically assisted OC steering system.

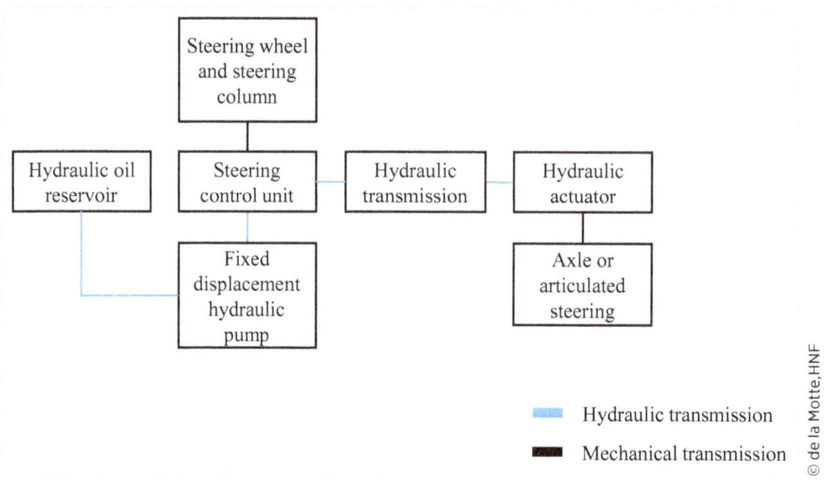

FIGURE 2.27 Closed center-load sensing steering system.

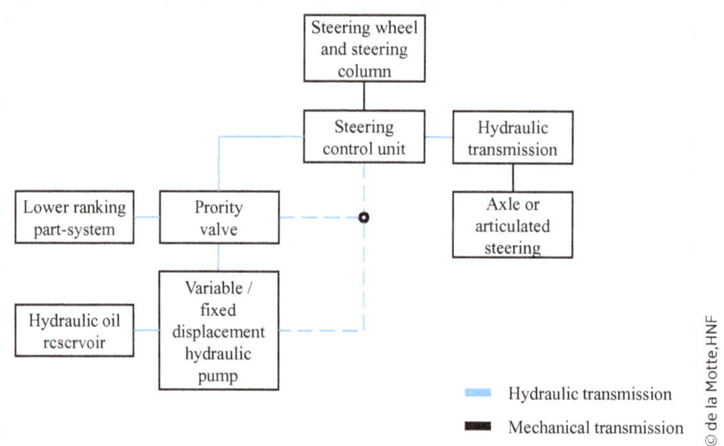

piloted three-way flow control valve (priority valve) realizes the supply to the SCU with first priority. The SCU needs to be a closed center (CC) unit with load sensing (LS) port. It gets charged on demand by providing a LS signal to the priority valve and the lower ranking part-system get always the remaining rest of the total pump supply.

The type of pump can either be a constant displacement pump or a variable displacement pump. Such CC-LS systems enable the user to reduce the fuel consumption and save energy. That helps to bring the TCO to a lower level.

All SCUs have a rotary directional control valve for leading the oil flow internally. OC SCUs are returning the oil flow from the pump to tank during neutral position (no steering, see Figure 2.28). At CC-LS SCUs, the inlet port is internally locked (see Figure 2.29). The oil flow is led via the priority valve to the lower ranking part-system.

The directional control valve forwards the oil flow into the gerotor set during clockwise or anticlockwise steering and from the gerotor set to the steering actuator port L or R during normal steering (also called servo operation). As one of its functions, the

FIGURE 2.28 OC steering system direction control valve + gerotor set in neutral position.

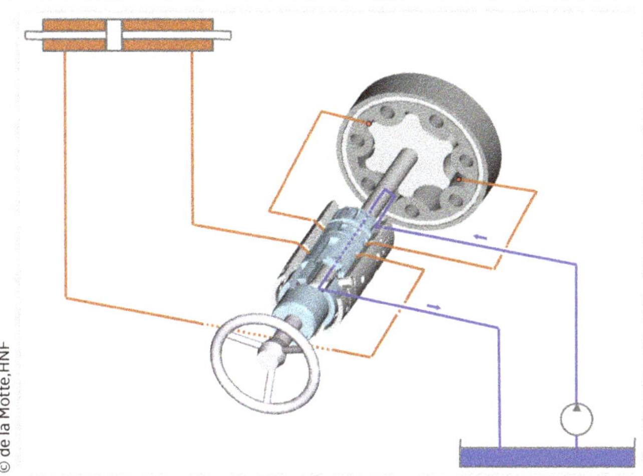

FIGURE 2.29 CC-LS steering system directional control valve and gerotor set with priority valve in neutral position.

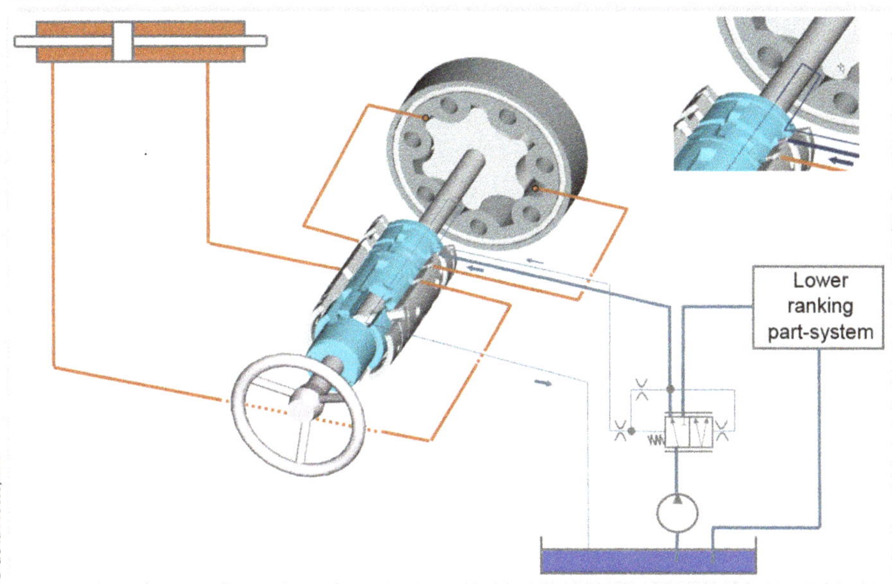

gerotor set meters the oil flow proportional from the (steer supply) hydraulic pump in relation to the steering wheel speed to the steering actuator (see Figures 2.30 and 2.31).

In a second function, the gerotor set works as a manually driven pump for manual steering (also called emergency operation) when the oil supply to the SCU fails. During manual steering, the turning effort on the steering wheel is 10-30 times more compared with the turning effort at normal steering.

The steering capability of a machine is defined by maximum allowed effort on the steering wheel ring by legal (EC machinery directive 2006/42/EG and EU 167-2013)

FIGURE 2.30 OC steering system directional control valve + gerotor set-normal steering.

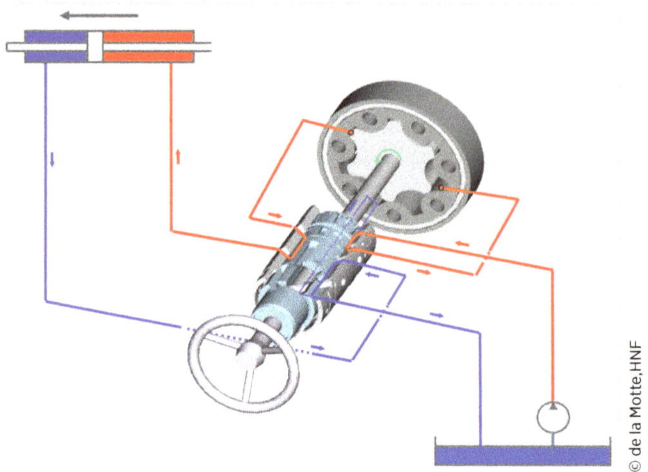

FIGURE 2.31 CC-LS steering system directional control valve + gerotor set with priority valve-normal steering.

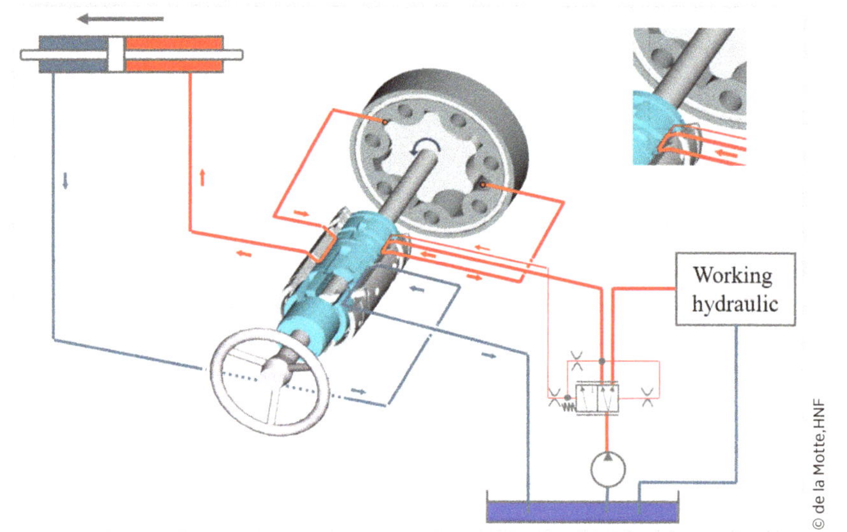

and/or standard requirements (ISO5010 or ISO10998). These limits shall not be exceeded, while driving through a test circle.

For the safety of the steering systems, valves are required, for example, check valves and pressure relief valves, which are typically integrated in a SCU. There is an inlet check valve in the P-port for preventing strong kickbacks on the steering wheel and to avoid sucking of air in the event of manual steering due to a broken P-line. A suction check valve (also called manual steering check valve) is required between the T-area and the P-area inside behind the inlet check valve. For limiting the input pressure and protecting the steering system supply pump, a pressure relief valve is required. The work ports must also be secured by pressure relief valves (also called anti shock valves) and parallel to these are anti-cavitation valves required for preventing vacuum pressure in the work ports L or R. According to legislation, the pressure setting of the anti-shock valves must be that much higher than the pressure relief valve setting; that the anti-shock valves will never open during normal steering operation. The mentioned valves can either be integrated in the SCU or can be separately installed in the steering system. Figure 2.32 shows symbols of a complete equipped SCU.

Related to the application requirements and steering comfort of Ackermann or articulated steered mobile working machines, load reactive (R) or non-load reactive (NR) SCUs are used. R means that the steer cylinder ports (work ports) are connected internally of the SCU to each other via the gerotor set (see Figure 2.32). That provides a feedback from the steered wheels on the axle to the steering wheel and the operator gets a similar steering feeling as known from automotive applications. R SCUs are only used for mobile working machineries with front axle Ackermann steering because they provide the right kinematics for self-centering to straight ahead while driving. Mainly fast driving on-road applications like agricultural tractors and some municipal machines have R SCUs.

At the NR SCUs are the cylinder ports always locked when the steering wheel does not get turned by the operator. Feedback from the wheels or articulated frame to the steering wheel is not possible due to this. These SCUs can be used for all applications. But they are needed to use for applications with articulated or rear axle steering or

FIGURE 2.32 Schematics of steering unit with single and double gerotor set.

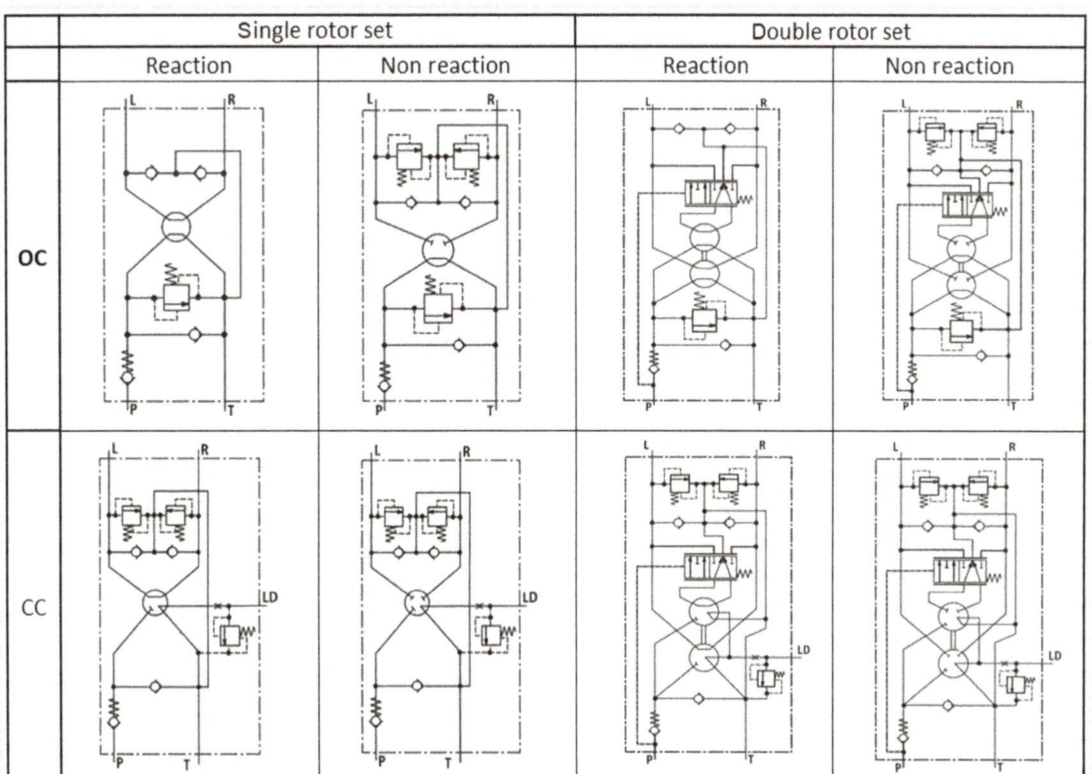

multi-axle steering because such machineries don't have a good self-centering to straight ahead. Mainly off-road-used applications like construction machines (except motor graders), combines, and harvesting machineries or material handling machineries have NR SCUs.

OC as well as CC-LS SCUs are available on the market either with R or with NR feature.

The agricultural and construction machinery market of medium heavy machines required in the beginning of the 1990 a cost-effective steering system that could realize manual steering according to legal requirements without using emergency oil supply from a back-up pump. Besides an existing SCU with two gerotor sets, two additional principals were introduced: a system with displacement amplification and a system with displacement reduction. The system with displacement amplification works with a gerotor set displacement required and sufficient for manual steering. A bypass to the gerotor set is adding a specific oil flow to the displacement of the gerotor set while normal steering. When the oil supply from the pump fails, the gerotor set displacement works only for manual steering. This requires a lower effort for turning the steering wheel during manual steering compared with the full amplified displacement.

The system with displacement reduction works with the gerotor set displacement as it is required and sufficient for normal steering. During manual steering, gerotor set chambers are automatically unloaded through an internal bypass loop so that the displacement gets reduced by half. Due to this, the steering wheel turning effort is reduced by half.

At the end of the 1970s, SCUs with double gerotor set were developed for the same purpose as abovementioned. They are working with two gerotor sets for normal steering. One gerotor set gets automatically unloaded on an internal bypass loop. When the oil supply fails, only one gerotor set works for manual steering. Due to this principal, it is possible to reduce the displacement by different ratios in accordance of the combined gerotor sets. That enables the use of such SCUs for medium-to-heavy machineries without any emergency oil supply from a back-up pump.

A special version of the steering units with double gerotor set was developed for offering a simple solution for realizing two steering ratios for normal steering. The operator gets the opportunity for selecting the ratio depending on the operation requirements. For example, it gets possible for driving on-road with four turns lock to lock and while working with a front loader or at the field head end switching to a steering ratio with two turns lock to lock.

This feature is realized by shifting the displacement active by using a solenoid activated directional pilot valve. Due to the generation of pilot pressure from the input of the steering unit, it also gets possible to use the displacement reduction when the oil supply to the steering unit fails. Figure 2.33 shows the principal schematic of a steering unit with active piloting of one of the double gerotor sets.

Modern agricultural machines as well as other applications have nowadays two parallel equipped steering systems-the main steering with the steering wheel and SCU for all operations and in parallel the secondary with a GPS- or joystick-controlled steer-by-wire system for off-road operations. If the main steering system is equipped with a R SCU, the reactive path needs to be locked while using the secondary steering.

FIGURE 2.33 Schematic of steering unit with double gerotor set for active shifting displacement and displacement reduction [55].

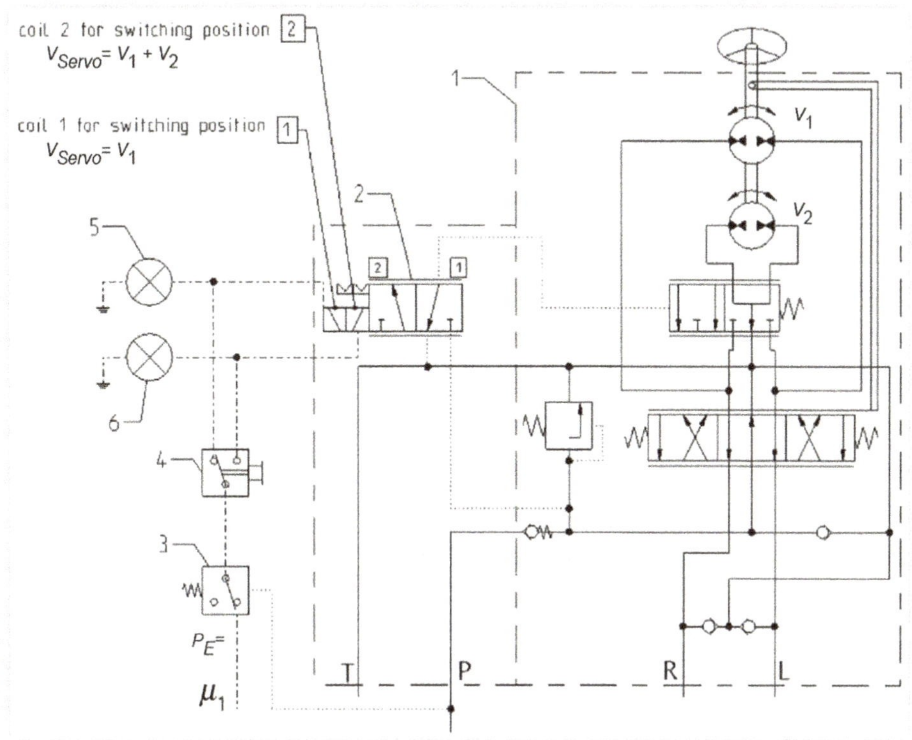

Several solutions with additional valve functions for this purpose are present in the market. The steering wheel must always have priority when the operator is turning it by law and the locked reactive path must be released immediately. Functional safety requirements have to be considered for this. An optimal solution for realizing such a request is a NR ⇔ R switchable SCU solution that works with all functions in both modes. This option can also be used for adapting the machine optimal on the operation requirements, for example, during fast on-road driving with R mode and slow off-road with NR mode.

2.4.2 Design and Functions of Semi-Active and Active Steering Systems

With hydrostatic steering systems presented in Section 2.4.1, it is possible to control mobile working machines reliably and precisely. Due to the wide range of applications for mobile working machines, there are different and sometimes contradictory requirements for their steering systems. The example of a tractor is a good illustration for this. For transport travel drives at high-speed and high vehicle-mass, an indirect, stiff steering behavior with a low gear ratio is advantageous, in order not to stimulate the trailer to uncontrolled movements by steering movements. When working in the field or at low-speed loading activities, direct and high translating steering behavior has a positive effect on the operator's workload and the productivity of the process.

FIGURE 2.34 Schematic of a NR ⇔ R switchable steering unit.

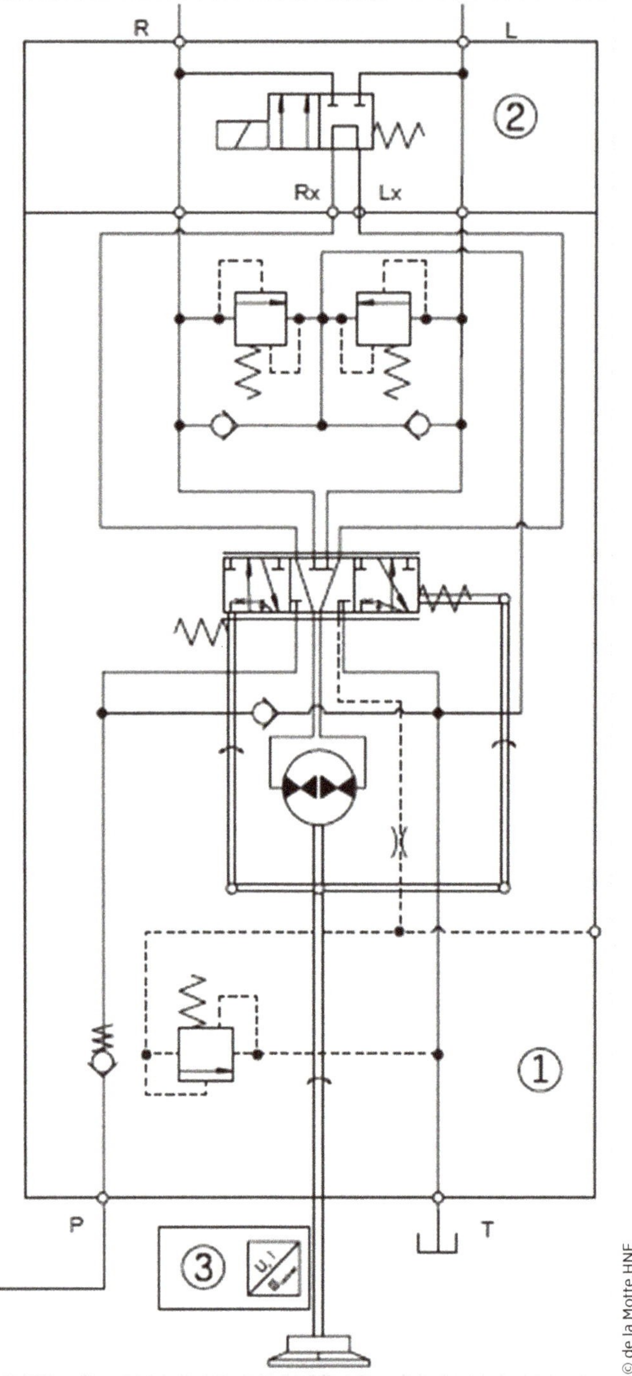

Conventional hydraulically assisted steering systems can't keep up with these requirements. Therefore, passive steering systems get extended by a semi-active or active part. In the following, the structure and implementable steering functions are described.

2.4.2.1 **Structure:** Basic core components of semi-active steering systems are illustrated in Figure 2.35.

Semi-active steering systems consist of a SCU, a parallel valve configuration (Valve block), some sensors, and an electric control unit (ECU) as well as an external set-point generator. Parallel to the SCU, the valve configuration provides an additional oil flow for steering. Set points are generated by sensors and transmitted to the ECU, which processes the set points and controls the valve configuration accordingly.

Typical semi-active steering system representatives are electro-hydraulic steering systems, also called add-on steering.

Active steering systems, especially superimposed steering, basically have a similar structures as semi-active steering systems. However, the active and passive parts of superimposed steering systems work simultaneously. Furthermore, this allows more complex steering functions to be implemented.

The hydraulically assisted SCU works in case of an error at the semi-active or active part as fall-back level.

In the case of the other representatives of active steering systems, called steer-by-wire steering systems, the passive steering component is no longer required. The mobile machines are only steered via the active part. Figure 2.36 illustrates the steer-by-wire system structure exemplarily.

Steer-by-wire steering systems basically consist of a "steering transmission," some sensors, a SCU as well as various external set-point generators, like Joysticks or electronic steering wheels. By using these external set-point generators, set points are generated and transmitted to the ECU. This device processes the set points and controls the

FIGURE 2.35 Structure of semi-active/active steering systems.

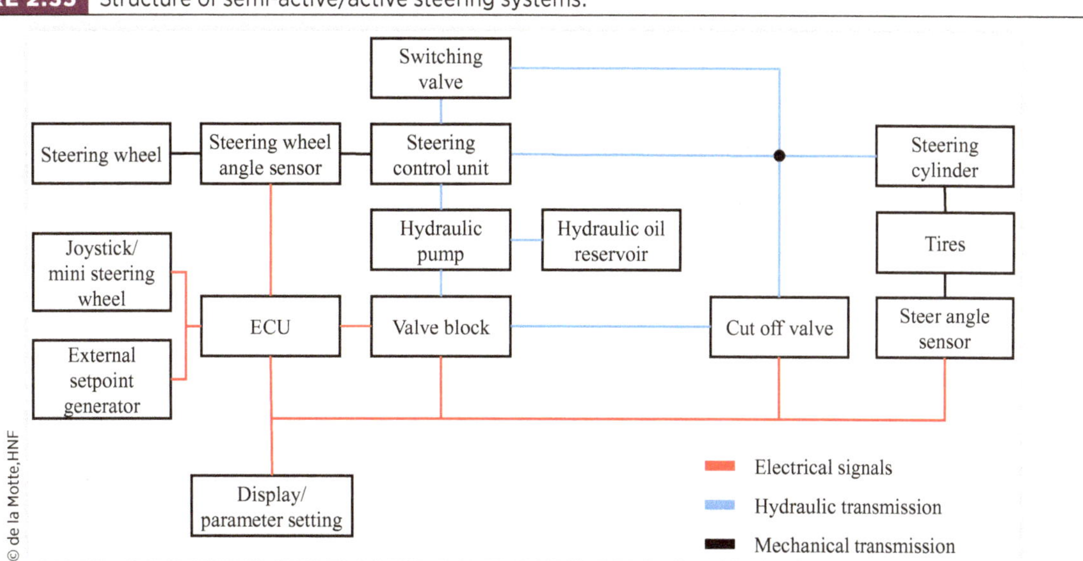

FIGURE 2.36 Structure of a steer-by-wire steering system.

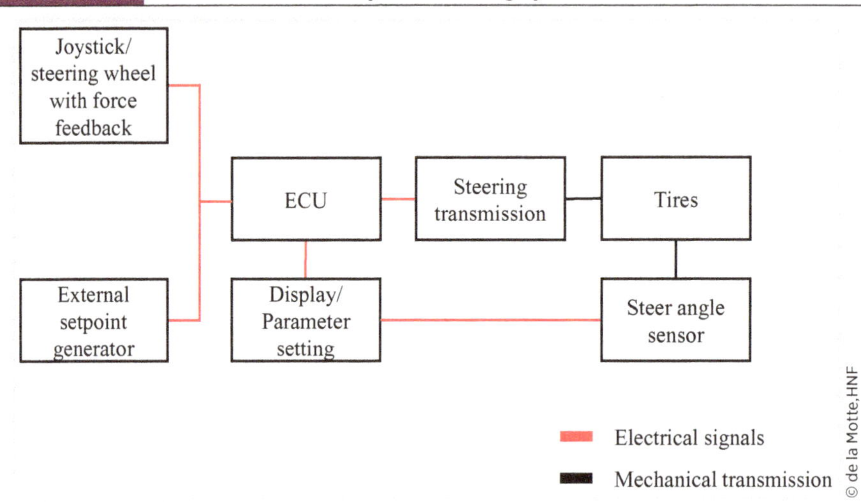

"steering transmission" accordingly. The "steering transmission" can either be an electrohydraulic or an electromechanical conversion. In case of an error, the system falls back on a so-called back-up steering transmission.

2.4.2.2 **Steering Functions:** With the semi-active and active steering systems listed in Table 2.2, the following steering functions can be realized.

2.4.2.3 **Parameterizable Superimposition:** At low vehicle speeds, the electrohydraulic valve configuration provides the biggest possible additional oil flow for steering to achieve direct steering response and reduce the driver's steering work. On the other hand is indirect steering behavior desirable for travel drives at high speed, that is, no additional oil flow from the valve configuration is required. Steering is performed exclusively with the hydrostatic SCU.

TABLE 2.2 Steering functions.

	Semi-active steering system	Active steering system	
Steering function	Electrohydraulic steering	Superimposed steering	Steer-by-wire
Parameterizable superimposition	○	●	●
GPS tracking	●	●	●
Joystick steering	●	●	●
Reverse driving system	●	●	●
Twelve o'clock steering wheel position	○	●	●
Stable straight ahead drive	○	●	●
Driving dynamics (steering assistance)	○	◑	●

2.4.2.4 **12 O'clock Steering Position:** Due to the design, hydrostatic SCUs have some slip on the steering wheel. This leads to minor further turning of the steering wheel when moved to the steering end stop. In this case, a defined 12 o'clock steering wheel position gets lost. This means the steering wheel angle and the wheel angle are no longer in line or in relation to each other. This behavior can be compensated by the electrohydraulic valve configuration by supplying an exactly measured additional oil flow.

2.4.2.5 **Stable Straight Ahead Drive:** If external forces are working on the steered wheels, for example, driving sideways on a slope, that may lead to little direction deviations. This undesired wheel-angle movement can be compensated by the supply of an exactly measured amount of oil to the related steering cylinder side by the electrohydraulic valve configuration.

References

1. Pischinger, S. and Seiffert, U. (Eds.), *Vieweg Handbuch Kraftfahrzeugtechnik (Handbook Automotive Technology)*, 8th edn. (Wiesbaden: Springer Vieweg, 2016).
2. Hoepke, E. and Breuer, S. (Eds.), *Nutzfahrzeugtechnik: Grundlagen, Systeme, Komponenten (Commercial Vehicle Technology: Basics, Systems, Components)*, 8th edn. (Wiesbaden: Springer Vieweg, 2016).
3. Kutz, M. (Ed.), *Mechanical Engineers' Handbook, Volume 3: Manufacturing and Management*, 4th edn. (Hoboken, NJ: Wiley, 2015).
4. Directive 2006/42/EC of the European Parliament and of the Council of 17 May 2006 on Machinery, and Amending Directive 95/16/EC (Recast).
5. "Regulation for Emission from Heavy Equipment with Compression-Ignition (Diesel) Engines," United States Environmental Protections Agency (EPA), https://www.epa.gov/regulations-emissions-vehicles-and-engines/regulations-emissions-heavy-equipment-compression, last view March 13, 2018.
6. Kunze, G., Goering, H., and Jacob, K., *Baumaschinen: Erdbau- und Tagebaumaschinen (Construction Machines: Earthmoving and Open-Cast Mining Machines)* (Wiesbaden: Springer Vieweg, 2002).
7. Renius, K.T., *Tractors: Two Axle Tractors, in CIGR Handbook of Agricultural Engineering, Volume III: Plant Production Engineering* (St. Joseph, MI: ASABE, 1999), 115-184.
8. Cottin, D., *Entwicklung von Landmaschinentragwerken (Development of Agricultural Machinery Structures)* (Frankfurt: VDMA Verlag, 2013).
9. Cottin, D., *Geschweißte Landmaschinentragwerke (Welded Agricultural Machinery Structures)* (Frankfurt: VDMA Verlag, 2013).
10. Knechtges, H. and Renius, K.T., "Gesamtentwicklung Traktoren (Development of tractors)," in: Frerichs, L. (Ed.), *Jahrbuch Agrartechnik (Yearbook Agricultural Engineering) 2016* (Braunschweig: Institut für mobile Maschinen und Nutzfahrzeuge, 2017), http://publikationsserver.tu-braunschweig.de/get/64169.
11. *MAN Bus and Truck, Grundlagen der Nutzfahrzeugtechnik LKW und Bus - Lehrbuch der MAN Academy (Basics of Commercial Vehicle Technology Truck and Bus - Textbook of the MAN Academy)* (Bonn: Kirschbaum, 2016).

12. Pahl, G., Beitz, W., Feldhusen, J., and Grote, K.-H., *Engineering Design - A Systematic Approach* (London: Springer, 2007).

13. *VDI-RICHTLINIE 2221: Methodik zum Entwickeln und Konstruieren technischer Systeme und Produkte (Methodology for the Development and Design of Technical Systems and Products)* (Düsseldorf: VDI-Verlag, 1993).

14. *VDI-RICHTLINIE 2222: Methodisches Entwickeln von Lösungsprinzipien (Methodical Development of Solution Principles)* (Düsseldorf: VDI-Verlag, 1997).

15. ISO/IEC/IEEE 15288: Systems and Software Engineering - System Life Cycle Processes, Geneva, 2015.

16. Linde, H. and Hill, B., *Erfolgreich Erfinden – Widerspruchsorientierte Innovationsstrategie für Entwickler und Konstrukteure (Successful Inventing - Contradiction-Oriented Innovation Strategy for Developers and Designers)* (Darmstadt: Hoppenstedt, 1993).

17. Gausemeier, J., Ebbesmeyer, P., and Kallmeyer, F., *Produktinnovation – Strategische Planung und Entwicklung der Produkte von morgen (Product Innovation - Strategic Planning and Development of Tomorrow's Products)* (München: Hanser, 2001).

18. Bonvoisin, J., Halstenberg, F., Buchert, T., and Stark, R., "A Systematic Literature Review on Modular Product Design," *Journal of Engineering Design* 27, no. 7 (2016): 488-514, doi :10.1080/09544828.2016.1166482.

19. Bendsoe, M. Ph. and Sigmund, O., *Topology Optimization - Theory, Methods, and Applications* (Berlin: Springer, 2004).

20. Fleczoreck, Th., Effzienzbewertung von Antrieben mobiler Arbeitsmaschinen am Beispiel eines Mähdreschers (Efficiency Evaluation of Drives of Mobile Machines Using the Example of a Combine Harvester), Dissertation, Technische Universität Braunschweig (Aachen: Shaker, 2013).

21. Frerichs, L., Hanke, S., Steinhaus, S., and Trösken, L., "EKoTech - A Holistic Approach to Reduce CO_2 Emissions of Agricultural Machinery in Process Chains," SAE Technical Paper 2017-01-1929, 2017, doi:https://doi.org/10.4271/2017-01-1929.

22. Hanke, St., Trösken, L., and Frerichs, L., "Development and Parameterization of an Object-Oriented Model for Describing Agricultural Process Steps," *Landtechnik* 73, no. 2 (2018): 22-35, doi:10.15150/lt.2018.3179.

23. Wu, X., "Entwicklung einer Simulationsmethode für maschinelle stoffgebundene Arbeitsprozesse (Development of a Simulation Method for Mechanical Material-Bound Work Processes)," Dissertation, Technische Universität Dresden, 2017.

24. Gruening, T., Kunze, G., and Katterfeld, A., "Simulating the Working Process of Construction Machines," *Bulk Solid Europe*, Glasgow, 2010.

25. Schramm, F., Bührke, J., and Frerichs, L., "Untersuchung von Verschleiß in der Bodenbearbeitung mit der Diskreten Elemente Methode (Investigation of Wear in Soil Tillage with the Discrete Elements Method)," in: *Tagung Land. Technik 2018* (Düsseldorf: VDI, 2018), 111-117.

26. Albers, A., Düser, T., and Ott, S., "X-in-the-loop als integrierte Entwicklungsumgebung von komplexen Antriebssystemen (X-in-the-Loop as an Integrated Development Environment for Complex Drive Systems)," *8th Conference Hardware-in-the-Loop-Simulation*, Kassel, 2008.

27. Brinkschulte, L. et al., "MOBiL – Eine auf mobile Arbeitsmaschinen optimierte Prüfmethode (MOBiL - A Test Method Optimized for Mobile Machines), 6th Conference Hybride und eneergieeffiziente Antriebe für mobile Arbeitsmaschinen (6th Conference Hybrids and Energy Efficient Drives for Mobile Machines)," in: *Proceedings, Karlsruher Schriftenreihe Fahrzeugsystemtechnik* (KIT Scientific Publishing), Vol. 50, 173-194, 2017.

28. Hicks, J., *Welded Joint Design* (Cambridge: Woodhead Publishing, 1999).

29. Klein, B., *Leichtbau-Konstruktion - Berechnungsgrundlagen und Gestaltung (Lightweight Construction - Calculation Bases and Design)* (Wiesbaden: Springer, 2013).

30. Wächter, M., Müller, C., and Esderts, A., *Angewandter Festigkeitsnachweis nach FKM-Richtlinie - Kurz und bündig (Applied Strength Verification According to FKM Guideline - Short and Sweet)* (Wiesbaden: Springer, 2017).

31. Hobbacher, A.F., *Recommendations for Fatigue Design of Welded Joints and Components* (Cham: Springer International Publishing, 2016).

32. Köhler, M., Jenne, S., Pötter, K., and Zenner, H., "Rechnerische Lebensdauerabschätzung," in: *Zählverfahren und Lastannahme in der Betriebsfestigkeit (Calculated Lifetime Estimation in Counting Procedure and Load Assumption in the Operational Stability)* (Berlin: Springer, 2012).

33. Siebenpfeiffer, W. (Ed.), *Leichtbau-Technologien im Automobilbau: Werkstoffe - Fertigung – Konzepte (Lightweight Technologies in Automotive Engineering: Materials - Manufacturing - Concepts)* (Wiesbaden: Springer, 2014).

34. Looman, J., *Zahnradgetriebe, Grundlagen, Konstruktionen, Anwendungen in Fahrzeugen (Gear Drives, Basics, Designs, Applications in Vehicles)* (Berlin: Springer, 1996).

35. Ansdale, R.F., "Differential Locks and Limiting Devices," Automotive Eng., January 1963

36. Bekker, M.G., *Theory of Land Locomotion: the Mechanics of Vehicle Mobility* (Ann Arbor: University Michigan Press, 1962), 2nd print.

37. Wong, J.Y., *Terramechanics and Off-Road Vehicle Engineering: Terrain Behaviour, Off-Road Vehicle Performance and Design* (Amsterdam: Butterworth-Heinemann, 2009), ISBN:978-0-75068-561-0.

38. Lewis, R. and Olofsson, U., *Wheel-Rail Interface Handbook* (Cambridge, Whoodhead Publishing Limited, 2009), ISBN:978-1-84569-412-8.

39. Pacejka, H.B., "Tyre Factors and Vehicle Handling," *International Journal of Vehicle Design* 1, no. 1 (1979): 1-23.

40. Ferhadbegovic, B., "Entwicklung und Applikation eines instationären Reifenmodells zur Fahrdynamiksimulation von Ackerschleppern (Development and Application of a Transient Tire Model for Driving Dynamics Simulation of Agricultural Tractors)," Dissertation, Technical University of Munich, 2008.

41. Pacejka, H.B., *Tyre and Vehicle Dynamics* (Oxford: Butterworth-Heinemann, 2002).

42. "The Hohenheim Tyre Model," https://reifenmodell.uni-hohenheim.de/en/home, last update September 20, 2017.

43. "CTI Cosin Tire Interface - API Reference and User's Guide," https://www.cosin.eu/wp-content/uploads/cti.pdf, Revision: 2020-2-r22237, last view May 27, 2020.

44. Hirschberg, W., Rill, G., and Weinfurter, H., "User-Appropriate Tyre-Modelling for Vehicle Dynamics in Standard and Limit Situations," *Vehicle System Dynamics* 38, no. 2 (2002): 103-125.

45. Zumkley, H., "Reifenparametrierung aus Fahrversuche mit einem Ackerschlepper unter besonderer Berücksichtigung des Hohenheimer Reifenmodells (Tire Parameterisation from Driving Tests with an Agricultural Tractor with Special Consideration of the Hohenheim Tyre Model)," Dissertation, University of Hohenheim, 2017.

46. Pacejka, H., *Modeling of the Pneumatic Tyre and Its Impact on Vehicle Dynamics Behaviour* (Delft: TU Delft, 1988).

47. Svendenius, J. and Wittenmark, B., "Brush Tire Models with Increased Flexibility," *European Control Conference (ECC 2003)*, Cambridge, UK, September 1-4, 2003, 1863-1868, doi:10.23919/ECC.2003.7085237.

48. Renius, K.Th., *Traktoren. Technik und ihre Anwendung (Tractors. Technology and Its Application)*, 2. Auflage (München: BLU-Verlag, 1987).

49. "General Terminology for Traction of Agricultural Traction and Transport Devices and Vehicles," ANSI/ASAE S296.5 W/Corr., December 1, 2003 (R2013).

50. Bock, G., "Feldversuche über die Zugfähigkeit von Ackerschlepperreifen (Fleet Tests on the Traction Capability of Tractor Tires)," *Grundlagen Landtechnik* 3 (1952): 88-100.

51. Söhne, W., "Die Kraftübertragung zwischen Schlepperreifen und Ackerboden (Power Transmission between Tractor Tyres and Arable Land)," *Grundlagen der Landtechnik* 3 (1952): 75-87.

52. Steinkampf, H.: Zur Methodik der Rollradien- und Radschlupfmessung (The Methodology of Rolling Radius and Wheel Slip Measurement), *Grundlagen der Landtechnik*, 21, no. 2 (1971): 40-44.

53. Grecenko, A., "Measurements of the Draught Forces from Off-Road Vehicles," in: *Zemedelska Technika* 24, no. 11 (1978): 643-660.

54. Pichlmaier, P., "Traktionsmanagement für Traktoren (Traction Management for Tractors)," Dissertation, Technical University of Munich, 2012.

55. HE14365 of the Hydraulik Nord Fluidtechnik GmbH & Co. KG., Ludwigsluster Chaussee 5, 19370 Parchim.

56. "Bosch Rexroth Projekthandbuch Hydrostatische Lenksysteme (Project Handbook Hydrostatic Steering Systems)," R961009559.

CHAPTER 3

Power Train

Swen Bosch, Prof. Danilo Engelmann, Dr.-Ing. Gerhard Geerling, Edwin Heemskerk, Dr.-Ing. Torsten Kohmäscher, Fabrizio Panizzolo, Dr.-Ing. Christian Pohlandt, and Prof. Dr.-Ing. Heinrich Steinhart

The main task of a mobile working machine is to fulfill a working process in the best possible way. Thus, on the one hand, the machine needs a traction drive to be mobile while on the other hand, the machine needs function drives to perform the work functions. Furthermore, there is a need to power additional auxiliaries to support the operator at operator's working place, like an air conditioner, or to support the working process technology with the power to control and measure systems. These auxiliaries are powered by a mechanic, hydraulic, or electric energy supply. Typical voltage levels for electric auxiliaries and for the batteries used in the machines are 12 or 24 V.

Power and torque of the primary energy provision are commonly supplied in a narrow speed range and need to be adjusted to match the need of the application. Various kinds of system solutions and gearboxes for the traction drive are available in the market and provide the required spread of transmission ratio for maximum tractive force as well as maximum ground speed. The power for the working functions has to be taken simultaneously from the primary energy provision and therefore make the power control a challenging task.

The following chapter will start with a description of the possible primary energy provision possibilities for mobile working machines (3.1). Before describing the traction drive (3.3) and the function drives (3.4), the basics of power conversion (3.2) for mobile working machines are presented.

3.1 Primary Energy Provision (By Danilo Engelmann and Christian Pohlandt)

To power, the function drives and the traction drive of the power train, a primary energy provision is needed. This energy is provided by a primary energy carrier that is further transformed by a prime mover to supply the functions and drives. A conventional power train is powered by a combustion engine followed by a gearbox as a prime mover and a fuel tank representing the energy carrier. An electric power train in comparison is powered by an inverter fed electric machine, representing the prime mover, and a battery carrying the energy. The comparison of a conventional power train versus an electric power train regarding the different energy carriers and prime movers is shown in Table 3.1.

3.1.1 Primary Energy Carriers

Nowadays, liquid fuels are most often the primary sources of energy in mobile machines that are burned in internal combustion engines. Due to the availability, subsidized low costs and the high energy density of the fuel, this form of energy is widespread. The most commonly used fuel in mobile working machines is diesel and, in a few applications, gasoline or gases, methane or petroleum gas stored in liquid or gaseous forms. The characteristics that fuel must have, are specified by norms as seen in Figure 3.1. This provides a uniform basis for the engine developers to design the engines for the fuels without any loss of efficiency.

Besides, electrical energy stored in batteries is used as a further primary energy source. Common battery types used for applications are lead-acid batteries, nickel-metal-hydride batteries (NiMH) and lithium-ion batteries (Li-Ion).

Lead-acid batteries are used as a starter battery to initially power the internal combustion engine, but it is also widespread in fork-lift trucks to supply the power train. To take advantage of the lead-acid battery weight, it is directly used as a counterweight and increases the stability of the forklift with a low center of gravity, see also Section 6.3.

The basic structure of a lead-acid battery consists of two oppositely polarized electrodes. The active material of the negative electrode is lead. The material of the positive electrode is lead oxide. Diluted sulfuric acid is used as an electrolyte to facilitate the chemical reaction. A separator splits the two electrodes and thus prevents a short circuit. To increase the capacity, several pairs of electrodes are usually connected in parallel.

TABLE 3.1 Comparison of different energy carriers and prime movers between conventional and electric power trains.

Conventional power train		Electric power train	
Energy carrier	**Prime mover**	**Energy carrier**	**Prime mover**
Fuel tank	Combustion engine	Battery	Inverter fed electric machine

FIGURE 3.1 Ragone plot for different energy sources useable in mobile machinery. [1].

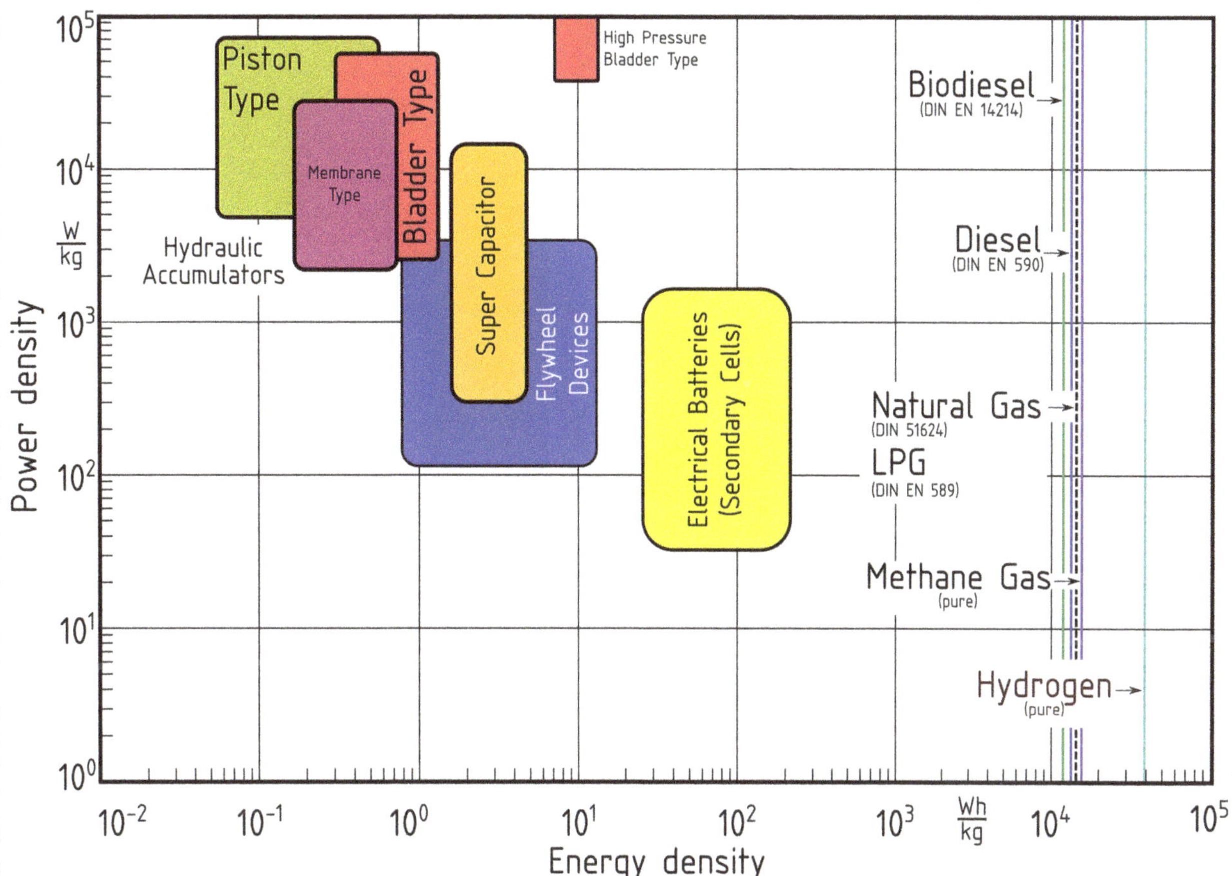

Compared to lead-acid batteries, a NiMH battery consists essentially of a hydrogen-storing metallization. The positive electrode active material is nickel-hydroxide. The active material of the negative electrode is a metal-hybrid alloy. As electrolytes diluted potassium hydroxide is mostly applied to NiMH batteries.

Nowadays, Li-Ion batteries are becoming more developed gaining attention for installation in mobile working machines. It should be noted that a Li-Ion battery behaves differently than lead-acid batteries or NiMH batteries. In a Li-Ion battery, although the Li-Ion is the active element, it also forms the active material of the electrodes themselves. Therefore, there are a variety of ways to combine active materials for the two electrodes, which can donate or absorb lithium ions. Typical active materials are graphite or lithium alloys for example. The positive electrode active consists of lithium iron phosphate or mixtures of cobalt-nickel manganese. The main aspects of combining suitable materials are the stability of the chemical reaction, safety, and lifetime.

The chemical reaction in a Li-Ion battery is principally generated by ion exchange between the two electrodes. Since lithium is the lightest metal and reacts very strongly with water, electrolytic anhydrous solvents are used in combination with salts. The separator is built with polyethylene or polypropylene.

Besides the above-mentioned batteries as primary energy sources, mobile machines with cable-guided energy supply are used in special applications of mining and logistics. An example is the port equipment which is used to load and unload ships.

The different energy sources can be characterized by gravimetric energy and power density or normalized with a volumetric base. Such a comparison of the different energy storage technologies allows a Ragone plot as shown in Figure 3.1.

Originally, the Ragone plot was created to compare the performance of different batteries. The gravimetric form of a graph shows the energy density in Wh per kg compared to the power density measured in W per kg and makes it possible to compare the different energy sources in terms of their weight. Since the ratio of mass and volume can differ, especially for gases, it is necessary for an appropriate selection of energy storages and therefore, the drive design should always consider both gravimetric and volumetric values.

The gravimetric and volumetric reference values must also be known. For example, the specified calorific values refer only to the fuel mass itself, i.e., without the mass of the tank system, similar to the electrical storages, the referenced mass values here vary between the mass of the individual basic cells and the stack. Different reference measures would distort the choice of storage.

Fuels for internal combustion engines are shown with lines on the scale of the Ragone diagram above, while the other storage forms are represented by areas. The line shape of the fuels comes from the calorific value prescribed in a narrow spread band in the corresponding norm of the fuel and thus determines the point on the axis of energy density. If no specific norm is referred to, the standard conditions for the substance are assumed (298.15 K = 25°C [77°F] and 1 bar [14.5 psi]), e.g., for pure Methane in Figure 3.1 The specific power density in the area under consideration is not determined by the fuel itself but by the process of the energy converter. For the remaining energy storage units, the power output of the energy converter in the driveline cannot specifically exceed that of the energy storage unit. Hence, there is a powerful tool to select energy stores if knowing how to use the Ragone plot.

3.1.2 Characteristics of Prime Movers

The main task of a drive system is to provide propulsion power (P_e) for the drivetrain and the work function. However, in comparison with passenger cars, the traveling speed required for mobile working machines is significantly smaller.

$$P_e = T \cdot \omega = T \cdot n \cdot 2\pi \tag{3.1}$$

However, a high torque should always be available in this small speed range. In addition to the absolute value of the torque, the course of the torque characteristic curve is also an important criterion. Thus, the following criteria can be defined for a characteristic curve of a prime mover.

- High torque at low speed, to ensure stable acceleration and high breakaway torque because the work processes of mobile machines often require a lot of torque at very low speeds.
- Stable working range through torque rise to generate stable operating points. The characteristic must reach the maximum at the earliest possible point.

Then with growing speed, the gradient of the characteristic curve should be negative. This means that the torque increases with decreasing speed, i.e., if a load jump is caused to the machine, the machine automatically compensates the load jump. For this behavior a sufficient torque rise is essential. Only when the maximum torque has exceeded the machines have to adjust the transmission.

- Constant range of nominal power is advantageous if the characteristic curves follow the corresponding power hyperbola after the nominal power has been reached. This allows the maximum performance of the machine to be maintained even under load fluctuations by adapting the transmission.

Figure 3.2 displays different characteristic curves. Especially the first (I) and second (II) characteristic curves are suitable for use with mobile working machines.

Characteristic curve number "I" has maximum torque at low speed with increasing speed, it shows a negative gradient so that a stable operating point beyond the maximum can be set. At nominal power, the characteristic curve fits the power hyperbola. It fulfills all points as explained above.

The characteristic curve number "II" is also suitable. It allows us to start from standstill with the maximum torque up to the nominal power, then it also fits at a wide speed range to the power hyperbola. A stable operating point can be set in this speed range of maximum power. In the speed range with low power, speed control can be used to maintain precise control of the process speeds.

FIGURE 3.2 Various possible characteristic curves of prime movers with the same nominal power blue: diesel engines; green: electric units; and red: spark-ignited engines.

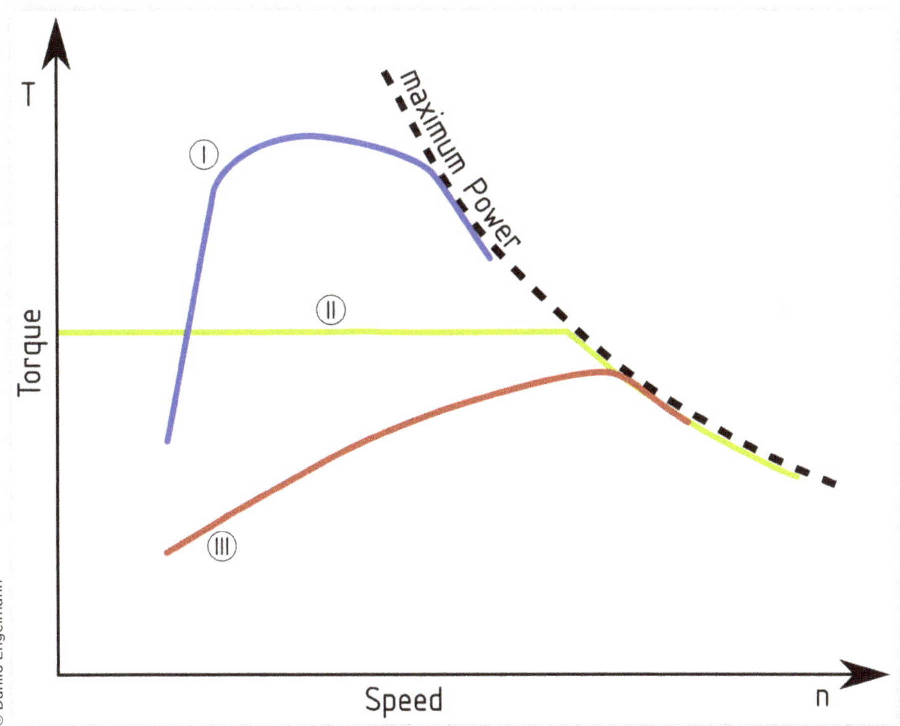

Characteristic curve number "III" is less suitable for use in mobile machines, the maximum torque is reached at the same point together with the nominal power. Only a small speed range is provided with a negative gradient. Stable operating points are difficult to achieve when the load varies at full power. Then the transmission must adapt its ratios, which often means a big loss of process speed.

From the characteristic curves in Figure 3.2, curve "III" is often used in applications such as sportive passenger cars application or smaller handheld machines and is typical for gasoline engines. Curve "I" and "II" are suitable for use in mobile machines. Curve "I" is a characteristic for diesel engines and "II" for electrical machines Therefore they are explained in more detail in the following.

3.1.3 Combustion Engines

Different types of combustion engines vary in different thermodynamic cycles. Especially the diesel engines are predestined for use in mobile working machines due to the shape of their characteristic curve which fits well for machine use and the high energy capacity of liquid fuels provides a long time for the machine to operate. There are also other engines such as gasoline or gas engines but mostly in indoor applications.

A typical characteristic curve of a diesel engine for mobile working machines is shown in Figure 3.3. Diesel engines, like all internal combustion engines, require a minimum speed (n_{idle}). With increasing speed, the torque will also increase until the

FIGURE 3.3 Normalized characteristic curve of a turbocharged 5.7 L six-cylinder engine for heavy-duty.

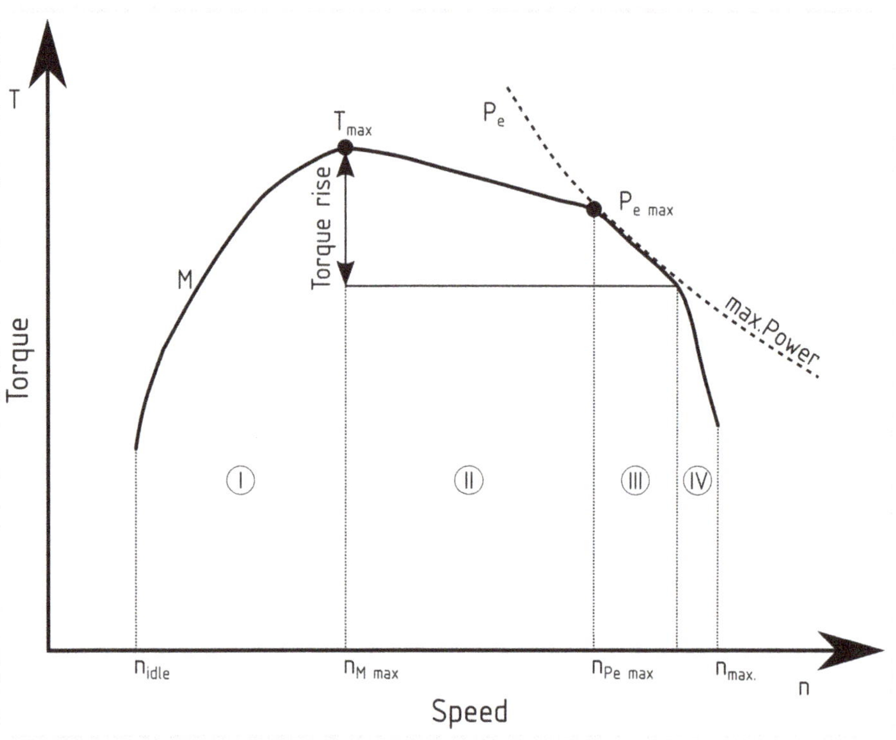

point of the maximum torque (T_{max}) is reached in speed range "I." In range "II," the torque curve decreases and touches the hyperbola of maximum power. This indicates the nominal power ($P_{e\,max}$) and corresponding speed ($n_{Pe_{max}}$), in the datasheets. In range "III," it is tried to keep the torque touching the maximum power hyperbola as long as possible to achieve a constant power range (see also Figure 3.4), the decrease of torque is compensated with augmenting speed so the output power remains the same. The last range "IV" is characterized by a rapidly decreasing torque and high fuel consumption. Therefore, operation in this range is avoided.

Figure 3.3 additionally contains the "torque rise," as mentioned in Section 3.1.2 before it is an important point for the use of these engines in mobile machinery. It is the distance of the torque maximum to the torque of the last useful operating point of the engine, normally the end of the constant power range. The torque increase is usually expressed as a percentage to compare the machines with each other. Values in the range and greater than 25% are common in mobile machines, bigger values are tried to achieve.

To enable the economical operation of a machine, the engines must have low consumption of fuels. If the absolute gravimetric fuel consumption ($\dot{m}_{fuel}$) is related to power (P_e) performed by the engine, the specific consumption (b_e) is obtained.

$$b_e = \frac{\dot{m}_{fuel}}{P_e} \tag{3.2}$$

The use of the specific consumption concerning the emitted physical work is particularly useful for mobile machines, as this makes the consumption analysis independent of the distance traveled by the machine because machines do not necessarily have to drive to perform physical work. If the lower heating value (H_u), also known as net calorific value, and the density (ϱ_{fuel}) of the used fuel are known, the efficiency can be determined. Thus, the specific consumption enables an assessment and comparison of the engines with each other (see Equation 3.3).

$$\eta_e = \frac{1}{b_e \cdot H_u} = \frac{P_e}{\dot{m}_{fuel} \cdot H_u} = \frac{P_e}{\varrho_{fuel} \cdot \dot{V}_{fuel} \cdot H_u} \tag{3.3}$$

The specific consumption is indicated with the characteristic curve and power in the datasheets of the engines (Figure 3.4). It changes with the load and speed, it takes a minimum value at the maximum torque. At higher speeds, there are negative effects on the engine working process and the torque drops slightly, which affects the specific consumption to increase.

In addition to the function of supplying power to the powertrain, the engines also have other functions (see Figure 3.5). By the clever arrangement of the aggregate, it can be achieved that the centers of gravity (COG) with its lever could reach a natural counterweight, so additional counterweights can be substituted in lifting machines such as wheel loaders. On tractors, the engines are even a structural component of the supporting structure of the machine, allowing a very compact design.

3.1.3.1 **Components of the Internal Combustion Engine and Exhaust Gas After Treatment:** Internal combustion engines need air from the environment in the combustion chambers to burn the fuel. Today, turbochargers are used for supercharging, this provides more air for the engine as a comparable naturally aspirated

FIGURE 3.4 Characteristic curves of torque and power with a normalized specific consumption of a turbocharged 3.0 L four-cylinder engine for medium-duty.

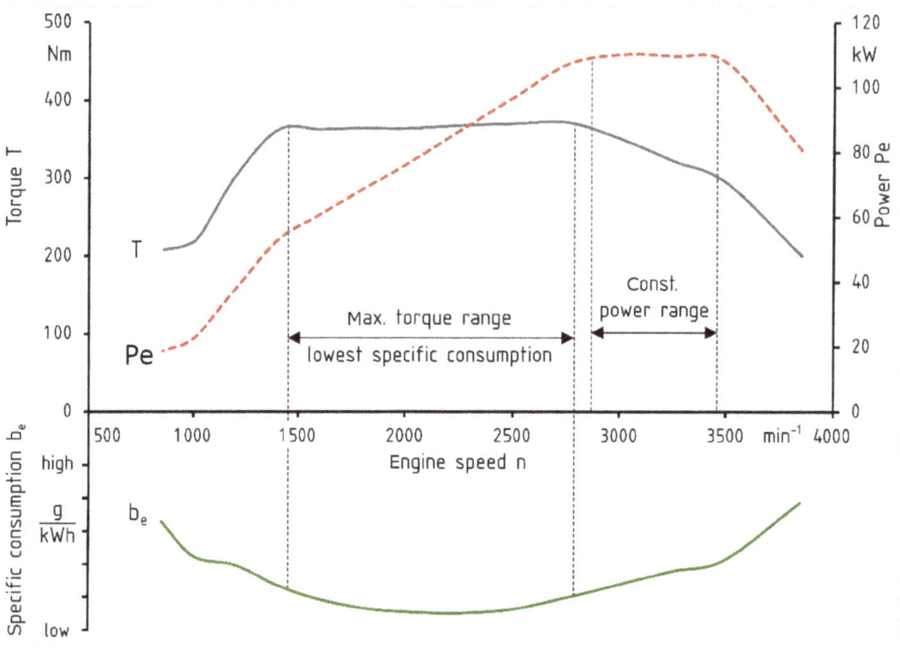

FIGURE 3.5 Arrangements of the engine in different types of machines.

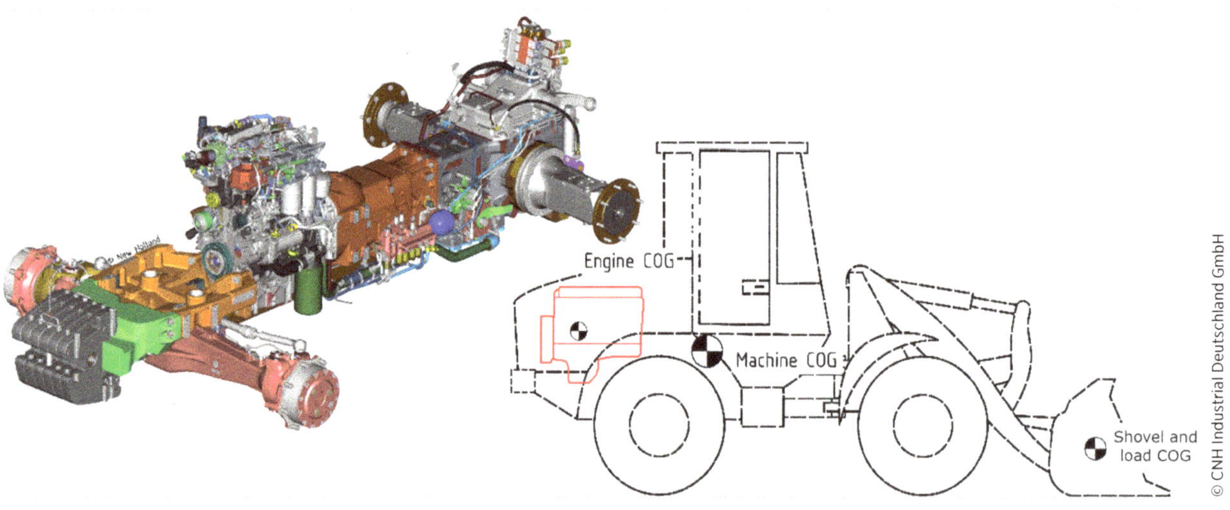

engine and allows the engine to deliver a higher power output. The engines usually have a single-stage but for more powerful engines a two-stage arrangement is present (2) and (3), like in Figure 3.6. The air heats up during compression in the turbochargers, a high outlet temperature would harm the emissions and performance of the engine if this air would be directly supplied to the cylinder. Therefore, it is cooled in the intercooler (4).

Depending on the operating point, the air is mixed with cooled exhaust gas (6) before it enters the combustion chamber. This measure, called exhaust gas recirculation

(EGR), reduces the NO_x emissions the engine caused by peak combustion temperatures in the combustion chamber. Especially at a partial load of the engine, a high admixture of 30% and more is possible, which reduces emissions considerably. The recirculation rate may not be too high either, otherwise, the combustion can be negatively affected, and the emission of unburned hydrocarbons and particles increases. The admixture is controlled by the valves (5) and (7), the relative displacements of valves between the exhaust system and the intake system are controlled by a model-based operating strategy to avoid false pressure ratios during transient operation.

After the combustion process, the exhaust gas leaves the cylinders via the manifold and the turbine of the high-pressure and low-pressure turbocharger (2) and (3). The turbocharger turbines use the enthalpy present in the exhaust gas to drive their compressors on the intake side. Afterward, the exhaust gas hits the DOC (8). Although this catalyst oxidizes the unburned exhaust components, it requires a certain temperature to operate, i.e., the light-off temperature. Therefore, this component is always located near the outlet manifold. If the catalyst temperature is too low, e.g., during cold starts, some fuel can be injected (19) which is oxidized in the DOC converter and thus heats it.

Then the exhaust gas reaches the DPF (9) where combustion particles are filtered. By measuring the pressure difference before and after the filter (11), information about the condition of the particles contained in the filter can be obtained. If the particles accumulate, the pressure difference increases. The regeneration must take place even if it does not take place continuously. Regeneration means the thermal decomposition of the particles trapped in the filter. A certain temperature is necessary for this, and here too, as with DOC, the filter can be heated by fuel injection (19) during the cold start.

The exhaust gas leaves the DPF and the nitric oxides are measured (12). A certain amount of aqueous urea solution is injected into a mixing line (16) correspondingly. Well mixed, the exhaust gas then enters the SCR reactor (17) where the urea is converted into ammonia, which preferentially converts the nitric oxides to nitrogen and water. Since ammonia is harmful to health and to avoid environmental damage, ammonia must be prevented from escaping into the environment, this is done with the trap catalyst (18) and continually monitored (12) at the end of the SCR system.

3.1.3.2 **Emission Standards:** The exhaust gas after-treatment shown in **Figure 3.6** is state of the art and has constantly evolved in line with new exhaust gas legislation. The current emission laws are EU Stage V and EPA Tier 4 in the U.S. EU Stage V was established by Regulation [2] and released in September 2016. Exhaust emission guidelines prescribed by the law reveal how much emissions an engine may emit in the prescribed test procedures. There are different performance classes into which the machines are allocated. Depending on the performance class, the corresponding quantities of emissions are then permitted (**Table 3.2**).

The exhaust emission levels are measured on the engine test bench with a special test cycle the nonroad transient cycle (NRTC) (**Figure 3.7**). The NRTC is a transient cycle developed by the U.S. Environmental Protection Agency (EPA) together with the European Union. NRTC testing is included in numerous tests of emission standards for the mobile machinery since the EU Stage III and IV, the EPA Tier 4, and Asia with Japanese 2011/13 regulations.

FIGURE 3.6 State of the art heavy-duty engine for mobile machinery with exhaust gas treatment. © John Deere.

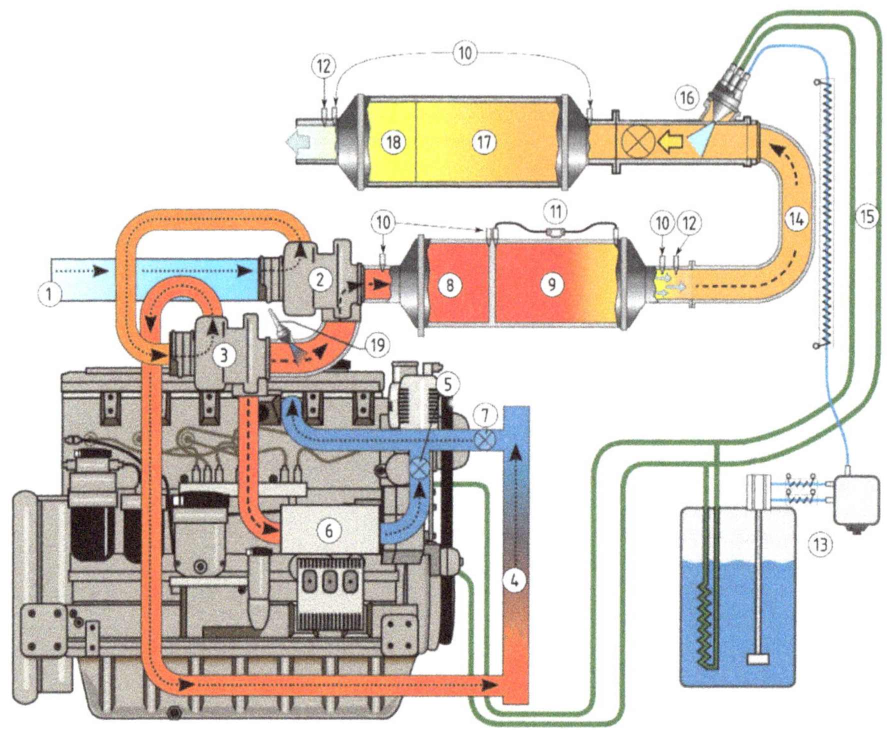

TABLE 3.2 EU Stage V emission standards.

Power P_e [kW]		$0 < P_e$ < 8	$8 \leq P_e$ < 19	$19 \leq P_e$ < 37	$37 \leq P_e$ < 56	$56 \leq P_e$ < 75	$75 \leq P_e$ < 130	$130 \leq P_e$ < 560	$P_e > 560$
Last valid stage	Stage	-	-	IIIA	IIIB	IV	IV	IV	-
	CO [g/kWh]	-	-	5.5	5.0	5.0	5.0	3.5	-
	HC [g/kWh]	-	-	7.5	4.7	0.19	0.19	0.19	-
	NOx [g/kWh]	-	-			0.4	0.4	0.4	-
	PM [g/kWh]	-	-	0.6	0.03	0.025	0.025	0.025	-
Stage V	CO [g/kWh]	8.0	6.0	5.0	5.0	5.0	5.0	3.5	3.5
	HC [g/kWh]	7.5	7.5	4.7	4.7	0.19	0.19	0.19	0.19
	NOx [g/kWh]					0.4	0.4	0.4	3.5
	PM [g/kWh]	0.4	0.4	0.015	0.015	0.015	0.015	0.015	0.045
	PN [1/kWh]	-	-	$1 \cdot 10^{12}$	$1 \cdot 10^{12}$	$1 \cdot 10^{12}$	$1 \cdot 10^{12}$	$1 \cdot 10^{12}$	-

FIGURE 3.7 Nonroad Transient Cycle (NRTC) as normalized speed and torque profiles.

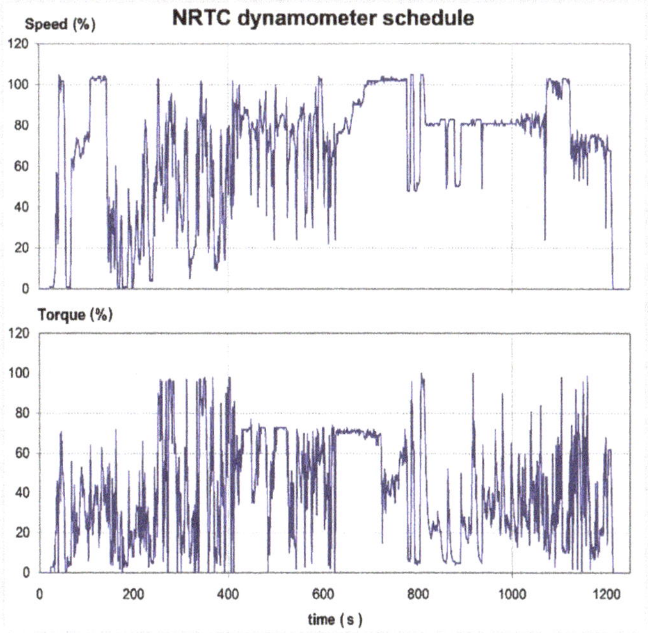

Reprinted Contains public sector information licensed under the Open Government Licence v3.0.

For each engine testing, the NRTC has to be performed twice. The first time with a cold start and the second time with a warm start. The final emission values are determined by the weighted averages of 10% for the cold start test and 90% for the test with the warm engine for the EU Stage V and in ratio 5%–95% for the EPA Tier 4. Another difference is that the number of particles is not considered for the Tier 4 standard, which otherwise prescribes comparable values for the emissions. As one can see, the emission standard EU stage V and EPA Tier 4 are comparable but not identical. The comparison and development of both standards are shown in Figure 3.8.

FIGURE 3.8 Comparison of EPA Tier and EU stage emission levels over the years 1996–2019.

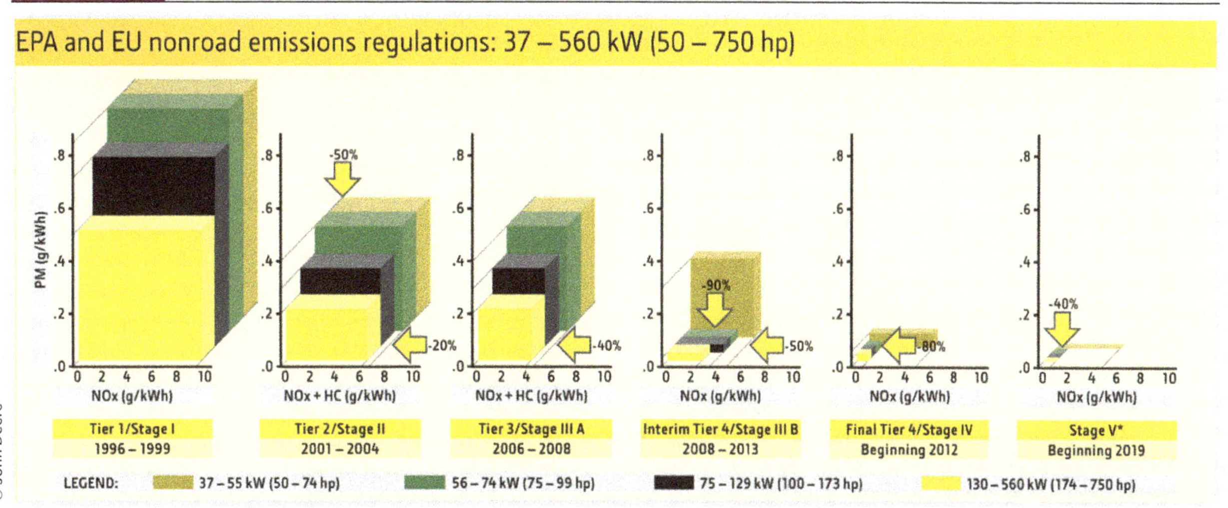

3.1.4 Electric Storage

The battery as primary energy storage provides the required electrical power in the function of a DC voltage source. The battery voltage depends on the cell type, the number of cells installed, and their interconnection. The cell itself is always a galvanic element consisting of an electrode pair, electrolyte, a separator, and sealed housing. It is available in the market in different shapes, e.g., round cells, pouch cells, or Hardcase cells according to **Figure 3.9** and with different chemistry like Li-Ion or NiMH. Several single cells connected in serial or parallel build a cell module. Therefore, the characteristic of a battery consists of nominal voltage, power, and other mechanical dimensions. The specific energy density of a Li-ion battery is about 150 Wh/kg. The typical voltage of a Li-ion cell is about 3.6 V and decreases depending on the duration and intensity of the load to a minimum allowable voltage level of about 2.8 V.

Figure 3.10 shows the measured voltage curve of a cell module removed from a high-voltage battery. It is a single cell module with a nominal voltage of 48 V, consisting of several cells connected in series. The discharge characteristic diagram shows the voltage curve over the capacity of the cell module at a constant discharge current load over a longer period. It can be seen that with increasing discharge current, the module voltage drops and consequently the available capacity of the cell module decreases

FIGURE 3.9 Comparison of different cell-shapes: round cells, hardcase cells, pouch cells.

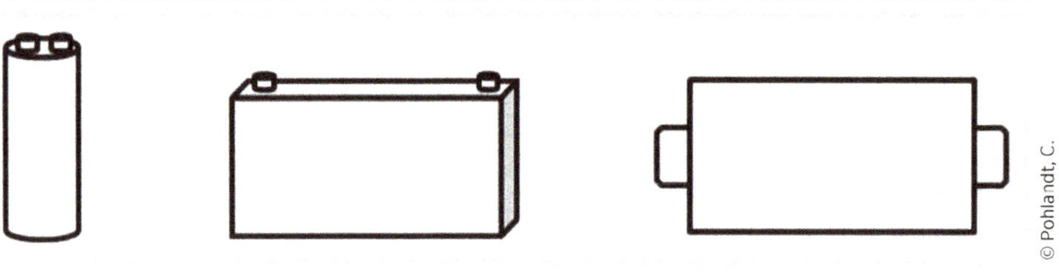

FIGURE 3.10 Measured discharge curve for a 48 V cell module [3].

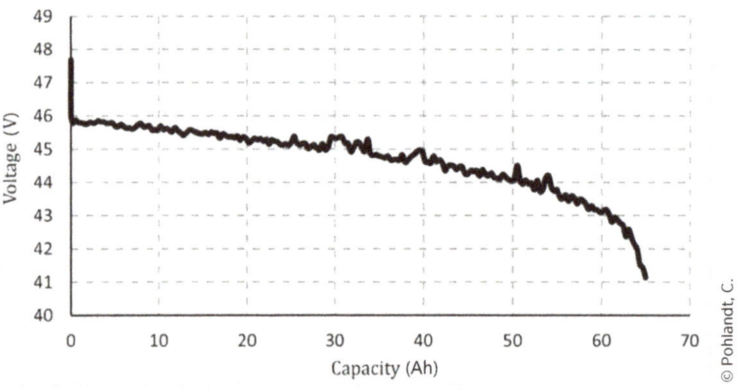

continuously. A higher discharge current would cause the voltage gradient to fasten and thus too limited usable capacity. The gradient of the discharge curve is depending on the battery technology and chemistry. For example, a discharge curve of a NiMH cell is extremely flat compared to those of a Li-Ion cell.

Besides the above-mentioned energy, power density, and discharge characteristics, there are several other typical requirements for rechargeable batteries such as:

- Operating and environmental temperatures
- Life cycle and lifetime
- Charging time for fast charging applications
- The self-discharge rate of a cell
- Acquisition costs
- Safety including charge and discharge protection circuits

Since the lifetime of a battery is a crucial indicator of customer satisfaction and warranty, more specific considerations are given. The lifetime is defined as a limited value according to the usable capacity guaranteed by the supplier. Often a usable capacity of 80% referred to the installed battery capacity at the beginning of life is defined as the end of life criteria and means that the lifetime of the battery is reached according to the desired application. The main functionality of the battery is still given but the remaining capacity is below the defined limited value. Hence the battery can still be used in other applications, e.g., as a stationary power supply till the end of life criteria for this second life application is reached. The lifetime can be specified either by a time value in years or by a value of cycles representing an application load profile. For example, today's Li-Ion batteries are designed for a lifetime of up to 12–15 years or more than 3,000 cycles. To be able to operate the battery safely, reliable and for a long time, an additional battery management system is needed. A management system fulfills the following basic functions:

- Safety, Control, and Supervision
- Power Management
- Thermo Management

Besides the Li-Ion technology, the lead-acid battery is the most commonly used type of battery in mobile working machines. Important reasons for this are the high level of safety and very good recyclability. Compared to other battery technologies, the price of a lead-acid battery is rather low, at least not because of its mass used as a starter battery. However, this battery type also has some disadvantages. Thus, the service life under high cyclic loads tends to be low and the specific energy density is only about 30–40 Wh/kg.

3.1.4.1 **Lead Acid Battery:** The charged lead battery (Pb-Battery) consists of a negative electrode whose active material is lead (Pb) and a positive electrode coated with lead oxide (PbO_2) as the active material. Diluted sulfuric acid is used as the electrolyte. A separator separates the two electrodes from each other and prevents a short circuit. To achieve large capacities, several electrodes are connected in parallel. This structural design is referred to as a plate set and shows the advantage that electrodes on both sides of the respective plate can be exchanged. The corresponding reaction equation is:

$$Pb + PbO_2 + 2SO^{2-}{}_4 + 4H^+ \rightarrow 2PbSO_4 + 2H_2O \tag{3.4}$$

When discharged, the electrode pair becomes lead sulfate ($PbSO_4$) and the amount of water in the sulfuric acid increases. The energy is released by discharging the battery. With an energy supply, the battery is charged and the processes in the overall reaction equation are reversed.

3.1.4.2 **Nickel Metal Hydride Battery:** A nickel-metal hydride battery (NiMH Battery) consists essentially of a hydrogen-storing metal alloy. The positive electrode active material is nickel hydroxide (NiO(OH)). The negative electrode active material is a metal hydride alloy. Inside a NiMH battery, the only hydrogen is exchanged between the electrodes during charging and discharging. Therefore, the transport medium is aqueous potassium hydroxide (K^+ OH^-). When fully charged, the hydrogen is stored in the negative electrode. During discharging this hydrogen is oxidized. The resulting hydrogen ions H^+ react with the OH^- ions of the potassium hydroxide to form water H_2O. Nickel is reduced from NiO(OH) to $Ni(OH)_2$ at the negative electrode.

The corresponding reaction equation is:

$$2NiO(OH) + 2\text{metal } H \rightarrow 2\text{metal} + 2Ni(OH)_2 \tag{3.5}$$

3.1.4.3 **Li-Ion Battery:** A Li-Ion battery behaves differently than lead or NiMH batteries. In a Li-Ion battery, while the Li-Ion is the active element, it also forms the active material of the electrode itself. Thus, there are a variety of ways to use an active material for the two electrodes that include or incorporate lithium ions. Typical active materials of the negative electrode are, e.g., lithium metals, lithium alloys, lithium oxides, and graphite. The positive electrode active material is selected differently depending on the manufacturers. The issues to consider when choosing the right material are chemical stability, safety, and durability requirements. Besides, the specific energy and power of the battery are essential selection criteria. Thus, e.g., lithium cobalt dioxide (LCO), lithium iron phosphate (LFP), lithium nickel cobalt aluminum oxide (NCA), and cobalt-nickel manganese mixtures (NMC) are available. The use of different active materials allows a variety of possible combinations representing all functional Li-ion batteries. The chemical reaction in a Li-ion battery is based in principle on the ion exchange between the electrodes. Lithium is the lightest metal and reacts very strongly with water. As the electrolyte, therefore, anhydrous solvents are used in combination with salts, e.g., lithium hexafluorophosphate ($LiPF_6$). The positive electrode active material is, e.g., lithium cobalt dioxide (LCO). The negative is graphite.Hence, the reaction equation is:

$$Li_{1-x}CoO_2 + Li_xC \rightarrow LiCoO_2 + C \tag{3.6}$$

The electrolyte contains lithium ions. The connectors are made of metal thin films of copper for the negative electrode and aluminum for the positive electrode. The separator is a polyethylene (PE) or polypropylene (PP).

3.2 Basics of Power Conversion (By Swen Bosch, Torsten Kohmäscher, Fabrizio Panizzolo, and Heinrich Steinhart)

Every mobile working machine has the final objective to fulfill the task for which the vehicle was originally designed for. Therefore, it needs the presence of a more or less complex transmission for the traction drive in-between the output shaft of the engine and the tires (or tracks).

This chapter will introduce the physical principles that find application for energy conversion in the field of mobile working machines. The focus is on the traction drive but the basics discussed previously are also important for the function drives. Based on basic knowledge, components and system solutions are introduced as well as relevant equations and assumptions discussed. Table 3.3 provides a high-level overview of the applied physical principles for the continuous variation of output speed: mechanic, hydrodynamic, hydrostatic, and electric. Continuous variable transmissions (CVT) are essential for the smooth and dynamic speed variation of mobile working machines in loaded conditions. The information in the table was reviewed and updated from an earlier overview provided by Renius in 2005 [4].

In the examples, shown in the pictures, symbols are used that are common in the community or defined in standards or norms. Hydraulic symbols are normed in the ISO 1219 [5] and mechanic transmission symbols follow the suggestion of Geimer, Renius, and Stirnimann [6, 7] (Figure 3.11).

TABLE 3.3 Important physical principles of continuously variable transmission for traction drives in vehicles. Modified from [4, 8].

Type of CVT			Principle of energy transmission	Ratio/speed control	Application	Efficiency
1	Mechanic		Traction forces within friction contact	The radius of traction force	Relevance for cars and MWM	Excellent
2a	Hydrodynamic		Mass forces at pump and turbine wheels (friction when the lock-up clutch is applied)	Automatically - dependent on load	Important for cars and MWM	Poor (excellent)
2b	Hydrostatic		Hydrostatic forces at pump(s) and motor(s)	Displacement of the units	Important for MWM	Moderate
3	Electric		Electromagnetic forces in air gaps of the motor(s) and generator(s)	Frequency of magnetic fields and load	Important for cars - gaining in MWM	Moderate to excellent

FIGURE 3.11 The *Renius* symbols for mechanic transmissions [7].

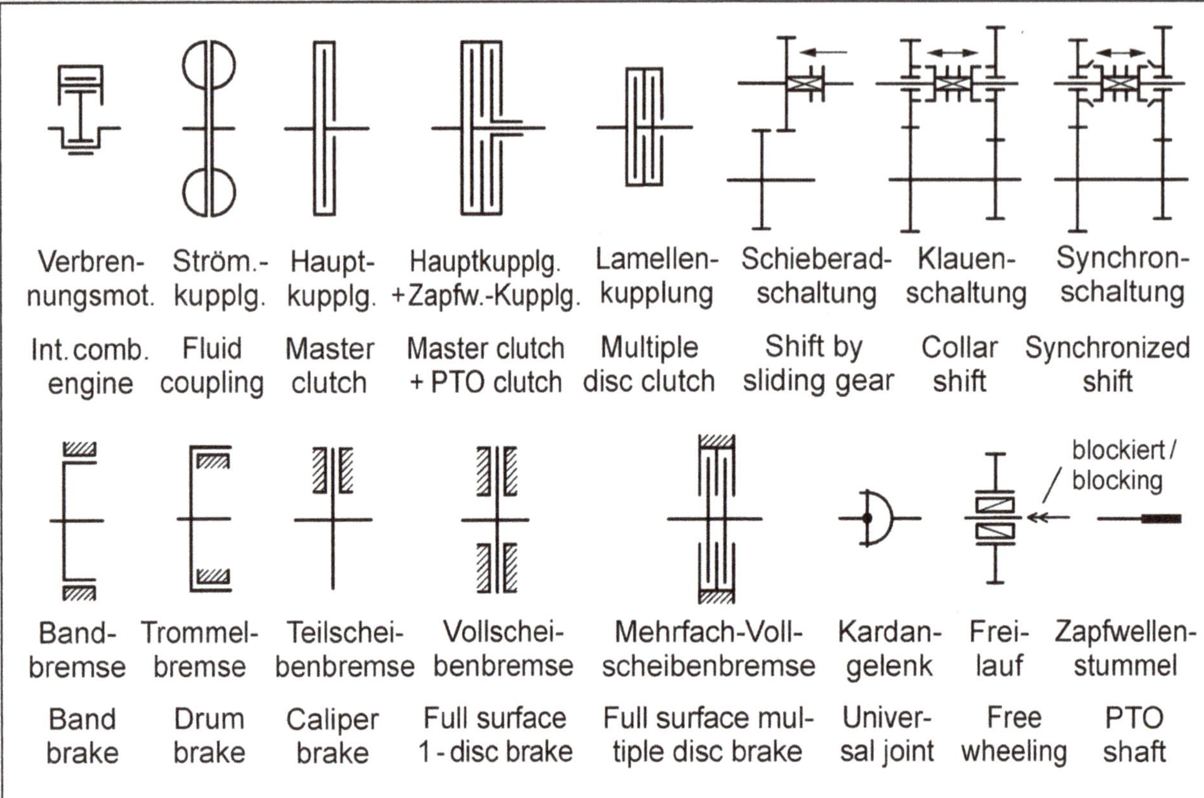

3.2.1 Mechanic Energy Conversion

The main role of a transmission is to convert the prime mover power $P_{PM} = T_{PM} \cdot \omega_{PM}$ to the required power for vehicle operation $P_v = T_v \cdot \omega_v$ in which the vehicle torque T_v corresponds to the vehicle tractive effort F_Z combined with the required vehicle speed v_v. If we consider each mechanism of transmission as a black box with a known input power P_{in} the input torque T_{in} and the input angular speed ω_{in} and with a known output power P_{out} the output torque T_{out} and the output angular revolutions ω_{out} it is possible to write the equation of mechanism efficiency η as:

$$\eta = \frac{P_{out}}{P_{in}} \quad \text{with } P_{in} = P_{out} + P_{losses} \tag{3.7}$$

and with P_{losses} as the representation of the internal losses of the mechanism.

Now, we can define the transmission ratio τ for each mechanism as the ratio between output and input angular speed:

$$\tau = \frac{\omega_{out}}{\omega_{in}} \tag{3.8}$$

Accordingly, we can define the mechanical transmission conversion ratio μ as the ratio between the output and input torque:

$$\mu = \frac{T_{out}}{T_{in}} \tag{3.9}$$

Substituting the two ratios within the equation of mechanism efficiency, we conclude:

$$\eta = \frac{P_{out}}{P_{in}} = \frac{T_{out} \cdot \omega_{out}}{T_{in} \cdot \omega_{in}} = \mu \cdot \tau \tag{3.10}$$

What reported above represents the fundamental equation of transmission involved mechanisms. At this point, it will be clear while, within a drivetrain, is needed the presence of many different transmission mechanisms having each one his distinct assigned task of:

1. Convert the 2 above-defined power-related factors μ and τ according to the following possible alternative approach:

   A. In a discontinuous way adopting a mechanical gearbox having a finite number of transmission ratios amplified than by a fix reduction ratio utilized for an axle (or a final drive unit in case of the tracked vehicle).

   B. Continuously adopting one of many existing CVT. (see next Section 3.3.3 for pure hydrostatic transmission and Section 3.3.4 for power splitting transmission)

   It is important to recognize at this point that the torque converter (TC) should also be considered belonging to a particular CVT family contributing as well as to power conversion. The chain drive variator could also contribute as a mechanism to convert power when properly integrated within a gearbox.

2. Cut, when required by the operator, for vehicle controllability reason, power transmitted flow via the presence of a dog-clutch or a master clutch; in other words, by de-clutching prime mover from the gearbox.
3. Engage, thanks to the presence of a dry or wet clutch or by a synchronizer and according to the selected transmission's layout, those shafts involved on power hand-over during each change of mechanical ratio inside the selected transmission.
4. Transmit the mission task required to power up to the wheels (or tracks) while passing through the different gearbox mechanism's shafts arriving than up to the output of the axle and connected wheels (or arriving up to the output of the final drive unit in case of the tracked vehicle). Gearbox shafts and Cardan shafts are providing their contribution as well.

Transmission Components: The following chapters go into the details of the components that are essential for transmissions of mobile working machines. Symbols, basic equations as well as relevant design considerations are introduced for gear stages, planetary gear sets, clutches, and two variants of mechanical variators [10].

3.2.1.1 **Gear Stage.** Gear stages consist of two meshing gear wheels that are connected to an input and output shaft. The gear stage is described by a fixed gear ratio i that is determined by the ratio of the tooth count for the input gear z_{in} and output gear z_{out}. Equation (3.11) describes how the output speed n_{out} of a gear stage is calculated from input speed n_{in} and the gear ratio i. Tooth counts for gears with external teeth are defined as positive values while the tooth count for gears with internal teeth are put in with negative sign. Equation (3.12) shows the calculation of input torque T_{in} from output torque T_{out} and gear ratio i. While there are no losses to consider in the speed calculation for the gear stage, the torque calculation has to consider losses or efficiency η.

$$i[-]=\frac{\omega_{in}}{\omega_{out}}=\frac{n_{in}}{n_{out}}=-\frac{z_{out}}{z_{in}}\Rightarrow n_{out}=\frac{n_{in}}{i} \tag{3.11}$$

$$T_{in}=\frac{T_{out}}{i}\cdot\eta^{w} \quad \text{with } w=\frac{P_{out}}{|P_{out}|} \quad \text{and} \quad \eta\in\,]0;1[\tag{3.12}$$

Depending on the actual direction of the power flow, the input torque can be calculated based on output torque T_{out} and efficiency η. Regenerative power flow is considered to have a positive sign (w = 1) and non-regenerative power flow a negative sign (w = −1). With that assumption, input torque T_{in} will be decreased (relative to ideal calculation) by regenerative power flow and increased by non-regenerative power flow. In most cases, the power flow will be non-regenerative and power losses will reduce the usable power from the source to the place of use. However, in HMT applications the regenerative power flow is part of the normal operating conditions and needs to be considered during the analysis.

Load-dependent losses in bearings and the meshing of gear teeth, as well as load-independent losses that are caused by splashing and squeezing of oil or by ventilation of air in the housing, are combined to the over-all losses and reduce the efficiency of gear stage. References and details about specific losses are available in Kohmäscher [8, 11].

3.2.1.2 **Planetary Gear Set.** A standard planetary gear set consists of a sun gear A, a ring gear B with internal gearing, a single or multiple planet gears Pl which are supported by the planet carrier C (**Figure 3.12**). Three planetary wheels are a typical arrangement as they offer outstanding self-centering of the sun gear as well as optimal power distribution. Sun gear, ring gear, and carrier are in concentric arrangement while the planet gears are meshing in between sun and ring gear. To be able and assemble the

FIGURE 3.12 Schematic of planetary gear set with the introduction of planetary members [8].

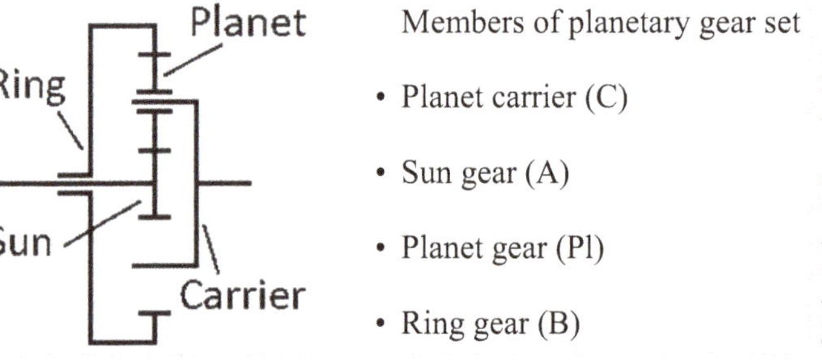

planet gears the tooth count of the ring gear and sun gear has to be dividable by the number of planet gears. The toothed module of a sun gear, ring gear, and planet gear must be the same since these are in continuous meshing. The approximate tooth count of the planet wheel can be derived from the tooth count of the sun gear and ring gear as follows – $z_{Pl} = (z_B - z_A)/2$.

The standard ratio of a planetary gear i_0 is the central element for the calculation of speeds and torques for the planetary members. It represents the speed ratio from sun gear A to ring gear B with the carrier shaft blocked and can be calculated using the tooth count for sun gear z_A and ring gear z_B. Tooth count for outer gearing (e.g., sun gear and planet gears) is defined as positive while tooth count for inner gearing (e.g., ring gear) is defined negatively. In Equation (3.13), the signs for the gearing and all tooth counts are considered with their positive or negative numbers.

$$i_0 = \frac{z_B}{z_A} \tag{3.13}$$

While the torque ratios for the planetary members are determined by the standard ratio i_0, the speeds follow the fundamental Equation (3.14). In HMT applications, variation in power flows in a planetary gear set is indicated by variation in the speed ratio of the relevant planetary members. The equation introduces two facts: (1) speeds ratios of planetary members follow a clear principle and (2) with a constant speed in one planetary member (e.g., the carrier) the speed of the other two members can only vary according to Equation (3.14).

$$n_A - i_0 \cdot n_B = n_C \cdot (1 - i_0) \tag{3.14}$$

A planetary gear set can be applied with two given input speeds to create a single output speed. One of these input speeds can also be 0 rpm which can be seen in the speed reduction of a wheel hub where the ring gear is part of the fixture housing. As the second option, the planetary gear set can be applied with a single given input speed to create two output speeds. These output speeds need to follow Equation (3.14) but their speed ratio is controlled by restrictions outside of the planetary gear set, e.g., the hydrostatic transmission in an HMT application. As a second example of single input speed, the differential for axles is introduced. This abnormal planetary gear set operates with a standard ratio of $i_0 = -1$ and the planet carrier C as the member for the single input speed. The speed ratio between the sun gear and ring gear—left and right side of the axle—is given by the driving direction (straight, left, or right turn) and the ground conditions.

Based on the torque equilibrium for the planet gears, the torque ratios between the sun gear A, ring gear B, and planet carrier C can be determined using the standard ratio for the planetary gear set. Equation (3.15) describes how the torques of the planetary members can be determined from a single known torque value in the planetary gear set.

$$T_C = (1 - i_0) \cdot T_A = (1 - 1/i_0) \cdot T_B \tag{3.15}$$

The resulting power losses are considered with the standard efficiency η_{PG} in Equation (3.16) and are dependent on the power flow in the gear meshing.

$$\frac{T_B}{T_A} = -i_0 \cdot \eta_{PG}^{w} \quad \text{as well as} \quad \frac{T_C}{T_A} = i_0 \cdot \eta_{PG}^{w} - 1 \quad \text{and} \quad \frac{T_C}{T_B} = \frac{1}{i_0 \cdot \eta_{PG}^{w}} - 1 \tag{3.16}$$

The efficiency coefficient w needs to be determined based on the direction of power flow in the meshing of the wheels and not the contribution of the rotating planet carrier. The direction of the power flow is analyzed by looking at the sun wheel in Equation (3.17).

$$w = \frac{T_A \cdot (\omega_A - \omega_C)}{\left|T_A \cdot (\omega_A - \omega_C)\right|} \tag{3.17}$$

There are numerous sources of power losses in a planetary gear set that can be separated into load-dependent and load-independent loss as in the section for the simple gear stages. Some more detailed information on the individual sources for power losses is summarized by Kohmäscher [11].

There are a couple of graphical methods for the analysis of planetary gear sets that are known to the community. The Kutzbach Plan by German Prof. Kutzbach is focused on the inside of the planetary gear set and support the planetary design.

The introduced method of the ladder diagram in Figure 3.13 is focused on the system outside of the planetary gear and allows for a simplified analysis of HMT. Ramm provides an example of the comparison of the different graphical methods in Ramm [12].

FIGURE 3.13 Introduction of a ladder diagram for planetary gear sets.

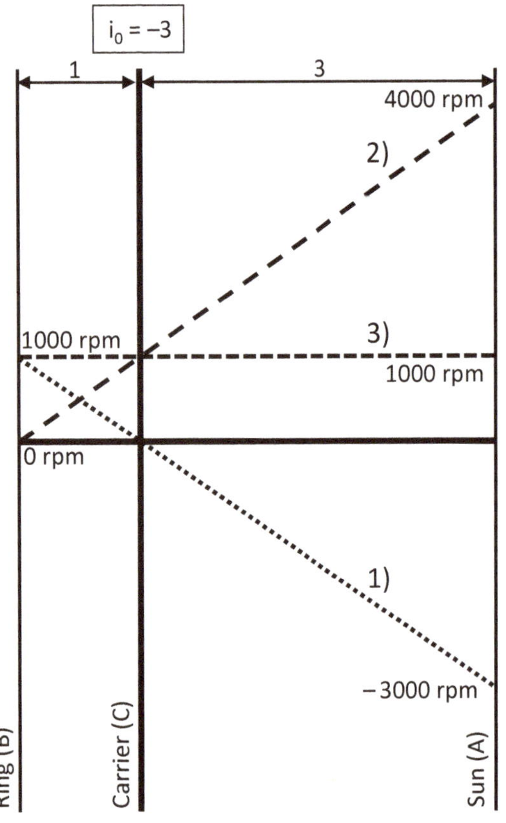

Introduction to ladder diagram:

- Horizontal line for 0 rpm
- Vertical line for the Carrier (C) as central element for planetary speed relation
- Positioning of vertical lines for sun (A) and ring (B) on the right and left side of the carrier – the distance needs to copy the standard ratio i_0 of the planetary
- Positioning of the vertical lines represent the ratios in planetary gears
- Positions above or below the horizontal "0 rpm-line" stand for speed values of the connected planetary member
- Valid operating conditions of the planetary gear set are represented by a straight line through the ladder diagram

Examples:

1. Constraints: nC = 0 rpm and nB = 1000 rpm
   → Result: nA = –3000 rpm
2. Constraints: nC = 1000 rpm and nB = 0 rpm
   → Result: nA = 4000 rpm
3. Constraints: nC = nB = 1000 rpm
   → Result: nA = 1000 rpm (blocked)

TABLE 3.4 Possible configurations and transmission ratios of planetary gear sets, based on [9].

Schematic	Configuration			Ladder diagram	Ratio	
	Input (1)	Output (2)	Blocked (B)		$i = n_1/n_2$	Range
 1 2	Sun	Carrier	Ring	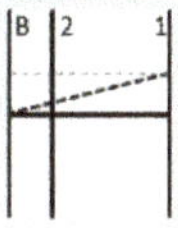 B 2 1	$1 + i_0$	$2 < i < \infty$
 1 2	Sun	Ring	Carrier	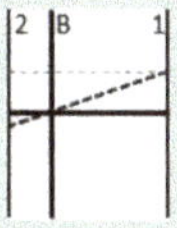 2 B 1	$-i_0$	$-\infty < i < -1$
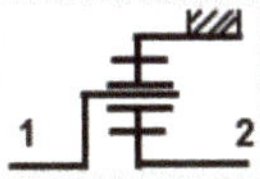 1 2	Carrier	Sun	Ring	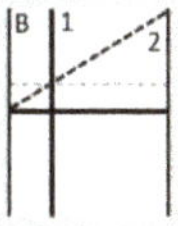 B 1 2	$\frac{1}{1+i_0}$	$0 < i < 0.5$
 1 2	Carrier	Ring	Sun	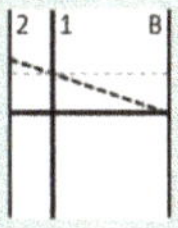 2 1 B	$\frac{1}{1+1/i_0}$	$0.5 < i < 1$
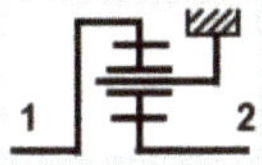 1 2	Ring	Sun	Carrier	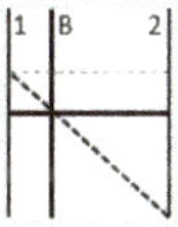 1 B 2	$-\frac{1}{i_0}$	$-1 < i < 0$
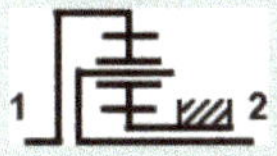 1 2	Ring	Carrier	Sun	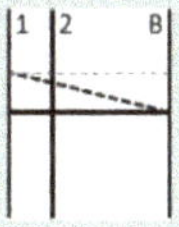 1 2 B	$1+\frac{1}{i_0}$	$1 < i < 2$

Planetary standard ratio $i_0 = -z_B/z_A$.

When one of the three planetary members is blocked, the planetary gear set is applied with a fix gear ratio for single power flow. With exchanging input and output shaft and blocking each member once, this results in six possible configurations. Only two out of the six possible configurations lead to a reversal of the rotational direction of input versus output speed (see highlighted ranges in Table 3.4). The table provides information on a possible schematic and introduces the ratio calculation together with the possible range for the ratio as well as a visualization using the introduced ladder diagram.

If the complete planetary gear set is blocked—e.g., when the sun gear is connected to the carrier via a clutch—all members turning with the same speed and direction, a direct mechanical gear is possible.

The standard gear ratio i_0 of a planetary gear set depends on the tooth counts for sun gear and ring gear but also on the design of the planet gear. Table 3.5 provides an overview with equations for the calculation of the standard gear ratio i_0 for the possible designs of the planet wheel as well as for the variation of inner and outer gearing on sun gear and ring gear. The standard gear ratio i_0 will become positive when sun gear and ring gear have both inner or outer gearing or when a double planet is introduced with both planets supported by a single carrier member. Make sure to only use absolute values for the tooth counts in the equations in Table 3.5. A positive standard gear ratio i_0 is reflected in a ladder diagram by positioning both lines for sun and ring gear on the same side of the carrier.

TABLE 3.5 Standard ratio of planetary gear set with compound planet gear.

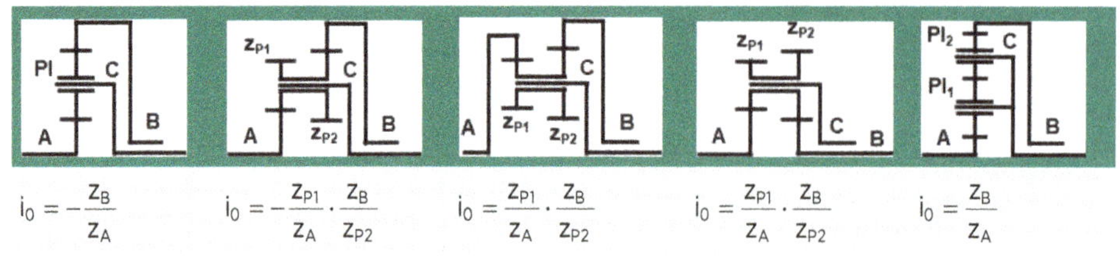

$i_0 = -\frac{z_B}{z_A}$	$i_0 = -\frac{z_{P1}}{z_A} \cdot \frac{z_B}{z_{P2}}$	$i_0 = \frac{z_{P1}}{z_A} \cdot \frac{z_B}{z_{P2}}$	$i_0 = \frac{z_{P1}}{z_A} \cdot \frac{z_B}{z_{P2}}$	$i_0 = \frac{z_B}{z_A}$

All tooth counts (z_A, z_B, z_{P1}, z_{P2}) need positive values for these equations.

3.2.1.3 Wet Disc's Clutch.

During each change of the ratio of mechanical transmission, the clutches involved during hand-over will have to face a dynamic working phase while sliding under load. The direct consequence of this transitory phase will result in a dissipation of energy and consequent heat generation at facing and slipping surfaces. This means that while designing a clutch not only will the static torque capacity have to be taken under consideration but also the thermal capacity which will play a fundamental role in dimensioning all sliding clutches utilized inside a mechanical transmission. Unlike dry clutches, currently utilized for full mechanical stepped transmissions (like those utilized for small tractors below 60–70 kW), wet clutches are largely utilized instead for mobile working machine transmissions and this implies that the friction surfaces will have to be lubricated (by a forced lubrication system) and most important to be cooled by the presence of mineral oil available on transmission sump [13, 14, 15, 16, 17, 18].

In Figure 3.14 the design of a heavy-duty clutch and the feeding with cooling oil, passing first through a channel machined on the clutch shaft, passing then through interposed clutch gear and exiting finally through clutch drum holes radially positioned just after the outer clutch disc's diameter, is shown. From the picture, the design of clutch actuation can be noticed as well as the force clamping generation system and the presence of the piston chamber receiving appropriate oil flow rate through dedicated clutch shaft channel, filled in by the oil provided by the actuation of the selected proportional controlled valve. Appropriate sliding seals will allow to secure and maintain on the clutch actuation system the selected pressure-controlled profile during clutch closing and once clutch will be engaged. Those seals are interposed between transmission main casting (being said sliding seals steady and radially in contact on their own external cylindrical surface with casting machined surface) and being instead seal's axial faces sliding against properly machined shaft grooves available at the clutch shaft extremity. A back-up plate is then inserted and secured with a span ring. A hub gear with outer diameter splines is inserted into the splines of clutch discs equipped in this case of teeth on the inner diameter. The counter-discs and hubs gear are free to increase in speed or rotate in the opposite direction as long as no pressure is present in that specific clutch.

The majority of wet clutches utilize today for clutch linings resin impregnated cellulose fibers, shortly called paper while sintered bronze lining has practically initiated to be dismissed for more than 20 years for mobile working machine applications.

3.2.1.3.1 Clutch Transmittable Torque. Considering a uniform pressure distribution on the annular friction surface of wet clutch disc's by simple passages it will

FIGURE 3.14 Clutch assembly and main components.

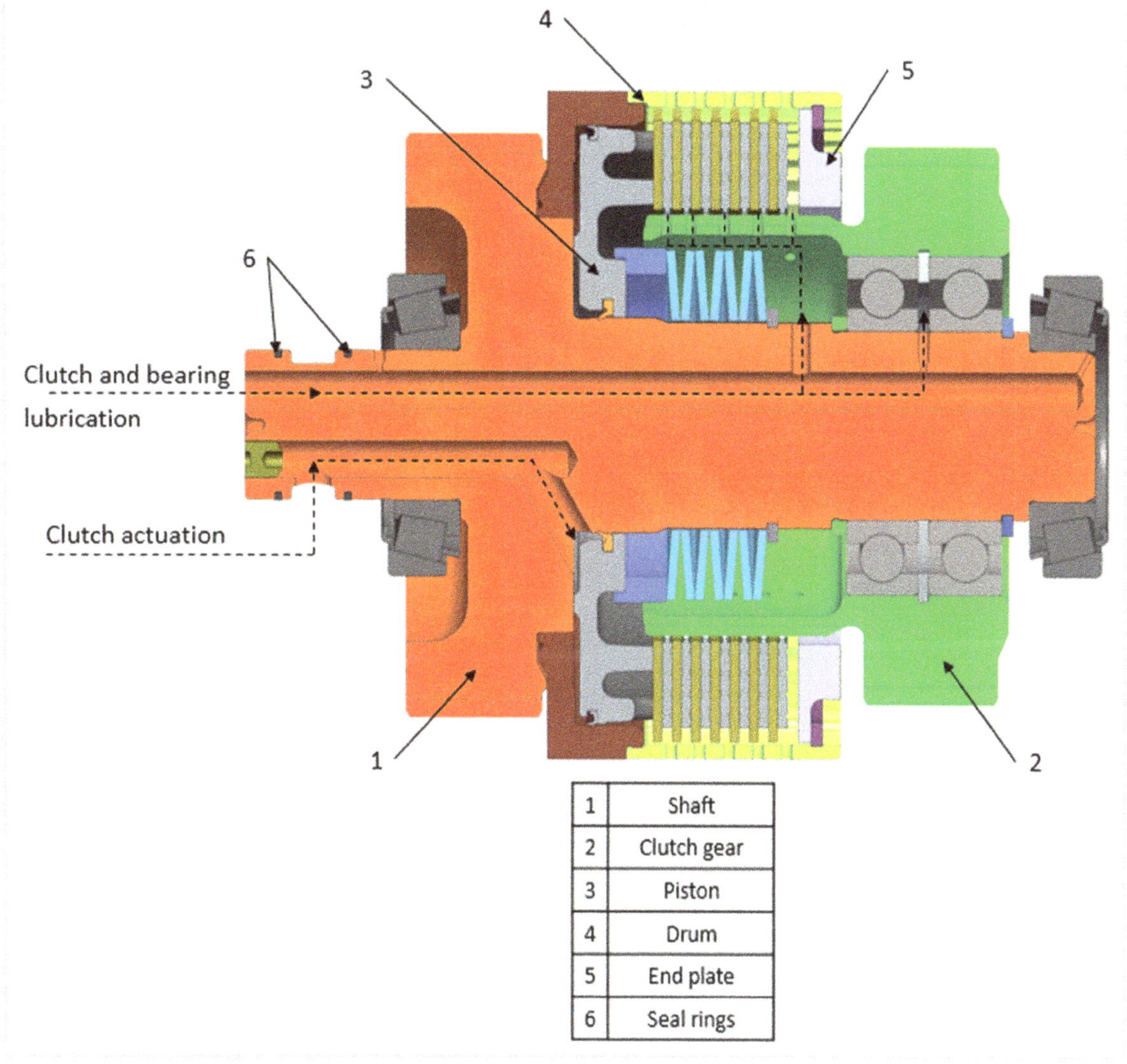

1	Shaft
2	Clutch gear
3	Piston
4	Drum
5	End plate
6	Seal rings

be possible to write the following formula providing the wet clutch transmittable torque as:

$$T = s \cdot R_m \cdot p \cdot A \cdot \mu \tag{3.18}$$

where

s is the number of sliding surfaces of each clutch package
A is the actuating piston surface [mm^2]
μ is the lining's friction coefficient (that could be either static or dynamic)
R_m need to be expressed in mm
p is the pressure in N/mm^2
T is the torque in Nm

$$R_m = \frac{2}{3}\left(\frac{R_{EXT}^3 - R_{INT}^3}{R_{EXT}^2 - R_{INT}^2}\right) \tag{3.19}$$

Depending on shifting condition and currently torque required for vehicle mission execution it is fundamental to design clutches and select appropriate linings for thermal capacity. By utilizing computer-assisted simulation programs it will be possible to predict during shifting condition lining's (as well counter-disc's) temperature raising and distribution profile either radially wise, on each lining and counter-disc, as well as axially wise along with the clutch package. It will be possible then to compare simulation provided lining and counter-disc temperature profiles with corresponding maximal allowable temperature level recommended by friction/lining manufacturers for selected friction material adopted for currently involved clutch design. It will be possible to predict currently dissipated power and involved energy as well as corresponding heat to be absorbed via the corresponding minimum required oil cooling flow. Having as simulation target the prediction of lining disc and a counter-disc temperature it will be necessary to have as input simulation data following information:

1. Vehicle specifications; all drive-train involved component's inertias and reduction ratios (from the engine to the axle and connected wheels or, in case of a tracked vehicle to output final drive and connected track).
2. The current working condition and vehicle speed as well as a mission required force or torque in terms of external loads.
3. Transmission specifications in terms of clutch cooling flow rate available from transmission cooling/lubrication system for all clutches, knowing in advance which are the geometrical dimensions (size) of clutch considered under thermal capacity investigation and knowing as well which are the physical characteristics of selected and investigated friction lining. For an explanation of the clutch thermal dimensioning process here following are reported the simulation results of four consecutive shift engagement and disengagement where oil flow is supposed equally distributed among friction pairs (lining disc's and counter-disc's).

From the attached pictures it is possible to see the outcome of a simulation. One can see the clutch lining's and counter-disc's temperature distribution as a consequence of available (designed) clutch oil cooling flow distribution, axially wise, across the entire clutch package. On the way to help the designer in post-processing simulation outcomings, the results are presented through the format reported here below.

3.2.1.3.2 Clutch Thermal Dimensioning Simulation Input Data. A plot that shows the working conditions of the clutch over time (**Figure 3.15**) is the first input data. It is composed of three sub-plots. The first represents the rubbing pressure on the first friction pair (next to the clutch piston), on the last (by the clutch endplate), and in the central clutch friction pair. The second subplot shows the oil pressure in the piston chamber and also the piston displacement. The last subplot shows the rotational speed of gear, drum, and the relative speed.

Additional two plots containing the lubrication information, namely the actual oil flow over time (**Figure 3.16**) and the mean oil fraction on friction surfaces over time and radius (**Figure 3.17**).

FIGURE 3.15 Clutch working conditions.

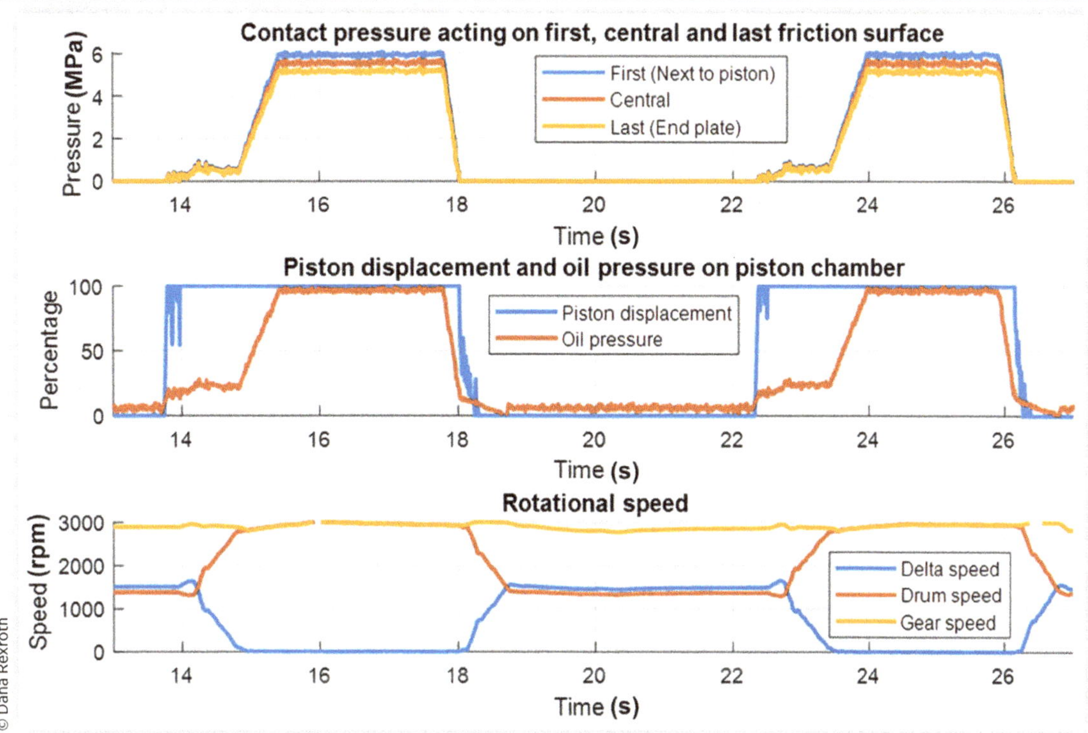

FIGURE 3.16 Overall oil flow into the clutch.

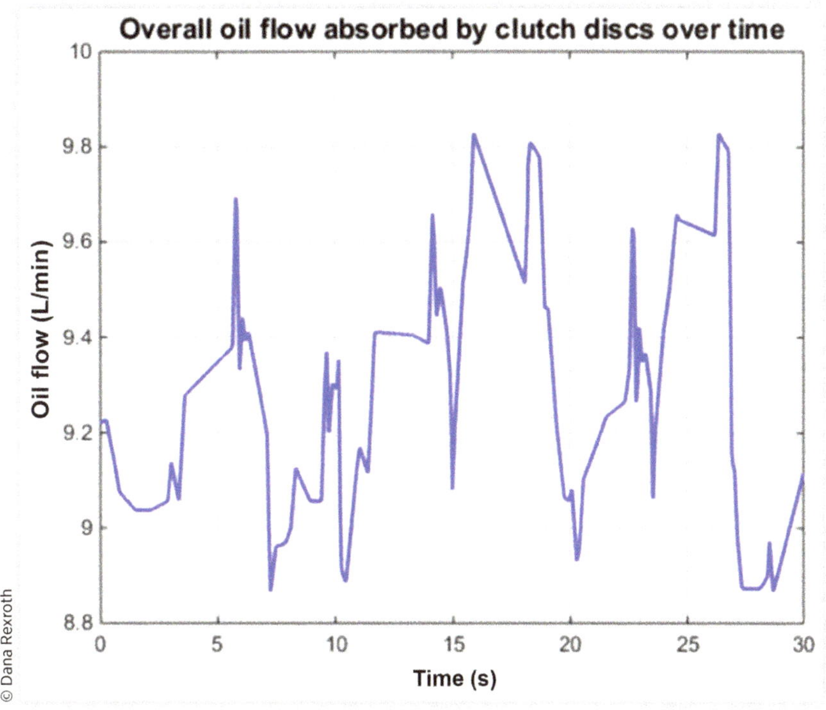

FIGURE 3.17 Mean oil fraction among clutch discs.

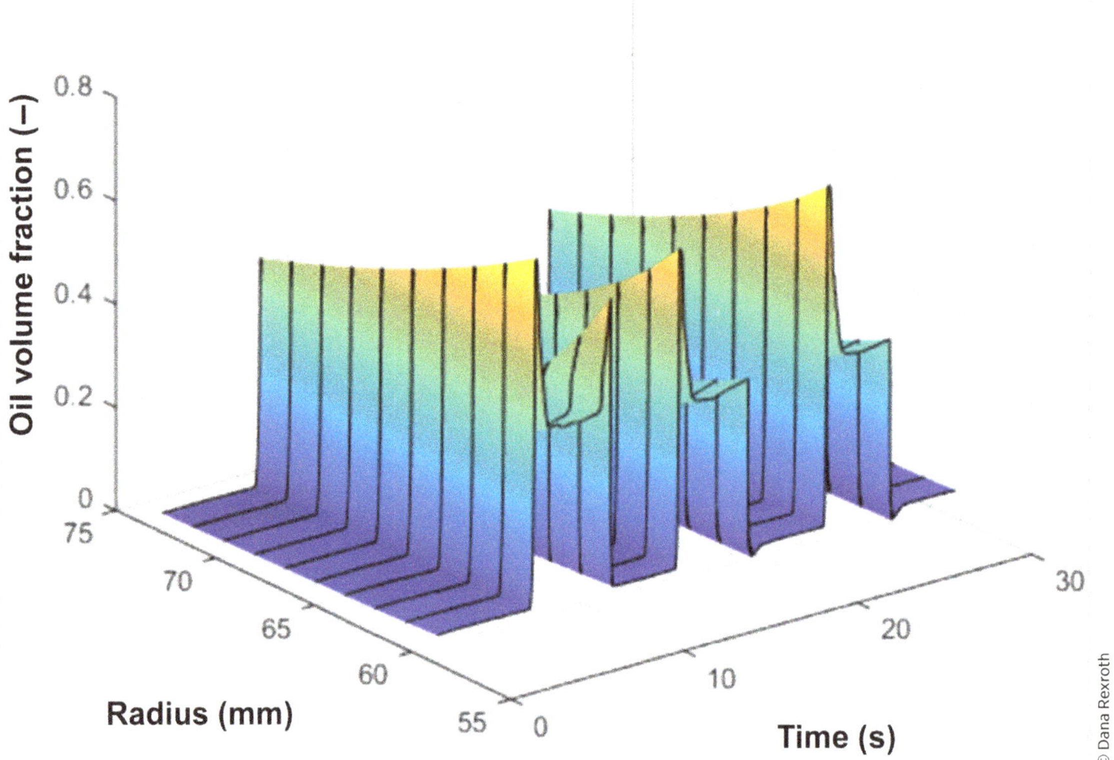

3.2.1.3.3 Clutch Thermal Dimensioning Simulation Output Data. Very interesting for direct visual and quick interpretation will be the 3D plot which depicts the thermal status of the clutch at the hottest instant (Figure 3.18). The separator discs are numbered starting from the piston toward the endplate. In this case (oil flow is assumed equally distributed among friction pairs), the hottest disc is the second. The friction discs are a lot colder than the separator discs. In the figure, those friction discs are not visible.

Of strong interest for the designer will be the possibility to see the following 3D plot which shows the thermal history of the hottest separator disc, as a function of radius and time (Figure 3.19). The maximum temperature reached into the clutch is also reported in the 3D plot title.

Interesting for the designer is also the following two plots, where the first (Figure 3.20) shows the thermal history of the hottest point into the clutch and also the oil temperature next to that element. The second (Figure 3.21) shows the slip torque transferred from the clutch during the whole simulation time. The torque is plotted only during the slipping phase, since the method used to calculate it is reliable only if the clutch delta speed is not zero.

FIGURE 3.18 Clutch thermal status at the hottest instant.

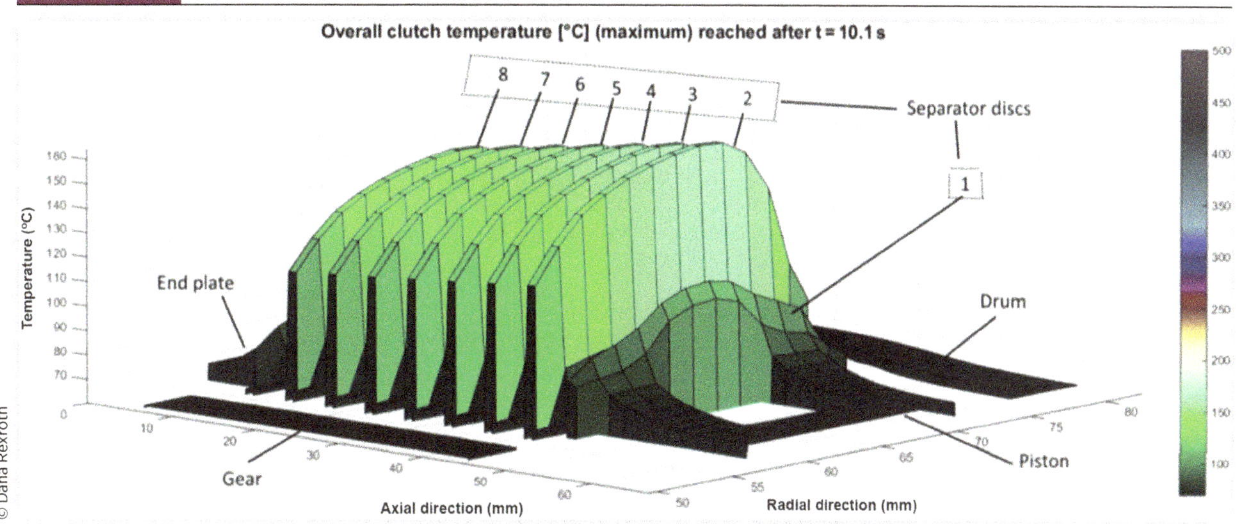

FIGURE 3.19 Thermal history of the hottest separator disc (number 2).

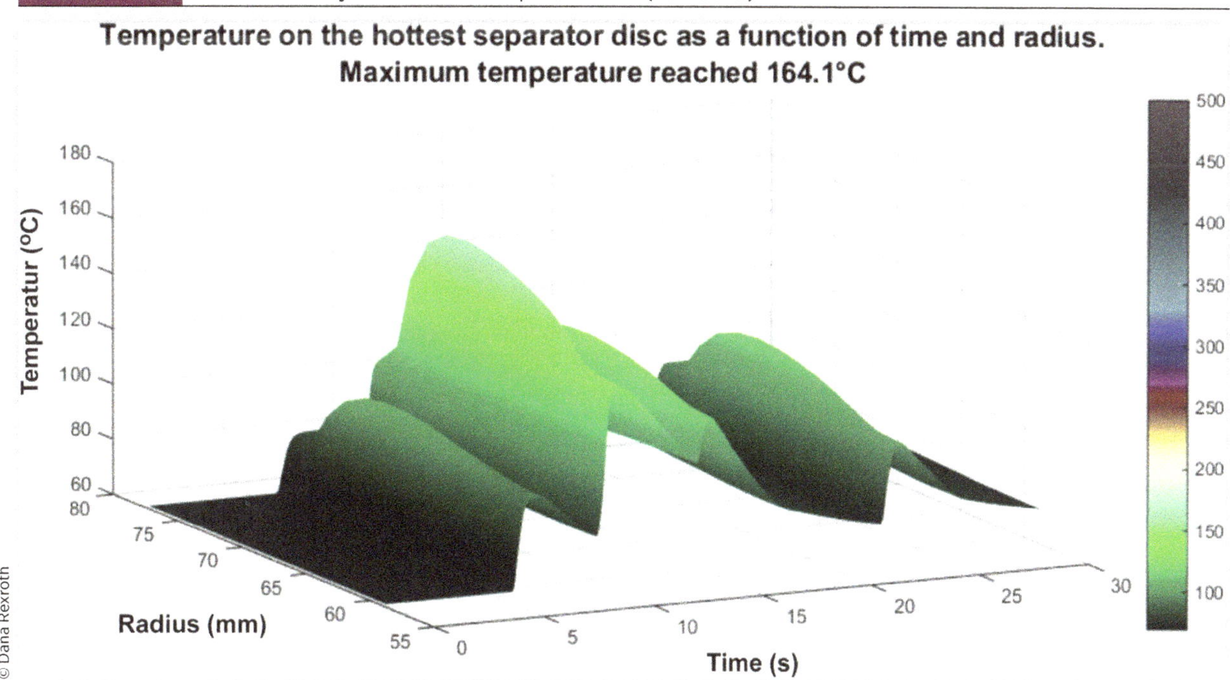

Interesting as well for the suitability confirmation or not of simulation selected counter-disc's thickness, to withstand the heat generated without distortion, is the following plot which shows the worst case of thermal stresses acting on the hottest separator disc. Notice that the most stressed point is at the outer radius, see Figure 3.22.

FIGURE 3.20 Temperature profile of the hottest point into the clutch.

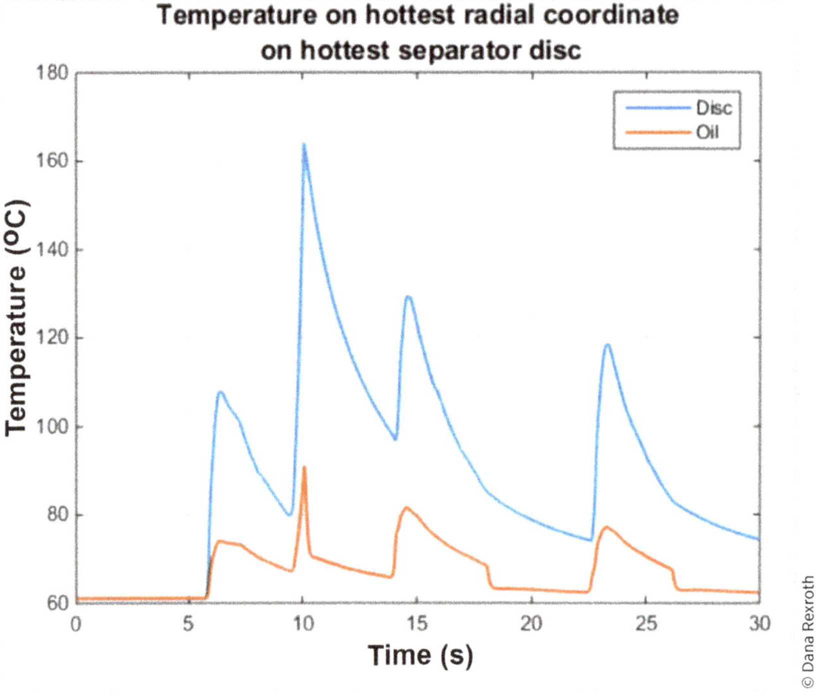

FIGURE 3.21 Torque transferred during shifts.

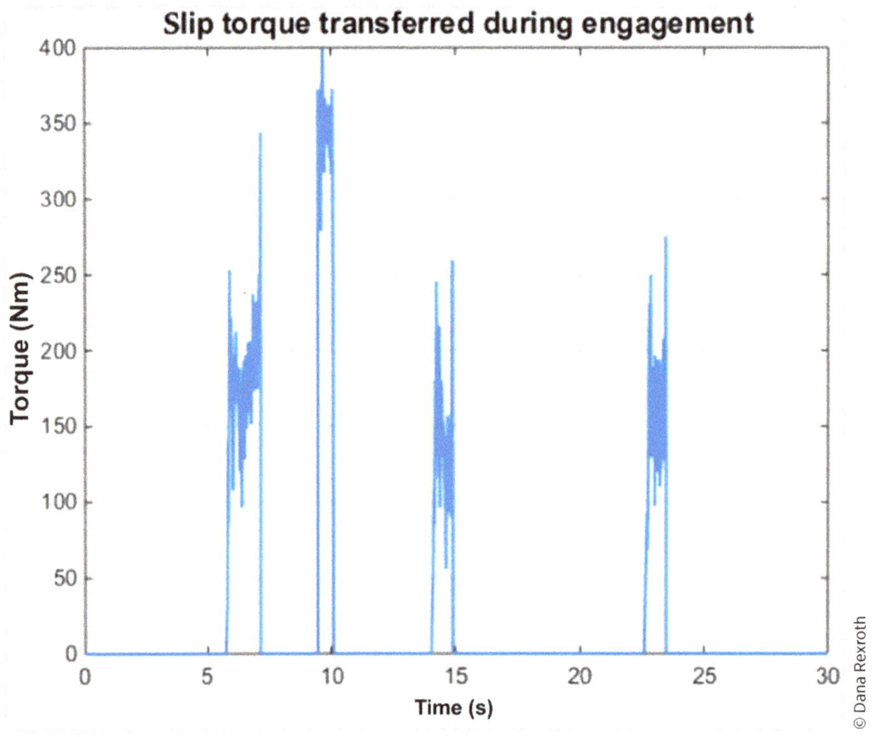

FIGURE 3.22 Worst case of thermal stresses acting on the hottest separator disc.

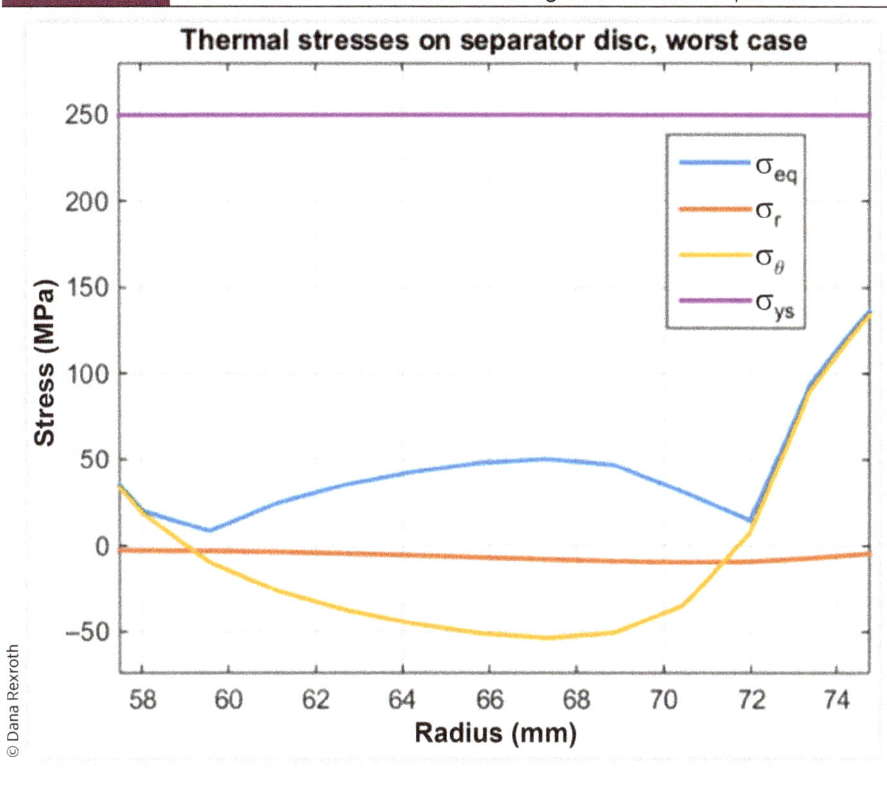

Finally, a plot that contains the specific power and a table reporting the specific energy and maximum specific power generated during simulated shifts (Figure 3.23 and Table 3.6).

Now comparing the results of the simulations with allowable friction material physical properties limits, it will be possible to judge the entire clutch design system and take next design modification steps if needed. The friction discs manufacturer

FIGURE 3.23 Specific power generated during the four simulated shifts.

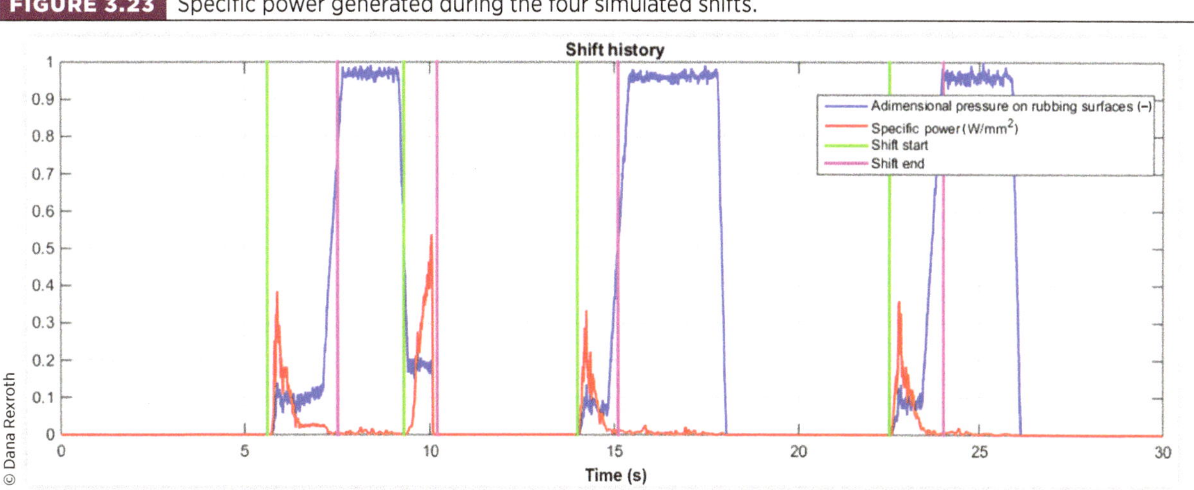

TABLE 3.6 Maximum specific power and total specific energy generated for each shift.

Shift number	1	2	3	4
Max specific power [W/mm^2]	0.385	0.537	0.335	0.359
Specific energy [J/mm^2]	0.119	0.167	0.088	0.104

recommends, for selected lining, to not overcome 1.21 J/mm^2 of specific energy and 0.810 W/mm^2 of specific power generated during a single shift.

Furthermore, it is also recommended to stay below 250 °C (482 °F) for lining surface temperature in the usual maneuvers (otherwise for the oil could start the thermal cracking phenomenon) and never overcome 300°C (572°F) at the rubbing interface, otherwise thermal damage on linings and discs could occur.

3.2.1.3.4 Clutch Thermal Dimensioning Conclusion. The presented case is a normal not severe working cycle for a clutch, with four asynchronous engagements and disengagements. The oil flow was equally distributed among friction pairs after having passed through the clutch gear. With the help of the simulation it was found that the mean oil volume fraction was always below one, more or less around 0.6 when the clutch was closed and cold, around 0.4 when the clutch was closed and hot and while the clutch was opened the volume fraction was around 0.05. During this normal not severe working cycle, the maximum instantaneous temperature reached was in the second separator disc, near to the piston, equal to 164.1°C (327.4 °F), corresponding to a temperature increase of 104.1°C (219.4°F) concerning the transmission oil sump temperature. The hottest instant happened during the asynchronous engagements and disengagements, where the highest specific energy was freed (0.167 J/mm^2), far below the allowable recommended limit.

FIGURE 3.24 Chain drive variator assembly.

3.2.1.4 **Chain Drive Variator:** This power conversion mechanism is made by utilizing a properly designed and flexible as much as possible steel chain running between two axially adjustable pulleys, respectively integrated on-chain variator input and output shafts, resulting in two variable chain contact diameters. Each pulley consists of an axially moving taper disc and, oppositely positioned on the same shaft, a specular taper disc rotating rigidly with the shaft itself. Clamped between the two pulley discs run the chain transmitting by friction the torque from input to output variator shafts. As mentioned, the two specular taper surfaces belonging to each pulley can be relatively moved axially in a continuous way changing accordingly, in a continuous way, chain to pulley contact diameter. Since chain has a constant length, if, e.g., the pulley integrated on the input shaft will be hydraulically actuated to reduce space between the two specular discs, increasing in such way the pulley to chain contact radius, the opposite will occur on the variator output shaft pulley; in such a way the chain variator mechanism will allow for an infinite (continuous) number of transmissions ratios. With this variator, see Figure 3.24, it will be possible to achieve a max speed

FIGURE 3.25 Chain drive efficiency map.

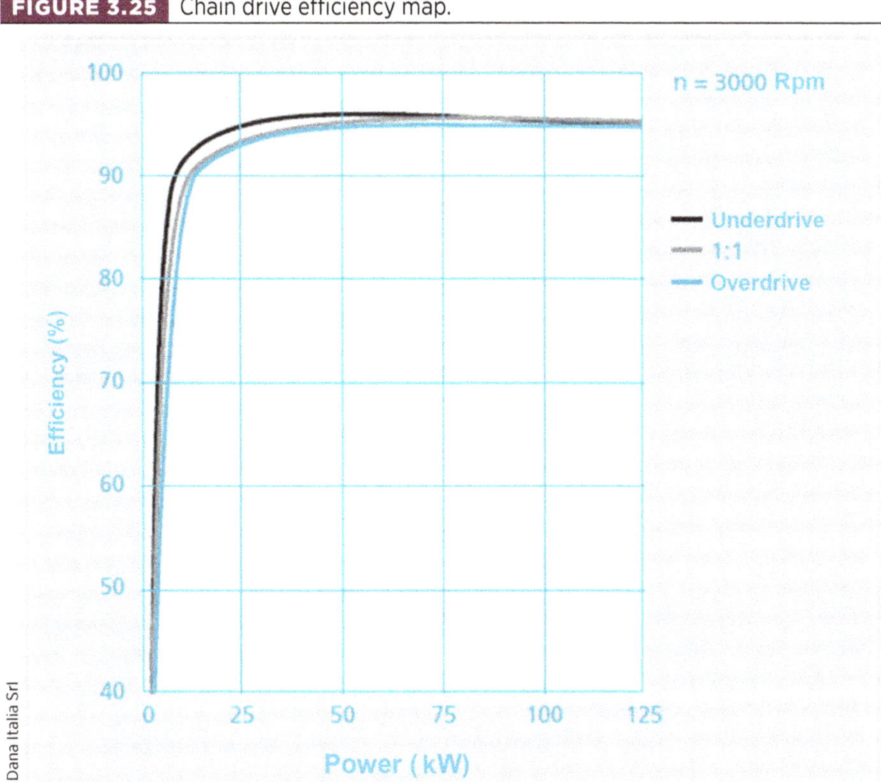

ratio of up to 2.45 in reduction and up to 0.4 in multiplication while showing a good variator efficiency level always around 95% within the operational working interval, see the efficiency map in Figure 3.25.

The clear physical limit of this mechanism is related to the fact that torque can be transmitted only by friction limiting in such a way power utilization ranges between 30 to 85 kW for compact tractor's applications. It is anyhow important to note that the variator alone will not be in the condition to cover the required tractor's transmission spread if not mandatory and rationally combined with an appropriate epicyclic gear train.

3.2.1.5 **Toroidal Variator:** After having been adopted for automotive and aeronautical applications, the full toroidal variator has been successfully introduced in the last 10 years on the drivetrains of small lawnmowers, garden, and recreational vehicles and only recently introduced in serial production for the drivetrain of small and compact tractors. The main advantage brought-up by the introduction of the toroidal variator on mobile working machines are since when properly integrated within a drivetrain, toroidal variator act as a continuous variable-ratio traction drive exhibiting simultaneously a high efficiency, a high power density and high reliability without the need of maintenance, being all internal mechanical components lubricated for life by a specially developed traction fluid. Full toroidal variator on his most simple configuration consists of two specular discs machined on their inside surface with a toroidal shaped groove both resulting in the so-called cavity. The two opposite and concentric grooves are clamped over a series of three rollers. Typically, the toroidal variators are arranged in pairs of cavities simultaneously clamped over the rollers, resulting in a compact assembly, without the need of adding-on any additional thrust bearing for transferring heavy axial loads to the transmission's housings. On each one of the two cavities, one disc performs the function of the input driving component while the other, facing the cavity disc act as an output-driven component. The transfer of the power from the driving to the driven discs is possible, thanks to the presence of rollers that are free to rotate around their supporting shafts as well, thanks to the traction fluid dynamically entrapped at each contact point between rolling component's surfaces. Tangential traction forces exchanged by rollers with the discs are possible thanks to the presence of particular synthetic and lubricating oil called traction fluid, having the property to improve consistently his viscosity by increasing the contact pressure between counter-rotating and mating components. Interesting to realize that this will occur without any metal-to-metal contact and slippage thanks, in fact, to particular traction fluid phase change of physical property. While moving up and down the rollers, so to change rollers position within toroidal cavities, the specular discs will rotate at different relative speeds one to each other. The change in the relative position of rollers versus discs within the cavity will result in a change of ratio inside the variator itself. Since rollers can continuously move, without power interruption during power transfer, the variator will result in a continuous variable transmission or CVT. Looking to the variator schematic, see Figure 3.26, it is possible to realize that when the rollers are positioned on the upper position of externally located

FIGURE 3.26 Full toroidal variator schematic.

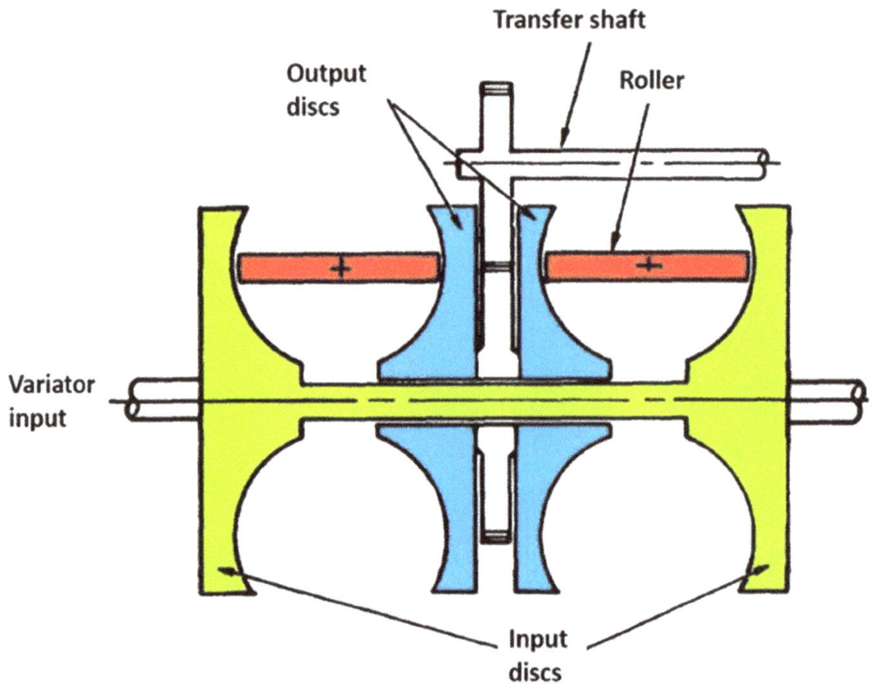

discs (corresponding to variator input discs) the output disc's speed will reach his maximum value and lowest transmitted torque.

When instead, the rollers are positioned on the lower position of variator externally located discs, the output disc's speed will be reduced resulting in a multiplication of the output torque. When the rollers will be kept on the central position of the cavities, the input and output discs of the variator will rotate at the same speed corresponding to the 1 to 1 (1:1) transmission speed ratio. For this type of full toroidal variator assembly, it will be possible to cover a speed ratio ranging from 1:3 minimum up to 3:1 maximum speed ratio corresponding in total to a ratio range or variator spread of 6. For most evolved variator design a ball ramp cam actuator system or a hydraulic cylinder actuator system will allow to continuously adjust the compression force on the rolling components across the entire speed and torque range. This continuous adjustment will contribute to optimizing not only variator efficiency in case of severe peak loads but will also allow us to keep approximately a high and constant value variator efficiency, independent from input power over the complete variator output working speed range, see Figure 3.27. Depending on mobile working machines' duties, the variator ratio range of 6 may not be sufficient to cover assigned vehicle tasks in terms of required torque versus vehicle speed. An intelligent and advantageous way to overcome this limit will be to integrate the full toroidal variator into multiple transmission layouts depending on considered applications. One of these possibilities could be covered while properly integrating an epicyclical gear set, resulting in one of the possible power-splitting transmission architecture design.

FIGURE 3.27 Full toroid variator efficiency for a 100 kW input power and 2,200 rpm input.

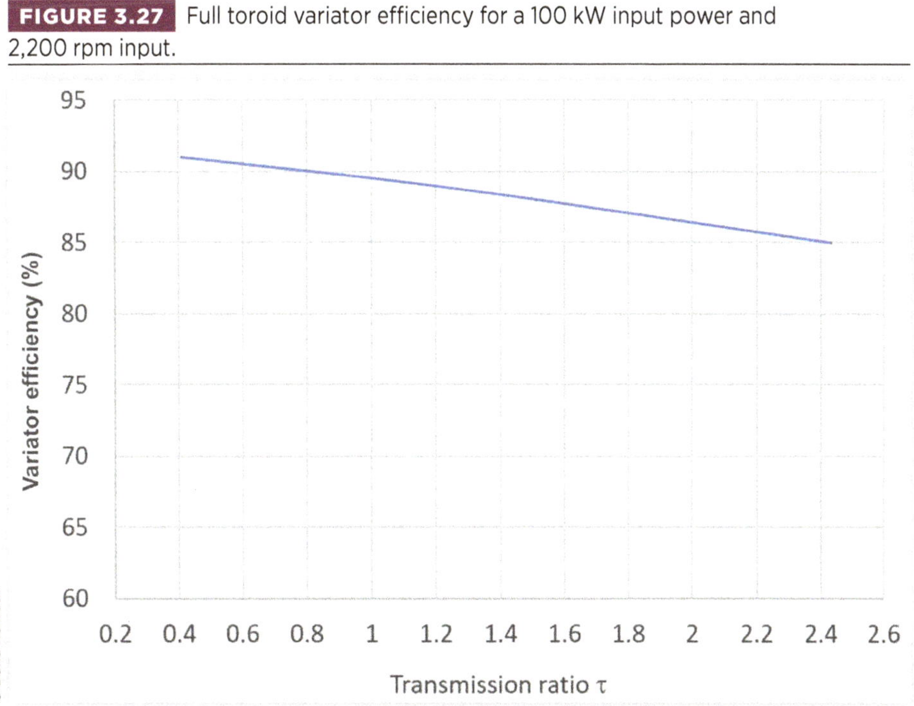

3.2.2 Hydraulic Energy Conversion

Mechanical energy can be converted into hydraulic energy and vice versa. Hydraulic pumps do the conversion from mechanical energy into hydraulic energy while hydraulic motors convert hydraulic energy into mechanical energy.

Furthermore, the hydraulic units are differentiated in terms of the type of energy conversion. Hydrodynamic units use the kinetic energy of a fluid. In a pump, the fluid is first accelerated and the reduction of speed in the motor creates the pressure built up according to Bernoulli's law. In mobile working machines, this principle is used in TCs. The basics of hydrodynamic elements can be found in Lechner [19].

Hydrostatic units work according to Pascal's principle and generate pressure in a "not moving" fluid (the fluid is only moving slowly). Detailed information on hydrostatic components and the working principles are described in Ivantysyn and Ivantysynova [20] and Murrenhoff [21].

3.2.2.1 Torque Converter: The TC device is playing a fundamental and unique role within mobile working machines applications because it makes it possible to convert simultaneously torque and speed in a continuous way. For this reason, it is still largely utilized. Mentioned previously (Section 3.2.1), torque and speed conversion represent common requirements, for mobile working machines, to fulfill the assigned task. TC belongs to the physical principle of hydrodynamic energy conversion devices. The main assembly constructive components are three, respectively:

1. An impeller or pump P is the driving component connected to the prime mover.
2. A turbine T that is the driven component connected to the TC output shaft.
3. A reactor component R can transfer reaction torque to the chassis.

The possibility to increase and convert the input torque is directly linked to the introduction of the reactor component that, while deviating the fluid flow and reacting/transferring reaction torque accordingly to the chassis, will allow summing on the TC output shaft both reactor torque and impeller input torque increasing, therefore, the output torque. Looking at Figure 3.28 it is possible to realize a typical schematic section of a single stage turbine version of TC with centripetal flow through the turbine called also TRILOK converter, currently largely utilized for mobile working machines. Single-stage TC means that there is only an impeller (pump) and only a turbine component and between those, radially positioned toward the inner radius of the fluid annulus, is interposed the reactor component. His tridimensional casting is rigidly connected to the hollow shaft bracing reactor to the transmission housing. Figure 3.28 gives evidence of the complex tridimensional path followed by the fluid flow. Looking at the design principle of the three mentioned components, the reactor belongs with some approximation to an axial flow turbomachine while the impeller and turbine belong to radial flow turbomachines.

For a selected geometry (blade's contour) of each of the three TC components, there will be only an optimum operation point, called also design point, where the efficiency η of the TC will reach its maximum level. Under this condition, the fluid flow will not face any impact losses at the input of each following in the path component. In other

FIGURE 3.28 Typical section of the torque converter.

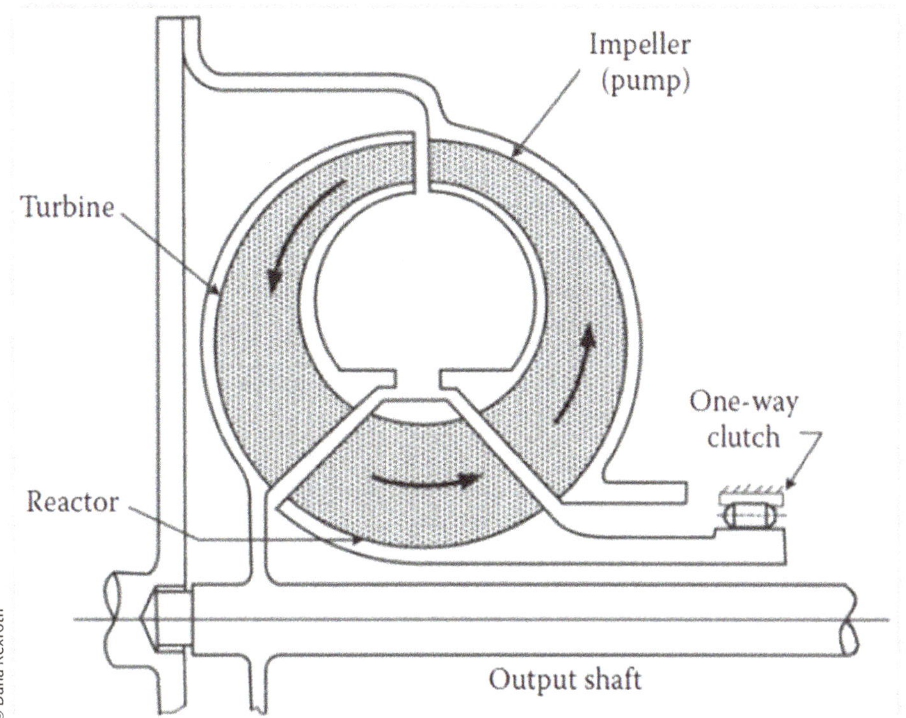

words, only under design condition the flow leaving each component will enter tangentially at the input section of the following one. Outside of the mentioned design point, input section impact losses, windage losses as well as leakage losses (due to blade gap between the output of a component and the input of the following one), together also with blade's friction losses will consistently change depending from out-of-design considered operating points. For such reason, the characterization of a TC can precisely and directly be done by experimental measurements. Once for a certain TC the characteristic curves will be available, it will be quite simple to utilize those curves for a specific design purpose supported also by similarity laws. For TCs, the equations provided at the input of Section 3.2.1 are still valid, so the TC efficiency η_{TC} can be written as follow:

$$\eta_{TC} = \frac{P_T}{P_P} = \frac{T_T \cdot \omega_T}{T_P \cdot \omega_P} = \mu \cdot \tau \tag{3.20}$$

where

T_T is the turbine output torque [Nm]
T_P is the pump input torque [Nm]
ω_T is the output angular speed [rad/s]
ω_P is the input angular speed [rad/s]

with

$$\mu = \frac{T_T}{T_P}$$

$$\tau = \frac{\omega_T}{\omega_P}$$

From hydraulic similarity laws it will be possible to arrive, after some simple passages, to the following important equation helpful to compare TC characteristic:

$$T_P = \epsilon \cdot n_P^2 \cdot D^5 \tag{3.21}$$

where

n_P is the impeller or pump revolutions [rpm]
$\epsilon = f(\tau)$ is a TC characteristic comparison function [$Nm \cdot s^2/m^5$]
D represents the TC fluid annulus output diameter [m].

At first approximation for a TRILOK type TC and considering impeller revolution as constant the turbine measured torque characteristic curve drop linearly from the so-called stall ratio μ_{Stall} to $T_T = 0$ where $\tau = 1$. If we let the impeller, driven from the engine, rotate at a constant revolution speed along with constant input torque, then the TC impeller will be driven by the constant input power. In such a condition, we would consider that the turbine torque will drop linearly from T_{Stall} to zero. Thus, the TC output power curve will be parabolic and accordingly, the efficiency will be a parabolic curve as taking η into account with P_P = constant. In reality, efficiency curves look a little bit different from a parabolic shape. If we look now to non-dimensional characteristic's curves, see **Figure 3.29** of a TRILOK TC, we can see stall point S, the design point M, where the efficiency reaches its maximum, and the point C called also the lock-up point where turbine torque is equal to pump torque and reactor torque is equal to zero. Starting from the point C, thanks to the presence of the free-wheel, interposed between reactor shaft and chassis (transmission casting), the entire reactor wheel will no longer deviate the fluid flow and will start to free-wheeling vanishing the scope for which the reactor was interposed between turbine and pump. From this point on, the TC characteristic will be the same as a hydrodynamic coupling and will exhibit a linear shape for efficiency characteristics.

The interval in the abscissa between speed conversion ratio τ at stall point S ($\tau = 0$) and lock-up point C ($\tau \approx 0.75$ in Figure 3.29) is defined TC conversion field [19]. Conversion characteristics of a TC can be modified while changing the turbine blade's inclination in such a way that the TC device can become more reactive resulting in higher multiplication of output torque. As a consequence of change, on mentioned characteristic behavior also the efficiency curve will change.

Generally, a more reactive TC will show his max efficiency point towards lower speed conversion ratio values, a reduction in conversion field amplitude, and a reduction in the level of max efficiency as well. For low-speed conversion ratio values, when exactly the torque is amplified, the efficiency is low and there is a consistent dissipation of power resulting in heat generation. For this reason, TC functioning in this field can work only during short transitory for a short time otherwise the fluid temperature will rise and could be seriously excessive. To mitigate and keep under control temperature increase a dedicated cooling system is always present on the TC of mobile working machines having cooling characteristics fully defined and established by the application severity and by the feedback coming from end-users.

FIGURE 3.29 Torque converter nondimensional characteristics.

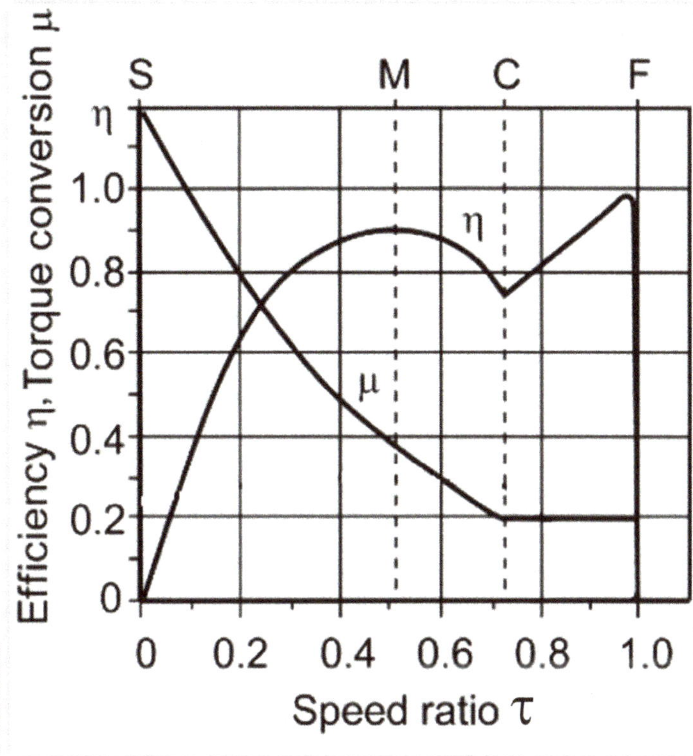

Performance characteristics of a TC are completely described by four parameters:

- Speed conversion ratio
- Torque conversion ratio
- Efficiency as already mentioned plus
- A capacity factor K that normally is provided by TC manufacturers.

The capacity factor (K) is an indicator of the ability of the TC to absorb or transmit torque. As mentioned earlier in this chapter, the capacity is closely related to the size and geometrical shape of the blades. For this reason, K could be utilized as a factor giving an indication of TC size-related performances. The capacity factor K is defined as:

$$K = \frac{n_P}{\sqrt{T_P}} \tag{3.22}$$

in $[rpm/(Nm)^{1/2}]$.

During measurements, done by TC manufacturers to characterize the hydrodynamic converting device, the pump input revolution will be kept being constant. The speed conversion ratio, torque conversion ratio, and pump absorbed torque are measured on a sufficient number of operative points necessary to draw, with sufficient accuracy, characteristic's curves, as shown in Figure 3.29.

If we are now interested in executing the engine-TC matching procedure, e.g., for a particular application in mobile working machines, the following procedure must be followed:

1. First, we have to select the engine speed, we are interested in, to evaluate the TC performance. Just remember: the engine speed is the same as that one of the TC pumps.
2. From the full load engine diagram, it is possible to see the available torque at the considered engine revolution.
3. The K-factor can be calculated from the described equation, knowing the engine torque and the speed available at the impeller.
4. Entering in the TC's characteristic diagram (curves) it will be possible to detect the torque conversion ratio μ and the speed conversion ratio τ, corresponding to the current capacity factor K in this selected operating condition.
5. Knowing that $T_T = \mu \cdot T_P$ and $n_T = \tau \cdot n_P$ both, TC output torque and speed, can be calculated.

Once the TC output speed and the torque for each of the considered operating points will be available it will be possible to predict the vehicle performances while having previously identified the most suitable mechanical gearbox design, in term of the needed spread, to cover vehicle task and a minimum number of transmission ratios needed to cover said spread, if a continuous variable transmission such as TC or chain variator, taken alone, could not sufficiently fit for the purpose.

3.2.2.2 **Hydrostatic Energy Conversion:** Hydrostatic power is determined by a hydrostatic pressure difference Δp combined with a hydrostatic flow Q. Hydrostatic energy can be stored in an accumulator that can be charged with the hydrostatic flow against a precharged second chamber that is filled with the inert gas nitrogen separated by an elastic diaphragm, an enclosed bladder, or a floating piston. Capacity for energy storage is limited while the charging and release rate of hydrostatic power is high.

$$P_{Hydr} = Q \cdot \Delta p \tag{3.23}$$

Hydrostatic units, i.e., pumps and motors are applied to convert between mechanical and hydrostatic forms of energy. Pumps are designed to consume mechanic power and generate hydrostatic power while motors consume hydrostatic power to generate mechanic power. Hydrostatic units without active elements can shift between pumping and motoring operation based on the external load conditions.

3.2.2.2.1 Hydrostatic Pump. Axial piston units in swashplate design are most commonly used in hydrostatic traction drives of mobile working machines. The main advantages of the swashplate principle are the compact design for a pump that can stroke in a positive and negative direction as well as the simple implementation of a mechanical drive through. Pumps can be stacked so that the number of required power take-offs for traction drive and work functions are reduced. Pumps can be mounted directly on

FIGURE 3.30 Swashplate pump with electric proportional displacement control [22].

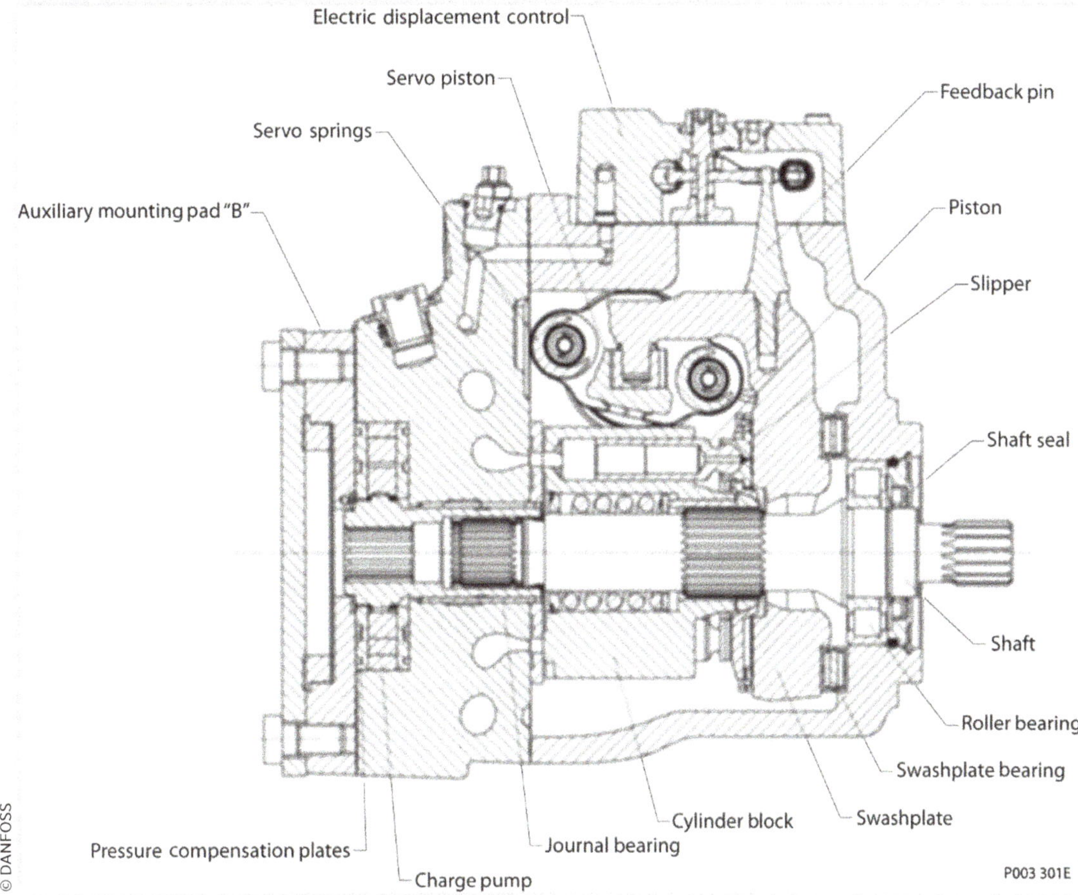

the diesel engine or via an additional splitter gearbox that might increase the pump speed or provide multiple mounting pads.

Figure 3.30 introduces a cross-sectional view of an axial piston pump in swashplate design with electric-proportional displacement control. The main components are the shaft and the cylinder block that holds the nine pistons with slippers that are the interface to the swashplate. The swashplate is centered in a neutral position by the pre-loaded servo springs and is stroked by the electric displacement control that receives the actual position of the swashplate via the feedback pin and controls pressure into the pressure chambers of the servo piston. The charge pump is positioned in the back of the pump and provides the servo system with pressure.

This whole package is assembled in a robust housing, which will be filled with oil in operation and is therefore sealed off with a shaft seal towards the input shaft. The swashplate bearing is essential to the smooth operation of the servo control.

3.2.2.2.2 Motor. The limited design space around wheels, axles, and gearboxes inside the vehicle frame requires compact and partly integrated solutions for the motor design.

FIGURE 3.31 Motor designs to provide solutions for various kinds of mobile working machines.

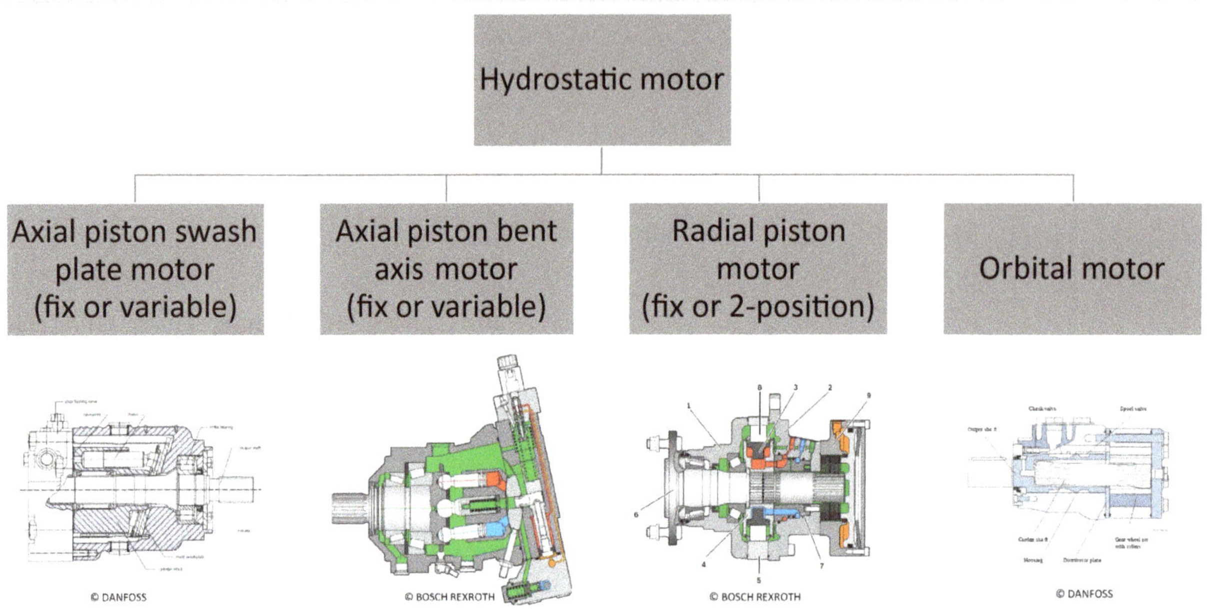

Therefore, a wide variety of motor designs are available for the traction drive of the various machines that all have their advantages and disadvantages. Figure 3.31 provides an overview of the motor types that are utilized in hydrostatic traction drives. While the axial piston units are mostly connected to a mechanical gearbox for speed reduction to the wheel, radial piston and orbital motors might be utilized directly at the wheel. Bent axis motors are the most frequent choice when a variable motor is required to perform the requested machine function. Radial piston units frequently have the option to switch off some of the radial pistons to realize a 2-position option.

Figure 3.32 introduces a cross-sectional view of an axial piston motor in bent axis design with electro-proportional displacement control. An electric signal commands a displacement which is converted into servo pressures by electric proportional control (8). These servo pressures move the differential servo piston (1) together with the connected valve segment (2) and bearing plate (3) until the commanded position is reached and fed back by the feedback spring (6). The sectional view shows the motor in maximum displacement. The minimum displacement is determined by the mechanical minimum displacement limiter (9). The motor shaft holds the optional speed ring (10) and is positioned in the housing with tapered roller bearings (4). The loop flushing system consists of the loop flushing shuttle spool (7) and the loop flushing relief valve (5).

For high-performance drive trains solutions with more advanced control systems, variable axial piston motors in bent axis, and swashplate design are most commonly used. Axial piston motors in bent axis design tend to provide efficiency advantages and allow for higher rotational speed and corner power. Hydraulic motors are available as fix motors that cannot change displacement, as 2-position motors

FIGURE 3.32 Bent axis motor with electric proportional displacement control [23].

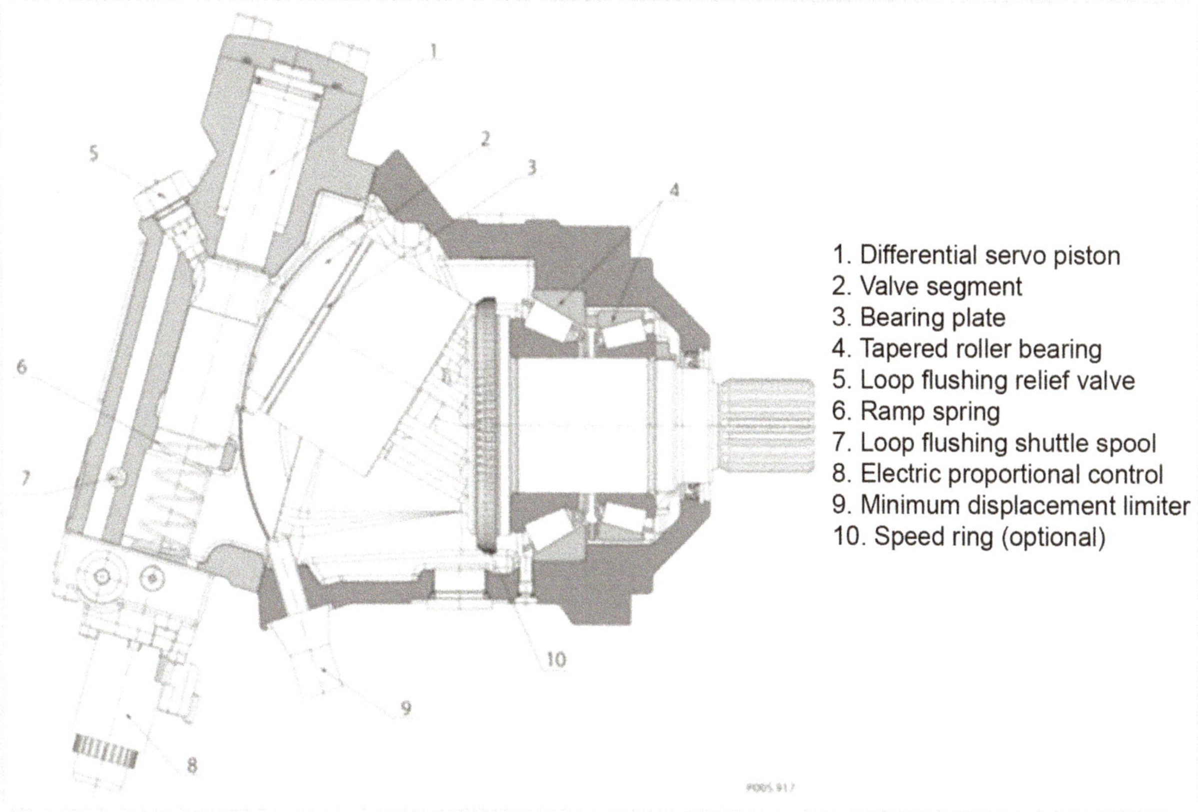

1. Differential servo piston
2. Valve segment
3. Bearing plate
4. Tapered roller bearing
5. Loop flushing relief valve
6. Ramp spring
7. Loop flushing shuttle spool
8. Electric proportional control
9. Minimum displacement limiter
10. Speed ring (optional)

that can change their displacement from minimum to maximum and finally as variable motors, that can vary their displacement continuously between a minimum and maximum value.

3.2.2.2.3 Torque. The relation between mechanical and hydrostatic variables is described for pumps (Index P) and motors (Index M) by the torque T, the actual displacement V and the pressure difference Δp in Equations (3.24) and (3.25). The hydro-mechanical efficiency η_{hm} increases the input torque to the pumping unit and it reduces the output torque of the motoring unit.

$$T_P = \frac{\Delta p \cdot V_P}{2 \cdot \pi \cdot \eta_{hm}} \quad \text{with } \eta_{hm} \in \left]0;1\right[\tag{3.24}$$

$$T_M = \frac{\Delta p \cdot V_M \cdot \eta_{hm}}{2 \cdot \pi} \quad \text{with } \eta_{hm} \in \left]0;1\right[\tag{3.25}$$

3.2.2.2.4 Flow. Another relation connects the rotational speed of the shaft n with the hydrostatic flow Q of the hydrostatic system in the Equations (3.26) and (3.27).

The volumetric efficiency η_{vol} reduces the flow output of the pumping unit (Index P) and it increases the flow consumption of the motoring unit (Index M).

$$Q_P = n \cdot V_P \cdot \eta_{vol} \quad \text{with } \eta_{vol} \in]0;1[\tag{3.26}$$

$$Q_M = \frac{n \cdot V_M}{\eta_{vol}} \quad \text{with } \eta_{vol} \in]0;1[\tag{3.27}$$

Both sets of equations use the actual displacement of the hydrostatic unit – V_P and V_M – as a key variable. The actual displacement depends on the physical maximum displacement and the actual swivel angle in comparison to the maximum swivel angle as in Equation (3.28).

$$V_P = V_{P,max} \cdot \frac{\alpha_P}{\alpha_{P,max}} \tag{3.28}$$

$$V_M = V_{M,max} \cdot \frac{\alpha_M}{\alpha_{M,max}}$$

3.2.2.2.5 Hydrostatic Transmission. In a hydrostatic transmission, pump and motor are connected via the two high-pressure ports to provide a continuously variable transmission ratio. For the analysis, it can be assumed that the flow of the hydrostatic pump matches the flow of the hydrostatic motor (Equation 3.29) and both hydrostatic units are loaded with the same delta pressure in the system (Equation 3.30).

$$Q_P = Q_M \tag{3.29}$$

$$\Delta p_P = \Delta p_M \tag{3.30}$$

The substitution of flows and delta pressures results in speed and ratios for the hydrostatic transmission that are based on the efficiency assumptions from Equations (3.24) to (3.27). Equation (3.31) shows the dependency of the motor speed from the displacements of pump and motor but also from the volumetric efficiencies. The ratio of the hydrostatic transmission is described by the speed ratio of the pump input shaft to the motor output shaft. Equation (3.32) shows the dependencies of the input torque to the pump – again it is the displacements and the hydro-mechanical efficiencies of the components.

$$n_M = n_P \cdot \frac{V_P}{V_M} \cdot \eta_{P,vol} \cdot \eta_{M,vol} \tag{3.31}$$

$$T_P = T_M \cdot \frac{V_P}{V_M} \cdot \frac{1}{\eta_{P,hm} \cdot \eta_{M,hm}} \tag{3.32}$$

3.2.2.2.6 Losses. As indicated in the earlier equations, hydro-mechanical and volumetric losses influence the overall efficiency of hydrostatic components and systems. Understanding the characteristics of the losses that cause the reduction inefficiencies allows an optimized system design. Hydro-mechanical losses lead to the over-all torque loss T_S and result from viscous and Coulomb friction as well as compression losses.

TABLE 3.7 Overview of pumping and motoring efficiencies [8].

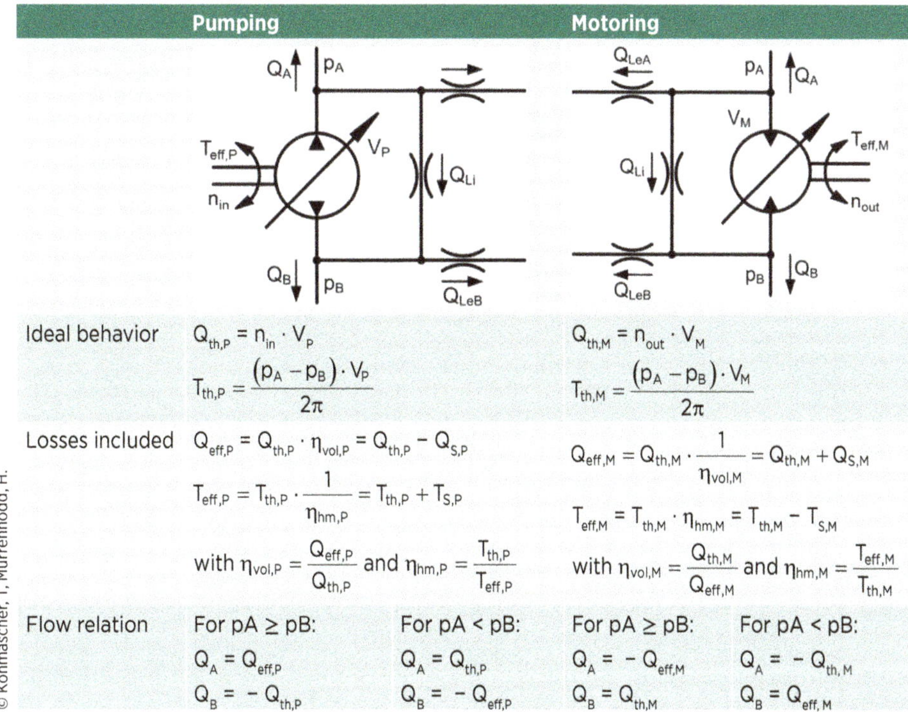

	Pumping		Motoring	
Ideal behavior	$Q_{th,P} = n_{in} \cdot V_P$ $T_{th,P} = \frac{(p_A - p_B) \cdot V_P}{2\pi}$		$Q_{th,M} = n_{out} \cdot V_M$ $T_{th,M} = \frac{(p_A - p_B) \cdot V_M}{2\pi}$	
Losses included	$Q_{eff,P} = Q_{th,P} \cdot \eta_{vol,P} = Q_{th,P} - Q_{S,P}$ $T_{eff,P} = T_{th,P} \cdot \frac{1}{\eta_{hm,P}} = T_{th,P} + T_{S,P}$ with $\eta_{vol,P} = \frac{Q_{eff,P}}{Q_{th,P}}$ and $\eta_{hm,P} = \frac{T_{th,P}}{T_{eff,P}}$		$Q_{eff,M} = Q_{th,M} \cdot \frac{1}{\eta_{vol,M}} = Q_{th,M} + Q_{S,M}$ $T_{eff,M} = T_{th,M} \cdot \eta_{hm,M} = T_{th,M} - T_{S,M}$ with $\eta_{vol,M} = \frac{Q_{th,M}}{Q_{eff,M}}$ and $\eta_{hm,M} = \frac{T_{eff,M}}{T_{th,M}}$	
Flow relation	For pA ≥ pB: $Q_A = Q_{eff,P}$ $Q_B = -Q_{th,P}$	For pA < pB: $Q_A = Q_{th,P}$ $Q_B = -Q_{eff,P}$	For pA ≥ pB: $Q_A = -Q_{eff,M}$ $Q_B = Q_{th,M}$	For pA < pB: $Q_A = -Q_{th,M}$ $Q_B = Q_{eff,M}$

Volumetric losses in pump and motor lead to the over-all flow loss Q_S of the system, which results from internal leakage Q_{Li} (between high and low-pressure side) and external leakage Q_{Le} (from high and low-pressure to the case). Flow losses (summarized in the volumetric efficiency η_{vol}) reduce the motor output speed relative to the ideal system while hydro-mechanical losses (summarized in the hydro-mechanical efficiency η_{hm}) plus the torque demand from the charge pump increase the required input torque to the pump over the ideally calculated.

Table 3.7 defines the pumping and motoring conditions of hydrostatic units. Starting with the ideal behavior, the losses and efficiencies are considered and finally, the calculated flows are assigned to the ports A and B of the hydrostatic unit [24]. This consideration is supportive when setting up a simulation model of the hydrostatic transmission.

3.2.2.2.7 Loss Modeling. Power losses due to flow and torque losses in the hydrostatic units greatly influence the overall efficiency of the hydrostatic transmissions. These losses are dependent on speed, system pressure, and displacement of the individual hydrostatic unit - pump or motor. Accurate modeling of the loss behavior is the first step towards modeling optimization of the complete system. As a starting point, precise measurements of components are essential to provide accurate data to create high-quality simulation models of existing components. More and more, system-level simulation is the basis for component selection, control development, and system optimization.

For the modeling of the loss behavior of hydrostatic transmissions, multiple principles were introduced in literature [11, 24]. After some attempts of modeling the losses

FIGURE 3.33 Datapoint and polynomial for flow loss (left) and torque loss (right) at 50% displacement [8, 24].

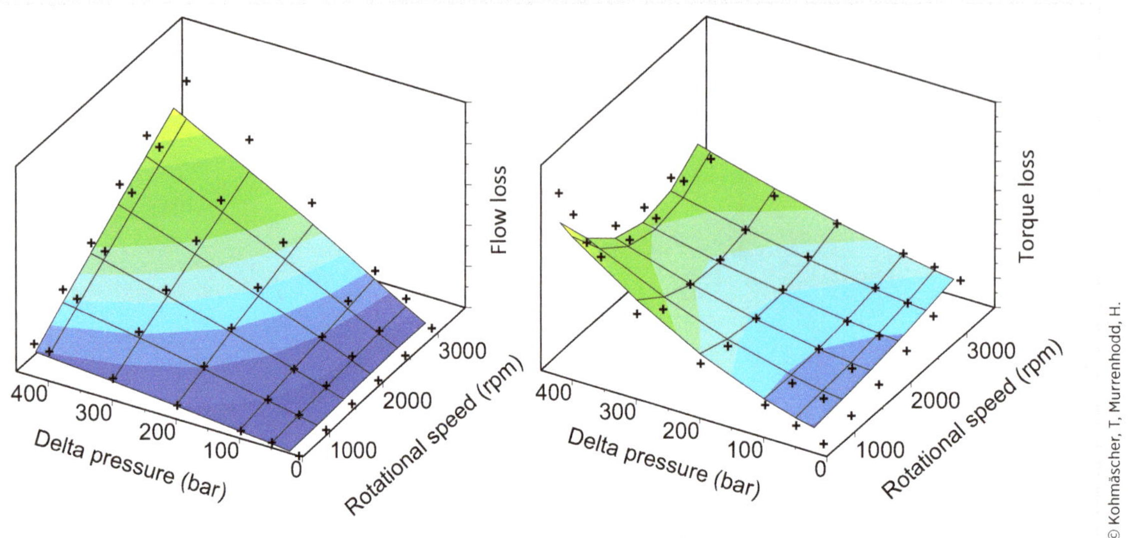

based on physical assumptions, the polynomic approach "Polymod" by Ivantysyn and Ivantysynova demonstrated superior performance in the modeling but also in the flexible integration into existing toolchains. It provides a purely mathematical, polynomial reproduction of the components loss behaviors in terms of flow loss Q_S and torque loss T_S - dependent on displacement V_i, speed n and pressure Δp.

$$Q_S = \sum_{i=0}^{f_1}\sum_{j=0}^{f_2}\sum_{k=0}^{f_3} K_{Q,kji} \cdot n^i \cdot V_i^j \cdot \Delta p^k \tag{3.33}$$

$$T_S = \sum_{i=0}^{f_1}\sum_{j=0}^{f_2}\sum_{k=0}^{f_3} K_{T,kji} \cdot n^i \cdot V_i^j \cdot \Delta p^k \tag{3.34}$$

Figure 3.33 displays exemplary the measured operating conditions of an axial piston unit at 50% displacement dependent on speed and pressure difference as well as the surface of the calculated polynomial of the loss model. The left diagram shows the flow losses while the right diagram shows the torque losses of the hydrostatic unit.

3.2.3 Electric Energy Conversion

3.2.3.1 **Fundamentals:** For the operation of mobile working machines, energy needs to be provided to the different electrical components of the machine. As the components are different in their operation principle, the demand for electrical energy differs not only in the amount but also in the way it has to be provided to the respective component. Electrical quantities as current or voltage can be divided into DC and AC quantities (DC from *direct current*, AC from *alternating current*). While DC quantities are completely described by their amplitude (e.g., 24 V battery voltage), AC quantities are characterized also by their frequency (e.g., the 400 V/50 Hz low-voltage grid in Europe). At this, 400 V

FIGURE 3.34 DC (left), single-phase AC (middle) and three-phase AC quantities (right).

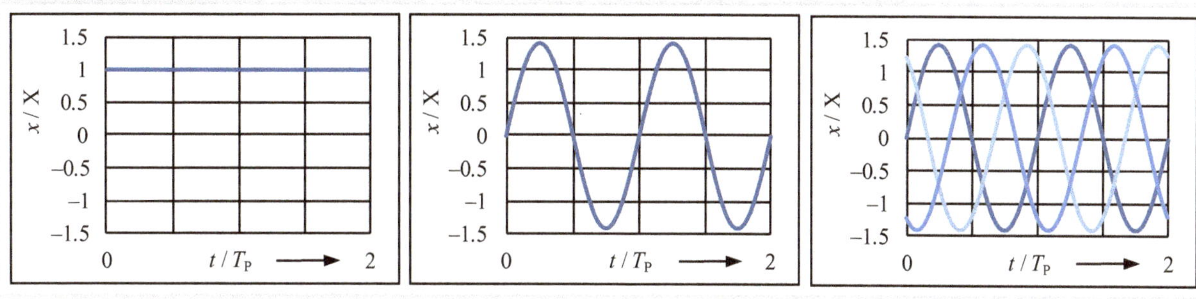

is the *root mean square* (RMS) value of the voltage. A sinusoidal quantity can be described as

$$x(t) = \hat{x} \cdot \sin(2\pi f \cdot t + \varphi) = \hat{x} \cdot \sin(\omega \cdot t + \varphi), \tag{3.35}$$

where

- f is the frequency
- $\omega = 2\pi f$ is the angular frequency
- φ is the phase angle

At sinusoidal quantities, the relation between the RMS value X and the peak value $\hat{x}$ is given by

$$\hat{x} = \sqrt{2} \cdot X. \tag{3.36}$$

The RMS value equals a DC value with a magnitude of X and is defined as the root of the sum of the squared sample values during a period with the duration T_P:

$$X = \sqrt{\frac{1}{T_P} \int_{t}^{t+T_P} (x(t))^2 \, dt}. \tag{3.37}$$

Especially in the field of electrical drives, not single-phase, but symmetrical three-phase AC quantities are required. They can be described as

$$\begin{aligned} x_1(t) &= \hat{x} \cdot \sin(\omega \cdot t) \\ x_2(t) &= \hat{x} \cdot \sin\left(\omega \cdot t - \frac{2 \cdot \pi}{3}\right) \\ x_3(t) &= \hat{x} \cdot \sin\left(\omega \cdot t - \frac{4 \cdot \pi}{3}\right) \end{aligned} \tag{3.38}$$

Figure 3.34 is showing the DC, single-phase AC and three-phase AC quantities in the time domain.

Usually, a conversion between DC and AC, AC and DC, DC to higher or lower DC voltage or from AC to AC with a different frequency and/or voltage is necessary. At this, it can be differentiated between the conversion types listed in Table 3.8. In many cases, the converters can realize an energy flow in both directions, and by this named "bidirectional." Otherwise, the classification is done according to the direction of the main energy flow.

TABLE 3.8 Conversion types.

Conversion	Symbol		Example
DC → DC (DC/DC-converter)	or	DC/DC	Voltage change from 24 V battery voltage to e.g., 5 V supply voltage for signal electronics
DC → AC (inverter)	or	DC/AC	Operation of a three-phase AC machine fed from a DC link
AC → DC (rectifier)	or	AC/DC	DC power supply fed from an AC grid
AC → AC (AC/AC-converter)	or	AC/AC	Variable speed drive fed from a three-phase grid

3.2.3.2 **Power Electronics:** Except for some special applications, the electrical energy conversion is usually realized by power electronics. Due to enormous progress in the development of power semiconductors over the last decades, power electronics are offering a reliable, flexible and efficient way of electrical energy conversion. With this, they are forming the link between the energy source and the load.

The conversion is realized by applying power semiconductor components like, e.g., diodes and transistors as valves. To avoid high losses and to achieve high efficiency of the power electronics, it is the aim to operate the valves either in their *ON* state or in their *OFF* state and with this as switches (hence e.g., the name *switched-mode power supply*). Thereby, the losses in the linear operation range can be reduced to a minimum. For a detailed view of loss mechanisms in power, electronics see Rashid's *Power Electronics Handbook: Devices, Circuits, and Applications* for an example [25]. In Figure 3.35, a typical system setup including an energy source, power electronics, and a load is depicted. As can be seen, power electronics do not only consist of the power semiconductors but necessarily also of control as well as of driver circuits.

Power semiconductors can be divided into three groups: not controllable (diodes), able to be switched on but not able to be switched off (e.g., thyristors) and able to be switched on and off like MOSFETs (metal-oxide-semiconductor field-effect

FIGURE 3.35 Structure of a typical setup.

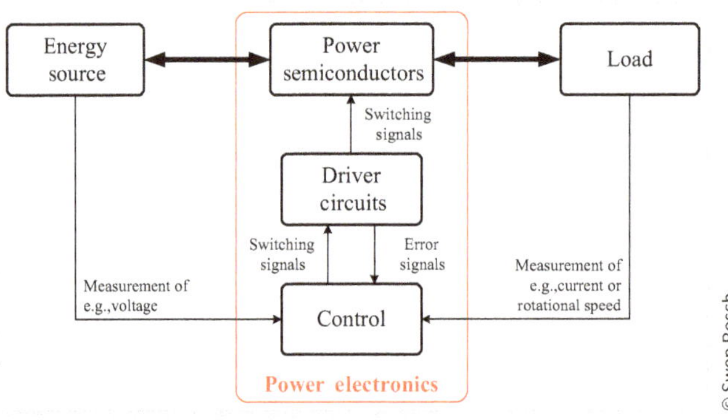

transistors) and IGBTs (insulated-gate bipolar transistors). At this, the MOSFETs and IGBTs are widespread for many applications.

Driver circuits are realizing the interface between the control and the power semiconductors. As the control is implemented on a microcontroller (μC) or a digital signal processor (DSP), which are working at a voltage level of 3–5 V, the driver circuit also adapts the voltage level of the switching signals to the required level of usually 12–15 V. Besides, the driver circuit is monitoring the correct function of the power semiconductors and is providing error signals in the case of an error so that the control can react and bring the system in a safe state or at least to avoid further damage.

The **control** is processing the input signals, which could be signals from current, voltage, or rotational speed sensors and also from a higher-level control system, providing reference values. As periphery, the μC or DSP is having a communication interface, analog-digital converters (ADCs) for the measurement signals and a pulse width modulation (PWM, see also Section 4.2.2) module, which is needed for the generation of the switching signals and therefore it is an elementary part of the most switched power electronic devices. Anyway, it has to be mentioned that there are control methods that do not need a PWM module, e.g., hysteresis control [26] or model predictive control [27]. In contrast to the PWM control, the switching frequency is not constant due to the working principle of these control methods.

For the following considerations regarding the PWM module, T_{Sw} is the switching time. Common switching frequencies $f_{Sw} = 1/T_{Sw}$ are located in the frequency range between 4 kHz and 20 kHz, which equals switching periods of 250 μs and 50 μs. In the PWM module, a counter is included, which can be incremented or decremented in every system clock (which is about 50 MHz or more, if no prescaler is used). The counter *CTR* can be implemented as an up-, down- or up-down-counter, depending on the application. For the explanations based on **Figure 3.36**, an up-down-counter is used. The example is based on the current control of an ohmic-inductive load with an inner voltage as a load (which could be a DC machine) using a so-called *single-phase half-bridge inverter.* At this, some simplifications are made: all electronic components are treated as ideal,

FIGURE 3.36 Single-phase half-bridge inverter: control/hardware structure and PWM operating principle.

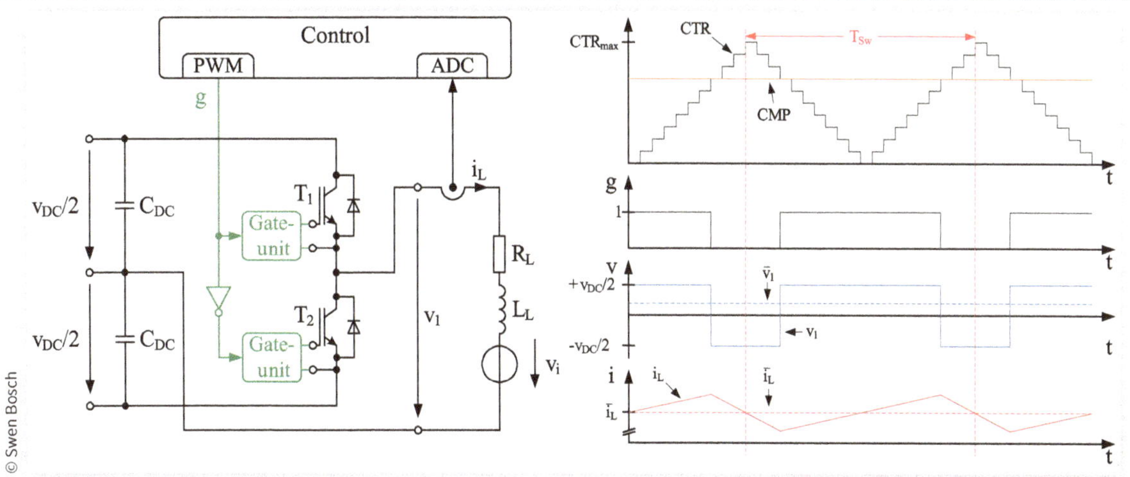

which means that there is no voltage drop on current-carrying power semiconductors as diodes or switches and that the time the switch needs to open or close is zero. Furthermore, v_{DC} has to be higher than the v_1. Besides the capacitors, the converter consists of two IGBTs as switches and two diodes. The two IGBTs are switched complementary to each other; hence the signal of the lower IGBT is inverted. If the upper IGBT T_1 is closed, the voltage between the load and the midpoint of the DC link v_1 equals $+v_{DC}/2$; otherwise, the lower switch T_2 is closed and v_1 is $-v_{DC}/2$. With this, the mean voltage within one switching period can be set between $+v_{DC}/2$ and $-v_{DC}/2$.

The PWM module is comparing the value of the PWM compare register *CMP* with the actual counter value *CTR* every increment. The compare value *CMP* is provided by the control and is updated always when the counter reaches its maximum. As long as the counter value *CTR* is below the compare value *CMP*, the switching function g, which is the switching signal provided to the driver circuit, is high, otherwise, it is low. The driver circuit is switching on and off the switches according to this signal. As it can be seen in Figure 3.36, a mean voltage (blue dashed line) within one switching period T_{Sw} can be provided to the load, but the time value of the voltage is either $v_1 = +v_{DC}/2$ or $v_1 = -v_{DC}/2$. This switching between two (or more, depending on the circuit) voltages is characteristic for switched-mode power electronics. As can be seen in the following, most of the switched-mode power electronic devices are also including energy storage elements like capacitors or inductors, which are necessarily needed for the computation of the current and for smoothing voltages or currents. Referring to the example, this would be the load inductance L_L.

In the following, an overview of the most widespread power electronic converters will be given. One of the most important passive converters (passive means not-actively switched, e.g., by a control) is the three-phase diode rectifier, as shown in Figure 3.37a, to generate a DC voltage from a three-phase AC source. If a diode rectifier is connected to a grid, line filters, often inductors L_f, have to be used to reduce harmonics of the grid side current to meet the grid standards. A DC capacitor C_{DC} is used to smooth the load voltage v_{DC}.

For the operation of three-phase electrical machines as variable speed drives, a three-phase DC-AC converter as depicted in Figure 3.37b can be applied. An overview of other applicable DC-AC converter types is given in Leonhard's *Control of Electrical*

FIGURE 3.37 (a) Three-phase diode rectifier and (b) three-phase inverter.

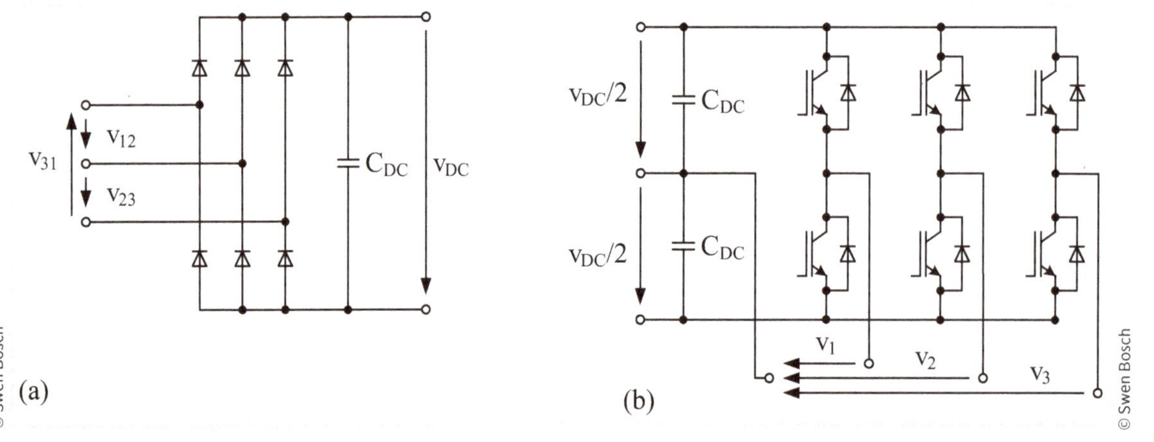

Drives [28]. The depicted converter is based on three-phase legs, which are build up in the same way as the single-phase half-bridge inverter already explained above. As the voltage of each phase leg can be varied within the voltage range between $+v_{DC}/2$ or $-v_{DC}/2$, the phase-to-phase voltages vary within v_{DC} and $-v_{DC}$ (assuming the same simplifications as for the single-phase half-bridge inverter). With this, it is possible to generate a three-phase AC voltage with variable frequency and voltage.

3.2.3.3 **Variable Speed Drives:** Variable speed drives and their control are a broad field and discussed in detail in Leonhard [28] or Doncker [29]. This chapter will focus on three-phase AC induction machines (IM) and synchronous machines (SM) as the most widespread machine types used for variable speed drives. The theory is presented based on rotational machines, but is also suitable for translational machines since the fundamental working principle of both machine types is the same. Furthermore, the theory of field-oriented control (FOC) will be explained based on a permanent magnet synchronous machine (PMSM).

In general, all electrical machines are consisting of a moving and a stationary part, named rotor and stator, respectively. The stators of an IM and an SM are largely the same and are based on a lamination package with slots. As illustrated in **Figure 3.38a**, three-phase symmetrical windings are placed in the slots. The electrical angle between the windings is always $2 \cdot \pi/3$ (120°), while the mechanical angle can be an integer part of the electrical angle. A current flowing in winding a results in a magnetic field in the direction of the a-axis, the same applies to the windings b and c. From an electrical point of view, as shown in **Figure 3.38b**, the windings can be described as an inductance $L_{S,\mu}$, where the index μ represents the axis a, b, or c (the ohmic resistances $R_{S,\mu}$ are neglected). If a three-phase symmetrical voltage $v_{S,\mu}$ (compare Equation 3.38) is connected to the stator windings, this results in three-phase symmetrical stator currents in the steady-state. Multiplying the stator currents with the inductances $L_{S,\mu}$ results in the magnetic fluxes $\psi_{S,\mu}$. For the following considerations, the magnetic fluxes $\psi_{S,\mu}$ are projected to the orthogonal αβ-coordinate system as shown in **Figure 3.38c**.

As the direction of the α-axis is equal to the a-axis, it can be shown, that the magnetic flux in the direction of the α-axis results in

$$\psi_{S,\alpha} = \psi_{S,a} - \frac{1}{2} \cdot \psi_{S,b} - \frac{1}{2} \cdot \psi_{S,c}, \tag{3.39}$$

FIGURE 3.38 (a) Simple stator winding scheme of a three-phase machine, (b) electrical representation, and (c) αβ-coordinate system.

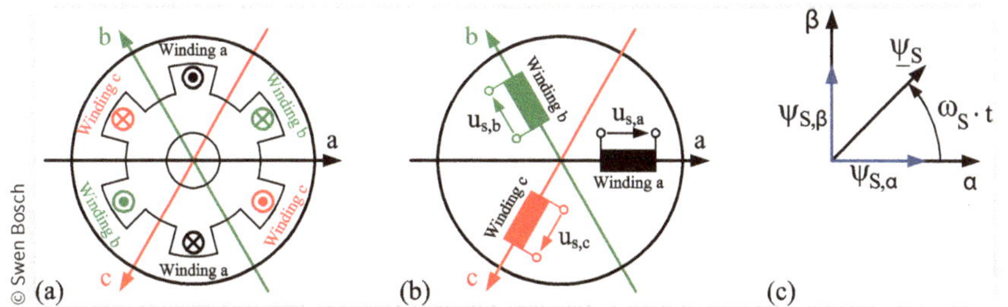

while the flux in the direction of the β-axis is given by:

$$\psi_{S,\beta} = \frac{\sqrt{3}}{2} \cdot \psi_{S,b} - \frac{\sqrt{3}}{2} \cdot \psi_{S,c}. \tag{3.40}$$

Insertion of the time depending flux by analogy with Equation (3.38) provides the time depending magnetic fluxes $\psi_{S,\alpha}(t)$ and $\psi_{S,\beta}(t)$ in the α- and β-axis:

$$\left.\begin{aligned} \hat{\psi}_{S,\alpha}(t) &= \frac{3}{2} \cdot \hat{\psi}_S \cdot \cos(\omega_S \cdot t) \\ \hat{\psi}_{S,\beta}(t) &= \frac{3}{2} \cdot \hat{\psi}_S \cdot \sin(\omega_S \cdot t) \end{aligned}\right\} \quad \underline{\psi}_S(t) = \psi_{S,\alpha}(t) + j\,\hat{\psi}_{S,\beta}(t) = \frac{3}{2} \cdot \hat{\psi}_S \cdot e^{j(\omega_S \cdot t)} \tag{3.41}$$

where $(3/2) \cdot \psi_S$ is the amplitude of the magnetic flux in the stator, which is rotating with the angular frequency ω_S. With this, $\psi_{S,\alpha}(t)$ and $\psi_{S,\beta}(t)$ can be represented as a rotating space vector $\underline{\psi}_S(t)$ in the complex plane. This transformation from the three-phase system to the αβ-frame is called *Clarke transformation*.

For most induction machines, the rotor is built as a squirrel cage rotor, which consists of a lamination package with slots, as shown in **Figure 3.39a**. The rotor slots are filled with aluminum. These so-called rotor bars are connected (short-circuited) to each other on both sides of the rotor (at machines with a power of more than 100 kW, copper instead of aluminum is used to reduce losses due to the higher electrical conductivity of copper). The rotating stator field is inducing a voltage in the rotor bars that is causing a rotor current. The combination of the current in the rotor bars and the moving stator field results in a force (known as Lorentz force). For rotational machines, this force is generating the torque. As can be seen from the theoretical descriptions, a force is generated only when there is a difference in the angular speed of the stator field and the angular speed of the rotor. This difference in the angular speed is called *slip* s and is the main characteristic of induction machines. The torque-slip characteristic of an exemplary induction machine is shown in **Figure 3.39b**. At this, T_i is the torque generated by the machine, T_B is the breakdown torque and s_B is the breakdown slip. In motor operation,

FIGURE 3.39 (a) scheme of a cut IM: stator and rotor lamination packages (bright grey), stator windings (blue, green, orange) and rotor bars (dark grey) (b) slip-torque characteristic (simplifying assumption $R_{S,\mu} = 0$).

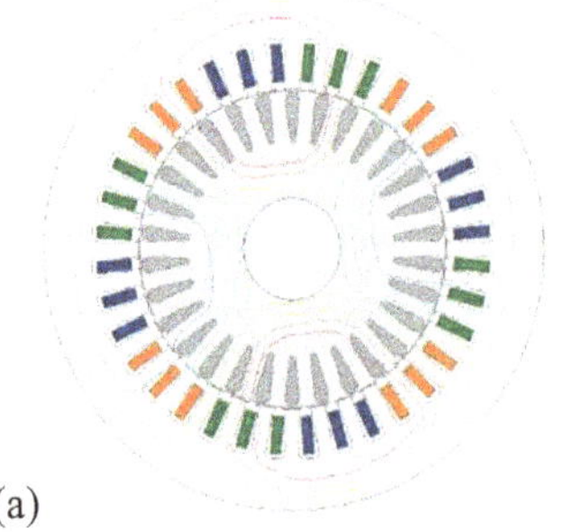

(a)

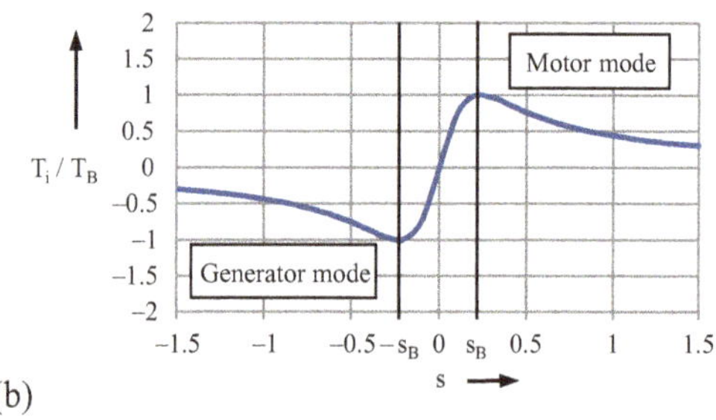

(b)

FIGURE 3.40 (a) scheme of a cut PMSM: stator and rotor lamination packages (bright grey), stator windings (blue, green, orange) and rotor magnets (green/red) (b) the alignment of the coordinate systems.

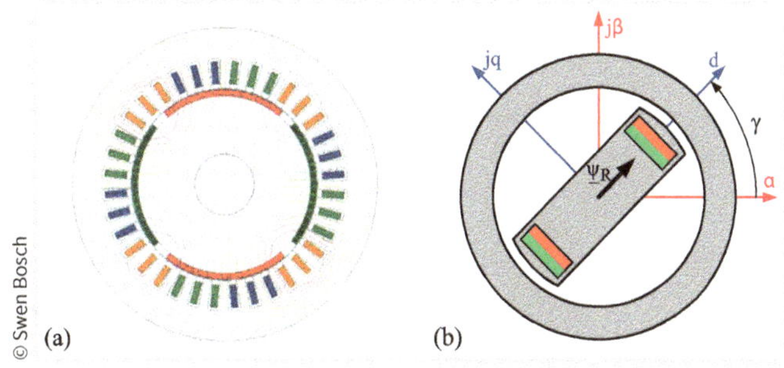

the machine is generating the maximum torque T_B at $s = s_B$. If the machine is burdened with a torque beyond the breakdown torque T_B, the machine stalls. In generator operation, the machine takes the maximum torque at $s = -s_B$.

For synchronous machines, it can be differentiated between two methods to generate a magnetic field in the rotor: electrically excited rotors and permanent magnet excited rotors. At an electrical excited motor, a DC current is supplied to the rotor either via slip rings or via a separate excitation machine with a rectifier. In the following, the focus will be on permanent magnet excited rotors, where permanent magnets are mounted in the rotor, as can be seen in **Figure 3.40a**. If a stator field is generated, the rotor will try to align with this field. At a moving stator field, the rotor will follow synchronously without slip, hence the name *synchronous machine*. The generated torque is depending on the angle between the rotor flux $\underline{\psi}_R$ and the stator flux $\underline{\psi}_S$. If the angle is zero, no torque is generated. If the angle equals $\pm\pi/2$, maximum positive or negative torque can be reached. As the stator field at inverter-fed variable speed drives can be controlled by the stator current, it is possible, to achieve a maximum torque for a given stator current. Due to this reason, the dq-coordinate system is introduced, where the d-axis is aligned with the direction of the rotor flux. As the dq-frame is rotating with the rotor, the d- and q-components appear as DC values, which simplifies the control significantly. The torque is influenced only by the field in the direction of the q-axis, while the field in the direction of the d-axis can be used for the so-called field weakening to achieve higher rotational speeds. The transformation from the αβ-frame to the dq-frame is called *Park transformation*.

3.2.3.4 Field-Oriented Control of a PMSM:

The field-oriented control is based on the dq-frame, which is rotating with the rotor. To adapt the stator flux to the dq-frame, the rotor angle γ has to be provided either by a sensor or a mathematical model of the machine. The rotation of the current space vector from the αβ-frame into the dq-frame is realized by using the Park transformation:

$$\underline{i}_{dq} = \underline{i}_{\alpha\beta} \cdot e^{-j\gamma} \triangleq \begin{bmatrix} i_d \\ i_q \end{bmatrix} = \begin{bmatrix} \cos(\gamma) & \sin(\gamma) \\ -\sin(\gamma) & \cos(\gamma) \end{bmatrix} \begin{bmatrix} i_\alpha \\ i_\beta \end{bmatrix} \qquad (3.42)$$

FIGURE 3.41 Overall current control structure and hardware setup.

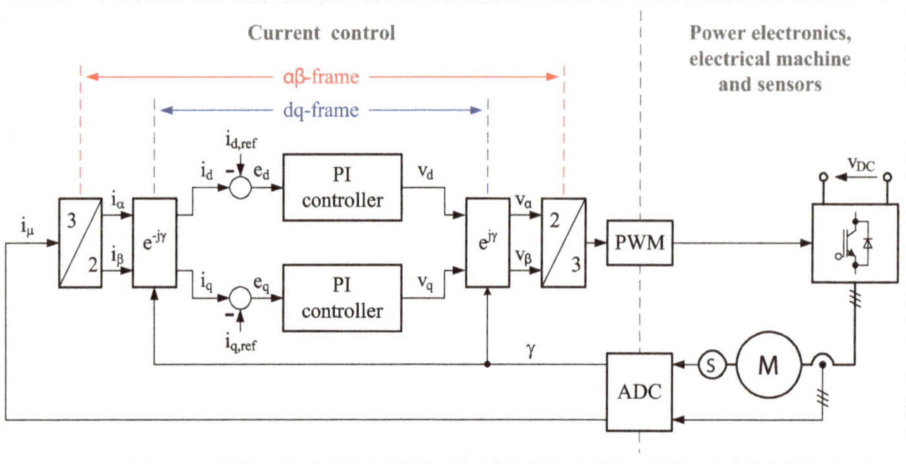

For the rotation of the voltage space vector back into the αβ-frame the inverse Park transformation is applied:

$$\underline{v}_{\alpha\beta} = \underline{v}_{dq} \cdot e^{j\gamma} \triangleq \begin{bmatrix} v_\alpha \\ v_\beta \end{bmatrix} = \begin{bmatrix} \cos(\gamma) & -\sin(\gamma) \\ \sin(\gamma) & \cos(\gamma) \end{bmatrix} \begin{bmatrix} v_d \\ v_q \end{bmatrix} \tag{3.43}$$

An overview of the field-oriented control including the power electronics as well as the electrical machine and the sensors is given by **Figure 3.41**. The control is providing the PWM signals to the inverter, which is fed by a DC voltage v_{DC}. The machine is powered by the three-phase output voltage generated by the inverter. The motor currents i_μ are measured and provided to the control. Besides, the rotor angle γ is determined by a sensor S. The currents and sensor signals are sampled by the ADC, before the following control algorithms are processed. Firstly, the currents are transformed in the αβ-frame by using the Clarke transformation and rotated by the angle γ into the dq-frame. Here, the control deviation $e_{d/q}$ is determined by subtracting the reference currents $i_{d,ref/q,ref}$ from the actual values $i_{d/q}$. The PI controllers are calculating the reference voltages $v_{d/q}$, which are rotated back to the αβ-frame and transformed to the natural frame by using the inverse of the Park and Clarke transformation. Finally, the PWM module is providing the switching signals for the inverter to generate the three-phase voltages for the electrical machine.

3.2.3.5 **Overview of an Exemplary Electrical System in a Mobile Working Machine:** An exemplary electrical system for a mobile working machine is shown in **Figure 3.42**. The main energy is provided by a combustion engine M, which is connected to a generator G. The three-phase voltage of the generator is rectified by an AC/DC converter and fed into the DC link, which is connected to the other electrical devices. For the operation of hydraulic pumps, air compressors or cooling fans, the DC voltage gets inverted for feeding the AC machines. For driving mobile working machines, a traction motor can be powered via an inverter. By feeding the DC link from a traction battery, also purely electric driving is possible. To protect the DC link from overvoltage, which can occur if too much energy is fed back from the traction motor while regenerative breaking, a braking resistor can be connected to the DC link to transform the energy into heat.

FIGURE 3.42 Exemplary electrical system in a mobile working machine; green arrows are showing the possible energy flows.

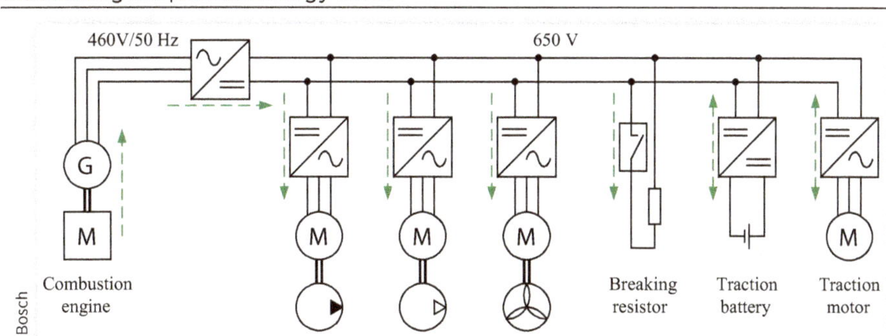

3.3 Traction Drive (By Torsten Kohmäscher and Fabrizio Panizzolo)

The main characteristic of a mobile working machine is determined by its traction and function drive. Since the primary energy provision, e.g., the combustion engine is not able to create a wide range of speed and torque the traction drive needs, transmissions are doing this task. The following subchapter will describe possible energy conversion systems for the traction drive.

It will start describing the mechanical gearboxes and their requirements for mobile working machines. To increase comfort and productivity hydraulic elements are implemented in the tractions drives. Therefore, transmissions with hydrodynamic elements are described as well as hydrostatic transmissions. The last group of described transmissions is power-split ones. They combine the high efficiency of mechanical gears with the continuous variable hydraulic transmissions.

3.3.1 Mechanical Gearboxes

Taking into consideration different off-highway market segments and roughly engine power below 100–130 kW there are many different vehicles developed to cover different application tasks. The market segment embracing all those different applications below engine power of 130 kW is worldwide recognized as a compact vehicle segment and it is practically dominated by the presence of the following hydrostatically driven mechanical transmissions:

1. Single-speed (ratio) transfer case
2. Single-speed (ratio) drop box, available with a specific drop from input to output shaft
3. Double speeds (rarely cases 3 speeds) drop box shiftable in stand-still condition

FIGURE 3.43 Drop box directly mounted at the input shaft on one axle.

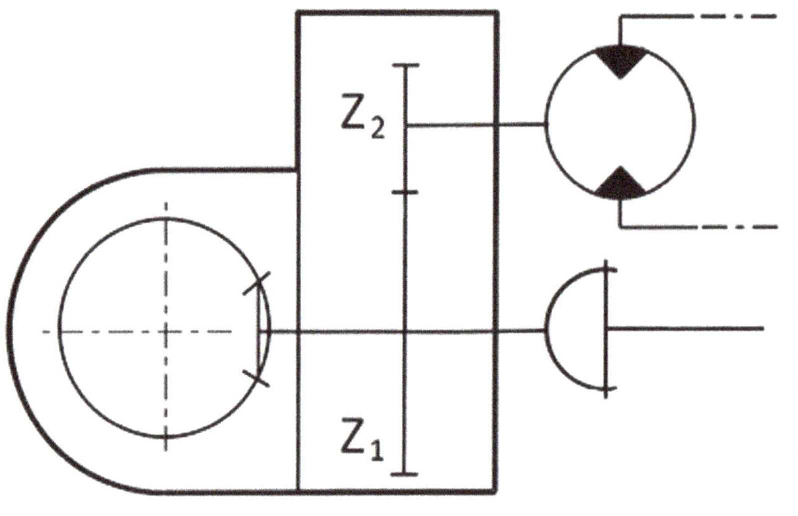

4. Double speeds (rarely 3 speeds) drop box shiftable while traveling with short term traction interruption like "shift on the fly" (SOF)
5. Double speeds (rarely 3 speeds) drop box shiftable under load without traction interruption like power shift transmission
6. CVT continuous variable speed transmission without any traction interruption

The level of complexity of the transmissions is growing depending from the main task and working cycles to be accomplished, max traveling speed required for considered vehicles, market required shifting quality, accepted shifting compromise level (as a trade-off between transmission costs and vehicle achieved performances), engine power as well different vehicle required performances and scaled-up vehicle's sizes.

Depending from different compact vehicle architectures/layouts all above-mentioned transmissions could be either directly flanged (rigidly connected) to one of the two axles or positioned in the middle of the vehicle, were more convenient between the front and rear axle while connecting the two driving axles with interposition of two Cardan shafts. Figure 3.43 presents a drop box directly flanged to the input of the axles while Figure 3.44 shows the middle vehicle positioned drop box.

3.3.1.1 **Single Speed Transfer Case and Speed Drop Box:** Those are the simplest execution of mechanical transmission design and foresee a chain of two or more connected gears resulting in a transfer case ratio suitable to achieve max required vehicle speed, having previously selected most appropriate total axle reduction ratio, and hydraulic variator torque conversion range (see this also as drivetrain ratio contribution) covered by maximum hydrostatic pump and motor displacement ratio.

FIGURE 3.44 Middle vehicle positioned drop box.

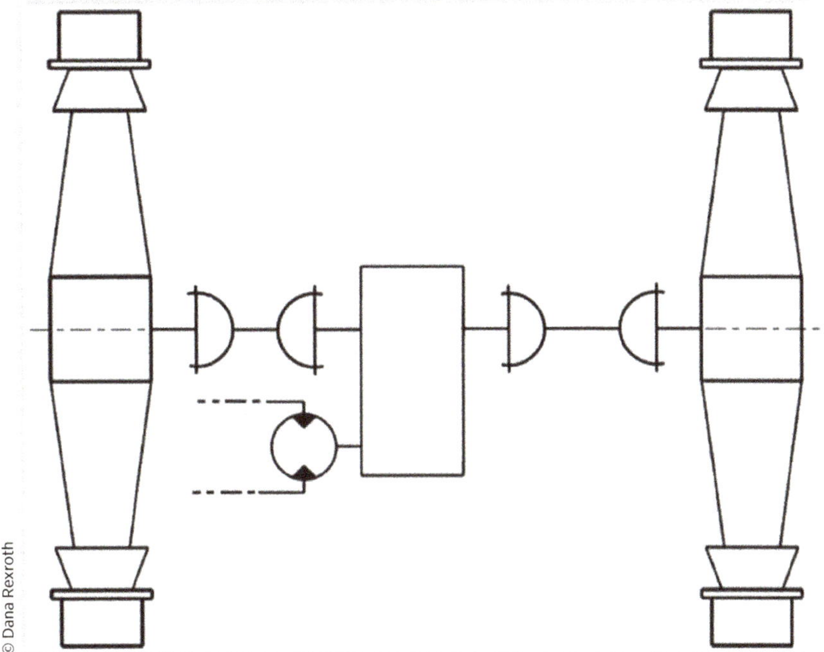

This type of transmission layout is currently utilized for an engine power up to 60 kW considering the availability of sufficient conversion range from the hydrostatic transmission. With the intent to cover different applications z_1/z_2 ratio interval is required between 0.97 (for 215 ccm displacement motor) and 6.4 (for 85 ccm displacement motor) depending on the size of the hydrostatic motor utilized. Corresponding maximum speed will result to be covered for an interval between 25–30 km/h (16–19 mph).

3.3.1.2 **Double Speed Stand-Shifting Drop Box:** When it is needed to travel at speed higher than 25–27 km/h (16–17 mph) and up to 35–40 km/h (22–25 mph) and when it is fundamental to change from site required working speed mode to traveling speed mode a double ratio drop box will be recommended. When it is also strongly required to optimize and keep under control, from a cost point of view, the sizes (and corresponding displacements) of hydrostatic units, then a double ratio drop box will significantly contribute, in terms of system integration, to the optimization of the entire vehicle drivetrain. In addition, when the application and the task to be performed will be fully covered by a first drop box ratio and only seldom it will be needed to move from a working site to another one then a second stand-still drop box ratio change, from first to second speed, will be an accepted feature for certain vehicles, considering also a cheaper cost advantage related to transmission control. It will be then possible to achieve maximum vehicle speed starting from zero speed, but this will be possible having previously selected suitable hydrostatic pump and motor size capable to fulfill, when combined with the drop box,

FIGURE 3.45 Double ratio standstill shifting drop box.

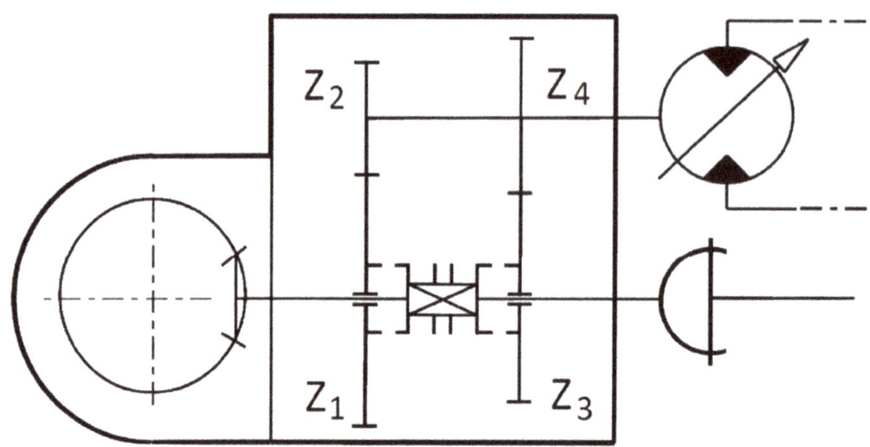

required performances without exceeding maximum recommended hydrostatic units allowable system working pressure. First transmission ratio z_1/z_2 and respectively second ratio z_3/z_4 required intervals to cover as many applications as possible are:

$$2.3 < z_1 / z_2 < 4.5 \tag{3.44}$$

and respectively

$$1.4 < z_3 / z_4 < 1.6 \tag{3.45}$$

For this drop box and for all possible applications to be considered, the displacement of the hydrostatic motor today never exceeds 110 ccm (**Figure 3.45**).

3.3.1.3 **Shift While Moving Drop Box (Shift on Fly: SOF):** When vehicles are equipped with engine power up to 70–80 kW this type of transmission layout is largely adopted on mobile working machines. The main design differentiating feature versus a stand-still shifting drop box is the possibility to shift (electronically and hydraulically assisted), from the first to the second ratio (and vice versa), under presence for a short term of traction cut. The magnitude transmission up-shift time is about 300 ms and about 450 ms for shifting down. A dedicated electronic control unit (ECU) will coordinate either the actuation of the drop box shifting mechanism as well dedicated sequence of change of the displacement of the hydrostatic units needed to synchronize the drop box input shaft together with here connected hydrostatic motor. Said shift will be possible if they adopt the following hardware system architecture:

1. Variable displacement pump
2. Variable displacement motor equipped with a special feature that will allow swash control plate to position it very precisely at a zero-degree angle (corresponding to the null displacement of the hydrostatic motor) during the shifting phase.
3. Availability of an input shaft speed sensor

FIGURE 3.46 Double ratios shift while moving drop box (please notice synchronizer presence).

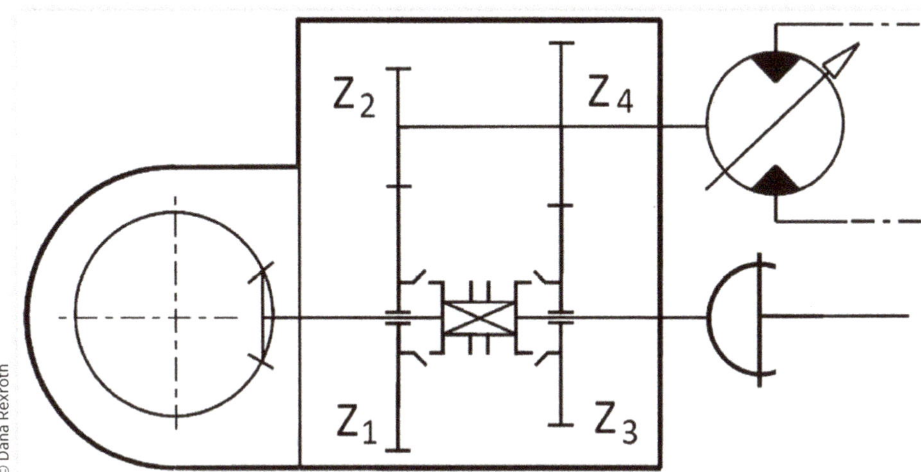

4. Shifting mechanism position sensor linked to synchronizer's shifting fork
5. A dedicated transmission control unit (TCU)

Despite a short-term traction interruption the entire shifting time will not exceed 300–450 ms and during this time the vehicle, particularly if approaching a steep gradient, will reduce a little bit his speed but then, thanks to electronic compensation, it will be possible to keep said speed drop within acceptable level without affecting driving comfort and quality while keeping anyhow vehicle behavior at an acceptable level. This compromise is currently well accepted by the mobile working machine market and this drop box layout is gaining more and more market share within compact vehicle applications. This achievement was possible thanks to the tailored system integration approach done starting from the engine to the wheels, leveraging on dedicated transmission control developed for the purpose. All considerations done on ratios for the drop box introduced above will still be valid for this particular evolved transmission configuration (Figure 3.46).

FIGURE 3.47 Hydrostatic motors summarizing drop box with double variable displacement hydrostatic motors.

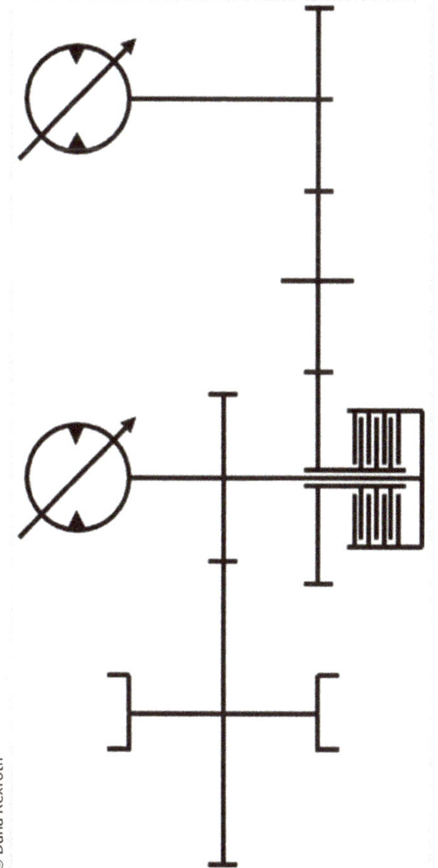

3.3.1.4 **Two Hydrostatic Motors Summarizing Drop Box:** This drivetrain architecture combines the utilization of CAN connected engine, drop box, and hydrostatic units (single variable displacement pump and double differentiated sizes of hydrostatic motors) (Figure 3.47). It has gained more and more popularity and presence within different compact mobile working machines market segments for engine power between 70–130 kW. The drop box design foresees, on highest-performing configuration, the presence of two variable displacement motors, in two different sizes each one properly connected to the transmission output

FIGURE 3.48 Hydrostatic motors summarizing drop box with one fix and one variable displacement hydrostatic motors.

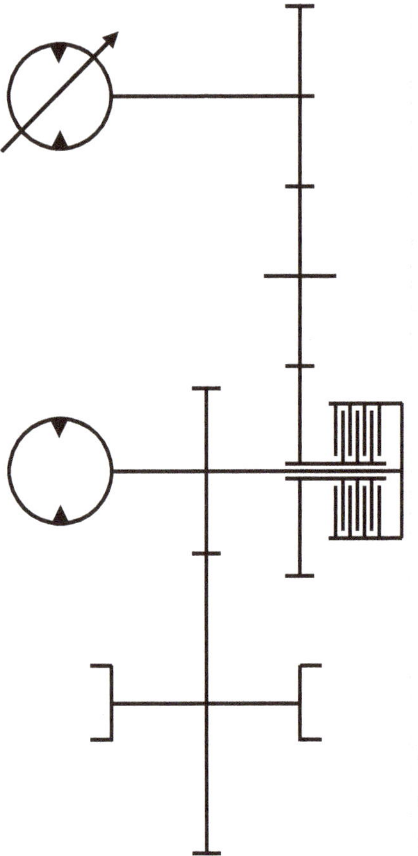

© Dana Rexroth

shaft via interposed gear's ratio. The optimized design of this drop box foresees that starting from a specified traveling speed, the bigger size hydrostatic motor will be disconnected from powertrain via a wet clutch. That specific disconnecting traveling speed has been selected on the way to avoid shifting maneuvers within operative intervals of speeds required for different tasks that will have to be accomplished by different vehicles. This disengagement will allow eliminating passive hydrostatic motor churning and friction losses that otherwise could not have been eliminated if the hydrostatic motor would not have been reduced to zero rotational speed and at the same time completely disconnected from the powertrain. When starting from stand-still both motors will provide their contribution to the propulsion of the vehicle while traveling the smaller size the hydrostatic motor will explore its entire speed and torque conversion ranges. It should be clear that here we are discussing an intelligent drive-train resulting in a CVT. The main advantage is the possibility to decouple engine speed from wheels speed either during normal working cycles as well as during long-distance traveling.

This system allows reducing fuel consumption and noise emissions, at relatively low/middle vehicle speed but, particularly during high speed traveling, the total drop box efficiency will be heavily affected by very high revolutions (and linked losses) generated by small size engaged hydrostatic variable displacement motor. As a potential improvement option, there is also the possibility to substitute the second smaller sized motor with a fix displacement motor. This is possible when vehicle performance requirements, (in terms of achieving a certain maximum speed at a certain gradient), are not so stringent up to force the designer to choose high conversion ranges for hydrostatic units. This change will result in an improved transmission efficiency from mid to high traveling speed (Figure 3.48).

3.3.2 Hydrodynamic Transmission

Particularly in the case of construction machinery, with the intent to execute the required task and have simultaneously full operational control, a skilled driver needs to dispose of a large range of torque variation (torque conversion) within relatively low vehicle speed range. This is the reason why discontinuous and multi ratios transmissions concepts have been adopted and are still largely utilized for different mobile working machines. During the operation of construction machinery it is usually required to face high peaks of traction demand (see also this as torque need from driving systems) at relatively low and particularly zero vehicle speed, e.g., on a wheel loader during pile entering condition. There is a hydrodynamic component that matches precisely with the request to allow high traction forces at zero speed and that, by its design principle, contribute as well with his own torque conversion multiplication factor while transitioning under load from low to zero vehicle speed. This fundamental component is the hydrodynamic TC. His first and main advantage is its self-adaption without the need of an operator control to sudden external and changing requests. The second advantage is the capability to stabilize the drivetrain in all different working conditions. One other distinct advantage of TC, as already mentioned, is to provide consistent and highest vehicle traction force at wheels (or tracks) at zero vehicle speed. This characteristic is mandatorily needed, e.g., by a wheel loader while entering in the pile with a front vehicle-mounted bucket. The TC has also the additional advantage to smooth the deceleration of the vehicle (seeing as a driving comfort advantage) when facing unexpected obstacles; the presence of fluid inside TC will additionally help in filtering, at certain levels, torsional vibration induced by the engine.

The main disadvantage of TC, this is the reason why in the current years the market is investing in alternative innovative transmissions, is mainly related to the fact that at zero speed condition and full available traction force, a cause of TC internal losses, the entire power coming from the engine will be dissipated resulting in a null TC efficiency. It is now clear that the presence of a TC in transmission will act as a continuous variable transmission contributing to reducing when adopted, the number of transmission ratios required to cover the entire transmission torque conversion range needed for a mobile working machine. In this way, the reduction in transmission complexity, linked costs, as well as less required installation space and most important the benefit of improvement of transmission total efficiency (now possible for the portion of mechanical transmission, thanks to the reduced presence of mechanical components) will be marginally offset, but at an acceptable level during normal operation. Introduced benefit on total transmission efficiency improvement will be anyhow completely canceled and reduced to value zero during vehicle stall condition (zero vehicle speed), but despite this, the integration of a TC within the general architecture of the transmission will allow now, as requested by the vehicle mission/duty, to have full traction in a stall condition. For those transmissions that will not foresee to integrate a TC as input driving component, it will be instead needed to cover the transmission spread of special mobile working machines by introducing a higher amount of mechanical transmission ratios up to 10–13 in the forward direction and up to 3–4 in the reverse direction.

As a common and fundamental requirement for mobile working machines, it is very important, during operation, to change the transmission ratio under load without interruption of vehicle movement and cut of traction force. There should be no interruption of transmitted power flow. This is one other distinctive design feature required for transmission foreseen to operate on mobile working machines such as a dozer while pushing on the front positioned blade, a wheel loader while digging and entering in the pile with the bucket, and other dedicated vehicles such as graders and scrapers. Mobile working machines operation effectiveness and productivity would have been seriously affected if the vehicle should have been forced to stop missing the said distinctive transmission feature. The need for the said distinct design feature has been accomplished by the consistent introduction, starting around from the year 1950 with a power shift transmission concept. This concept gained more and more application presence worldwide and it is still a concept largely adopted and utilized for different vehicles by construction machinery manufacturers. Such required shifting under load function has been fulfilled by the adoption as driving transmission component of multi-plates wet discs clutches and by the adoption, for clutch shifting control, of the hydraulically actuated and electronically modulated clutch system together with the adoption of clutches overlapping control strategy applied to both closing and opening clutches (involved on current transmission ratio change). The clutch overlapping phase will last for a very short time with a falling magnitude, depending on the vehicle's external condition, from approximately 350 m/s up to 800 m/s and in particular cases even more. During the overlapping time, the pressure and the corresponding torque transmitted by the clutch going to be closed will increase until the closing clutch will be capable to transmit the actual torque. During this interval, the outgoing clutch will stay engaged (loading engine by an additional torque). This will happen until the expiring of the overlapping time interval. At this moment the outgoing clutch will be opened while the other will continue to close increasing his pressure up to reach full engagement. During this handover, there will be, for a short term, a power dissipation on both clutches and consequently, those involved clutches will need to be cooled by forced oil flow passing through discs packages. The combination of clutches for the said overlapping phase will allow the vehicle to transmit, as required, traction and speed to the soil without interrupting vehicle movement and task fulfillment.

3.3.2.1 **Hydrodynamic Transmission (Layshafts Architecture):** The general architecture of hydrodynamic transmissions utilized on the mobile working machine is currently covered by two types of gearbox layouts the so-called parallel countershafts (layshafts) transmission design family and the planetary transmission design family, respectively. Both transmissions are available, depending on the applications, in a different but finite amount of transmission ratios covered by properly connected gears belonging to the different transmission shafts positioned inside the transmission case. The utilization and engagement of different available and sequenced transmission ratios will allow to reduce transmission output revolutions and, at the same time, to increase transmission output torque. Transmission output revolutions and output torque will be transmitted then to the axle or involved track driving components, contributing to the execution of vehicle expected tasks. For both transmission families, the following common driving components and transmission features, as well as transmission functions, can be detected on different transmission layouts available in the market:

1. TC as the main transmission driving input component. The lock-up feature is also more and more frequently utilized with the scope to improve, when possible, the total transmission efficiency.
2. A gearbox containing inside all different driving components such as shafts, bearings, gears, wet clutches, actuation, and feeding systems for clutches, hydraulic pump, forced lubrication system for all clutches and involved mechanical components, optional transmission output disconnecting device, sometimes as optional as well interaxle differential and differential lock. Attached outside of the transmission case there will be clutches control valve's block with corresponding hydraulic valves presence and finally transmission oil filtration system.
3. The transmission drop-case function is fully dependent on application segments. For example, wheel loader and industrial forklift trucks will need for

TABLE 3.9 Shifted gear versus corresponding engaged clutches.

Transmission speed		Activated clutches
Forward	4th	$\mathbf{FW_{HIGH}}$ **+ C2**
Forward	3rd	$\mathbf{FW_{HIGH}}$ **+ C1**
Forward	2nd	$\mathbf{FW_{LOW}}$ **+ C2**
Forward	1st	$\mathbf{FW_{LOW}}$ **+ C1**
Reverse	1st	R + C1
Reverse	2nd	R + C2

frame integration purposes, a cause of vehicle own architecture, a different drop distance from the transmission input shaft to the corresponding transmission output flanges. For that reason, long drop (LD), intermediate drop (ID), and short drop (SD) configurations of transmission applications platforms for mobile working machines have been developed by specialized manufacturers.

4. One or more power take-off (PTO functions) to drive different auxiliaries such as vehicle implement pumps, emergency steering system, and axle brake cooling flow recirculating pump systems are present as well normally on power shift transmissions available on the market.

There are a few specialized companies in the world producing and commercializing both countershafts and planetary transmission families even though planetary transmissions are mostly utilized for heavy-duty applications in construction and hauling vehicles.

Now, as a common example valid also for all other countershafts power shift transmission, the architecture of a power shift transmission utilized for a loader backhoe application is explained in more detail. It is a four-speed (ratios) forward and two-speed (ratios) reverse transmission (Table 3.9).

The transmission is composed of five main assemblies:

1. The converter and pump drive section including PTO for driving implement pump (see yellow marked components in Figure 3.49).
2. The input or directional clutches (green marked components).
3. The range clutches (brown marked components).
4. The output section (blue marked components).
5. The electronic transmission controls (showing integration on the gearbox see Figure 3.52).

3.3.2.1.1 The Converter and Pump Drive Section. The engine power is transmitted from the engine flywheel to the impeller through the impeller cover. This element represents the pump portion of the hydrodynamic TC and is the primary component that, by his centrifugal design principle, will radially direct oil flow picked-up at TC center and discharge it at the outer diameter facing turbine input blades. From here the oil flow moving toward TC centerline direction will allow the turbine itself to drive the transmission input shaft and directional clutches. Oil flow exiting from turbine blades

FIGURE 3.49 Powershift transmission section.

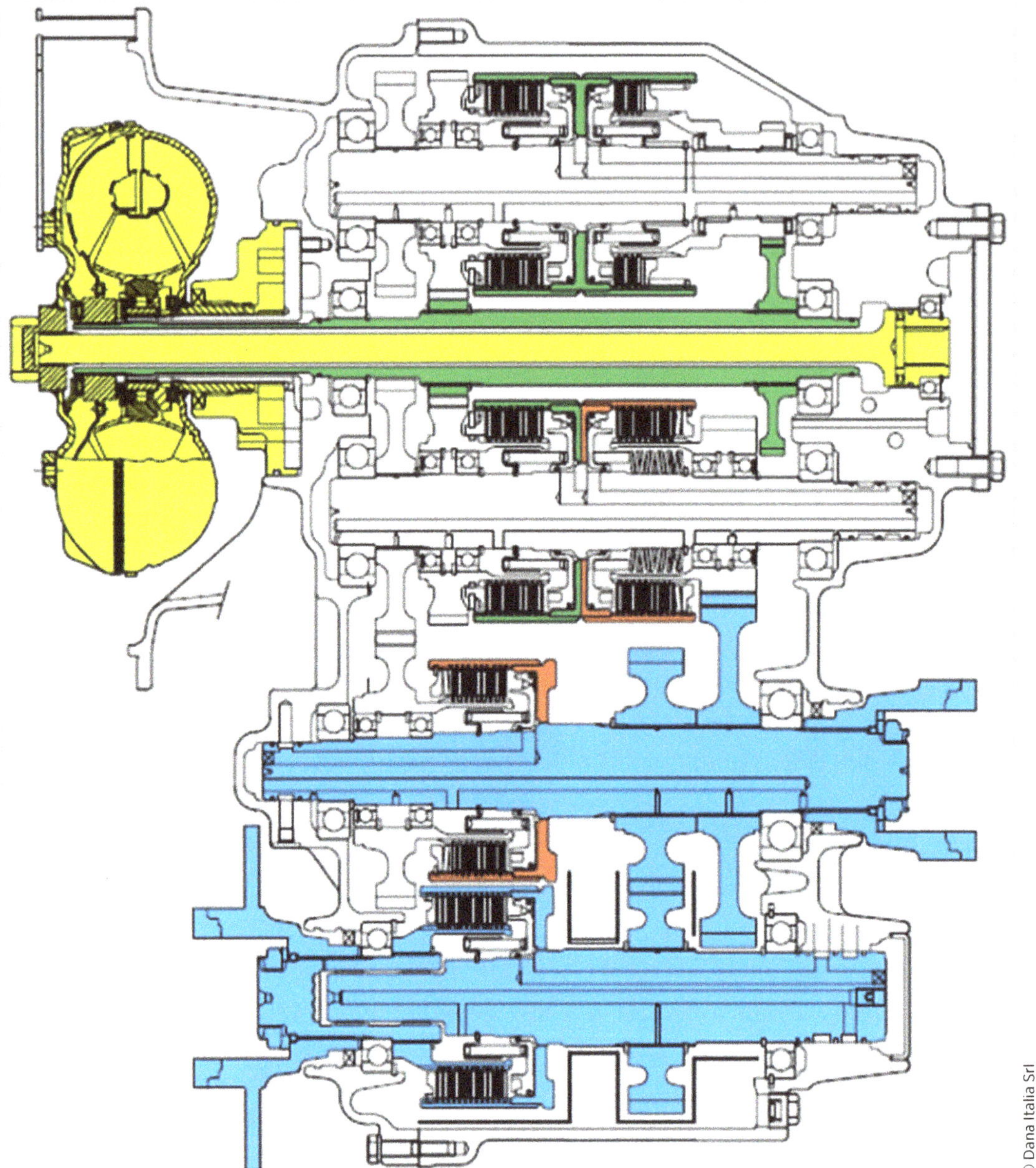

will face the stator blades (distinctive component available within TC). The stator while reacting to the transmission case will allow, to the oil flow, to change direction and to recirculate versus the pump input blades. The stator of the TC is located in-between and at the center of the inner diameters of the impeller and turbine elements. This recirculation will allow the converter to multiply the torque. The torque multiplication is a function of the blading (impeller, turbine and reaction member) as well as of converter output speed (turbine speed). The TC will multiply engine torque to its designed maximum multiplication ratio when the turbine shaft will be at zero speed (stall).

FIGURE 3.50 Schematic layout of presented four forward two reverse power shift transmission.

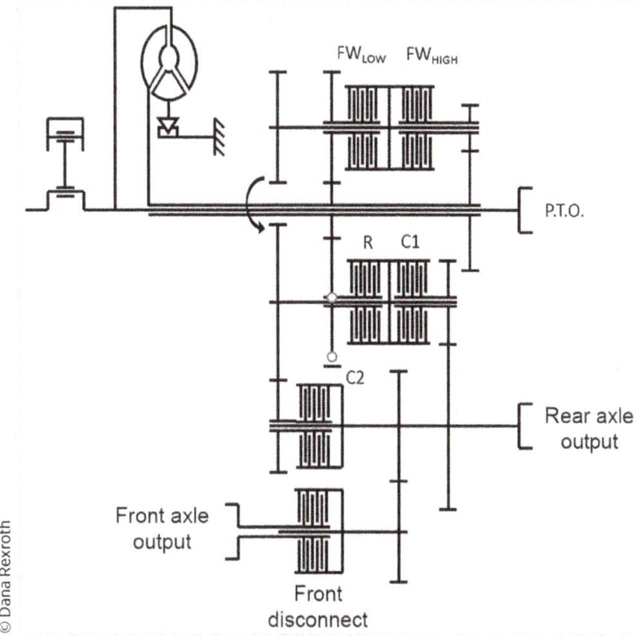

Therefore, it can be said that as the turbine shaft is decreasing in speed the torque multiplication is increasing. Rigidly connected to the impeller cover there is a splined shaft, coaxial to the turbine shaft, which drives the PTO positioned at the back of the transmission case. Since the shaft is connected to the center of the impeller cover, the PTO speed will be the same as engine speed.

The rear side of the impeller cover has a tanged drive which drives the transmission charging pump located in the converter housing. The transmission charging pump speed is also the same as the engine speed (Figure 3.50).

3.3.2.1.2 The Input or Directional Clutches. The turbine shaft driven from the turbine itself will transmit power to the forward or reverse clutches. These clutches consist of a splined drum with a bore to receive a hydraulically actuated piston. The piston is oil-tight by the presence of sealing rings. A steel counter-disc with external splines is inserted into the drum and rests against the piston. Next, a friction disc with splines at the inner diameter is inserted. Discs and counter-discs are alternated until the required clutch package is achieved. To engage the clutch, the control valve which is fitted on the side of the transmission will direct pressurized oil through tubes and internal casting's passages to the selected clutch shafts. Oil sealing rings are located on each clutch shaft. These rings direct the oil through a drilled passage into the shaft to the desired clutch. The pressure of the oil forces the piston discs and counter-discs against the backup plate. The counter discs with splines on the outer diameter clamped together with discs inserted instead on the hub will allow the entire clutch package to rotate rigidly. When the clutch is released a return spring will push the piston back and oil will drain back via the control valve into the transmission sump. The transmission has one reverse clutch and two forward clutches (forward low and forward high). This result, in combination with the two range clutches, in the transmission having four forward speeds and two reverse speeds. The engagement of the directional clutches is modulated, which means that clutch pressure is built up gradually. This will enable the unit to make forward, reverse shifts while the vehicle is still moving and will allow a smooth engagement of drive. The modulation is electronically controlled in the control valve block, via modulated valves.

3.3.2.1.3 The Range Clutches. Once a directional clutch is engaged, the power is transmitted to the range clutches (first or second). Operation and actuation of the range clutches are similar to the directional clutches. The engagement of the range clutches is also modulated to enable a smooth engagement. The modulation for these clutches is achieved using an orifice fitted in the control valve which is controlled electronically and which limits the oil flow to the clutch during shifts. In the range clutches, the plate before the endplates are dished to build up the clamping force of the clutch gradually.

3.3.2.1.4 The Output Section. With one range clutch engaged power is finally transmitted to the output shafts (Figure 3.51).

FIGURE 3.51 Range clutches and front output shaft disconnect clutch.

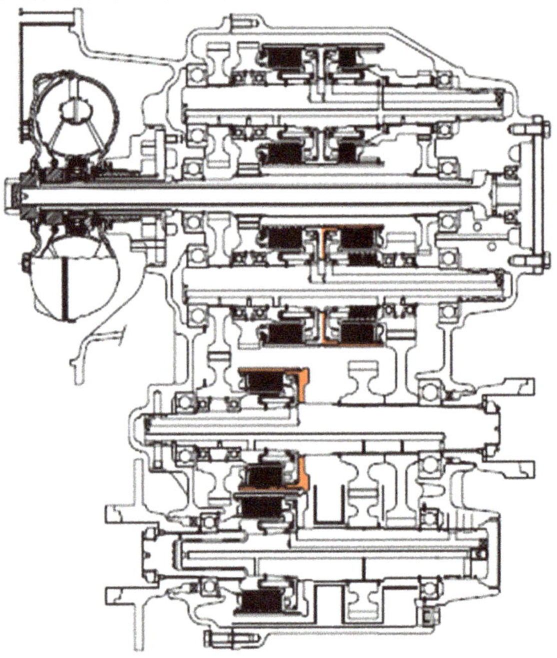

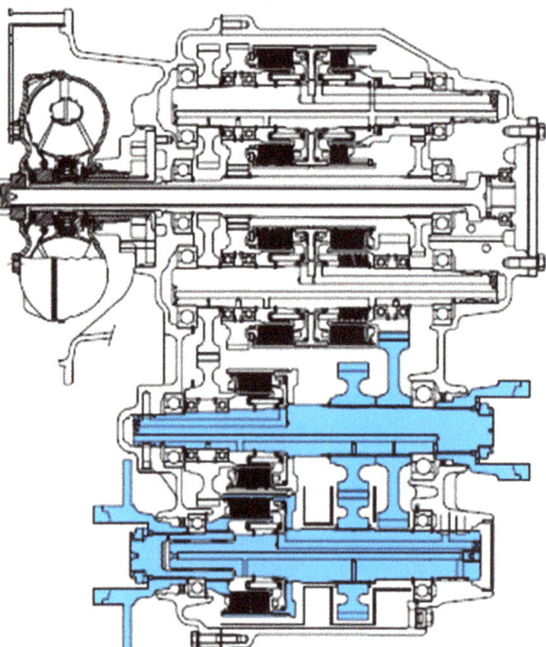

The transmission can have an upper output at the rear side of the transmission case and a lower output at the transmission front side. Output rotation of the rear upper flange is opposite the engine rotation when the forward clutch is engaged, while output rotation of the lower and the front output is the same as the engine rotation with the forward clutch engaged. Due to loader backhoe tires arrangement and difference in loaded radius between front and rear tires during an operation mode, the ratio between upper and lower output in the example is 0.951:1. The lower front output has an axle disconnecting clutch to enable the four-wheel drive. The clutch is similar to the other clutches except that it is actuated by an on-off valve without clutch modulation. The disconnect is controlled electronically. Without an electrical signal, the clutch is always engaged.

3.3.2.1.5 The Transmission Control (Refer to Hydraulic Diagram See Figure 3.53) The transmission is controlled by an electronic gear selector. This unit has a microprocessor that receives certain inputs (gear selector position, speed sensor, ...) which are processed and will give output signals to the control valve (Figure 3.52). The control block presented in Figure 3.53 has six solenoids, six shift spools, a pressure

FIGURE 3.52 Transmission control unit positioning.

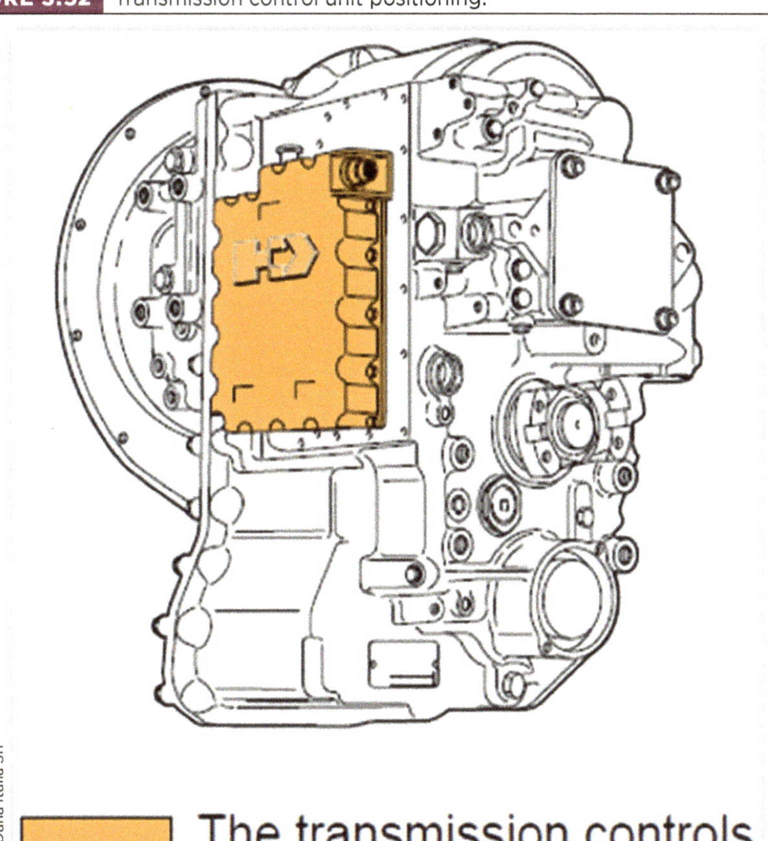

FIGURE 3.53 Presented power shift transmission hydraulic scheme (neutral, second clutch, and disconnect clutch engaged status) [30].

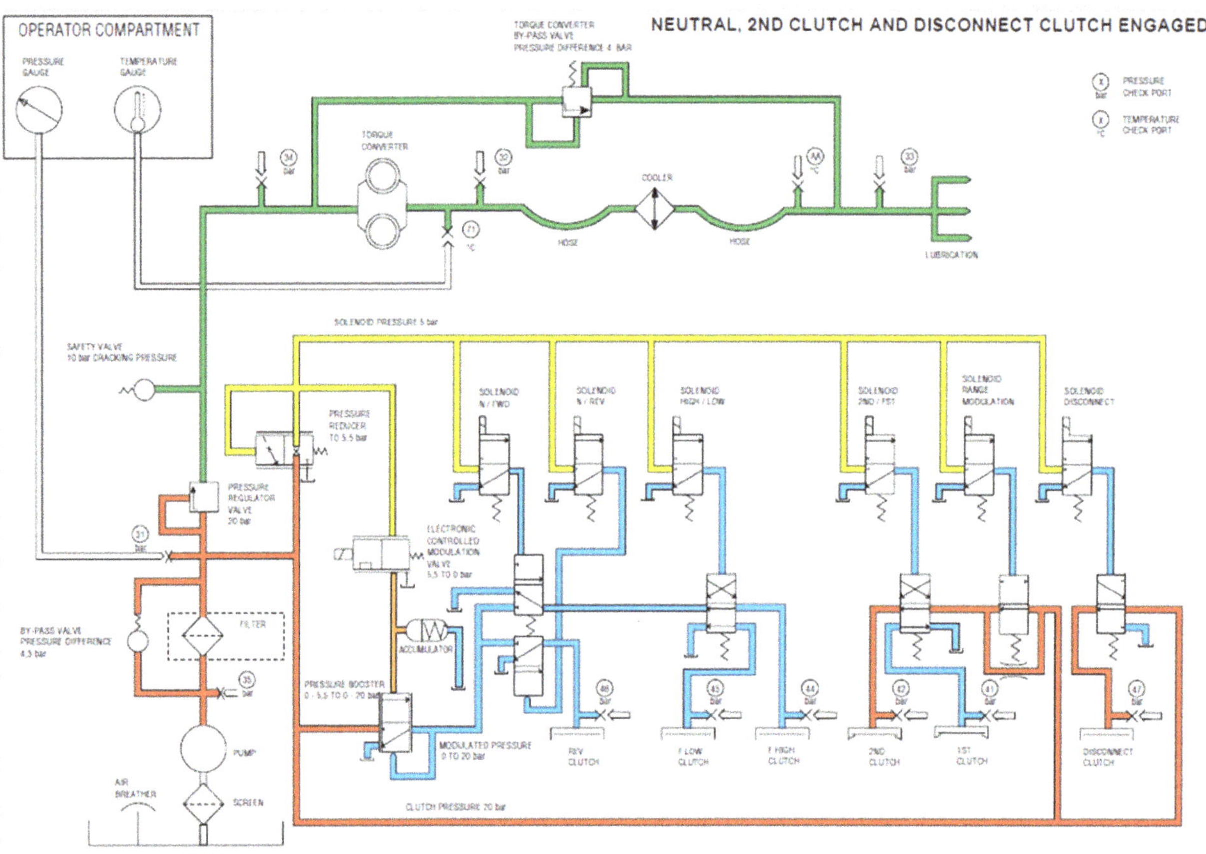

reducer valve, an electronically controlled modulation valve, an accumulator, a pressure booster and a speed sensor. For more details regarding the transmission hydraulic scheme see Figure 3.53 showing the following transmission status: NEUTRAL (with second clutch and disconnecting clutch engaged).

The Operation of the valve is as follows:

Regulated pressure 20 bar (290 psi) is directed to the shift spools, the pressure booster, and the pressure reducer valve. In the pressure reducer, the regulated pressure is reduced to 5.5 bar (80 psi). This reduced pressure is used as a supply for the solenoid and electronic controlled modulation valve. When activated, the electronically controlled modulation valve will give an output pressure curve from 0 to 5 bar (0–73 psi). This pressure curve is multiplied in the booster so that a curve from 0 to 20 bar (0–290 psi) is available for the directional clutches. Between the electronic modulation valve and the booster there is an accumulator to damp any hydraulic vibration. When the forward mode is selected the electronic modulation valve and the forward solenoid are activated. The pilot pressure of the forward solenoid will move the shift spool so that a forward clutch can be fed with modulated pressure. If the high/low solenoid is not activated the forward high clutch is engaged; if it is activated, the forward low clutch is engaged.

When reverse is selected the electronic modulation valve and the reverse solenoid are activated and the pilot pressure of the reverse solenoid will move the shift spool so that the reverse clutch can be fed with modulated pressure. The shift spools from forward and reverse are locked against each other with return spring in between; this makes it sure that only one direction can be selected.

Range shifting is selected as follows:

If the range solenoid (second/first) is not activated, regulated pressure is fed through the modulation shift spool and the second/first shift spool to the second clutch. If instead the range solenoid (second/first) is activated, the pilot pressure will move the shift spool so that the first clutch is fed. It has to be considered here that also range clutches have a modulation, which operates as follows: when the range is changed, the oil will flow through the modulation shift spool to the chosen range clutch momentary until the friction discs and counter-discs are closed against the dished plate. This condition corresponds to the so-called kiss-point (KP) clutch status where entire discs and counter-discs of the clutch package are completely facing one to the other (no space in between each disc versus counter-discs) but without transmitting yet any torque. At this moment the range modulation solenoid is activated. The pilot pressure will move the modulation shift spool so that the oil supply is fed through an orifice which is in the bypass of the valve.

The controlled and feed volume of oil is used to move and push the clutch piston gradually against the clutch dished outer plate until the clutch is fully closed. This will give a smooth buildup of torque. Instead, when the range modulation solenoid is released, the modulation shift spool returns to its rest position allowing full oil flow out of the clutch. The control valve block controls also the front axle disconnecting clutch. If the solenoid is not activated the full oil pressure is fed through the disconnecting valve shift spool to disconnect the clutch. If the solenoid is activated, the pilot pressure will move the disconnecting valve shift spool to block the oil supply to the disconnected clutch to engage it. The control valve needs also a speed sensor. This sensor will pick up upper output gear speed. This information is used in the electronic gear selector to determine shift logic. Since the sensor picks up the upper output gear speed, the signal will be in direct relation to the turbine speed if any directional clutch is engaged.

3.3.2.2 **Planetary Transmission:** For many years, planetary transmissions have been adopted to drive heavy duty mobile working machines. The reason why different, but similar, planetary architectures have been selected, instead of countershaft's transmission layout, is mostly related to the following advantages:

1. The input and output shafts of a planetary transmission are co-axials. This could be a strong drivetrain integration requirement for some particular vehicles.
2. Different selected planetary sets are providing to the designer the possibility to optimize gear's design from strength and transmittable power density point of view.
3. Inertias of selectable planetary transmission components involved in gear shifting is more favorable than corresponding heavier inertias faced instead with a countershaft's transmission design during up and downshifting phases.

As an introduction to the planetary transmission design the gearbox Voith, model 402 Certoplan II [10], has been selected here as a reference. This transmission design was originally and is still installed on a dozer tracked vehicle having an engine power of 147 kW. Full cinematically equivalent planetary transmission layouts are utilized for tracked dozer applications by mobile working machines manufacturers and are differing from presented gearbox only for chosen planetary gear set's ratios and for connected component's arrangement only on transmission output section. More detailed design information can be recognized looking to the connection by a clutch adopted, on the output transmission section, between the sun gear and planetary carrier, (alternative design currently available in the market for a Japanese dozer transmission). In this arrangement, the single available clutch is integrated on the last planetary set positioned at the output side of the transmission. As it can be realized from enclosed stick-diagram (see Figure 3.54) planetary transmission layout presented here foreseen the presence of TC followed by four planetary sets that could function either singularly or connected in series by two or three planetary sets to realize a full symmetric gearbox having the same three transmission ratios in forward as well in reverse. The selection of each speed is obtained hydraulically actuating under load one of the four brake packages and the single available first gear clutch or actuating simultaneously two brake packages according to the shifting actuation sequence reported on the Table 3.10 positioned below transmission stick-diagram. For complexity reduction and production standardization processes, the four planetary gear sets have been designed utilizing three planet gears each. For modularity purposes, three gear sets have been adopted having the same amount of teeth for sun, planets, and ring gear components, while for the fourth remaining gear set the same ring gear with the same teeth has been utilized for commonality and production standardization processes purpose and related advantages. The main

FIGURE 3.54 Schematic layout of three forward and reverse speeds planetary transmission.

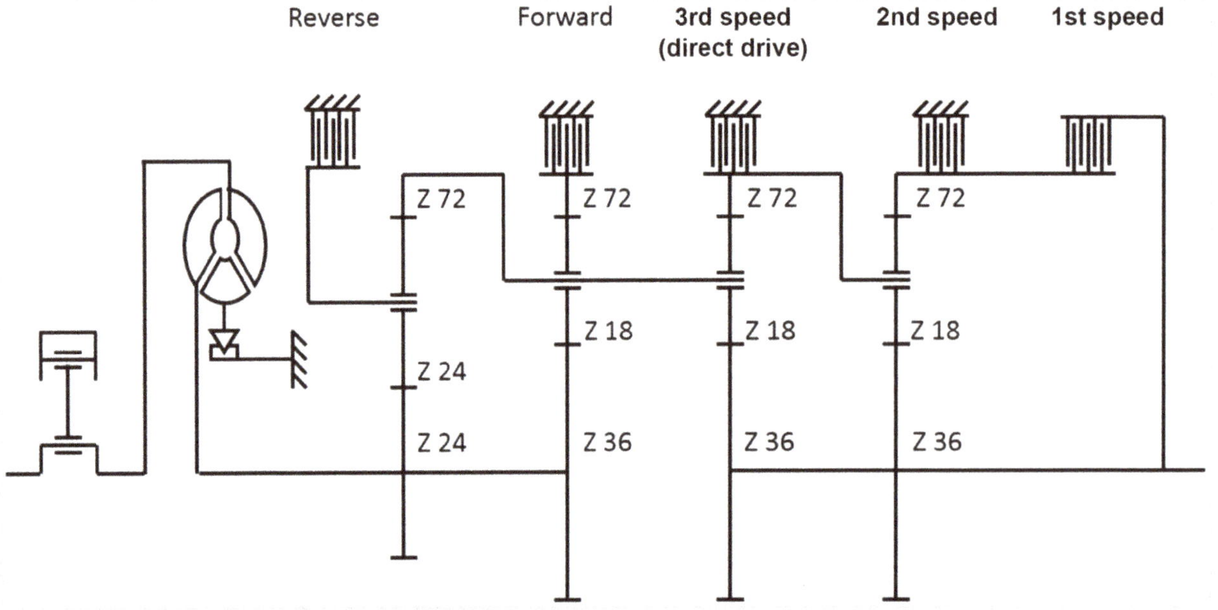

TABLE 3.10 Shifted gear versus corresponding engaged clutch and brakes; and symmetrical transmission ratios.

Activated components	Forward		Reverse		Ratio
	BF + C1	First	BR + C1	First	3
	BF + B2	Second	BR + B2	Second	1.667
	BF + B3	Third	BR + B3	Third	1

advantage of presented transmission architecture is the combination of the smallest possible packaging space required to install the transmission (being the transmission design very compact) and the manufacturing simplicity and related low costs resulting in a well-balanced system's integration project that has been strongly taken as reference by all vehicle manufacturers engaged on equivalent transmission development done since the introduction on the market.

Alternatively, it is possible to block the last epicyclic reduction carrier to the sun by a clutch and this will not introduce any cinematic change. This arrangement is present in the market as well and utilized for dozer applications.

3.3.3 Hydrostatic Transmission

Hydrostatic transmissions (HST) are an integral part of the traction drive of various mobile working machines in optimized variants and configurations. The HST consists of a variable pump that supplies hydrostatic flow and a motor that consumes this flow to provide mechanical power to its output shaft. In a closed hydrostatic circuit, the high-pressure ports of the pump and motor are directly connected. The complete pump flow is consumed by the motor and routed back to the pump. In open hydrostatic circuit design, the motor can be connected to the pump via a control valve. This valve controls the flow from the pump to the motor while the return flow from the motor is throttled back to the tank. This way, the pump flow can be split between multiple hydrostatic power consumers, but the return flow of the motor needs to be controlled by a specialized brake valve that prevents low-pressure events in the motor during dynamic operation.

Both circuit designs operate according to the flow coupling principle in which motor speed is determined by the available flow and motor displacement while the system pressure is determined by the current torque load on the shaft of the hydrostatic motor. Most HST on mobile working machines work according to the flow coupling principle. The pressure coupling is considered an alternative principle in open-circuit design which utilizes a pressure controlled high-pressure line with hydrostatic accumulator to provide power to hydrostatic motors. The deceleration energy of the hydrostatic motor can be stored in the accumulator. With the given system pressure, the torque of the hydrostatic motor is determined by motor displacement. A torque balance on the motor shaft between the created motor output torque and the external torque load on the motor shaft results in the turning direction and rotational speed of the hydrostatic motor. Hydrostatic flow is consumed based on motor displacement and the rotational speed of the motor. With the same turning direction of the motor, power can be stored in the accumulator by dynamically stroking the motor through neutral into the other direction. Due to component availability, the application of systems in pressure coupling is limited today to systems for hydrostatic energy regeneration like swing drives in hydraulic excavators or hydraulic brake assists in delivery trucks. Application in traction drives still needs to be developed.

3.3.3.1 **Closed Circuit HST:** The HST in closed circuit design and flow coupling provides the advantages of continuous variation of the machine's ground speed, jerk-free reversals of the driving direction, as well as 4-quadrant operation with wear-free deceleration by braking against the drag torque of the diesel engine. The components of the HST can be flexibly positioned on the machine so that the diesel engine can be positioned transverse to the driving direction to act as a counterbalance weight in wheel loaders or to provide room for processing equipment. Gearboxes can be added to the system design to increase the overall ratio of the traction drive. High-pressure levels of up to 500 bar (7,250 psi) are realistic in modern HST and are the reason for the high values of motor torque and power density. The maximum output torque of the motor is available from the initial machine movement without increased engine speed or an increase in power loss – a clear advantage of the HST over gearboxes with TCs.

Figure 3.55 introduces the schematic of a typical HST in closed circuit design with flow coupled hydrostatic units.

The pump (1) is connected to the motor (2) via two high-pressure lines in a closed hydrostatic circuit which is pressure protected by two pressure relief valves (10) that can be set to more than 450 bar (6,500 psi). The closed-circuit has to be precharged to a relatively low-pressure level around 25 bar (360 psi) to reduce air content in the oil and thereby increase the stiffness of the HST. The charge pressure relief valve (4) limits the pressure in the charging system for the HST. The charge pump (3) supplies cold oil to the loop through the filter (5) and the two check valves (6) that only open towards the low-pressure side of the loop. Hot oil is flushed out of the closed-circuit across the loop flushing valve (7) which connects to the respective low-pressure side when significant drive pressure is reached. The additional pressure relief valve (8) behind the loop flushing valve sets the pressure level on the low-pressure side of the HST. In the stand-still of the HST, the charging system provides pressure to both pressure lines of the loop and maintains a pressure level that is close to the pressure in the charging system. The temperature

FIGURE 3.55 Schematic of closed-circuit hydrostatic transmission in flow coupling [11].

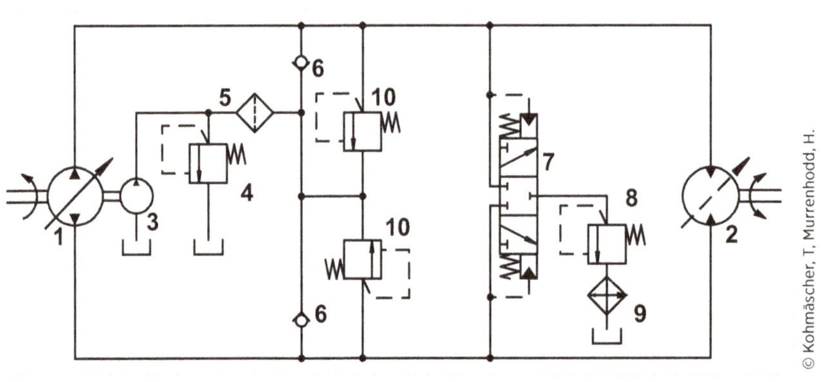

level of the flushed oil is reduced by the oil cooler (9). At least one of the hydrostatic units (pump [3] or motor [32]) needs to be variable to realize a CVT.

The direction of power flow through the HST is determined by the load on the motor shaft and therefore both hydrostatic units need to be able to operate in pumping and motoring mode [11]. The charging system and high-pressure relief valve (HPRV) are integrated into the pump while loop flushing is integrated with the motor to provide some additional cooling flow for the motor case.

The flow coupling of pump and motor is the base assumption for the calculation of the speed relationship in the closed-circuit HST. Hydrostatic flow Q in the closed-loop and system delta pressure Δp between the high and low side of the HST is the key characteristic values in the HST.

3.3.3.2: **Advantages of Hydrostatic Traction Drives:** Hydrostatic traction drives are well established in the markets for mobile working machines that aim for optimization of fuel economy, but not all applications can benefit in the same way. Table 3.11 summarizes the main advantages and disadvantages of hydrostatic transmissions that are applied in traction drives.

Advanced control strategies contribute to noise reduction and efficiency improvement by reducing the speed of the engine and pump for the optimized engine operation with an impact on the noise generated by the pump.

TABLE 3.11 Advantages and disadvantages of hydrostatic traction drives.

Advantages
High power density compared to electric drives
A simple implementation of rotatory and linear movements
Simple distribution of power to multiple consumers - e.g., motors or cylinders
Flexible positioning and packaging of the system installation
Very precise and step-less positioning independent of the load condition
Full functionality available at constant engine speed - given by work process
Power flow can be reversed to assist in machine deceleration or downhill control against drag torque of the diesel engine and thereby reduce brake wear
Component behavior can be selected to be load-dependent or load-independent
Separation of ground speed and engine speed for implementation of efficient system solutions due to ratio-controlled system behavior
Driving direction changed by stroking pump through neutral displacement without the need for mechanical reverse gear. Same top speed in forward and reverse possible.
Disadvantages
Volumetric and hydro-mechanical power losses through leakages and friction
Maintenance requirements for pressure fluid (e.g., filtration and cooling)
Environmental challenges by noise, risk of external leakages, and flammability
Relatively low total efficiency at maximum speed compared to mechanical transmission

3.3.3.3 Guidelines towards Loss Reduction: Losses in hydrostatic components and HST will impact the efficiency of the traction drive of mobile working machines. However, the loss behavior can be greatly influenced (improved) during the design phase of the control system and the component selection. The recommendations in Table 3.12 support the reduction of losses on the system level [4].

Further efficiency improvement can be achieved by the integration of the HST into a hydro-mechanical power-split transmission (HMT). Losses in the hydrostatic circuit are reduced by reduction of the hydrostatic power ratio and selection of smaller components or by speed reduction of the hydrostatic components.

3.3.3.4 Transmission Structures: The flexible combination of hydrostatic units together with the electronic displacement control of pump and motor enables the development of numerous transmission structures that can be optimized to fit the mobile working machine in scope (see Figure 3.56). The preferred solution will be selected based on machine-specific requirements such as the spread of transmission ratio, the control concept, or the limitations in design space on the machine.

The first differentiator for transmission variants is the control for pump (primary) and motor (secondary) displacements – it could be a variable pump with fix motor, it could be fix pump with variable motor, or it could variable pump with variable motor. With both components variable, the next question is whether the commands to the two components are sequenced, parallel, or combined in an optimized way.

To further extend the operation range of a hydrostatic transmission, as a second differentiator, mechanical gearboxes for merging power flows, or for shifting mechanical gear ranges. One single hydrostatic motor or multiple motors could be connected to a single gearbox. The control concept for the gearbox determines the performance of the complete system solution. The combination of a hydrostatic transmission with a multi-speed gearbox provides the advantage of a high spread in transmission ratio together with a beneficial efficiency profile. Radial piston motors can be shifted between high and low displacement and thereby allow a multispeed operation.

TABLE 3.12 Recommendations for the reduction of losses in a hydrostatic transmission system.

Recommendations for reduction of system losses
Utilize displacement variation of pump and motor (primary and secondary control)
Prefer hydrostatic components with high efficiency - e.g., 45° axial piston motors in bent axis design
Select at least the motor in bent axis design (not swashplate design) with the motor displacement larger than the pump displacement
Keep pump speed relatively low
Prevent frequent operation at pressure levels above 200–250 bar (2,900–3,600 psi) which will also be beneficial for product life
Include pressure limitation for the pump instead of high-pressure relief valves only
Optimize the charging system in flow and pressure
Plan for additional gear stages to keep motor speed low

FIGURE 3.56 Structures of continuously variable hydrostatic transmissions [4].

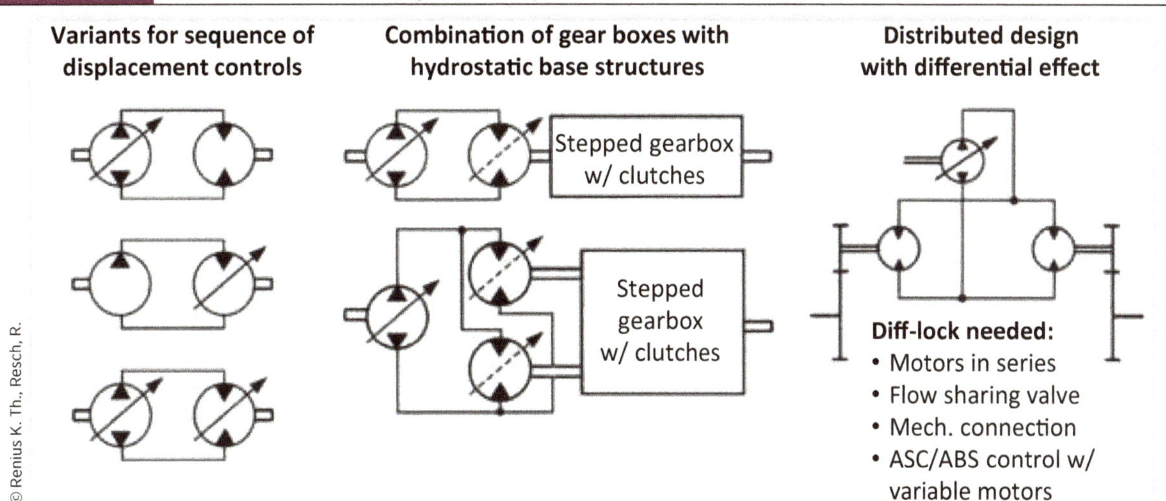

The third differentiator for transmission variants is the distributed design of hydrostatic transmissions which leads to systems with a single pump and multiple hydrostatic wheel drives, e.g., sprayer or aerial work platforms. The distributed system architecture with motors close to the wheels is difficult for the implementation of a multi-speed solution but allows for flexibility in the positioning of components. Pump flow is distributed flexibly between multiple motors by following the path of least resistance. This way, the distributed architecture includes a differential functionality between axles and wheels. Active locking of the differential feature is required to limit wheel slippage during slippery ground conditions and can be reached by flow sharing devices that work on the principles of throttling or displacement.

Electronic concepts require speed measurement on the individual wheels and torque reduction on the slipping wheels. This way the differential lock can be implemented without the need for additional components. In general, hydrostatic systems can be very different in the distribution of the machine.

3.3.3.5 **Pressure Limiter Valve:** When driving a machine with HST against a robust obstacle, system pressure will rise and can exceed the pressure setting of the high-pressure relief valves (HPRV). The HPRV will open and relieve the pressure into the charge gallery of the pump, which will cause huge power loss and put stress on the pressure fluid and the components. This power loss heats the oil and leads to two consequences - (1) the heat needs to be cooled out of the circuit afterward and (2) the oil might be damaged when this situation continues for a longer duration. Pressure limiter (PL) valves protect the hydrostatic circuit from high system pressure and heat insertion by actively de-stroking the swash plate angle of the pump when the pressure setting is reached.

To prevent interaction between PL valves and HPRV, the pressure limitation by the PL valves is set to be 30 bar (440 psi) below the setting of the HPRV. De-stroking of the pump will result in a pressure reduction without the generation of surplus heat and the associated power loss. The pump will be stroked back to the position that is required to compensate for the flow losses and maintain the pressure in the system. The machine

FIGURE 3.57 System diagram of the pump with electronic displacement control and pressure limiter (PL) valves.

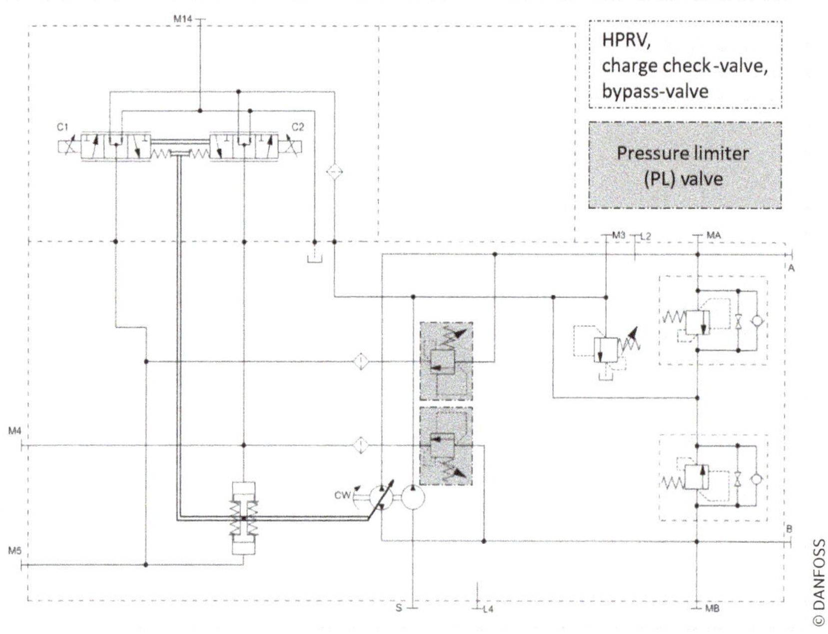

© DANFOSS

will stop driving but the pressure in the hydrostatic circuit will stay high and the maximum driving torque is transferred to the wheels. The implementation principle of the PL valves varies with the supplier of the hydrostatic pump. In the introduced schematic (Figure 3.57), the PL valve for side A will create pressure in the servo chamber for side B to actively force the pump back towards a neutral position. This is possible since the control pressure is in the range of 30 bar (440 psi) while the high-pressure level at PL setting can easily by above 400 bar (5,800 psi). The PL valve for side B will work on the servo pressure of side A in the same way.

3.3.3.6 **Pump Control Options:** The selected option for the displacement control of the pump in a hydrostatic traction drive determines whether the behavior of pump displacement is smooth and dependent on the current load conditions or robust and independent from the load. The electric-proportional control of pump displacement (EDC – Electronic Displacement Control) positions the swashplate proportionally to an electric input command (current – mA) without any influence from system pressure or engine speed – load-independent behavior. The actual angle of the swashplate is fed back into the pump control valve and the pressures in the servo chambers of the pump are modulated to maintain the commanded angle. This control provides a very robust positioning of the pump angel and thereby a very robust volumetric flow.

The electric-proportional pump control without mechanical feedback of the actual swashplate angle into the pump control valve (NFPE – Non-Feedback Proportional Electric) is load-dependent and applied in the smooth, automotive control of a hydrostatic traction drive. The pressure in the servo chambers of the pump is proportional to the electric input command – not the displacement of the pump. The resulting angle of the

FIGURE 3.58 Schematic of pump control valve of EDC (left) and NFPE (right) [22].

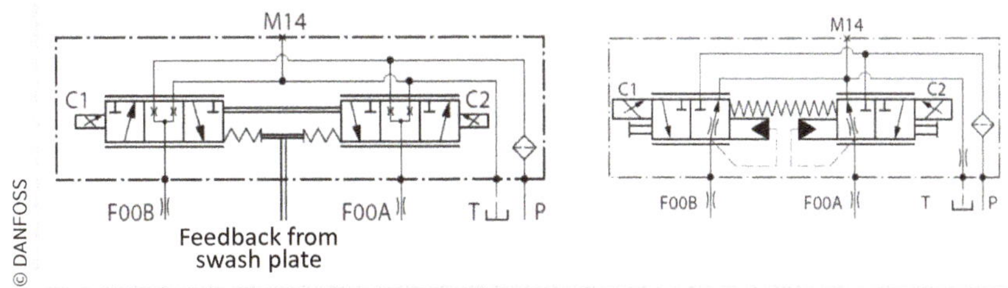

swashplate is a consequence of the counter-acting forces from the servo system of the pump, which is dependent on system pressure, pump speed, and pump displacement. With a constant current command, the pump might reach maximum flow at low system pressure. Rising system pressure reduces the pump displacement and thereby the flow. The flow might even drop to zero at the highest system pressure. The resulting envelope of flow and pressure can be interpreted as a limitation of the hydrostatic power output of the pump based on the electric input command.

Machines such as wheel loaders and telehandlers prefer load-dependent behavior in the traction drive to control wheel torque sensitively during digging, while combines and forklift trucks are normally equipped with a traction drive that provides load-independent behavior. The selection of the appropriate control concept is essential to the success of machine development.

Figure 3.58 introduces the hydrostatic schematics of the EDC (left) and NFPE (right) pump control options. The pressure routings to the servo cylinder sides A and B are labeled with F00A and F00B, while P and T refer to the charge pressure supply and the pump housing which is serving as tank pressure level for the pump control. C1 and C2 refer to the solenoids for the control of the pump.

The control valve of the electronic displacement control (Figure 3.58, left) is implemented as a 3/4-directional spool valve (3 positions/4 ports) which is actuated by two proportional solenoids from each end of the spool. The active solenoid transmits a force to move the spool and connect the corresponding chamber of the servo cylinder with the charge pressure. The pressure difference in the servo cylinder results in stroking of the pump, which ultimately leads to force equilibrium on the spool via the mechanical feedback to the control valve. This mechanical closed-loop control of the swash plate stays active continuously, keeping the pump at the correct displacement and compensating for varying control forces.

The non-feedback electric proportional control (Figure 3.58, right) utilizes two proportional pressure reducing valves which are sharing the central spring. The actuated solenoid provides a force to move the spool and the corresponding servo chamber is connected to the charge pressure. The pressure build-up in the servo chamber creates a force on the opposite side of the spool until the commanded pressure and thereby a force equilibrium is reached. Pressure and force are not influenced or modified by any other input or control loop than the current command to the solenoid [22].

3.3.3.7 **Motor Limits:** Hydrostatic motors are the power consumers of the HST and in that role, the hydrostatic motor can also contribute to the power management in the

FIGURE 3.59 Electric-proportional motor control, speed sensor and loop flushing [23].

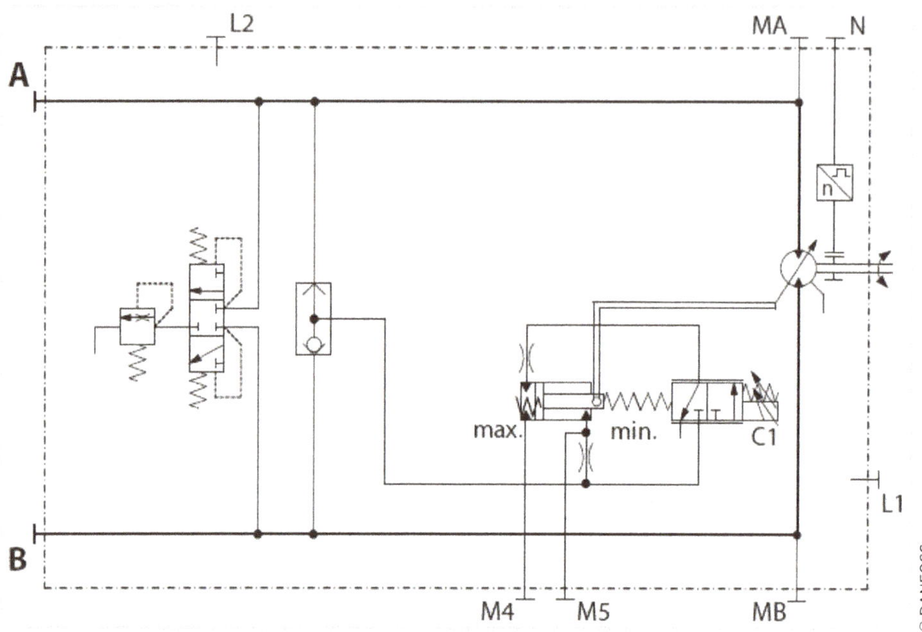

traction drive. The motor consumes pump flow to create a rotation in the output shaft of the motor. The working machine will react to the increasing motor speed with a resisting torque on the shaft. Dependent on the actual motor displacement, this torque will be balanced with an increased pressure level in the HST.

Figure 3.59 introduces the hydrostatic schematic of a state-of-the-art electric proportional motor with loop flushing and speed sensor on the output shaft. Information on the speed of the hydrostatic motor will be used in the mechatronic control system for the traction drive. The volume flow of oil from the kit leakage, the displacement control valve, and the loop flushing valve are collected in the housing of the motor before for cooling and lubrication before it is drained to the hydraulic tank of the machine via the ports L1 and L2.

The displacement of some hydrostatic motors can be reduced to zero which switches the motor off hydrostatically like pressing the clutch in mechanical transmission. In this condition, the motor will not generate torque or consume flow when spinning freely, so that innovative control concepts like anti-slip control, automatic gear shifts, or front-wheel assist can be implemented.

The capability of a motor is determined by the maximum torque and the maximum rotational speed – the corner power. While the maximum torque of the motor is limited by the physical displacement and the maximum allowed pressure, the speed is limited by various limitations in tribology, material selection, and motor design. In some motors, those result in a displacement dependent speed limitation.

The diagram in Figure 3.60 illustrates exemplarily the displacement dependency of the rated and maximum speed limit of a 110 cm^3 motor in bent axis design. The smaller

FIGURE 3.60 Rated and maximum speed [rpm] dependent on displacement [%] for 110 ccm bent axis motor [23].

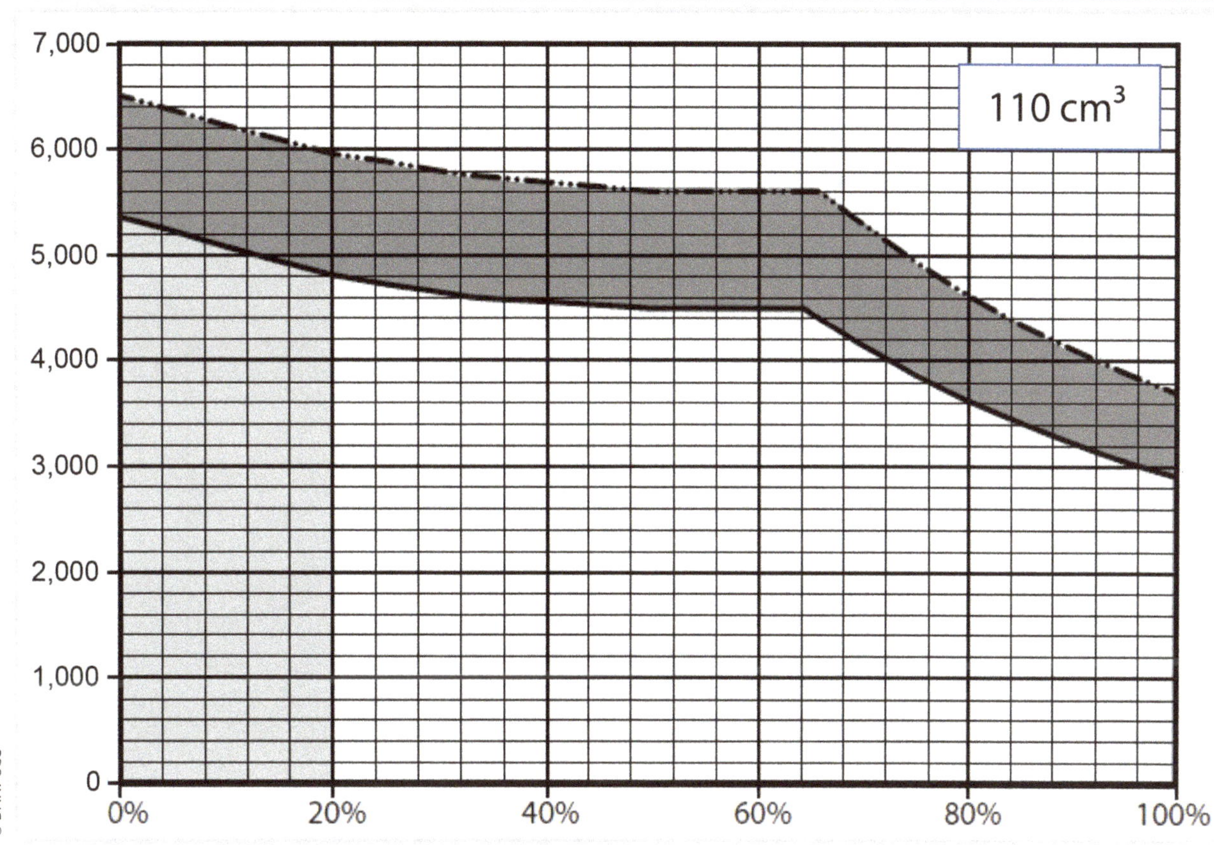

the displacement gets, the higher the allowed rated and maximum speed of the motor in the application becomes. In the grey area, the operation is only allowed for a limited time, during short over speed events at reduced pressure levels.

3.3.3.8 **Motor Control Options:** The displacement control of an axial piston motor in bent axis design is most frequently utilizing control pressure that is taken via a shuttle valve from the high-pressure side of the hydrostatic transmission (high-pressure controlled). The default position of the motor describes whether the motor is at minimum or maximum displacement without an active control signal. It has to be specified during the design process and it is an essential part of the safety concept and the intended fault reaction. In some motor designs, the default position can be modified by turning the end cap with the servo cylinder by 180°.

3.3.3.8.1 2-Position Control. In the simplest control concept, the motor can be commanded to a minimum and a maximum displacement using a hydraulic control pressure or an electric command. This change in motor displacement is completely independent of the actual load in the system and is used to provide 2-speed behavior of the traction drive without the need for a mechanical gear shift. Low gear at maximum motor displacement provides the high tractive force while the high gear at minimum

FIGURE 3.61 Control schematic of 2-position motor with electric command [23].

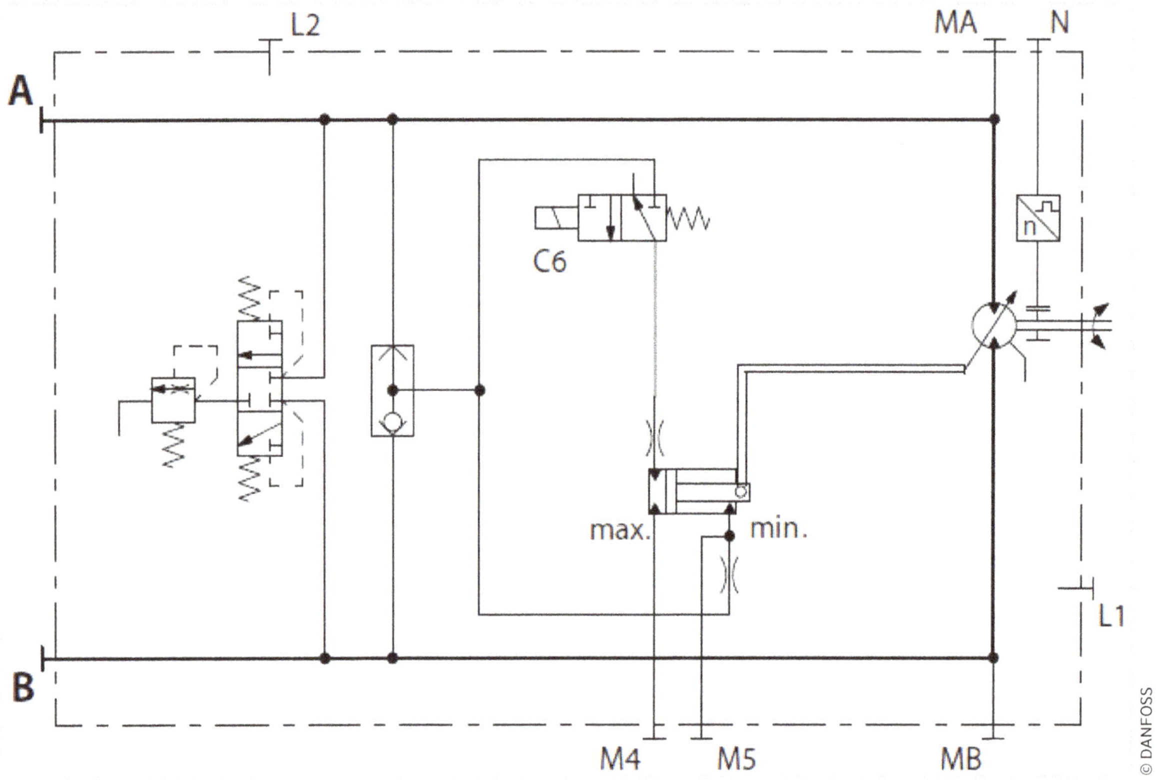

displacement allows for higher ground speeds. The allowed minimum displacement of the motor must be determined during the design process to prevent over-speed situations of the motor during normal operation and is directly dependent on expected engine speed and pump displacement. A mechanical hard stop will be adjusted to limit the minimum displacement to the determined value.

The electric 2-position control is realized by the ON/OFF solenoid C6 (Figure 3.61) which actuates the 2/3-directional valve that connects the piston side of the servo piston either to control pressure (ON = minimum displacement) or to pressure of motor case (OFF = maximum displacement). The ring chamber of the servo cylinder is always connected to the selected control pressure. For the hydrostatic 2-position control the solenoid C6 is replaced with a hydrostatically actuated valve.

3.3.3.8.2 Pressure Compensation Over-Ride (PCOR). Fix motors and 2-position motors are not capable of managing the load on the diesel engine, which is the prime mover in most mobile applications. This can be acceptable on machines with relatively high engine power, such as a combination, but can be very challenging for a machine with relatively low engine rating, e.g., a wheel loader. Since the motor is not directly connected to the engine, it can only adjust the load on the engine by modifying the system pressure in HST. Load dependent control concepts for the motor were developed to improve the performance and extend the potential range of applications. Assuming a

FIGURE 3.62 Control schematic of electric 2-position motor and PCOR that defaults to minimum displacement [23].

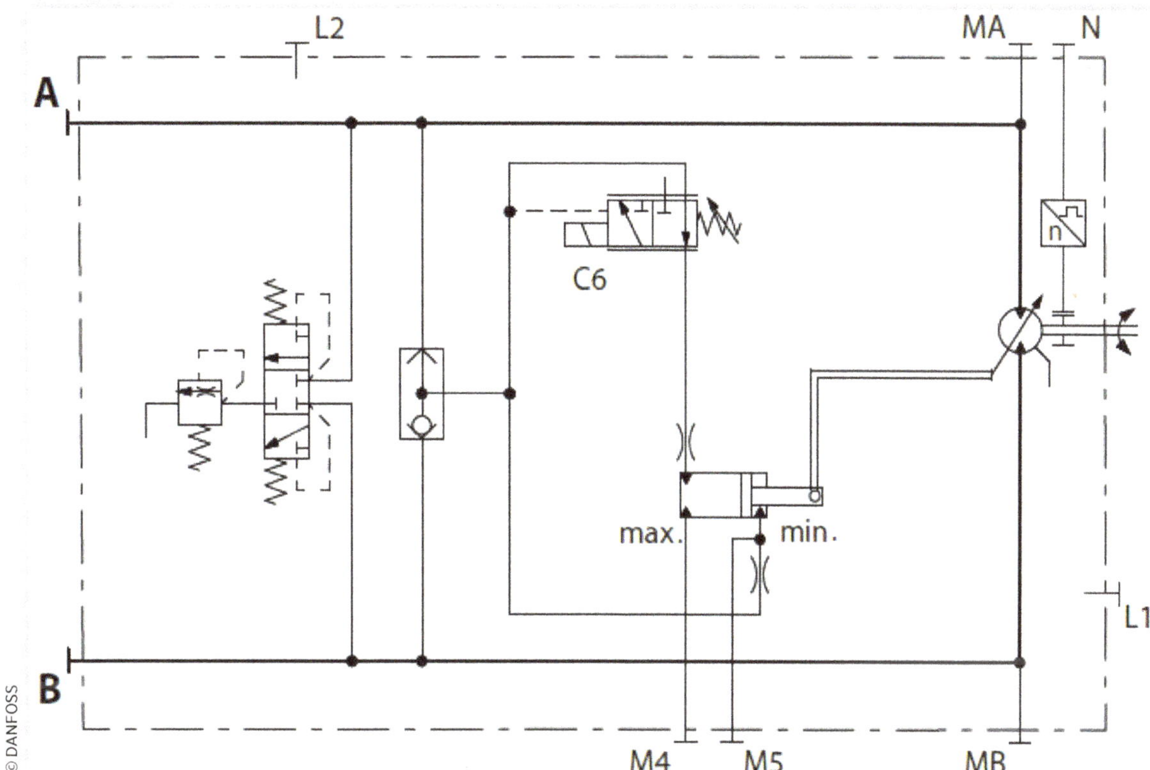

constant displacement of the pump, the displacement of the motor is modulated to keep the system pressure and thereby the torque on the engine shaft as close as possible to the desired operating point. System pressure is maintained constant by allowing for a variation of ground speed. This function is called "Pressure Compensation Over-Ride (PCOR)" since the primary command to the motor is overruled by a secondary control input, the pressure-based compensation. The schematic in Figure 3.62 introduces the pressure based PCOR function that utilizes the same spool that is connected to the C6 solenoid.

The PCOR becomes active when the primary control is commanding relatively small displacement and the system pressure rises about the setting of the pressure threshold. For a motor with 2-position control, this would be the minimum displacement for high ground speed. Coming from maximum ground speed on level ground and the motor at minimum displacement, the machine drives at low a system pressure of about 100 bar (1,450 psi). In this example, the machine drives into a slope that is growing continuously steeper and the system pressure rises due to the higher tractive force requirement at the wheels. The PCOR becomes active when the system pressure rises above the pressure threshold which could be set to 250 bar (3,600 psi) pressure. The motor displacement is continuously increased to keep the pressure close to the threshold, to provide the increased tractive force at larger motor displacement and the same pressure, and to reduce ground speed as a compromise for the increased traction force. Motor displacement can be modulated between the minimum and maximum displacement to maintain the intended system pressure – the available hydrostatic power can be fully utilized throughout the displacement range of the motor. The setting of the pressure threshold is optimized to consume the available engine power for the hydrostatic traction drive. The PCOR can only increase the displacement of the motor over the primary input command. If the motor is commanded to the high displacement by the primary command already, the PCOR feature might not become active at any time. It is possible to combine the PCOR with a 2-position control option and electric proportional displacement control as a primary motor command.

3.3.3.8.3 Brake Pressure Defeat (BPD). The brake pressure defeat (BPD) functionality is a supportive feature that comes together with the PCOR and is intended to limit the PCOR operation to the advantageous operating conditions only. The PCOR provides very beneficial performance while driving the machine is operated under heavy load. However, during strong deceleration or hydrostatic braking events, increasing motor displacement due to high system pressure would only lead to more deceleration and even higher system pressure. The BPD valve actively selects the pressure side as the supply pressure for the motor control pressure instead of a pressure-based selection using a hydrostatic shuttle valve. This prevents PCOR activity on high-pressure events that happen on the deceleration or braking side of the HST. The PCOR feature should always be applied together with a BPD valve in hydrostatic traction drives to prevent unintended or unexpected machine behavior.

Figure 3.63 introduced the hydrostatic schematic for 2-position control with PCOR and BPD value for the selection of the motor control pressure. The 3-way/2-position BPD

FIGURE 3.63 Control schematic of the electric 2-position motor with PCOR and BPD [23].

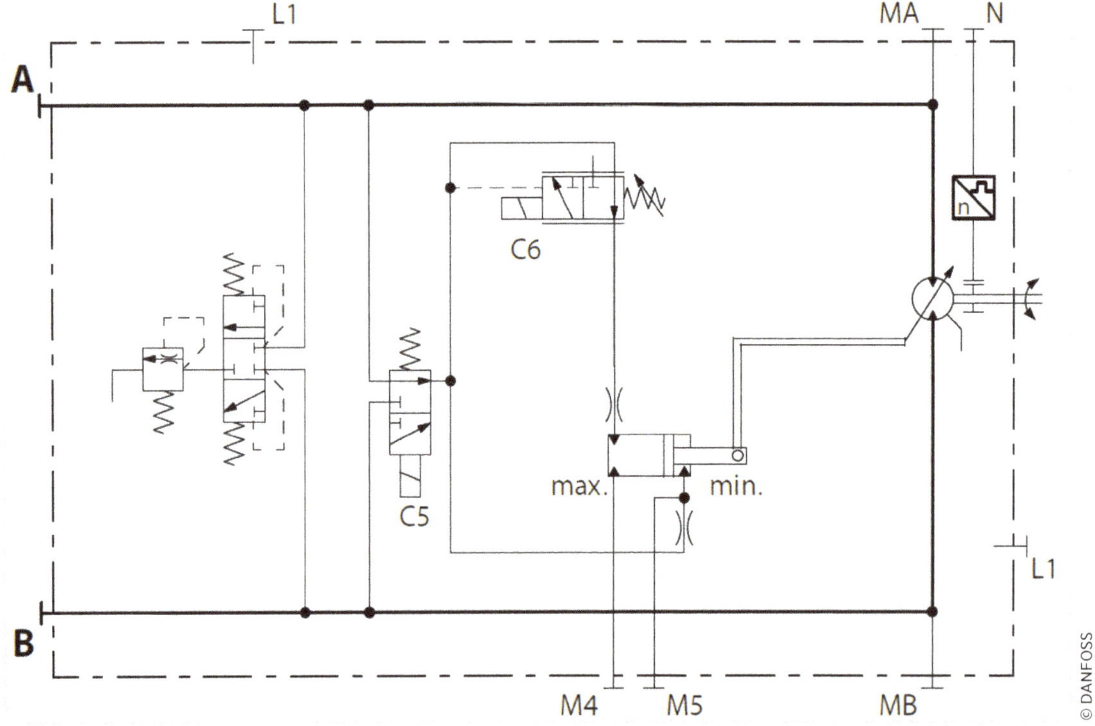

valve is switched by the solenoid C5 to select the supply pressure for motor servo control as well as the pressure signal to the PCOR valve. Dependent on the driving direction of the machine, the BPD valve is commanded to select the high-pressure side for normal driving operation. During braking events, the high-pressure of the HST changes and due to the BPD selection only the pressure in the low-pressure branch of the HST is available for motor control and the PCOR will not become active.

3.3.3.8.4 Electric Proportional Motor. With the electric-proportional control, the motor displacement can be command by an electric current input to a position between the minimum and maximum displacement. This is the most flexible control option for the motor and is applied for advanced control options on vehicle level, e.g., motor over-speed protection, electronic PCOR functionality, or shifting strategies for hydrostatic transmissions. The proportional motor control can also be designed with a pressure signal as the input command.

The electric proportional control of motor displacement is achieved by the proportional solenoid C1 which actuates a 3-way/2-position control valve (see Figure 3.64). The position of the servo cylinder is fed back to the opposite face of the directional valve via a feedback spring. When a displacement is commanded, the pressure in the piston chamber of the servo cylinder is increased and modulated between control pressure and tank pressure until the servo piston reaches the commanded positions and the force equilibrium is reached at the spool. The solenoid acts in the direction of increased pressure in the piston chamber while the feedback spring reduces the pressure level.

FIGURE 3.64 Hydrostatic schematic of electric-proportional motor control (w/o PCOR and BPD) [23].

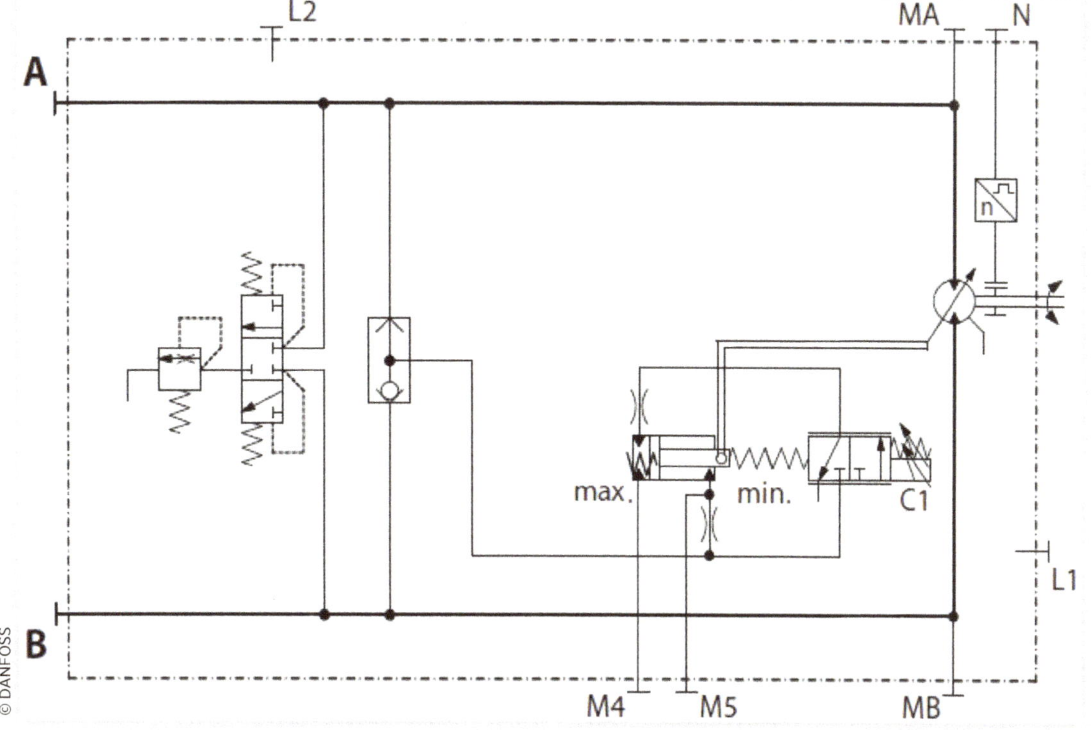

The ring chamber of the servo cylinder is always connected to high pressure. This way, the motor displacement is controlled independently of the pressure load in the HST. The default position without current command is maximum motor displacement in this example.

3.3.3.9 **Control of Hydrostatic Traction Drives:** The machine operator expects a powerful hydrostatic traction drive that loads the diesel engine heavily when needed and supports fuel-saving strategies when possible. Throughout the whole operation envelope, the diesel engine needs to be protected against overload (anti-stall feature) and too high rotational speeds (engine over-speed).

For the realization of the intended functionality, the system engineer can choose hydrostatic components with various characteristics. In the traditional automotive drivetrain controls for mobile working machines with HST, hydro-mechanical controls in a pump and motors were applied. Recent developments are more frequently applying components with electro proportional control inputs to implement fully mechatronic concepts. The components are expected to follow their input commands and the system solutions are embedded in the software completely. Adjustments in performance and dynamics of the traction drive are achieved by parameter modifications in the software. More advanced and more flexible control systems become possible and multiple optimized working modes for the various machines can be provided.

Fundamentally, hydrostatic traction drives can be split into "load-dependent" and "load-independent" categories. The selected control concept is solely dependent on the requirements of the machine application. In "load-dependent" traction drives, NFPE pumps are applied together with PCOR compensated electric-proportional motors. While in "load-independent" drive trains, the combination of the EDC pump with an electric proportional motor is beneficial.

3.3.4 Power-Split Transmission

Efficiency requirements for the CVT of today's mobile working machines cannot be realized by the implementation of purely mechanical, hydrostatic, or electric CVT. In power-split transmissions, the variable ratio of the CVT is combined with the high efficiency of mechanical gears to a multimode transmission. Planetary gear sets and gear stages are essential for splitting or merging of the power flows.

In hydro-mechanical transmissions (HMT), the hydrostatic transmission (HST) is implemented to provide the continuous variation of the speed ratio in the variable power branch. The limited conversion ratio of the HST can be applied in multiple HMT modes to provide the required spread in the overall transmission ratio. Switching between modes in an HMT frequently happens at synchronous speed conditions. The CVT capability of the HST is used to vary the transmission ratio continuously within each of these modes. As always in HST, the hydrostatic components can be positioned with high flexibility.

The basic idea of the power split principle is presented in Figure 3.65. The input power from the engine is split into a constant ratio mechanical path and a variable ratio path using a CVT unit. On the output side of the transmission, both power flows are

FIGURE 3.65 Principle of a CVT with external power split [4].

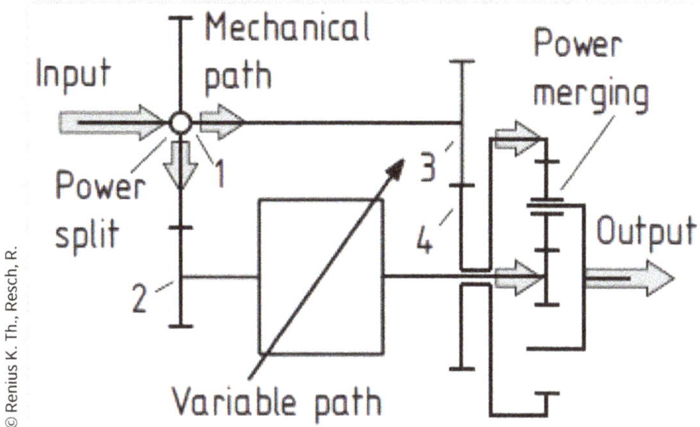

merged, and the output power can finally be used for the traction drive. Technical applications of power split transmissions in traction drives of mobile working machines can be found in three basic structures: (1) Input Coupled (IC-HMT) as in Figure 3.65, (2) Output Coupled (OC-HMT) which is a mirror image of Figure 3.65, and (3) Compound (CC-HMT) which utilizes a planetary gear set on the input and the output side.

The system efficiency of the power-split transmission typically ranges above the efficiency of the applied CVT due to the high efficiency in the mechanical path.

This shall be demonstrated by a typical example with realistic efficiency values for full power operation: with a power split in an HMT of 70% mechanical and 30% CVT power flow and related mechanical efficiency of 95% and CVT efficiency of 80% for an HST with swashplate units, the resulting HMT efficiency can be calculated to 90.5% [4].

$$\eta_{HMT} = 0.7 \cdot 0.95 + 0.3 \cdot 0.8 = 90.5\% \quad (3.46)$$

The efficiency of a power-split transmission is dependent on the losses in the variable and fix mechanical path. In an HMT, the losses in the variable path—which is the HST—on the hydrostatic power ratio but also on the speed in the hydrostatic pump or motor.

The design of power-split transmission allows for the possibility of circulating power, which is disadvantageous due to the possible negative impact on power losses. Regarding the IC-HMT of Figure 3.65, the output speed of the variable power path that drives the sun gear in the planetary can be reversed in the case of a hydrostatic or electrical CVT unit. This means that its power flow direction is reversed – in Figure 3.65 now from right to left. This power is superimposed to the system input power, it must be transferred by the mechanical path again to the right and it is thus called "circulating." If its value is kept low compared to the input power, total system efficiency can remain above that of a direct CVT. This strategy is therefore often applied in hydrostatic IC-HMT for tractors to use the HST in forward and reverse operation and thereby increase the speed range of the output in a single mode.

3.3.4.1 **Splitting and Merging of Power Flow:** In HMT, gear stages and planetary gears are used for splitting and merging the mechanical and hydrostatic power flows (see Figure 3.66).

FIGURE 3.66 Splitting (left side) and merging (right side) of power flow in HMT [11].

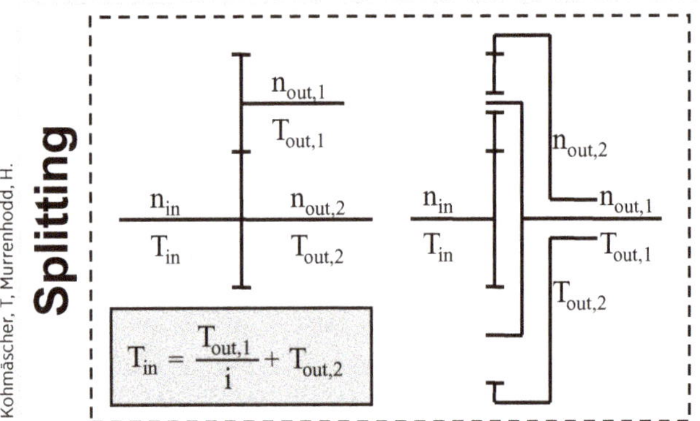

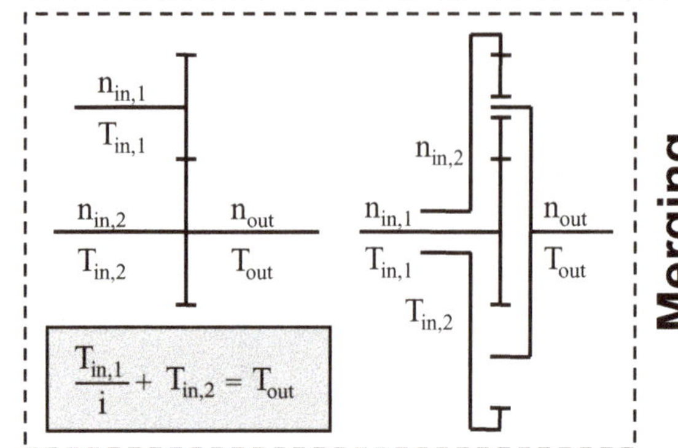

Application of gear stages for power splitting results in a fix speed ratio between the input speed n_{in} and the output speeds $n_{out,1}$ and $n_{out,2}$. Therefore, the power flows across the upper and lower shaft can only be varied by the transmitted torques $T_{out,1}$ and $T_{out,2}$. For merging two power flows the torques $T_{in,1}$ and $T_{in,2}$ are combined to the output torque T_{out}. Based on this interpretation, IC HMT uses a gear stage as the power splitting device is also called "torque split" HMT. Application of planetary gears for power splitting results in a fix torque ratio between the input torque T_{in} and the output torques $T_{out,1}$ and $T_{out,2}$. Therefore, the power flows through the two output members of the planetary can only be varied by the output speeds $n_{out,1}$ and $n_{out,2}$ that are dependent on the standard ratio i_0 of the implemented planetary. For merging two power flows the speeds $n_{in,1}$ and $n_{in,2}$ determine the output speed n_{out} while the input and output torques are again in fix ratio due to the planetary restrictions. Accordingly, OC HMT that applies a planetary as the power splitting device is also called "speed split" HMT.

3.3.4.2 **Power-Split Structures:** For the following introduction of power split transmissions, it will be assumed that the CVT is an HST so that the power-split transmission is referred to as a hydro-mechanical transmission (HMT). Three general structures of HMT are available to the designer of an efficient transmission: IC-HMT, OC-HMT, and CC-HMT. HMT modes can be designed to utilize either one basic structure, such as IC or OC, or multiple structures, such as OC combined with CC.

HMT is referred to as CVT which highlights the fact that the transmission ratio of HMT can be adjusted in many very small steps due to the high resolution of the ratio in the connected HST. Most implemented HMT for traction drive allows for a speed reduction until the machine stops and can hold its position even on steep slopes. The market refers to this feature as "powered ZERO," "powered NEUTRAL," or infinitely variable transmission.

3.3.4.2.1 Input Coupled HMT. IC HMT systems are characterized by having the first hydrostatic unit, which must have variable displacement, connected to the transmission input shaft via gears as shown in Figure 3.67. One member of a 3-member planetary is driven by the transmission input shaft (e.g., sun gear) while a second element is coupled

FIGURE 3.67 Schematic of Input Coupled HMT structure (torque split).

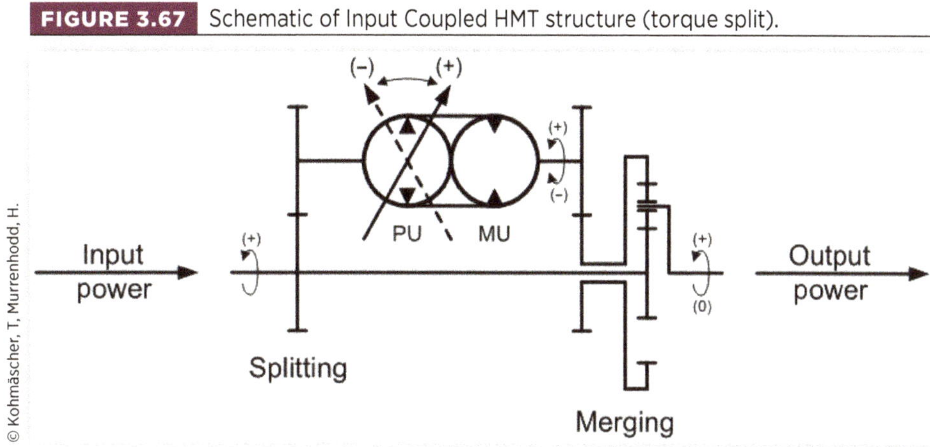

to the transmission output shaft (e.g., carrier). The second hydraulic unit, which is typical of fixed displacement, is connected to the third element of the planetary (e.g., ring gear).

The input coupled system is operated with the variable hydrostatic unit stroking from maximum negative displacement, over the center, to maximum positive displacement. As a result, the second hydraulic unit rotates from full negative speed, to zero, to full positive speed thereby changing the speed of the third planetary element (e.g., ring gear). Theoretical peak efficiency of the system is achieved with the first variable hydraulic unit near-zero displacements thereby transmitting no power hydraulically into the planetary. This typically corresponds to the middle of the output speed range where the hydrostatic power ratio and the speed of the motor unit are zero. In HMT applications today, the Input Coupled transmission requires a mechanical reverser to provide reverse vehicle operation.

Figure 3.68 introduces the left side of the ladder diagram for the IC-HMT. The earlier introduced concept is applied to review the conditions at the planetary gear set.

FIGURE 3.68 Ladder diagram and power flows for Input Coupled HMT.

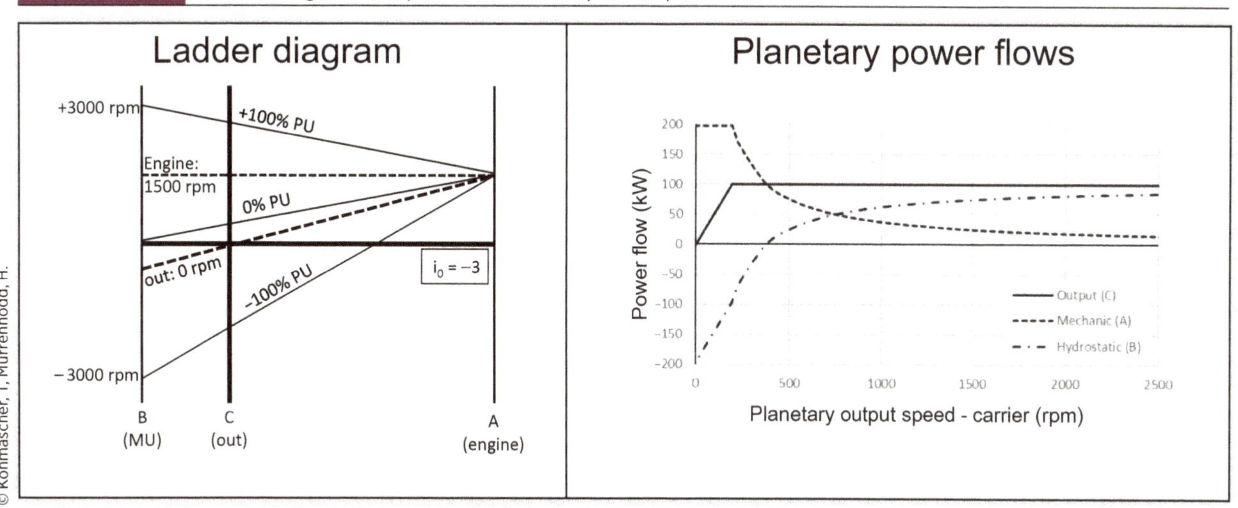

The IC-HMT in the schematic is analyzed for a planetary gear set of standard ratio $i_0 = -3$ in the ladder diagram - with the carrier (C) as output, the sun gear (A) as the constant input speed from the engine at 1,500 rpm, and the ring gear (B) for the input speed from the motor unit (MU) in the HST. In a first step, the horizontal line for the engine speed is put in at 1,500 rpm level. In the second step, three key conditions for the HST are added: (1) Pump unit (PU) at +100% displacement with the ring gear speed at +3,000 rpm, (2) PU at 0% with 0 rpm on the ring gear, and (3) PU at −100% with the ring gear at −3,000 rpm. The speed variation from +2,625 rpm to −1,875 rpm on the carrier output shaft can be determined graphically. Finally, the line for 0 rpm output speed on the carrier shaft is added to understand the lower limit for this IC mode in the HMT. It becomes clear that the PU can only be used to 1/6 of maximum negative displacement to not reverse the rotational direction of the carrier member which double-serves as the output shaft. For reversals, the IC-HMT requires a mechanical reverser gearbox.

On the right side, Figure 3.68 presents the mechanical and hydrostatic power flows through the planetary gear set. The bold line shows the output power in the carrier which is 100 kW maximum, but torque limited towards 0 rpm rotational output speed. The hydrostatic power flow (dash-dotted line) crosses the 0 kW-line at 375 rpm output speed which indicates the point of 100% mechanical power transmission and peak efficiency. Towards higher speed, both powers flow contribute to the output power which keeps the efficiency high. While at lower speeds the power circulates back through the hydrostatic power branch and the mechanic power flow rises above 100 kW to settle at 200 kW due to the torque limit in the system. The torque limit in the gearbox is caused by the setting in the HPRV of the HST in the IC-HMT. Close to stand-still, the hydrostatic power branch creates −200 kW circulating power which results in 0 kW output power at the carrier.

3.3.4.2.2 Output Coupled HMT. The full utilization of OC-HMT requires that both hydrostatic units are variable in displacement and connected to a 3-member planetary gear set. The planetary is coupled to the transmission input shaft (e.g., carrier) and the output shaft (e.g., sun gear) as shown in the schematic in Figure 3.69. The first hydrostatic

FIGURE 3.69 Schematic of Output Coupled HMT structure (speed split).

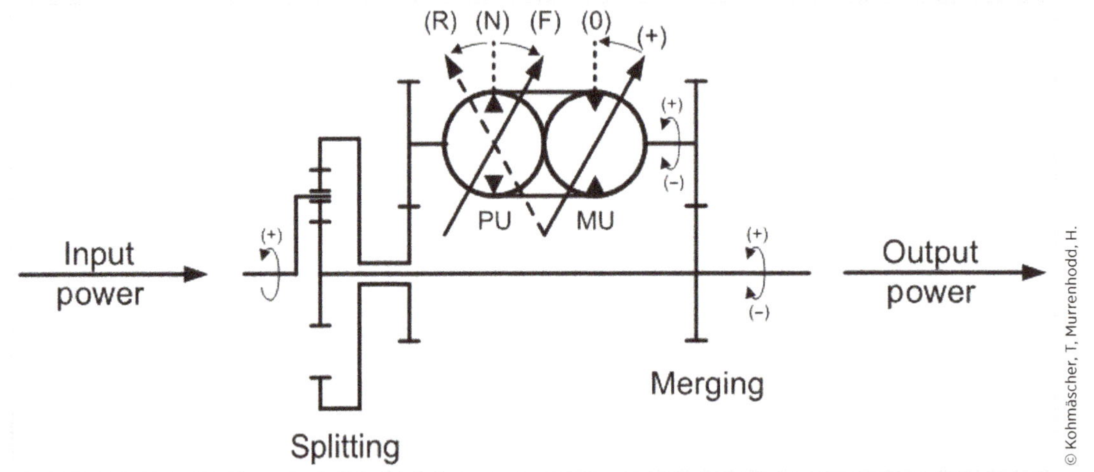

unit is then coupled to the third planetary element (e.g., ring gear) while the second hydrostatic unit is coupled to the transmission output shaft via a gear stage.

The OC system operates with the first hydrostatic unit, referred to as the pump unit, determining the direction of vehicle travel. When the pump unit is at zero displacements, the vehicle is at a standstill. Stroking the pump in one direction will produce forward driving while stroking it into the opposite direction produces reverse driving. The second hydraulic unit, referred to as the motor unit, strokes from maximum displacement to zero displacements after the pump unit has reached a maximum angle. The motor unit can rotate either direction to provide forward or reverse driving.

The theoretical peak efficiency of this system is achieved at maximum forward transmission speed when the complete engine power is transmitted mechanically. Power losses in the motor unit are partly dependent on the rotational speed and might become dominant at higher ground speeds. Due to the speed relationship of the planetary, the rotational speed of the pump unit will be reduced while driving forward but it will be increased when driving in reverse. This must be considered when designing the HMT to not overspeed the pump unit in reverse operation. Since the power loss in the pump is greatly dependent on the rotational speed, forward driving provides a significant efficiency benefit over reverse driving. Therefore, if equal efficiency characteristics in forward and reverse are required, a mechanical reverser would be required for the OC system as well.

The OC-HMT in the earlier schematic is interpreted in the ladder diagram of Figure 3.70 (left) assuming a standard ratio of $i_0 = -3$. The combustion engine supplies engine power at a constant speed of 1,500 rpm to the carrier (C). The sun gear (A) transmits mechanical power to the output and is connected to the motor unit (MU) while the ring gear (B) is connected to pump unit (PU) of the hydrostatic power path that provides power to the output via the motor unit (MU). In a first step, the horizontal line for the engine speed is put in at 1,500 rpm level. In the second step, three key conditions for the HST are added: (1) PU at 0% displacement and MU at 100% which blocks the sun gear and results in a stand-still of the machine, (2) PU at 100% and MU at 0% which

FIGURE 3.70 Ladder diagram and power flows for Output Coupled HMT.

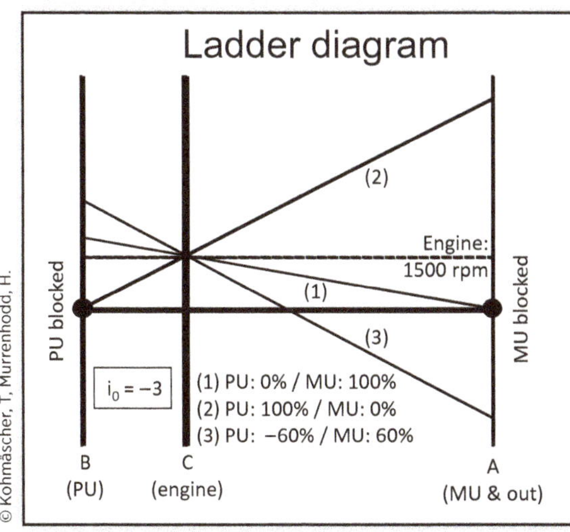

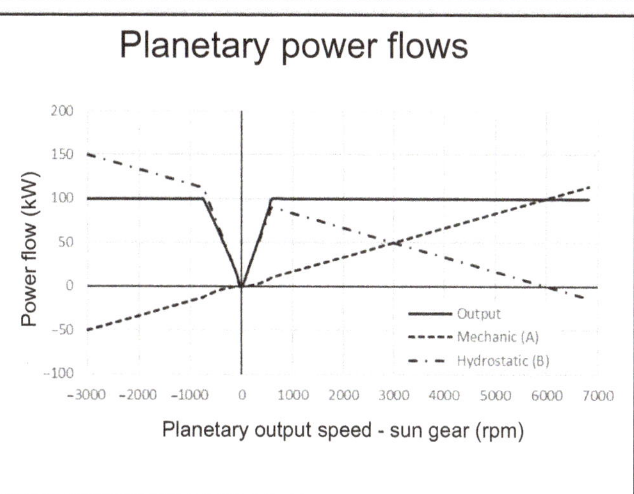

blocks the ring gear and leads to the maximum forward ground speed, and (3) PU at 60% and MU at 60% which results in the maximum reverse ground speed. With one hydrostatic unit at 0% and the other at 100% displacement in the conditions 1 and 2, the HST blocks the speeds at the planetary member that is associated with the unit at 100% displacement. In condition 3, the pump displacement is kept low in reverse operation to allow for a higher speed limit of the PU. For the visualization of the functionality of the OC-HMT in the ladder diagram, the straight lines for the HMT condition are pivoted around the fix input speed from the engine at the carrier.

Figure 3.70 shows on the right side the mechanical (dashed) and hydrostatic (dash-dotted) power flows through the planetary gear set concerning the output speed of the sun gear. The bold line shows the output power which is determined by the merged power flows from sun gear and HST and reaches a maximum value of 100 kW. The power is torque limited towards 0 rpm rotational output speed. The hydrostatic power flow is at 100% when starting to drive in forward and reverse operation. In forward operation, the mechanical power flow increases with output speed until it reaches 100% at maximum ground speed. Theoretically, it would be possible to increase the ground speed beyond this point but due to the circulating power flow, the efficiency drops quickly. Due to the different direction of the torque on the output shaft in reverse operation, the mechanical power flow circulates back to the planetary and the hydrostatic power flow rises above the power input. The circulation of power in reverse driving will result in reduced efficiency and therefore the OC-HMT might have to be equipped with a reverse gear.

3.3.4.2.3 Compound HMT. Compound HMT requires that both hydrostatic units are variable in displacement and connected to a 4-member compound planetary gear set as introduced in Figure 3.71. The planetary is coupled to the transmission input shaft (e.g., carrier 1 and sun 2 gear) and output shaft (e.g., ring 1 gear and carrier 2).

FIGURE 3.71 Compound HMT structure.

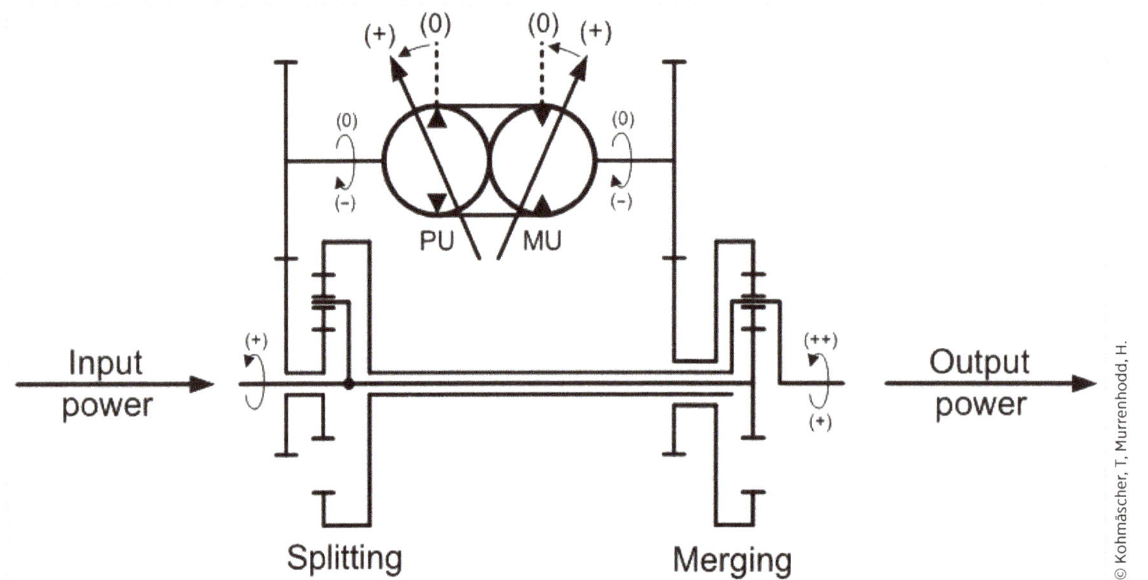

The hydrostatic units are coupled to planetary members that are not connected to the transmission input or output shafts (e.g., sun 1 gear and ring 2 gear). In the schematic, the 4-member planetary gear set looks very much like two planetary gears sets while industrial implementations might be significantly more compact. The schematic in Figure 3.71 emphasizes the advantages of the compound HMT that leads to the high-efficiency level – the mechanical power flow is transmitted via two shafts and the hydrostatic power ratio is strongly reduced.

At the low output speed condition of the compound HMT, the first hydrostatic unit (pump unit) on sun 1 gear (A1) is at zero displacement and high speed. The second unit (motor unit) on ring 2 gear (B2) is at maximum displacement and zero speed – blocked condition number 1. To increase the transmission output speed, the displacement of the pump unit is increased, hydrostatic flow is created, and its speed will decrease. Besides, the displacement of the motor unit is decreased thereby increasing its speed. At the high output speed condition, the pump unit will be at maximum displacement and zero speed—blocked condition number 2—while the motor unit will be at zero displacement and maximum speed.

Theoretical peak efficiency of the compound system exists at the blocked conditions number 1 and 2. At these two operating points, the engine power is being delivered purely mechanically through the planetary as no power is being transmitted hydraulically by the rotating kits. The power losses in the hydrostatic units are caused by the high pressure and the higher rotational speed in one of the units. The compound HMT is not applied down to zero vehicle speed due to efficiency concerns due to high levels of circulating power within the planetary gears.

The schematic of the compound HMT is interpreted and analyzed based on the ladder diagram in Figure 3.72 (left) assuming a standard ratio of $i_{01} = i_{02} = -3$. This time a 4-member planetary gear set needs to be visualized in the ladder diagram. The first planetary is created as before. The second planetary is added by matching the lines for the shared planetary members (C1 = A2 and B1 = C2) and adding the line for the free planetary member according to the standard ratio i_{02} for the second planetary (line for

FIGURE 3.72 Ladder diagram and power flows for Compound HMT.

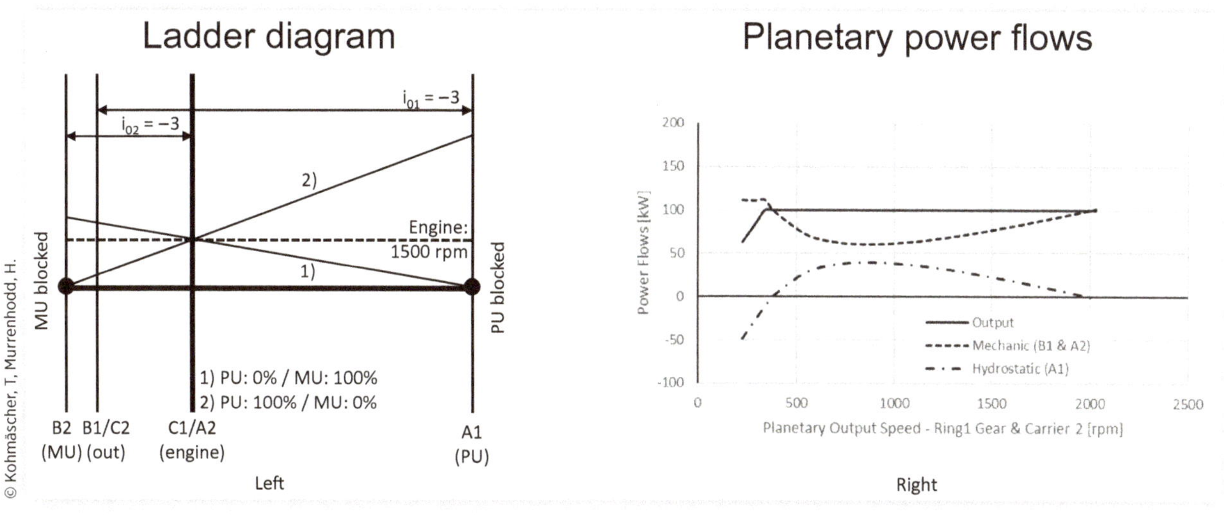

B2 to the left of the line for B1). The engine supplies power at a constant speed of 1,500 rpm to carrier 1 and sun 2 gear (C1 and A2). The ring 1 gear (B1) transmits mechanical power to the output via carrier 2, the sun 1 gear is connected to the pump unit (PU) that supplies the motor unit (MU) at the ring 2 gear (B2) while the output power is taken from carrier 2. In a first step, the horizontal line for the engine speed is put in at 1,500 rpm level. With the engine speed being fix in the analysis, the lines of specific conditions will pivot around the crossing of engine speed and C1/A2-line. In the second step, the two key conditions for the HST are added: (1) PU at 0% displacement and MU at 100% which blocks the sun gear (A1) of the first planetary to create 375 rpm output speed and (2) PU at 100% and MU at 0% which blocks the ring gear (B2) of the second planetary to create 2,000 rpm output speed. Since the line for the output shaft B1/C2 is positioned between the point of pivoting and the line for the ring gear (B2) of the second planetary which is blocked by the motor unit (MU), it cannot reach zero output speed.

Figure 3.72 also shows on the right side the mechanical (dashed) and hydrostatic (dash-dotted) power flows through the planetary gear set concerning the output speed. The bold line shows the output power which is resulting from the merged power flow and reaches a maximum value of 100 kW. As a second limitation, the maximum pressure in the HST becomes active in high-torque situations which is the case close to 0 rpm output speed. Compared to the hydrostatic power ratios in IC-HMT and OC-HMT, the hydrostatic power ratio in CC-HMT stays well below 50% in this example. The blocked conditions 1 and 2 can be identified by 0 kW hydrostatic power which is the case at 375 and 2,000 rpm output speed. Outside of the blocked conditions, power flows start to circulate back via the hydrostatic path which makes those ranges less interesting for HMT implementations. Especially towards stand-still, the hydrostatic power ratio rises quickly and impacts efficiency in a bad way. Therefore, CC-HMT modes are normally combined with an OC-HMT mode to start the machine from stand-still. The component requirements for OC-HMT and CC-HMT are very similar which makes this option attractive.

Kress [33], Renius [4], and Kohmäscher [11] investigated the power flows and efficiency characteristics of the introduced basic concepts in detail.

3.3.4.3 **Efficient and Flexible Hydrostatics:** The power losses in the HST greatly influence the overall power loss of the HMT solution. Amongst other things, the transmission designer needs to balance cost, development time, power losses, space claim, and production volume when sourcing the hydrostatic components or the HST that can be integrated into the transmission design. The market offers three general directions for the HST strategy:

- Standard hydrostatic pumps (swashplate) and motors (bent axis)
- Compact HST with swashplate pump and fix motor (bent axis or swashplate)
- Integration of highly efficient 45° bent axis units

The application of standard hydrostatic pumps and motors offers some unique advantages over the other solutions. The components are easily available, serviceable, exchangeable and can be applied quickly for any planned production volume. The high number of redundant components in bearings, shafts, and housings might cause high system costs. It should be possible to design all introduced HMT structures with standard

FIGURE 3.73 Transparent view of 45° bent axis rotating kit.

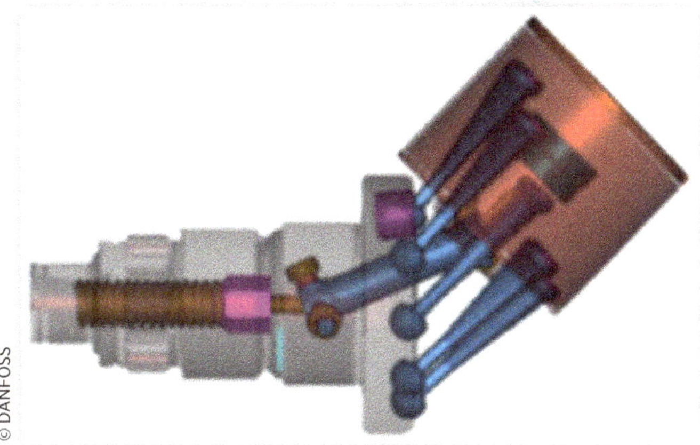

components – for OC-HMT and CC-HMT it might be best to utilize two bent axis motors for efficiency.

The effort for the implementation of compact HST into the HMT transmission design is slightly higher but it also offers some advantages – compactness, efficiency, and cost. Pump and motor are in the back-to-back arrangement and the compact HST comes pre-tested from the supplier to the gearbox assembly. Bosch Rexroth and Linde Mobile Hydraulics developed multiple configurations and sizes of this compact HST that were implemented in HMT solutions for many tractor applications at CNH and ZF. This option is most likely the preferred selection for transmissions series that run with medium to high production volumes.

Flexible integration of 45° bent axis units into yokes and compact HST can be considered the option with the highest effort but also with the highest potential for optimization, efficiency, and integration. The option was selected by Fendt, John Deere, and Claas for the development of tractor specific transmissions solutions and by the supplier ZF for the development of a wheel loader solution for the general market. Today, Danfoss Power Solutions is the only supplier that has experience with high volume production of 45° bent axis technology.

Large angle bent axis hydrostatics shown in Figure 3.73 offer significant efficiency advantages over the standard bent axis and swashplate type axial piston hydrostatics. The maximum stroking angle of 45° will increase the maximum torque output, power density, and efficiency without increasing the general size of the package. Additionally, the 45° bent axis technology has been designed to operate in a dry case environment with forced point lubrication to eliminate splashing losses from the rotating units.

This high component efficiency is the key to the selection of 45° bent axis technology for OC and CC HMT structures. These structures strongly depend on the performance of hydrostatic technology to meet the machine requirements.

3.3.4.4 **Yoke Technology:** The high stroking angle of the 45° bent axis kit technology requires the use of yokes to supply the hydrostatic units with flow and pressure. Figure 3.74 introduces a typical yoke design on the left. The solid main body is designed to hold and position the rotating kit of the hydrostatic unit while the upper and lower

FIGURE 3.74 Yoke design and a hydro module for independently controlled units.

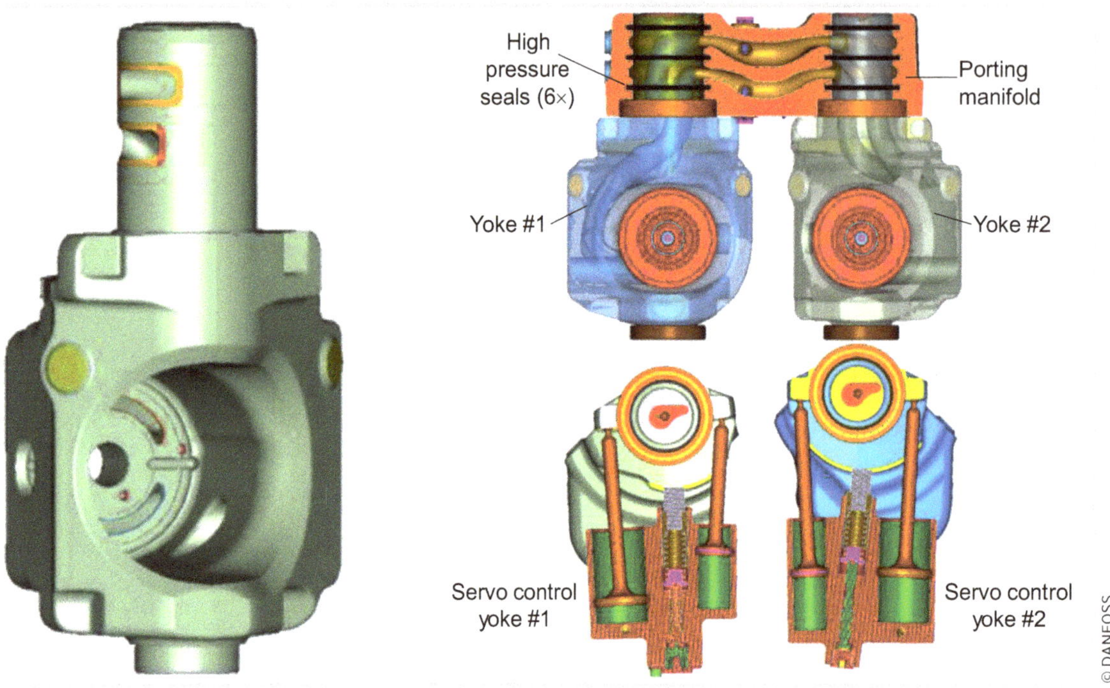

trunnions position the yoke in the manifold. Two of these yokes are combined to a compact HST with a manifold that also contains loop flushing, high-pressure relief valves, and the charging system – this is also referred to as the hydro module. Hydrostatic flow and pressure are routed towards the back of the rotating kit through channels inside that cast iron from the manifold. In total, three high-pressure seals are needed to securely keep the hydrostatic fluid contained inside the hydro module. Two independent servo controls allow for the individual positioning of the yokes.

The dual yoke, shown in Figure 3.75, can provide a similar function as the hydro-module shown before while significantly decreasing the number of components and thereby improving robustness. The dual yoke contains the necessary components to form a closed hydrostatic circuit including fluid porting passages, high-pressure relief valves, charge check valves, loop flushing valve, and servo control pistons. The dual yoke reduces the number of complex yoke castings from two to one. The inclination angle of the kits to each other is fixed. While the first kit is inclined at 0°, the other kit will be inclined at 45°. As the first kit increases displacement, the second will simultaneously decrease displacement.

Additional advantages can be derived from the single servo control system to position both hydrostatic units, the elimination of high pressure seals together with reduced complexity in the casting cores, the more compact flow channels to reduce pressure losses as well as the weight reduction. With pump and motor unit changing their displacement simultaneously, the maximum output torque of the dual yoke design is reduced right from the start. This needs to be considered and most likely compensated during the design of the transmission by adjustment of gear ratios or setting of pressure limitations [34, 35].

FIGURE 3.75 Dual yoke hydro module.

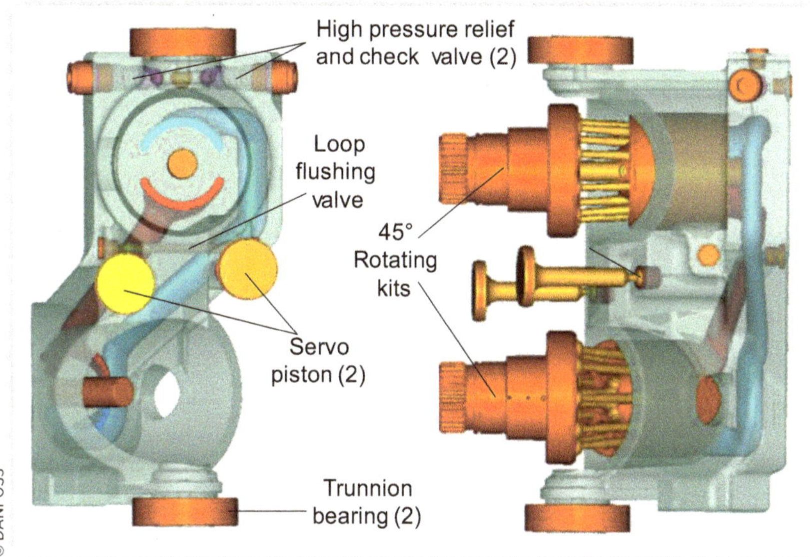

3.3.4.5 **HMT Implementations:** After the introduction of the three basic HMT structures - IC-HMT, OC-HMT, and CC-HMT - and the interpretation of HMT designs by ladder diagram, it will be very interesting to analyze HMT solutions in the market. The following paragraphs will introduce a couple of established HMT solutions in the tractor market as examples.

3.3.4.5.1 Example 1 (IC-HMT). Following the order of the basic HMT structures, the first example is an IC-HMT with two forward ranges and a single reverse range through a reverser behind the planetary - the John Deere "AutoPowr" for the 7010 and 7020 tractor series which was introduced in 2001 (see Figure 3.76). The HST in the HMT applies 45° axial piston units in bent axis design as variable pump unit and as fix the motor unit for superior system efficiency. The HPRV in the HST is set at 550 bar (8,000 psi), but typical pressures under heavy load are about 250 bar (3,600 psi).

The concept of the transmission applies the IC-HMT structure in two forward modes with the mode change at synchronous speeds of planetary members which requires a 4-member compound planetary gear set. The engine drives both, the variable pump unit (PU) and the left sun gear (A1) of the merging compound planetary. Its ring gear (B) speed is controlled by the fixed displacement motor unit which can rotate in both directions. Its carrier (C) output speed is used for range L which is activated by clutch KL to offer a maximum travel speed of 15 km/h while its output of the right sun gear (A2) is used for range H activated by clutch KH to offer 50 km/h as maximum travel speed. An extra-planetary gear set with double-planets for reverse operation up to 17 km/h also utilizes the carrier (C) output of the merging planetary but reverses the rotational direction by braking the ring gear with clutch KR.

In stand-still of the machine, low range is actuated by clutch KL to utilize the carrier (C) output speed, the PU is stroked into the negative direction and the ring gear turns in the opposite direction of the incoming engine speed so that hydrostatic power is

FIGURE 3.76 Example 1 - John Deere "AutoPowr" for 7020 tractor series [4].

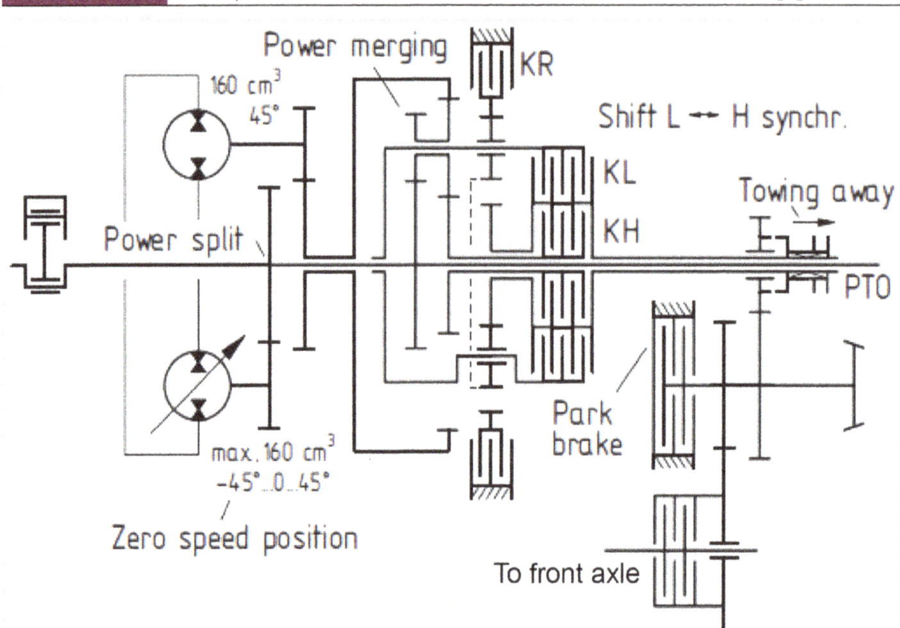

circulating back via the HST. At the lock-up of the ring gear, in about the middle of the range with the PU close to the neutral position and MU close to zero speed, the efficiency reaches its maximum value. When the PU is stroked into positive direction the ring gear speed has the same direction as the engine speed, circulation of power stops and the HST contributes to the output together with the mechanical power path. At the end of the low range, all planetary members rotate at the same speed and a synchronous mode change becomes possible by closing clutch KH and opening clutch KL. The direction of torque on the MU turns around and the high-pressure side in HST changes which again results in power circulation back through the HST. The PU needs to compensate its displacement slightly to keep the speed by making up for the changed loss behavior in the HST. In the high range, the PU starts to stroke towards negative displacement again, the output speed at sun 2 gear is increased and above the lock-up, the hydrostatic power flow adds to the output power. The reverse range is activated by the brake KR and works with the output shaft of the carrier (C) and works in the same way as the other ranges.

The transmission control is described in the paper of Mistry and Sparks explaining the hydraulic circuit in detail and electrical circuit in concept [36].

For the analysis of the transmission with the ladder diagram, first, the compound planetary gear set as the central element for merging of the power flows needs to be interpreted as two 3-member planetary gears. Figure 3.77 introduces the first planetary gear as the combination of sun 1 (A1) gear as the input from the engine, ring (B) gear as the input from the hydrostatic power branch and carrier (C) which is the output shaft for the low and reverse ranges. The second planetary connects sun 2 (A2) gear which is the output shaft for the high range to the ring (B) gear and the carrier (C). For the analysis in the ladder diagram, the standard ratios are assumed to be $i_{01} = -1$ and $i_{02} = -4$.

FIGURE 3.77 Example 1 - Interpretation of 4-member compound planetary as two 3-member planetary gears [11].

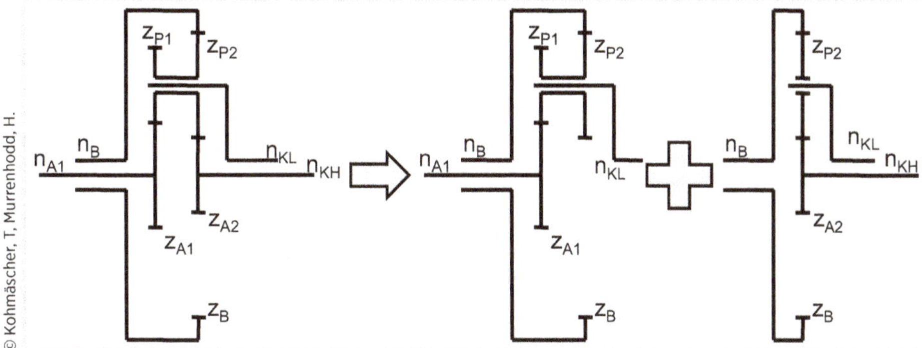

Figure 3.78 introduces the ladder diagram for the HMT in the first example. The first planetary is represented by the three lines on the left that are labeled "B," "C," and "A1" while the fourth line to the far-right side that is labeled "A2" was added for the second planetary.

In a first step, the engine speed level is highlighted on the A1-line with a black dot which will be the center for the pivoting for the analysis. In a second step, three key conditions for the HST are added: (1) PU at –100% displacement and MU at 100% which results in the stand-still condition for low range (see crossing with a line for the carrier) and maximum travel speed in the higher range, (2) PU at 0% displacement and MU at 100% which blocks the ring gear for purely mechanical power transmission, and (3) PU at 100% displacement and MU at 100% to reach the synchronous speed for all planetary

FIGURE 3.78 Example 1 - ladder diagram.

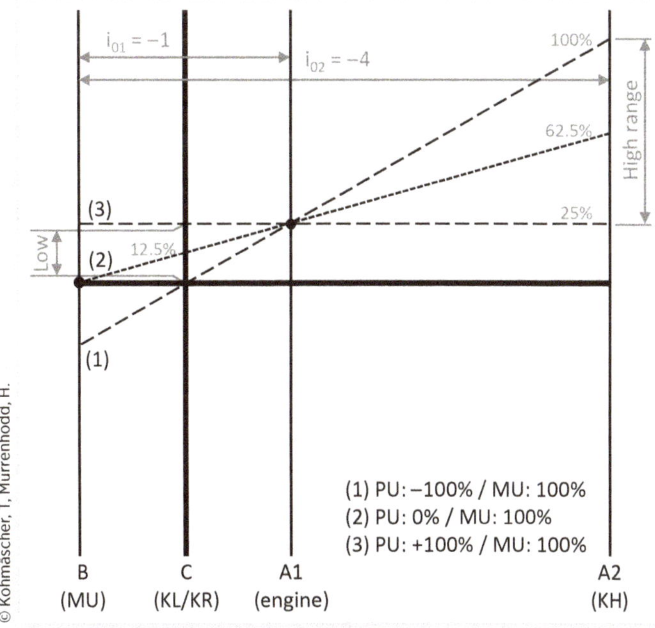

FIGURE 3.79 Example 1 - HMT efficiency and speed of planetary members vs. output speed [11].

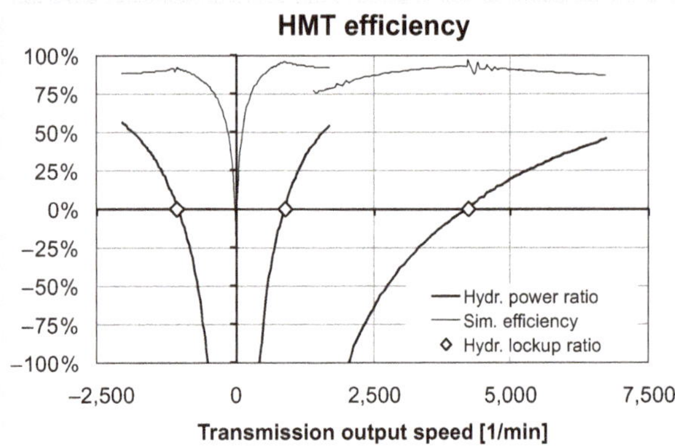

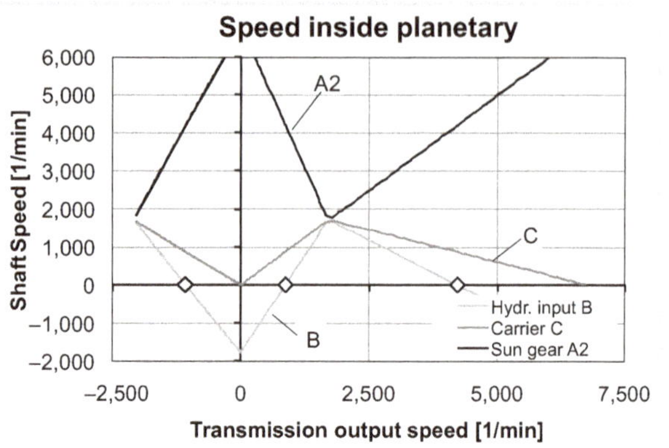

member and simplify the range shift. In a third step, the utilized speed ranges for a low and high range marked in the diagram.

Figure 3.79 introduces simulated system efficiency, hydrostatic power ratio for the HMT and the speeds of all planetary members assuming 1,800 rpm fix engine speed on sun 1 (A1) gear.

The curves in the left diagram present the efficiency and the hydrostatic power ratio with the three diamonds on the power ratio marking the lock-up points for the ring (B) gear (two in forward and one in reverse). The peak efficiency values can be identified at the lock-up points while the efficiency drops towards higher motor speeds (see B in the right diagram). Circulating power values are displayed as negative in the diagram and quickly rise beyond −100% which strongly influences efficiency. For the reverse operation, the additional planetary gear is added to the power flow and thus the efficiency is reduced compared to the "low" forward mode. The curves in the right diagram introduce the speed relation in the planetary gear set by displaying ring (B) gear for the hydrostatic input, carrier (C), and sun 2 (A2) gear.

3.3.4.5.2 Example 2 (OC-HMT). The second example is an OC-HMT with a single HMT range but two mechanical speed ranges down-stream of the HMT section of the transmission – the Fendt "Vario" for the Favorite 926 which was introduced in 1995 (Figure 3.80). It became the worldwide first in series produced power-split transmission for standard tractors in 1996 [4].

The mechanic input power from the engine is split by the planetary on the input side of the transmission and merged on the shaft that holds the motor unit and the mechanical gear connection to the sun gear of the planetary. The power characteristics follow the introduced behavior for the pure OC-HMT – hydrostatic power portion is 100% in the starting point for both driving directions. In forward operation, the hydrostatic power is reduced to zero at the top speed of the HMT range and it increases in reverse operation towards 150% when the machine reaches 50% of the top speed in forward operation. The HST in the HMT applies 45° axial piston units in bent axis design – a single pump unit that is connected to the ring (B) gear of the planetary and two motor units that share the shaft for the merging of mechanical and hydrostatic power flow.

FIGURE 3.80 Example 2 - Fendt "Vario" for Favorit 926 tractor series [4].

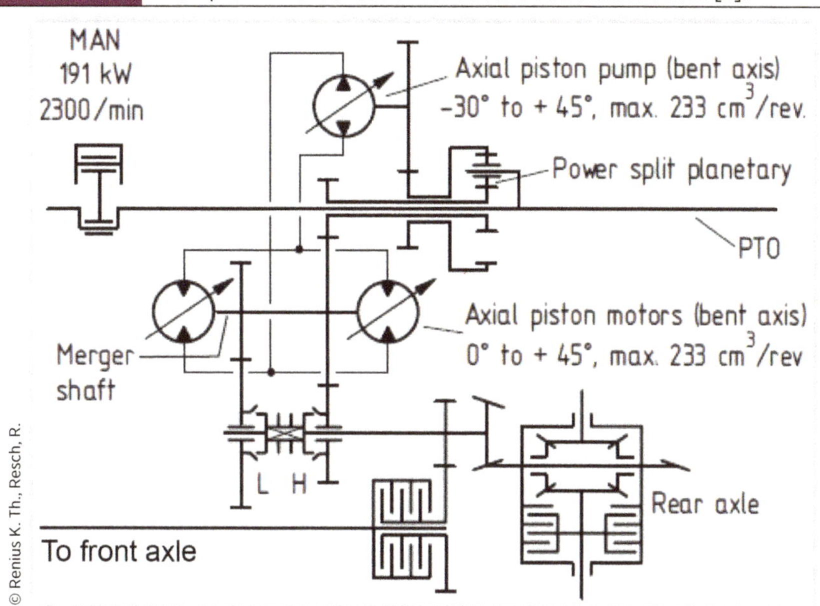

The HMT is combined with two conventional ranges – range "L" covers speeds up to 32 km/h (20 mph), range "H" up to 50 km/h (31 mph). Depending on the selected engine speed, these top speeds are near the lock-up point for 100% mechanical power transmission. The speed potential beyond the lock-up point is not used in the "Vario" implementation to prevent circulating power in the forward range. The "Vario" solution started with a shift-on-stop only range selection but the option for shifting while driving below 20 km/h (12 mph) was introduced in 2003. This shift needs to be supported by an adjustment of the ratio in the HMT since it can be executed at any time and is not happening at synchronous speeds of planetary members.

The Fendt "Vario" concept applies a simple mechanical structure but requires large hydrostatic units that provide high torque and rotational speed capability at outstanding efficiency values. In 1996 this kind of hydrostatic unit was not available in the market and therefore Fendt started to develop its own 45° variable bent axis units with support from Sauer-Danfoss (today: Danfoss Power Solutions). Due to the high-efficiency values of 95%–96%, it was even possible to reverse in the OC-HMT mode and accepting the circulating power flow through the HST. To compensate for the higher noise levels of the large hydrostatic units, Fendt had to isolate the HMT inside the transmission housing by elastic suspension as common for engines.

A detailed analysis employing a leader diagram is not required since the specific conditions for the implemented solution follow the characteristics that were introduced for the basic concept before.

3.3.4.5.3 Example 3 (CC-HMT): The third example is a combination of output coupled HMT with compound HMT modes – the John Deere "AutoPwr" for the 8030 tractor series which was introduced in 2005 (see Figure 3.81). As mentioned earlier, the compound HMT is well suited for high efficiency, but it requires an OC-HMT for starting

FIGURE 3.81 Example 3 - John Deere "AutoPowr" for 8030 tractor series [37].

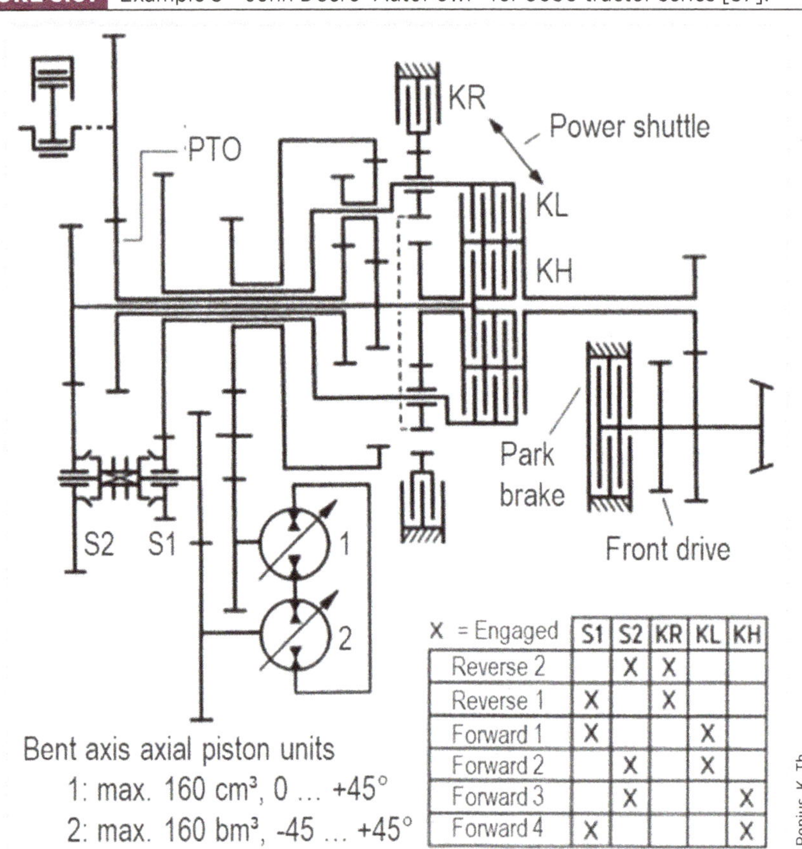

X = Engaged	S1	S2	KR	KL	KH
Reverse 2		X	X		
Reverse 1	X		X		
Forward 1	X			X	
Forward 2		X		X	
Forward 3		X			X
Forward 4	X				X

from a standstill. The transmission works with four HMT ranges that are shifted at synchronous speed conditions while implementing a reverser which is already known from the 7000 series [37]. The 4-members of the planetary gear set are sun 1 (A1) gear which is powered by the engine, sun 2 (A2) gear which can be selected as the output shaft, ring (B) which is directly connected to the hydrostatic motor unit (MU), and the carrier (C) which also can be selected as the output shaft. In the bottom right corner, Figure 3.81 presents the shifting logic for the four forward and two reverse HMT modes. The transmission utilizes three multi-disc clutches (KR, KL, and KH) and synchronized dog clutches (S1 and S2). As before, the disc clutches select the planetary member for the transmission output while the synchronizer clutches alternate the connection of the pump unit (PU) to the planetary gear set – either sun 2 (A2) gear or the carrier (C). The hydrostatic motor unit (MU) stays always connected to the ring (B) gear. Both hydrostatic units in the HST are variable down to ZERO displacement with the pump unit (PU) stroking in both – positive and negative – directions. In the first and third forward modes, as well as the first reverse mode, the hydrostatic pump unit (PU) is connected to the transmission output shaft (S1 → KL; S2 → KH; S1 → KR). Therefore, those modes are considered output coupled HMT modes. In the other modes, none of the hydrostatic units is connected to the input to the output shaft of the transmission. Those modes are considered compound HMT modes.

The interpretation of the 4-member compound planetary gear set in the center of this transmission as two 3s planetary gears is the same for example 1 – the ratios are assumed to be different. For the analysis in the ladder diagram, the standard ratios are assumed to be $i_{01} = -1.2$ and $i_{02} = -3.6$.

Figure 3.82 introduces the ladder diagram for the HMT in the third example. The first planetary is represented by the three lines on the left that are labeled "B," "C," and "A1" while the fourth line to the far-right side that is labeled "A2" was added for the second planetary.

In a first step, the engine speed level is highlighted on the A1-line with a black dot which will be the center for the pivoting for the analysis. In a second step, three key conditions for the HST are added: (1) PU connected to the carrier and at –100% displacement and MU at 0% which results in the stand-still condition for FWD1 and top travel speed in FWD4, (2) PU connected to the carrier or sun 2 gear and at 0% displacement and MU at 100% which blocks the ring gear for purely mechanical power transmission, and (3) PU connected to sun 2 gear and at 100% displacement and MU at 0% which results in lock-up point for sun 2 (A2) gear which is never utilized in the control concept of the transmission. In a third step, a special condition for the planetary gear set was added—all members turning at the same speed so that the planetary rotates as a block—as

FIGURE 3.82 Example 3 - ladder diagram.

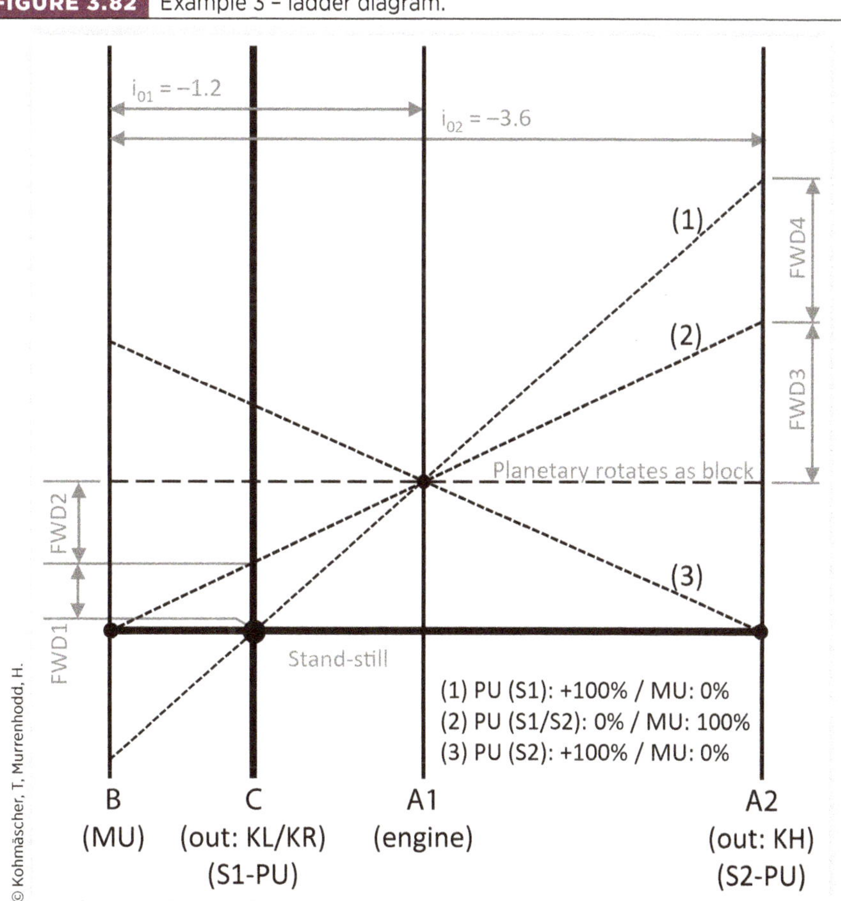

FIGURE 3.83 Example 3 - HMT efficiency and speed of planetary members vs. output speed (VW: forward; RW: reverse) [11].

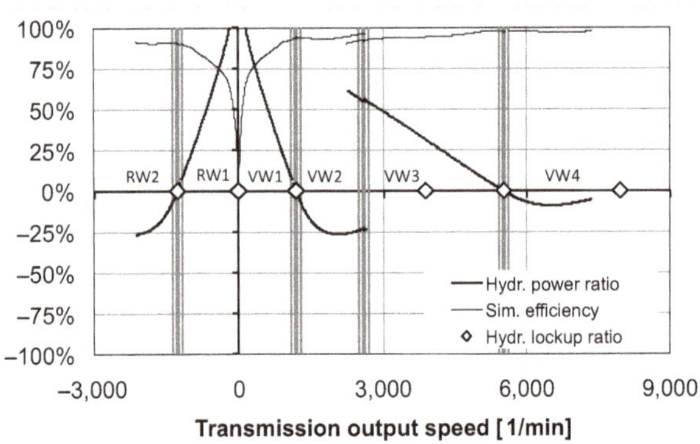

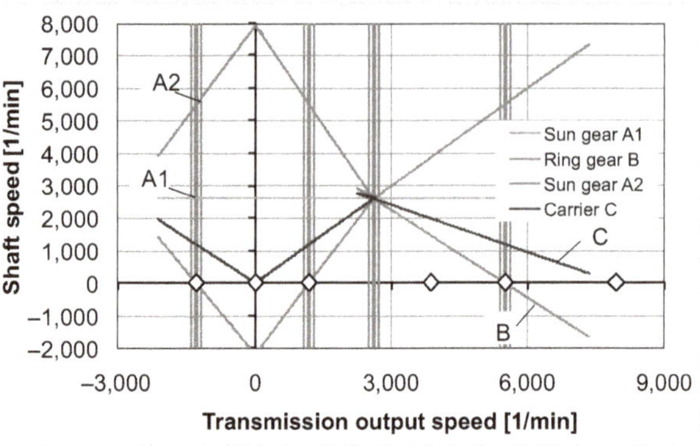

an alternative condition for a synchronous shift. In step four the utilized forward speed ranges FWD1, FWD2, FWD3, and FWD4 are highlighted in the diagram.

The curves in the left diagram of Figure 3.83 present efficiency and the hydrostatic power ratio. The diamonds mark the multiple possible hydraulic lock-up points of the HMT. The hydrostatic power ratio indicates the output coupled (straight lines) and the compound (curved) HMT modes. In this case, the hydrostatic power in the compound modes circulates back and reduces the maximum efficiency. Not all the mechanically possible hydraulic lockup points are utilized. The vertical bars indicate the three mode shifts in forward and the single-mode shift in reverse operation. The additional planetary gear, which is required for reverse operation, reduces the efficiency in reverse compared to the first two modes in forward operation.

The curves in the right diagram of Figure 3.83 introduce the speed variation of the first sun gear A1 (engine), ring gear B (MU), second sun gear A2 (S2 and KH shaft), and carrier C (S1, KL, and KR shaft). Mode shift from FWD2 to FWD3 takes place when all planetary members turn at the same speed – therefore no speed adjustment is required.

3.4 Function Drives (By Gerhard Geerling and Edwin Heemskerk)

Mobile hydraulic applications [38], like the implement function of a mobile machine, differ from other hydraulic applications, such as industrial hydraulics [39, 40] and aerospace [41], since the pressure and flow demand varies greatly over time and between different functions. Unlike other hydraulic applications, several functions are often supplied by one single pump and are operated in parallel. This means that the total installed power on the actuator side is generally considerably higher than the available pump power which is normally limited by the prime mover's power. This is possible because the actuators seldom require their maximum power at the same time and if they do the driver, as part of the control loop, can handle this lack of power or oil flow. The need for compact and cost-effective machines leads in many cases to a one pump hydraulic solution.

The requirements for the working hydraulics in mobile machinery is manifold so that different hydraulic systems have been developed for different applications.

These solutions have different characteristics that, depending on the machine, can be an asset or a drawback. Typically, the known solutions for working hydraulics differ in the following properties:

- Energy efficiency
- Load dependency
- Behavior during flow or power shortage
- System stiffness, damping
- Accuracy
- Complexity
- Expandability

In this chapter, the most common working hydraulics solutions for function drives in mobile machinery are described and the above characteristics are discussed based on typically machine applications the solutions are common in.

A typical mobile machine driveline setup with open-loop implement hydraulics is shown in Figure 3.84.

Such a system usually consists of:

- The pilot actuation (PA), in form of lever, hydraulic pilot control units, joysticks, foot pedals and switches. These can be mechanically connected to the main control valve or by hydraulic or electric lines. The hydraulic and electro-hydraulic pilot actuation uses oil from the pilot oil supply.
- The pump (P) that is driven by an energy source as a diesel engine (on the main shaft or the power take-off) or an electric motor.

FIGURE 3.84 Typical mobile machine system setup, i.e., of a wheeled excavator.

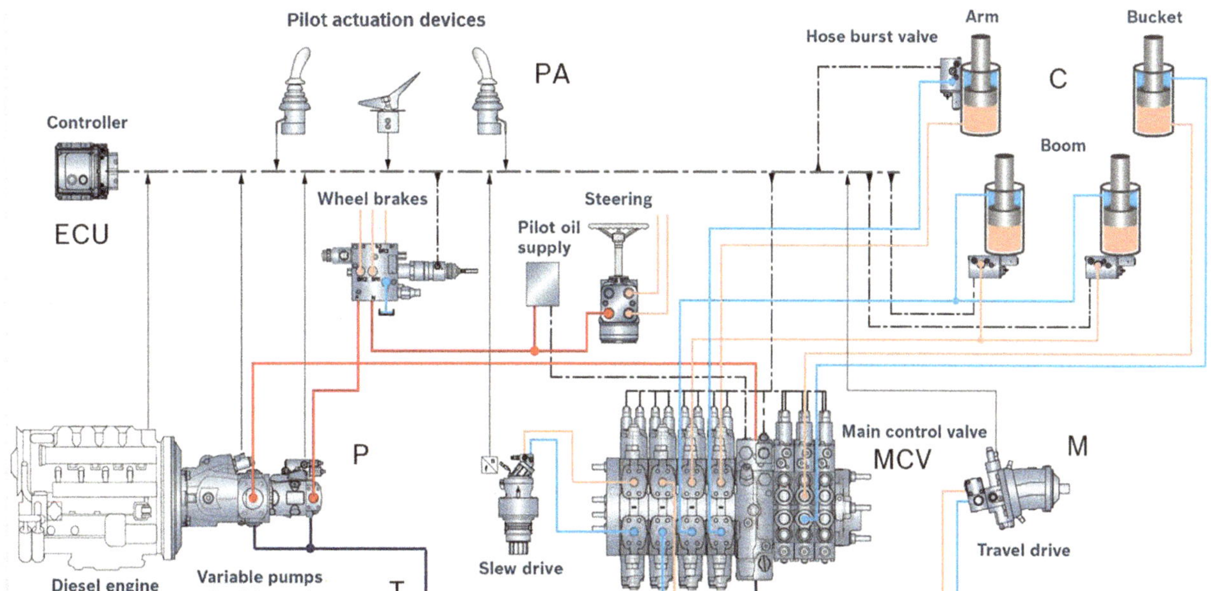

- The main control valve (MCV) that distributes the pump oil to the different consumers based on the driver request.
- The hydraulic actuators, cylinders (C) and motors (M) that transforms the hydraulic energy into the movement of the machine or attachment.
- The electronic control unit (ECU), also called the controller that runs the system software and drives the coils of the electro-hydraulic valves.
- An oil reservoir or tank (T) that contains the oil that is currently not in the system. Typically, the oil is filtered before entering or after leaving the oil tank. The oil cooling is often interlinked with the tank. Filtering and cooling is not part of this book.

Special safety valves, as the hydraulic wheel brakes or the hose burst valves that are only used in some machines and are not part of this book.

Hydraulic assisted and active steering systems are described in Section 2.4.

3.4.1 Basic Pilot Actuation (PA) Principles

The pilot actuation (PA, see Figure 3.84) has the task to transform the driver's wish into control of the hydraulic system. This pilot actuation is typically done with foot pedals, levers, knobs, and switches. Knobs and switches are normally used for machine settings (i.e., to set the engine speed) and to control functions of lower importance (i.e., to lower or raise the stabilizer on a wheeled excavator). Levers and foot pedals are used to proportionally and steady control of a machine function. The latter is usually done in one of the following three ways.

3.4.1.1 Mechanical Pilot Actuation: Compact machines and machines with functions that are operated from outside the cabin (i.e., on truck-mounted cranes) sometimes use mechanical actuation. Figure 3.85 shows a mechanical operated 4/3 way valve.

Due to the mechanical connection between the lever (3) and the spool (2) the operating directly controls the position of the spool. The springs (5) in the cap (4) enables a controllable proportional pilot actuation of the spool and ensures the return of the spool to its neutral position as soon as the lever is not held. Depending on the machine and valve design the lever either actuates one (left example in Figure 3.85) or two (right example) functions. Mechanical levers are very robust, cost-effective and allow a compact

FIGURE 3.85 Mechanical operated valve and 2-axis lever [42, 43].

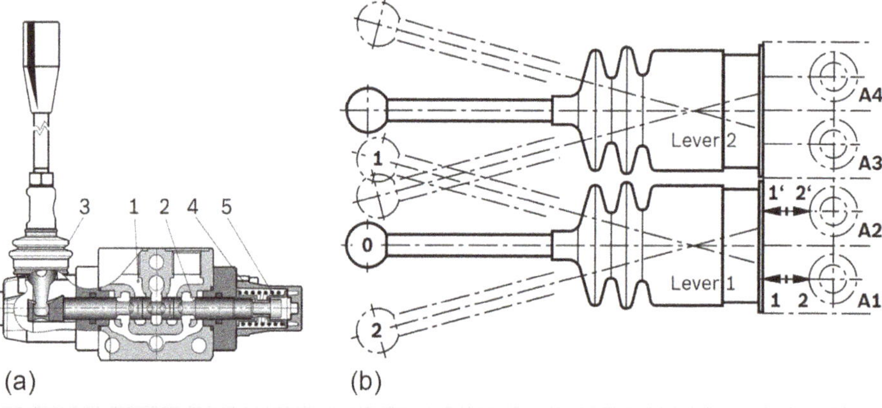

machine design, but it requires the manual pilot actuation to be directly on the control block or connected by a system of rods, cables and/or levers.

3.4.1.2 **Hydraulic Pilot Actuation:** For bigger machines, a mechanical lever leads to a complex mechanical design if the operator wants to control the machine from inside a cabin. For these machines, hydraulic pilot control units are a common solution as an input control element. Hydraulic pilot control units are mechanical operated pressure reducing valves that reduce a supply pressure (typically from a low-pressure source) to a pressure that is proportional to the lever angle. This pressure, usually called pilot pressure (p_{St}), is then connected with a hydraulic hose to the corresponding spool cap of the main control valve and used to displace a spool inside the main control valve.

Figure 3.86 shows a cross-section of a hydraulic pilot control unit. When not actuated, the control lever is held in zero position by the four return springs (8). The control ports (1, 2, 3, 4) are connected to the tank port T via the drilling (11). With deflection of the control lever (5) the plunger (9) pushes against the return spring (8) and the control spring (7). The control spring (7) firstly moves the control spool (6) downwards and closes the connection between the appropriate port and tank port T. At the same time the appropriate port is connected to the port P via the drilling (11). The control phase begins as soon as the control spool (6) has found its balance between the force of the control spring (7) and the force which results from the hydraulic pressure in the appropriate port (ports 1, 2, 3, or 4) [44].

FIGURE 3.86 Hydraulic pilot control unit, cross-section and hydraulic symbol [44].

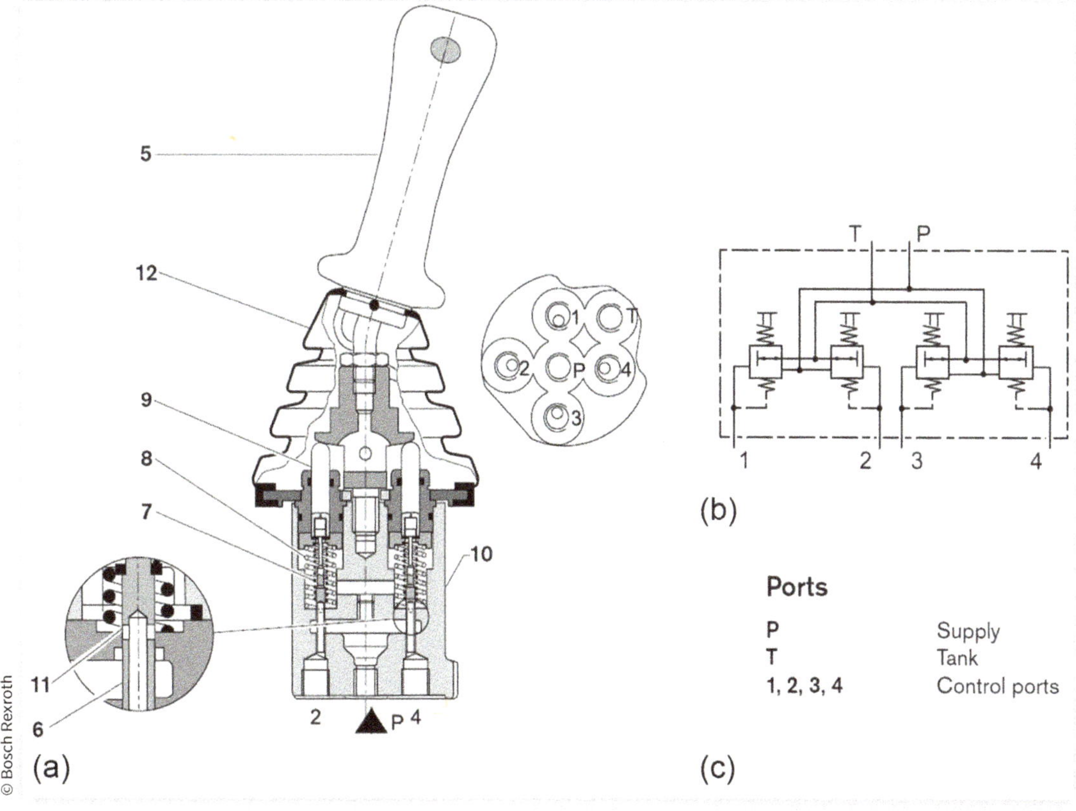

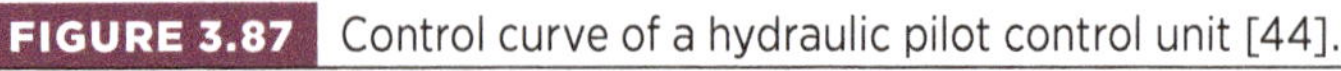

FIGURE 3.87 Control curve of a hydraulic pilot control unit [44].

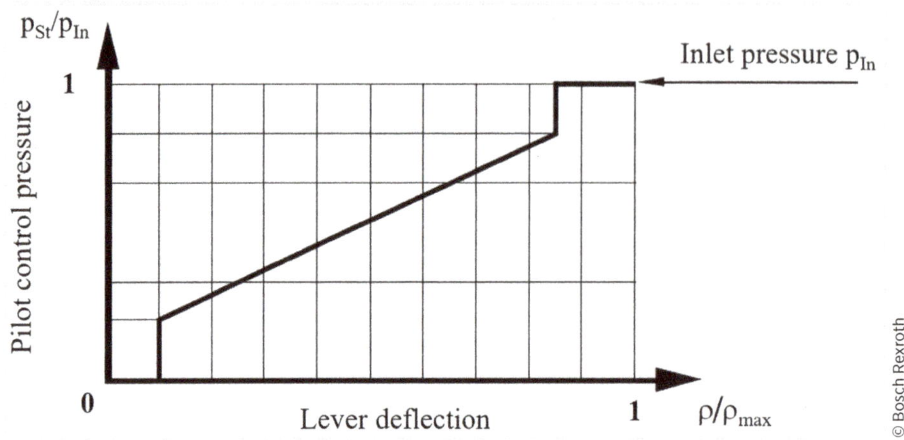

Through the interaction of control spool (6) and control spring (7) the pressure in the appropriate ports is proportional to the stroke of the plunger (9) and thus the position of the control lever (5). Figure 3.87 shows a typical control pressure curve with the lever deflection. In many cases, mobile machines are operated in a rough terrain that can lead to unwanted lever deflection while driving. To prevent this, the pilot control unit has a dead band of about 2°-3°. After passing this angle the control pressure increases proportionally with the lever angle. Just before the end of the lever angle, the maximum proportional value is achieved. If the lever is pushed further, the supply pressure is not reduced anymore and is forwarded to the control port. This is done, to ensure that a spool inside the main control valve stays at a full stroke and is not affected by flow forces caused by the hydraulic flow over the spool.

3.4.1.3 **Electro-Hydraulic Pilot Actuation:** A growing number of mobile machines are equipped with electronic remote controls (also called joysticks and electronic foot pedals). The reason for this differs from machine to machine. Typical reasons are the size of the machine (i.e., mining excavators), use of a radio remote control (i.e., knuckle boom crane) or automation (i.e., underground drilling). Figure 3.88 shows an electronic joystick and foot pedal. The design and its behavior are similar to a hydraulic pilot control unit, with the difference that it has an electrical output and not a hydraulic. When not actuated the operating element is held in a neutral position by the return springs (8). With deflection of the lever/pedal (1), plunger (5) pushes against return spring (8). The magnet (7) mechanically linked to the plunger (5) moves upward or downward while following the operating element actuation direction. The command value generated by the sensor (4) is proportional to the deflection of the lever/pedal.

There are different kinds of electronic joysticks and pedals. They vary in the sensor type (physical sensing principle) and the output signal. The most common outputs of electronic joysticks are current I [A], voltage V [V], pulse-width modulation (PWM) and controller area network bus (CAN bus). A more detailed explanation can be found in Chapter 4. This electronic signal is either directly connected to an electro-hydraulic pressure reducing valve or a magnetic coil to control a spool in the main control valve or first send to an electronic control unit (ECU) as shown in Figure 3.89.

FIGURE 3.88 Electronic joystick and foot pedal [45].

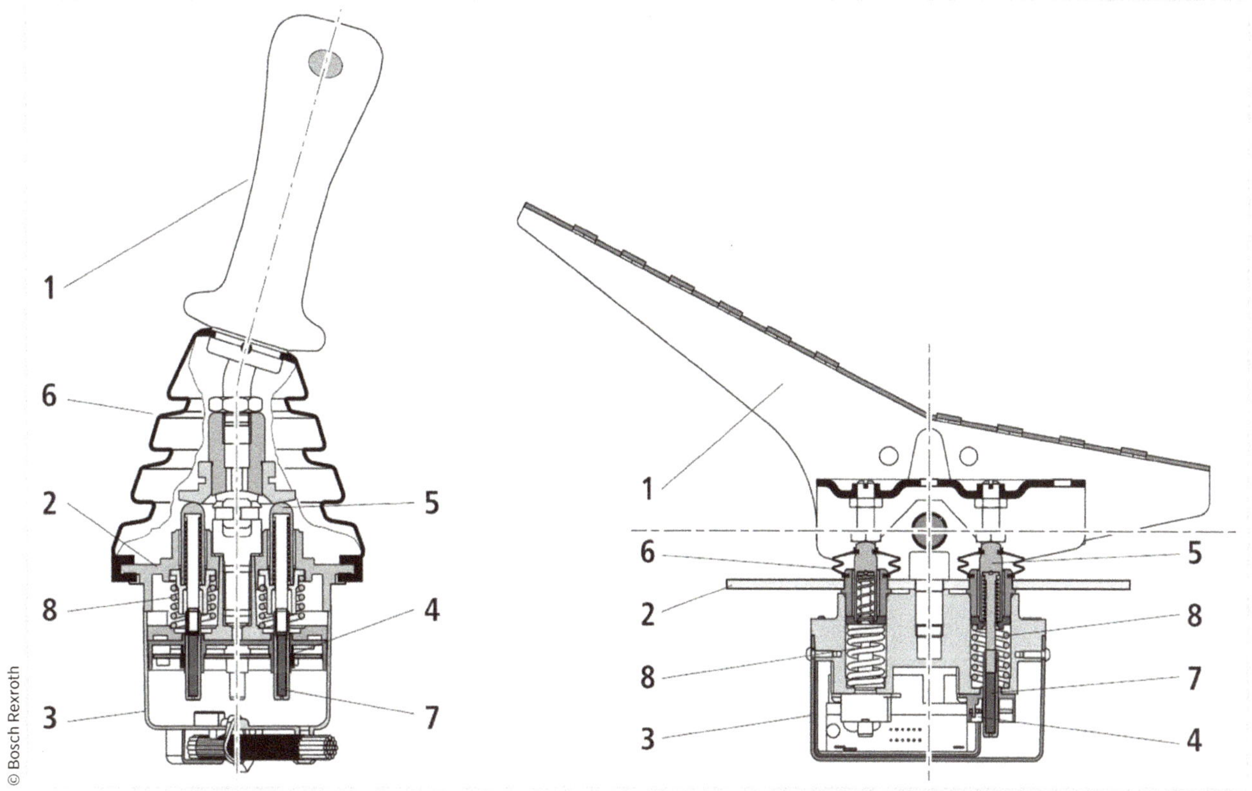

FIGURE 3.89 Electro hydraulic spool pilot actuation.

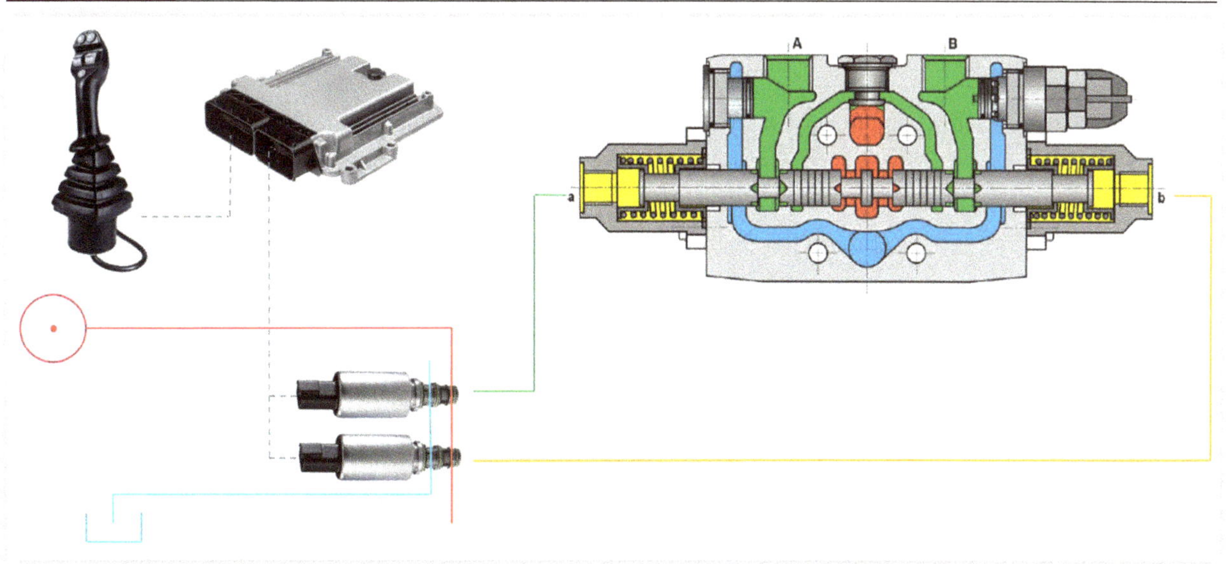

3.4.2 Basic Pump (P) Control Principles

The pump has the task to supply hydraulic oil flow or pressure to the system. Typically driven by a prime mover it translates rotational energy in oil flow (see Chapters 1 and 3). There are different kind of pump designs and they can be classified in different ways.

- **Axial piston vs. radial piston vs. gear vs. vane vs. screw (Figure 3.90)**

  The different design principles of a pump have an influence on the efficiency, dynamics, pressure level, and cost but are not so important to the principle system behavior and are not part of this chapter/book. For a detailed explanation of that, please refer to books on the fundamentals of hydraulic fluid power, e.g., Bosch [39], Murrenhoff [21], and Ivantysynova [20].

- **Constant vs. variable**

  The flow of a pump with a constant displacement volume (Figure 3.91, left) only depends on its rotational speed. The flow (Q_P) of a variable pump (Figure 3.91, right) depends on its rotational speed (n) and its variable displacement volume (V_g) which corresponds to the swashplate position respectively angle (α), see Equations (3.26) and (3.28).

  Variable pumps offer more flexibility but are generally more expensive.

- **Pressure vs. flow source**

  If a pump works as a flow source, it will supply a defined oil flow (Q_P) independent of the pump pressure (p_P). A pump working as a pressure source will supply oil flow at a given pressure (p_P) independent of the amount of oil flow (Q_P) that it is supplying. This behavior is accomplished by the pump regulator. Typically, a pump works either as a flow or as a pressure source, but there are solutions available where the pump can do both, depending on the need of the machine. Another common pump regulator for implement hydraulics is the horsepower P_P controller, which prevents the overloading of the prime mover (i.e., diesel engine), see Equation (3.23).

3.4.2.1 **Flow Control:** A constant pump normally works as a flow source. A variable displacement pump can also operate as a variable flow source with the help of a flow controller. Figure 3.92 shows an axial piston pump with a regulator with a pilot-pressure-related control that adjusts the displacement (between the minimal displacement $V_{g\,min}$ and the maximum $V_{g\,max}$) of the pump in proportion to the pilot pressure (p_{St}) applied at port Y. The oil is pumped from a reservoir, connected to the suction port

FIGURE 3.90 Pump design principles [39].

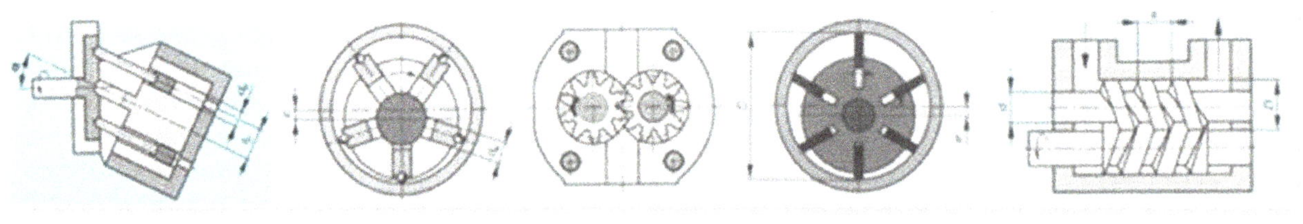

FIGURE 3.91 Hydraulic scheme of a constant (left) and a variable pump (right).

FIGURE 3.92 Variable displacement axial piston pump with a flow controller and its static system behavior [46].

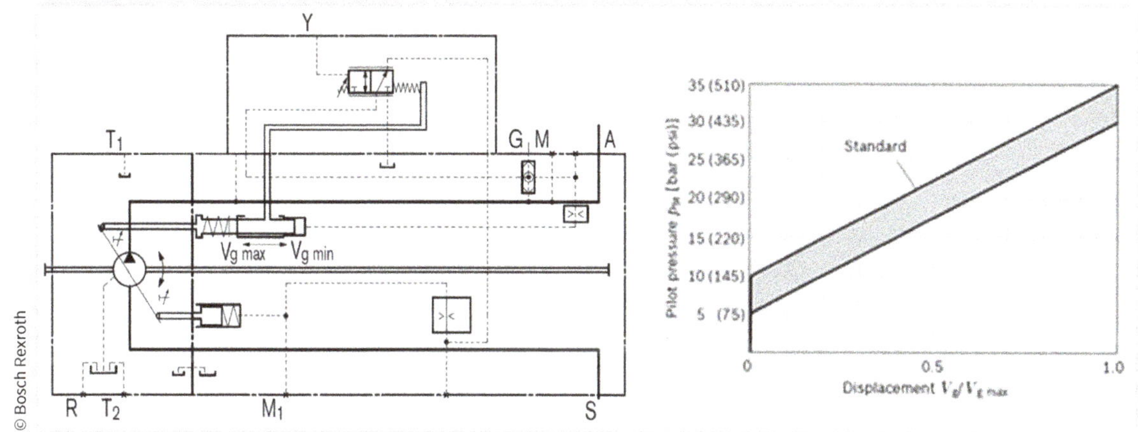

(S), to the pump outlet, A. Basic mechanical depressurized position without a pilot signal is the minimum displacement V_{gmin}. With increasing pilot pressure, the pump swivels to a higher displacement. The required control pressure is taken either from the load pressure or from the externally applied pilot control pressure at the G port. To ensure the control even at low operating pressure <30 bar (435 psi) the port G must be supplied with an external pilot control pressure of approx. 30 bar (435 psi). This is a typical controller used, i.e., in the positive control (see Figure 3.114). M and M1 represent measurement ports that are used for pump or machine initiation and advanced solutions. The internal leakage leaves the pump through the drain ports R, T_1, and T_2.

3.4.2.2 **Pressure Control:** Figure 3.93 shows a mechanical set pressure controlled variable axial piston pump, as used in constant pressure systems. The oil is pumped from a reservoir, connected to the suction port (S), to the pump outlet, A. The pressure control keeps the pressure in a hydraulic system constant within its control range even under varying flow conditions. The variable pump only moves as much hydraulic fluid as is required by the actuators. If the operating pressure exceeds the setpoint set at the integral pressure control valve, the pump displacement is automatically swiveled back (in the direction of V_{gmin}) until the pressure deviation is corrected. M and M1 represent measurement ports that are used for pump or machine initiation and advanced solutions. The internal leakage leaves the pump through the drain ports R, T_1, and T_2.

FIGURE 3.93 Variable displacement axial piston pump with a pressure controller and its static system behavior [46].

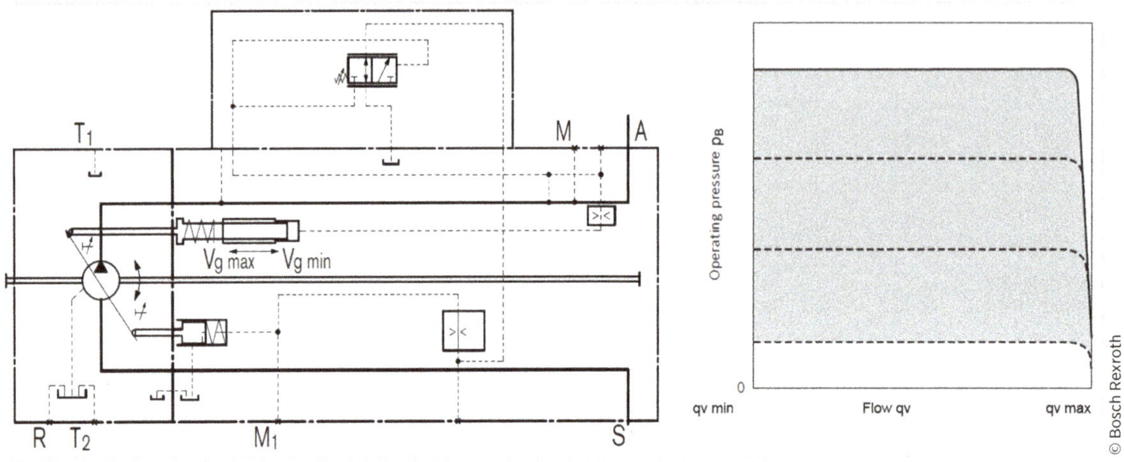

3.4.2.3 **Complex Controller:** Most modern hydraulic systems in mobile machines have more than one controller. Figure 3.94 shows a variable displacement axial piston pump with power control (LR), pressure cutoff (PCO) and load sensing (LS). This is a typical pump you would find in a multi-functional working machine using classic load sensing (LS).

The pump has three controllers that interact in different situations. The main controller is the load sensing controller (LS). This is a pressure controller with the specialty that it tries to set the pump pressure to the external pressure at the port X plus a given value defined by a spring in the LS regulator. In a hydraulic load sensing system (see Figure 3.119) the load pressure p_L of the actuator is connected to the X port.

FIGURE 3.94 Variable displacement axial piston pump with power control (LR), pressure cut-off (PCO) and load sensing (LS) and its static system behavior [46].

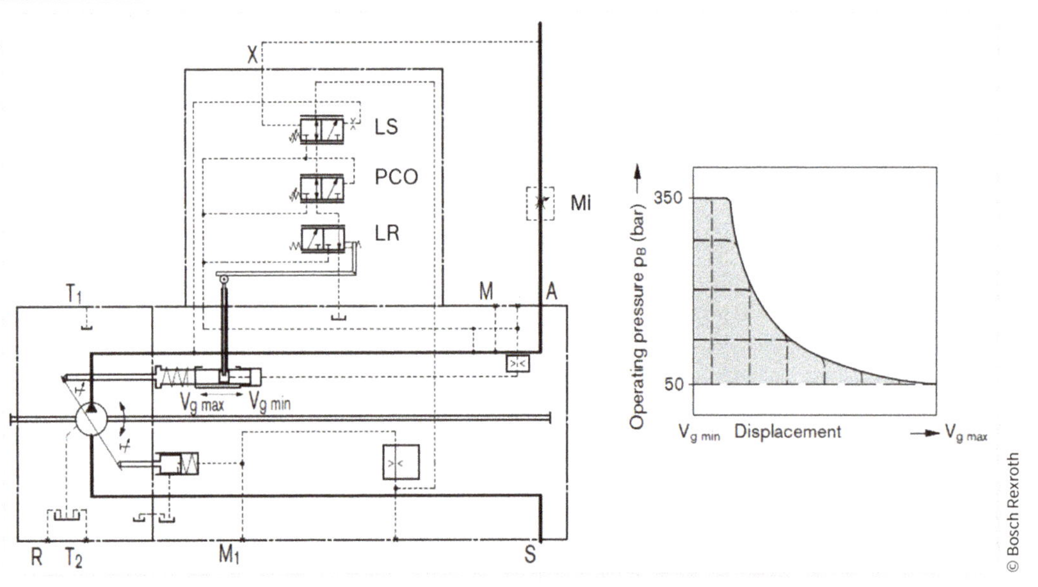

Mi represents the meter-in orifice of the main control valve. The LS regulator ensures that the pressure difference over the meter-in orifice Mi Δp_{Mi} is constant and approximately equal to the spring value in the regulator.

$$\Delta p_{Mi} = p_P - p_L \tag{3.44}$$

The opening of the meter-in orifice determines the flow to the actuator. As Δp_{Mi} is constant the flow to the actuator is proportional to the opening and independent of the load. If the differential pressure Δp_{Mi} increases the pump is swiveled back (towards V_{gmin}), and, if the differential pressure Δp_{Mi} decreases, the pump is swiveled out (towards V_{gmax}) until the pressure drop across the sensing orifice in the valve is restored. Additionally, the pump has a pressure cut-off controller (PCO). This controller limits the maximum pressure p_P the pump can supply. This controller is inactive most of the time and only intervenes when the pump pressure is too high to prevent damage to the hydraulic system or the mobile machine itself. The third regulator in Figure 3.94 is a horsepower regulator (LR). The power control (LR) regulates the displacement of the pump depending on the operating pressure so that a given power is not exceeded for a given pump speed. The precise control with a hyperbolic control characteristic provides optimum utilization of available power. The operating pressure acts on a rocker via a measuring piston and an externally adjustable spring force (defining the power setting) counteracts this. If the operating pressure exceeds the set spring force, the control valve is actuated by the mechanical rocker, the pump swivels back (in direction of V_{gmin}). The lever length at the rocker is shortened and the operating pressure can increase at the same rate as the displacement decreases without the power being exceeded. M and M1 represent measurement ports that are used for pump or machine initiation and advanced solutions. The internal leakage leaves the pump through the drain ports R, T_1, and T_2.

3.4.3 Basic Valve Principles

In mobile applications, the distribution of the pump oil flow to the different actuators (i.e., cylinder and motors) is mostly done by a central main control valve (MCV) as shown in Figure 3.84. The main control valves that are used in mobile applications today can be divided into roughly two categories, *open* and *closed center* valves. **Figure 3.95** shows an example of an open center (OC) valve. The characteristic of an OC valve is that if no function is activated the oil from the pump flows through the main control valve (through the bypass canal (CB)) to tank with relatively low resistance as shown in Figure 3.95 [47].

Figure 3.96 shows a proportional 6-3 way valve and its typical nomenclature. The valve has six ports, pump (P), tank (T), the cylinder/motor ports A and B (with A typically connected to the rod side of the cylinder) and the two ports of the center bypass channel (CB). If the main control valve has more than one valve section, the cylinder/motor ports get an indices i (i.e., A_1 or B_3).

The opening area between two ports are named after their flow direction, i.e., the opening area in the 6/3-way valve between the pump and the port A_1 is called A_{PA1}. The pilot actuation of the valve is named after the actuator port that is connected to the

FIGURE 3.95 Open center valve with six valve sections in monoblock design [47].

© Bosch Rexroth

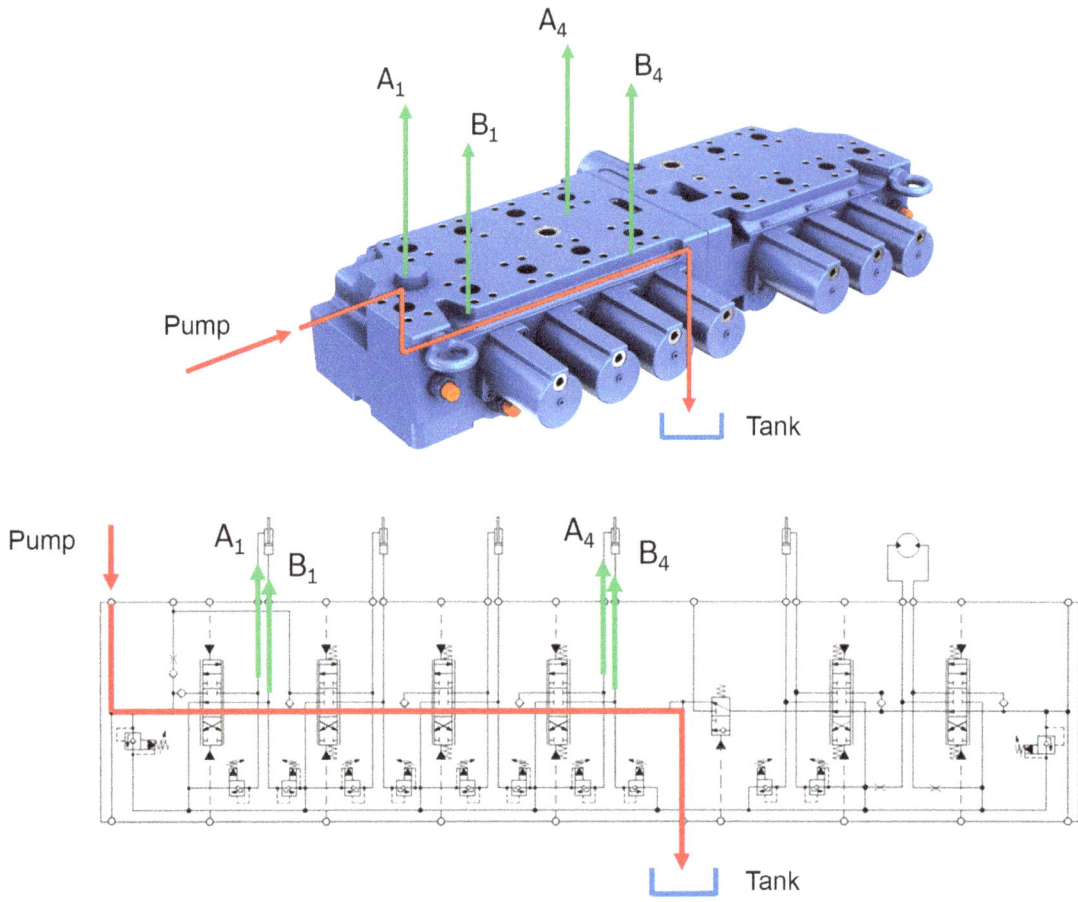

FIGURE 3.96 Nomenclature for a proportional 6/3 way spool.

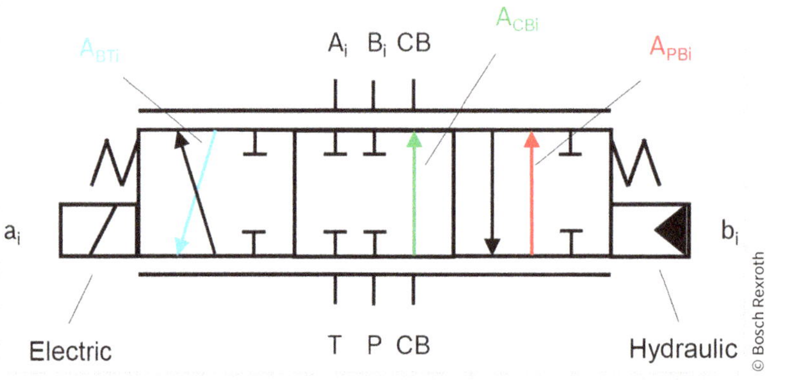

FIGURE 3.97 Closed center valve with nine valve sections in a sandwich design [48].

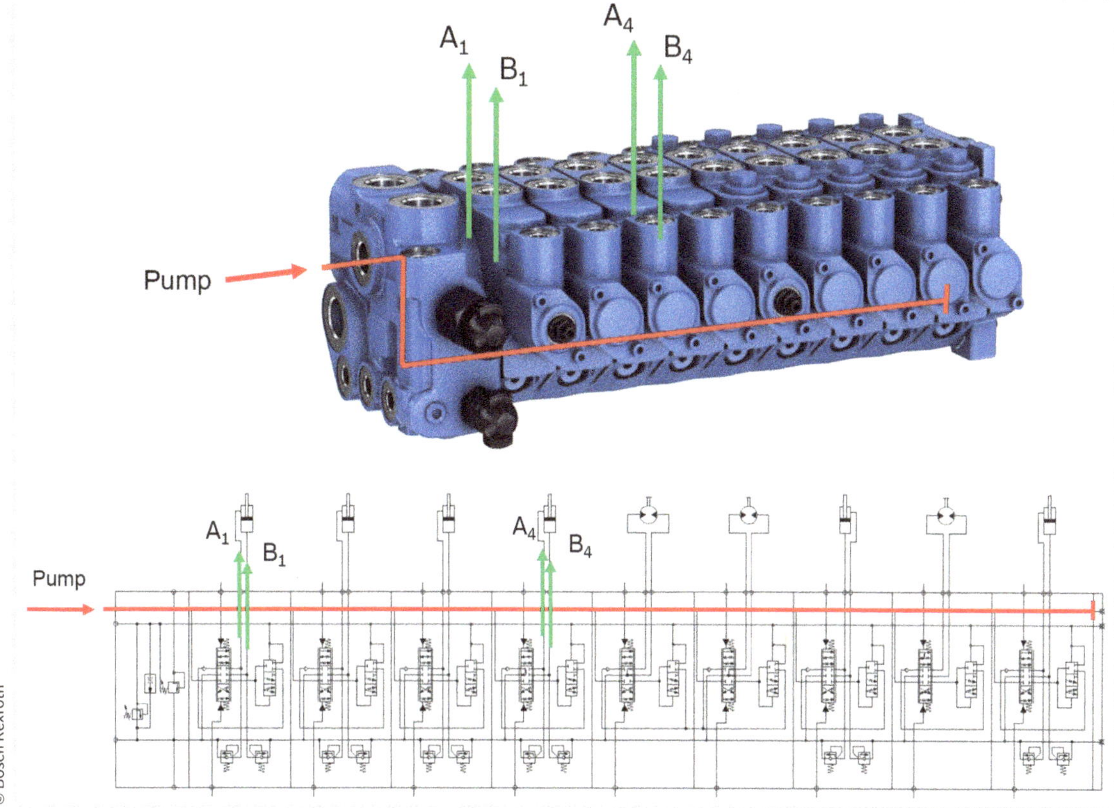

pump when activated. So, when b_2 is activated, valve 2 is opening a connection from the pump to the port B_2.

On the contrary, the closed center valve has no connection from the pump to the tank in a neutral position. As Figure 3.97 shows, the pump channel in a closed center valve is a blind alley.

To control two or more actuators with only one pump, special attention is needed to allow parallel movement. In case two actuators are only connected by 4/3-way directional valves to the pump supply (see Figure 3.98) and are opened in the same way (meter-in opening areas $A_{PA1} = A_{PA2}$) the actuator with the lowest load will receive all the oil flow for as long as the pressure drop Δp_{PAi} over the opening area A_{PAi} is lower than the load pressure difference.

$$Q_{PAi} = \propto_D \cdot A_{PAi} \cdot \sqrt{\frac{2 \cdot \Delta p_{PAi}}{\rho}} \quad \text{with} \quad \Delta p_{PAi} = p_P - p_{Ai} \tag{3.45}$$

To allow for a parallel movement, the operator needs to reduce the valve opening of the actuator with the lower load pressure until the pressure drop over the meter-in orifice compensates the load pressure difference.

Although the control quality requirement in parallel operation is very different depending on the machine, this system behavior is in most cases neither ideal nor

FIGURE 3.98 Simple open-loop hydraulic circuit with two 4/3 proportional valves [49].

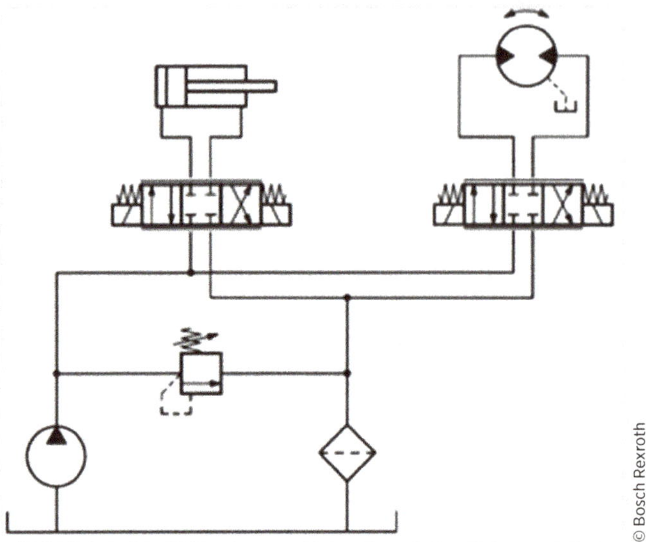

acceptable. Especially when the load pressures change quickly due to external forces (i.e., when hitting a rock in the soil while digging) or when a cylinder hits the end stop it is very difficult to control a parallel movement appropriately.

To better control the parallel movement in a system where more than one actuator is supplied by a pump at the same time, the following three basic principles are commonly used.

- Control of the parallel movement by valve sequence
- Control of the parallel movement with the help of pressure compensators
- Control of the parallel movement with the help of sensors and electronics

Some market solutions also use a combination of those principles.

3.4.3.1 **Control of the Parallel Movement by Valve Sequence:** One way to influence the flow behavior in case more than one actuator is supplied by one pump is the way the different valve axis is connected to the pump line. Depending on what sequence and in what form the valves are connected to the pump line, the behavior of the parallel movement is defined. Figure 3.99 shows the three typical ways that two open center valve axis can be connected to the pump line.

With a parallel connection (Figure 3.99, left) both actuators are connected to the pump line directly and in case of equal meter-in orifice opening areas ($A_{PA1} = A_{PA2}$) all the flow oil goes to the actuator with the lowest load.

On versions with a tandem connection, the pump flow for the second axis first goes over the first axis (Figure 3.99, middle). So the pump flow to the second axis is restricted by the center bypass opening area (A_{CB1}) if the first axis is activated. When the first axis is fully actuated, only the first actuator will receive oil from the pump. This setup is

FIGURE 3.99 Parallel, tandem and serial connection.

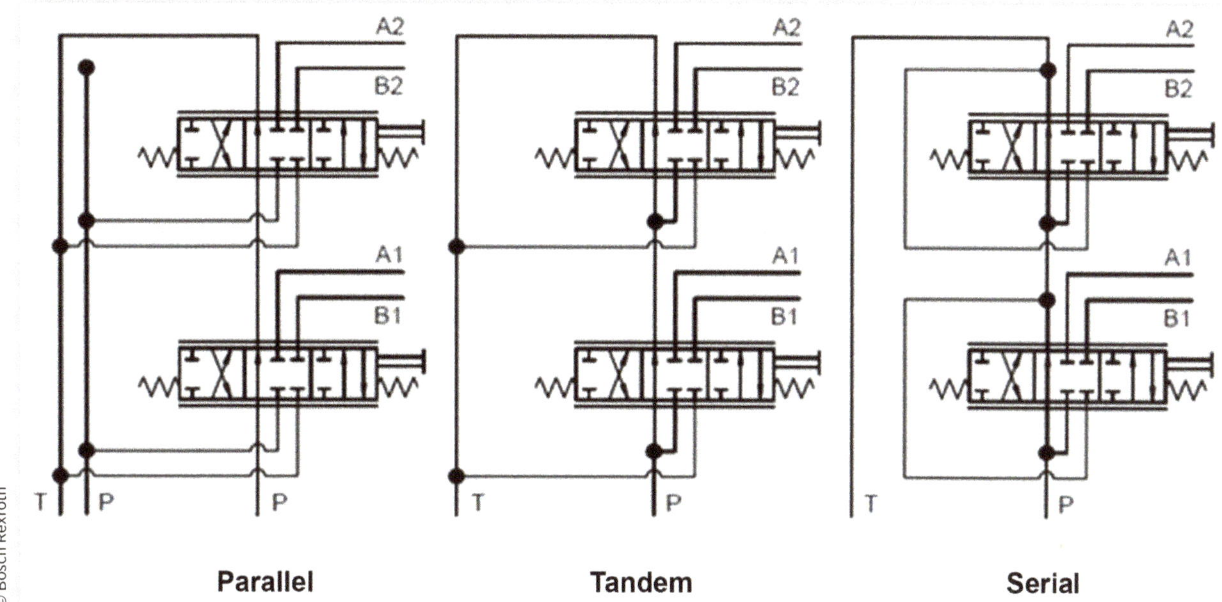

commonly used when certain machine functions need to have a priority concerning another function (i.e., the swing priority vs. the arm movement in crawler excavators).

With a serial connection (Figure 3.99, right) the second valve also receives no pump oil flow when the first valve is fully activated, but the return flow from the first valve can be used by the second valve. Today this is the least common connection between the three.

3.4.3.2 **Control of the Parallel Movement with Pressure Compensators:** A second common way to control the parallel movement of multiple machine functions connected to one pump is with the use of pressure compensators. These are additional self-regulating hydraulic elements restricting the flow from the pump to the cylinder/motor under certain conditions with the aim that all valve axis have the same pressure drop over their meter-in orifice A_{PA}. Two different principles of compensators are common today, the (traditional) load sensing compensator (LS, see Figures 3.100–3.102) and the flow sharing compensator (see Figures 3.103 and 3.104). Often these compensators are named individual pressure compensators to emphasize that the compensator is affecting only one axis and is not an inlet pressure compensators (also named un-loading or by-pass compensator) that is used in combination with load sensing systems to control the pump flow for the whole main control valve.

In a traditional (individual) LS pressure compensator the goal is to keep the pressure difference Δp_{PA} ($\Delta p_{PA} = p_{P'} - p_{LS}$) over the meter-in orifice A_{PA} of a valve axis constant. The pressure compensator is typically situated upstream between the pump and the meter-in orifice (see Figure 3.100) and comprises a spool (1) that is loaded on the side (area $A_{p'}$) with the pressure $p_{P'}$ in front of the valve opening and on the other side (area A_{LS}) with the load sensing pressure p_{LS} (with p_{LS} being the load pressure p_A in a quasi-static view) and a control spring (2). The pressure compensator is opened by the

FIGURE 3.100 Traditional (individual) upstream load sensing (LS) pressure compensator.

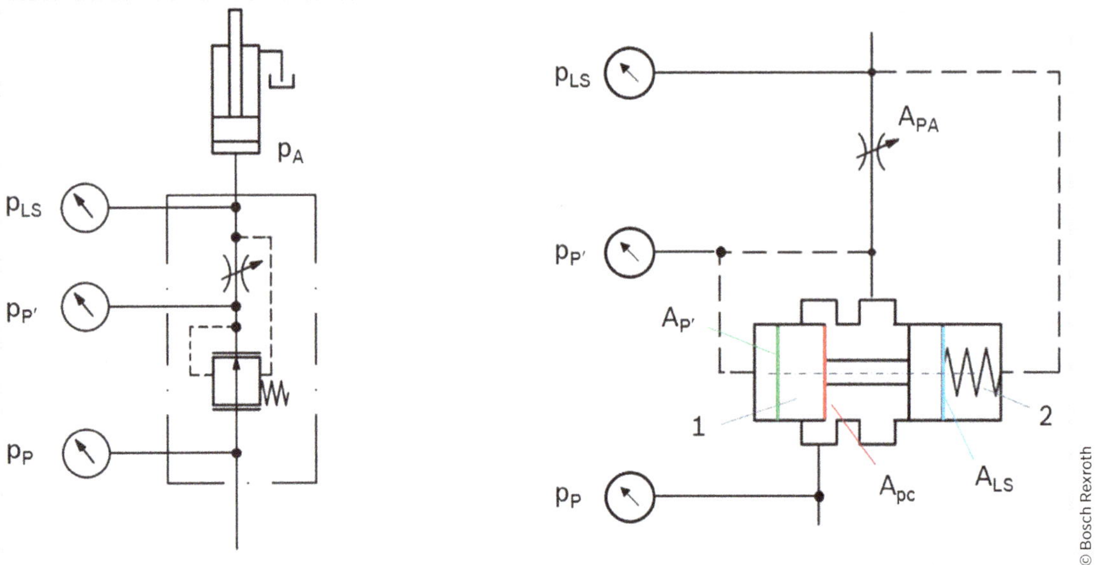

FIGURE 3.101 Traditional (individual) downstream load sensing (LS) pressure compensator.

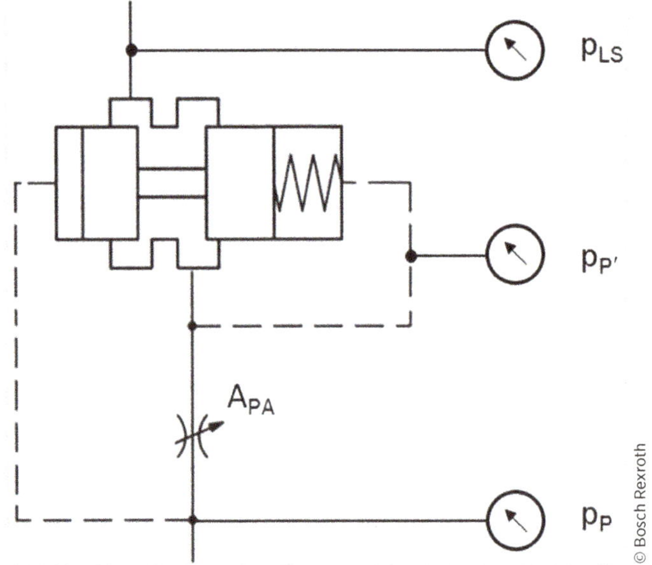

load pressure p_{LS} and the control spring (2). The pressure in front of the valve opening $p_{P'}$ closes the pressure compensator.

Figure 3.101 also shows a traditional LS pressure compensator, but in this version, the pressure compensator is behind the meter-in orifice (also called a downstream compensator in comparison to an upstream compensator). In both versions, if the pump pressure is lower than the load pressure, the pressure compensator spool will open as

FIGURE 3.102 Traditional (individual) upstream load sensing (LS) pressure compensator with load holding.

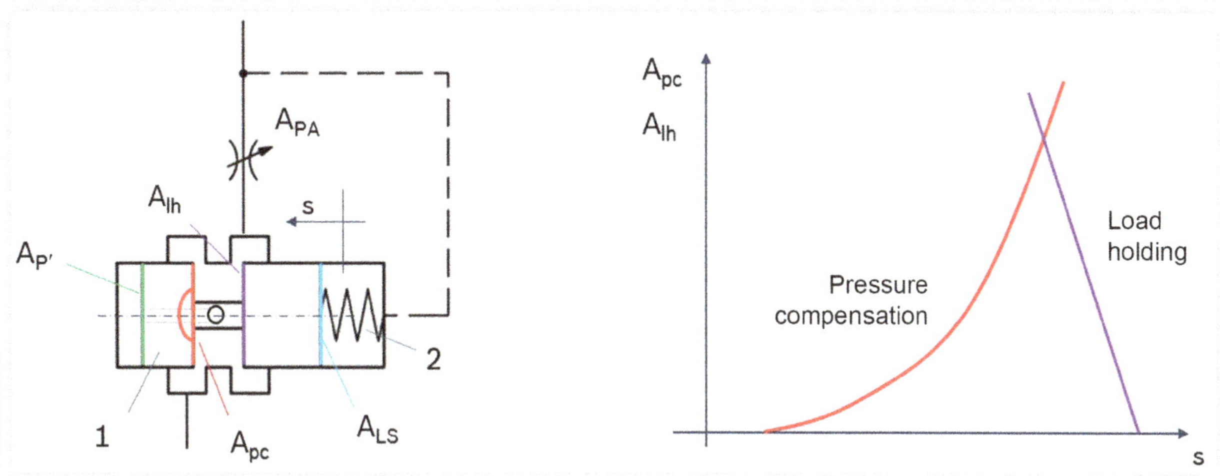

FIGURE 3.103 Downstream flow sharing LS pressure compensator.

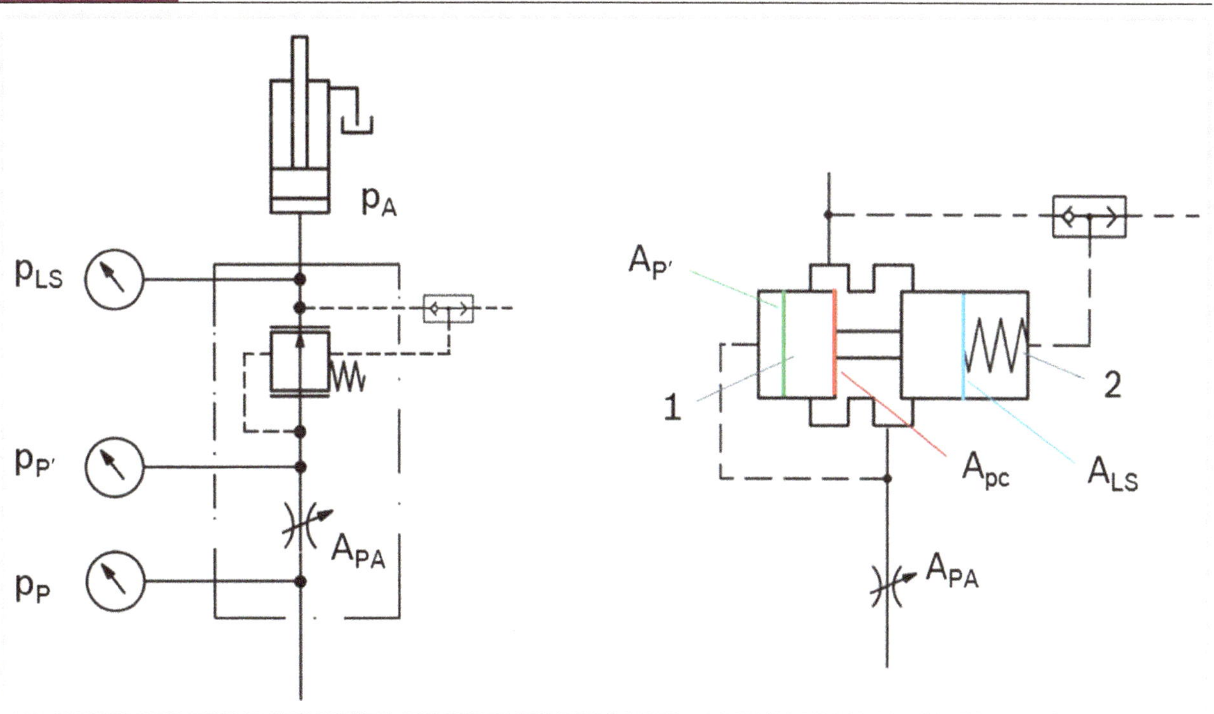

much as possible. In this case, the oil would also flow from the load to the pump. This can be prevented with a check valve before or after the meter-in orifice.

Figure 3.102 shows a pressure compensator that has the load holding function integrated. The difference between these versions is the additional control edge A_{lh} and the way the pressure $p_{P'}$ is applied to the pressure compensator (using drilling to connect the $A_{P'}$ side to the spool neck).

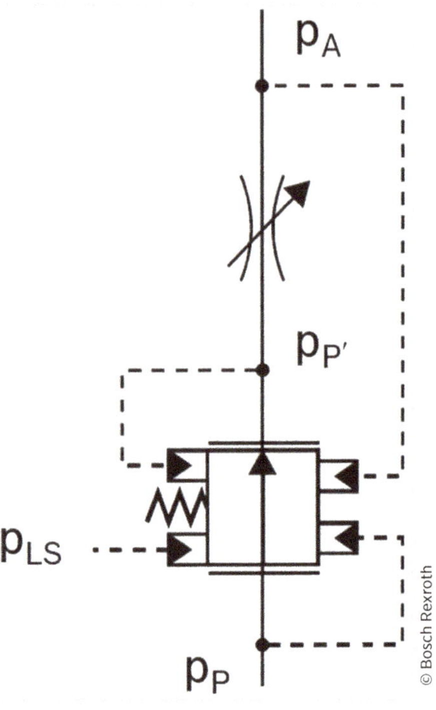

FIGURE 3.104 Upstream flow sharing LS pressure compensator.

Assuming i.e., that the pump pressure p_P is 100 bar (1,450 psi), the control spring (4) corresponds to 10 bar (145 psi) and the load pressure p_{LS} is 20 bar (290 psi) than the pressure compensator will find a spool position s, that will generate a pressure drop of 70 bar (1,015 psi) between the pump and the spot P′ right before the meter in orifice (force equilibrium on the pressure compensator).

$$p_{P'} \cdot A_{P'} = p_{LS} \cdot A_{LS} + F_{Sp} \tag{3.46}$$

The pressure drop the control edge A_{pc} (Figure 3.102 shows a common characteristic versus the spool stroke) of the pressure compensator needs to generate depends on the pressure difference between the pump and the load. The position of pressure compensator spool depends on this pressure drop and the flow to the actuator Q_{PA} (see Equation 3.45).

As the spring force F_{Sp} changes concerning the spool position, the Δp_{PA} over the meter-in also varies. The gradient of this change depends on the spring stiffness.

As soon as the pump pressure is smaller than the load pressure the compensator spool will fully open and with that close the load holding control edge (A_{lh} in Figure 3.102) and with that the connection from the pump to the load. This load holding function does not use a seat valve but a spool in a housing with minimal clearance. By this it is not completely leak free, in the sense of that there is a small leakage during load hold operation. Still this is for many applications sufficient.

Assuming that more than one function (cylinder or motor) is supplied with the same pump at the same time and that the system is under-saturated (meaning the pump cannot supply all the requested oil for all functions), the supply pressure will drop. As a result, the compensator spool at the highest load will eventually open completely. By this, that function will first lose speed and then even stop completely. Functions operated simultaneously at lower pressure levels will; however, move normally. Thus, this setup will priories the functions with the lower load pressure in case off undersupply.

This behavior is not ideal for all mobile machine types, so that a second pressure compensator design, known as the flow sharing pressure compensator, is also commonly used. This compensator uses the highest load pressure of all active functions to control the flow.

The downstream flow sharing LS compensator also consists of a spool with two control sides (1 in Figure 3.103) and a spring (2). One control side is connected with the highest load pressure p_{LS} and the other side of the pressure compensator is connected to the pressure $p_{P'}$ directly behind the meter-in orifice A_{PA}. In a single operation, the flow sharing LS compensator has hardly any effect on the flow to the cylinder or motor, only generating a small pressure loss (based on the spring rate). The pump solely defines the flow. However, when more than one cylinder or motor is activated the pressure compensators are active. They ensure that all meter-in orifices A_{PAi} have the same pressure difference over them, based on the force equilibrium on the pressure compensator spool, see Equation (3.46). As the highest active load pressure p_{LS} is the same for all pressure compensators, the pressure after the meter-in orifice $p_{P'}$ is also the same for all meter-in orifices. As the pump pressure is also the same for all functions, therefore the flow to all functions is purely dependent on the opening of their meter-in orifice. By this, all functions are given the same priority, which means that in undersaturation all functions will decrease in speed in the same ratio compared to their flow request. This flow sharing functionality can also be achieved by placing the compensator upstream (see Figure 3.104) of the directional valve, but today the downstream version is the most common [50].

3.4.3.3 **Electronically Control of the Parallel Movement:** The third way to control the behavior of hydraulic actuators as cylinders or motors in parallel movement is to adapt the opening of the valves according to the pump and load pressures. For these additional pressure, sensors are used and software defines the opening of the spool continuously. This is typically done with valves that have an independent meter-in and meter-out.

In a system with independent meter-in and meter-out, here shown in Figure 3.105 with four 2/2-way electrohydraulic controlled valves, the valve opening from the pump to the actuator (i.e., A_{PB}) can be constantly adapted concerning the pressure difference $\Delta p = p_P - p_B$. To do this the pump and the load pressures are sensed, i.e., with pressure sensors. If the pump pressure rises, i.e., due to a parallel actuator, the valve opening can be adapted to keep the flow to the actuator constant.

3.4.4 Well-Established Systems for Hydraulic Function Drives

Today's well-established hydraulic systems on the market can be obtained by the combination of the different pump controller and valve concepts described above (see Figure 3.106).

FIGURE 3.105 Independent meter-in and meter-out control of a cylinder.

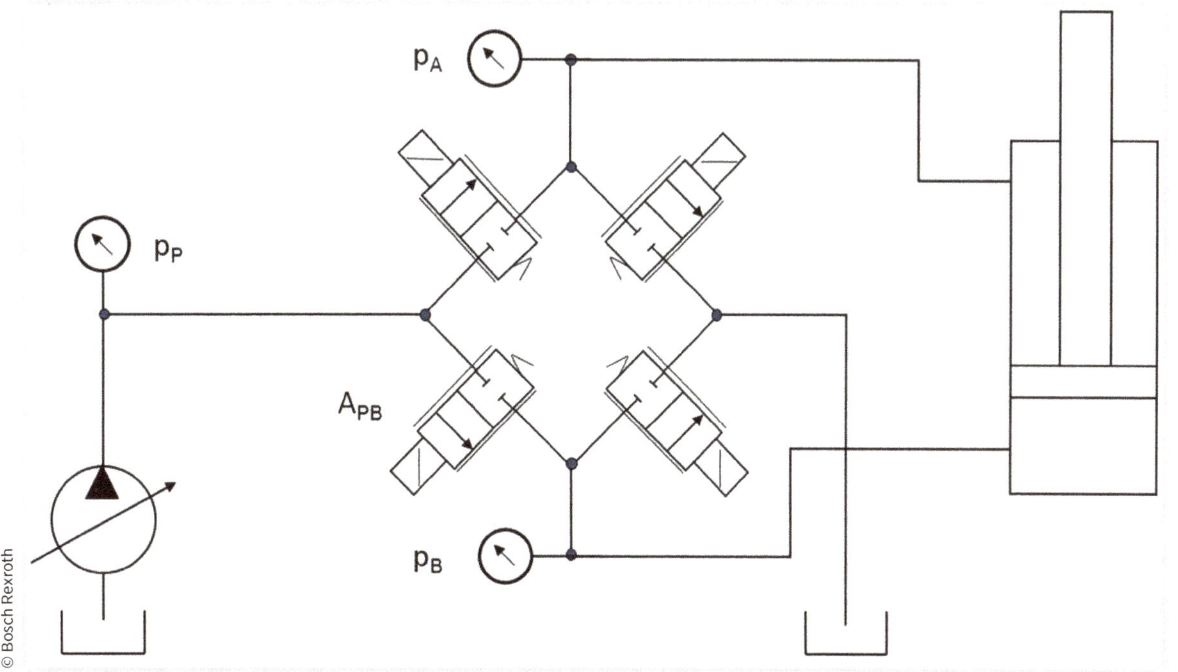

FIGURE 3.106 Common working hydraulics systems.

			valve			
			open center	closed center		
				w/o compensator	LS compensator	flow sharing compensator
pump	constant	pressure source+				
		flow source	throttle control / constant flow		LS with constant pump	
	variable	pressure source – controlled			eLS	
		pressure source – regulated	negative control	VBO / *const. pressure* (**) / *MPN* (***)	classic LS	LUDV
		flow source – controlled	positive control	*pump controlled* (*)	EFM	
		flow source – regulated				

LS

Note that the hydraulics solutions on mobile machinery sold today will in some cases have additional features included in the control systems to compensate certain flaws or to improve the machine's specific behavior under certain conditions.

In the next sub-chapters, the predominant hydraulic systems for function drives are described. In two categories, there are solutions available, but they are not as generic as the rest of the table. These are written in italics in Figure 3.106 and are described shortly below. Besides, descriptions can be found in the literature on hydraulic fundamentals, i.e., Murrenhoff [21] and Ivantysynova [20].

- * In some applications a pure **pump controlled** system is used, where the valve mainly acts as a switch. For example in bigger mining excavators, the hydraulic system has 3–10 pumps. These pumps are switched 100% to a certain function and supply oil only to one actuator at a time. More pump controlled actuators are described in Chapter 5.
- ** In industry [39, 40] and aerospace [41] applications quite a few hydraulic systems work with a **constant pressure supply** in which the movement of the actuators is done with 4/3-way valves that control the meter-out flow often close to the cylinder or motor. This setup is in many cases not very energy efficient but very precise in control. Some mobile applications, which focus on precisions and stiffness in their control, are equipped with such meter-out controlled constant pressure systems.
- *** In the last few years, research has grown on **multi-constant pressure systems or networks** (MPN) for mobile machines intending to significantly increase the energy efficiency of the overall system. The innovative STEAM (Steigerung der Energieeffizienz in der Arbeitshydraulik mobiler Maschinen (Increasing the energy efficiency in function drives of mobile working machines)) concepts is a perfect example in that direction [51, 52, 53] but also digital hydraulic concepts that use fast-switching valves in combination with constant pressure supplies [54, 55] show promising results.
- + A constant flow pump cannot work as a pressure source without additional components. Still, for the sake of the structural overview, it is kept as part of the table in Figure 3.106.

3.4.4.1 **Throttle Control:** Throttle control is the most common hydraulic system for function drives. A fixed displacement pump (P) as a constant flow source is used together with an open-center main control valve (MCV), also called a 6/3 way valve, to supply oil flow to the cylinders or motors, see **Figure 3.107**. **Figure 3.108** shows a cross-section of an open center (OC) valve and its corresponding hydraulic scheme.

When the valve is not activated, the oil from the pump (P) can flow through the central bypass channel (CB) across the valve section, through the filter (F) into the tank (T), keeping losses in standby minimal. When the spool (5) in one of the valve sections (1) of the main control valve (MCV) is moved from its neutral position (increase in pilot actuation (i) and spool stroke (s)), the central bypass notch A_{CB} on the spool begins to close, see **Figure 3.109**. By this, the oil flow from the pump to the tank is throttled and with that, the pump pressure p_P rises according to the orifice formula [21], Equation (3.45).

FIGURE 3.107 Open center (OC) throttle control.

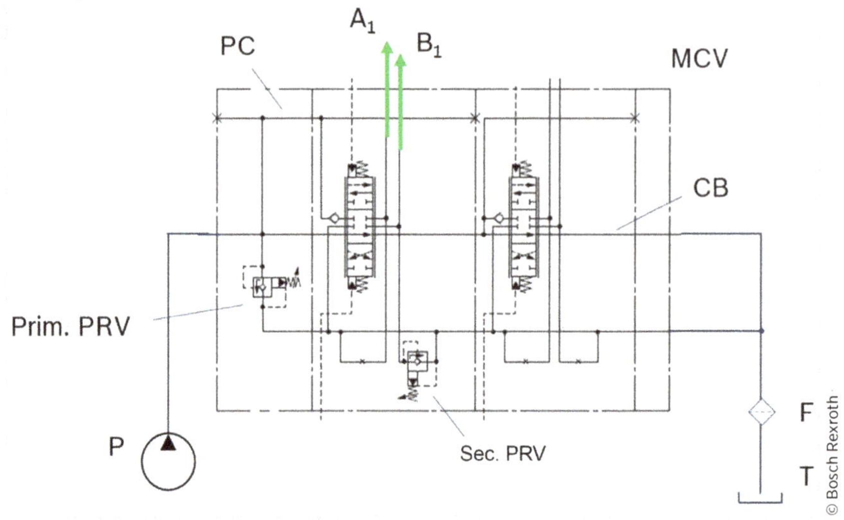

FIGURE 3.108 Hydraulically operated open center (OC) valve [56].

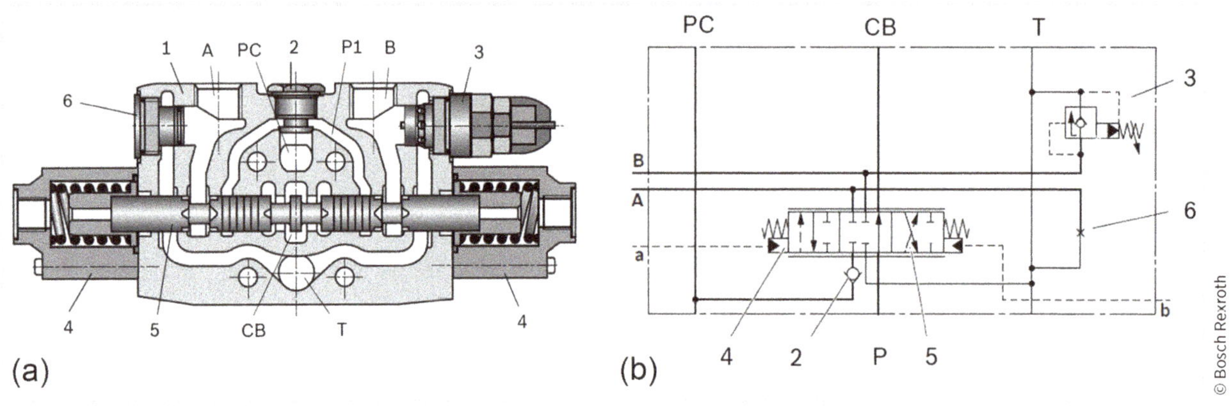

FIGURE 3.109 Example of opening areas vs. spool stroke in an open center (OC) system including flow and pressures.

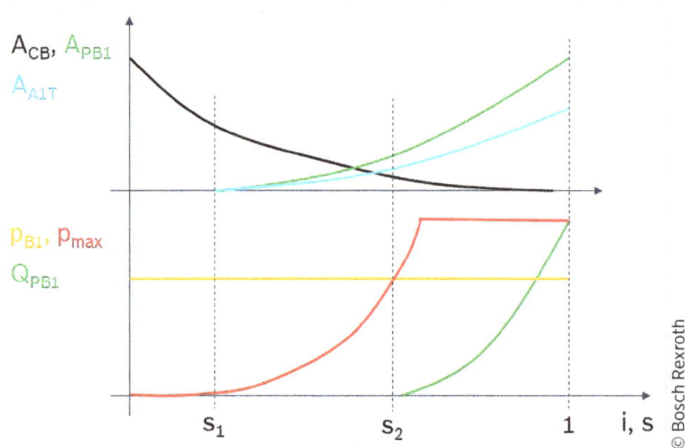

As long as the pump pressure p_P is lower than the load pressure (p_{B1}) the pump pressure rises in the relation of the closing of the center bypass area A_{CB}. This pressure is in front of the actuated valve section and the connected parallel channel (PC). In Figure 3.109 a pressure p_{max} shows the maximum pump pressure concerning the spool displacement in a single operation if no oil flows to the cylinder/motor. The maximum pump pressure is defined by the pressure set in the pressure relief valve in the inlet (prim. PRV). This graph shows the load dependency of open center throttle control. This load dependency and the pump pressure at the opening of the meter-in area A_{PB1} at s_1 is dependent on the machine function it is designed for.

Additionally, there is a check valve (2) between the PC and the P1 channel, also called the load-holding valve. This check valve only opens when the pressure in the PC channel is higher than the pressure in the P1 channel (that is equal to the load pressure). If there would be no load-holding valve, the implement, e.g., a boom cylinder, would lose oil via the center bypass channel to the tank and drop although "Lift" is activated for as long as the load pressure is not reached. As soon as the pump pressure exceeds the load pressure and the meter-in A_{PB1} is opened oil flows to the implement. Still, the oil will flow to the tank through the center bypass (CB) until the spool is fully displaced. The amount of oil flow to the tank is dependent on the spool displacement (defining the meter-in opening and center bypass area) and also on the load pressure. With the design of the control edges of the spool (the different opening areas), the load pressure sensitivity is defined. This load feeling gives the operator a pressure control at the start of the movement, which means that he/she can control the acceleration of the load, giving the system a smooth control with high damping. When fully actuated, the pump flow is completely directed to the implement with only small losses. In a parallel movement, the driver will have to find the right position of the joystick/pedal to get the wanted actuator reaction. An inexperienced operator might experience this pressure sensitivity as an inconsistency and it can then be regarded as a disturbance. A skilled/trained operator can use this information feedback from the system to his advantage and increase the machine's fine controllability. Furthermore, this load behavior reduces shocks to the driver as the oil flow takes the path of least resistance when an obstacle is hit. This load sensitivity comes with two major drawbacks. The open center throttle control system has a bad energy efficiency when used in partial actuation and it is difficult to do precise proportional flow control. The major benefit of the throttle control lies in the simplicity of the system setup, which leads to reliable and robust systems at a low cost.

Based on the above, the throttle control is widely used in mining excavators, wheel loaders, skid steer loaders, mini excavators and backhoe loaders (Figure 3.110).

FIGURE 3.110 Typical machines that use throttle control.

3.4.4.2 **Negative Control (NC):** For machines that are used full time, i.e., on 2–3 shifts in quarries or mining, energy efficiency is very important, as the fuel cost plays an important role in the total cost of ownership (TCO). To improve the energy efficiency compared to an open center throttle control but still keeping its controllability the open center control block can also be operated with a variable displacement pump. One common strategy to control the variable flow of the pump is negative control (NC), also called negative flow control, as shown in Figure 3.111.

In a negative control system (NC) the oil flow from the central bypass (CB) flows through an orifice (O_{NC}) before it goes to the filter (F) and ends in the tank (T), see Figure 3.111. This restriction leads to a backpressure p_{NC} in front of the orifice O_{NC} that is dependent on pump flow to the tank. This pressure p_{NC} is connected to the pump regulator to control the stroke of the pump. What is special about this solution is that an increase in the pressure p_{NC} leads to a de-swashing of the pump, hence the name negative control (see Figure 3.111, right). Typically, the pressure range of the pressure p_{NC} is 0–40 bar (580 psi).

When not activated, all the pump oil flows through the center bypass (CB) of the main control valve (MCV) to the NC orifice (O_{NC}). The backpressure p_{NC} in front of the orifice de-swashes the pump until Q_{min} is reached. This minimal oil flow (Q_{min} > 0 L/min) is used for flushing the system to cool/heat the oil and the hydraulic components (depending on the temperature) and to filter the oil. As soon as a joystick/pedal is activated, a spool (6/3-way, comparable to the spools in throttle control, see Figure 3.109) in the MCV starts to stroke and closes the CB. In consequence, the pump pressure rises, but as long as no oil flows to the cylinder, the pressure p_{NC} and with that the pump swash angle stays identical. Once the pump pressure exceeds the load pressure, the oil will flow to the cylinder and less oil will flow through the CB to the tank. This causes the pressure p_{NC} to drop and with that the pump to swivel out. This also happens when a second machine function is operated

FIGURE 3.111 Negative control (NC) system and pump swivel characteristics.

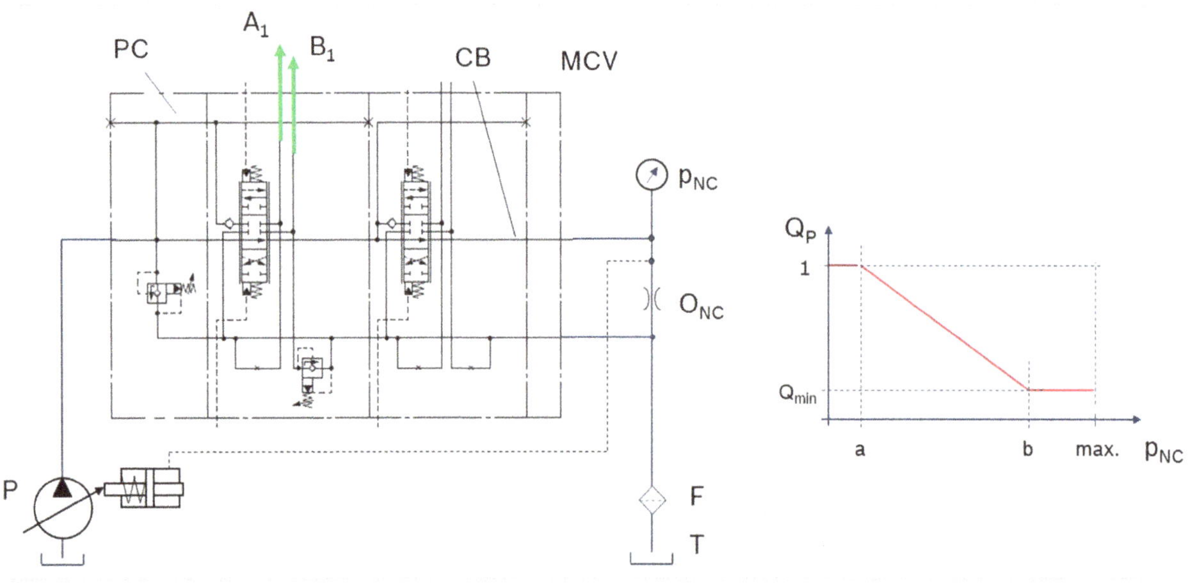

FIGURE 3.112 Common negative control (NC) applications.

simultaneously to a first (i.e., with a pilot actuation of 50%). In parallel operation, the resulting center bypass area A_{CBsum} is dependent on both individual CB areas A_{CBi}.

$$A_{CBsum} = \sqrt{\frac{A_{CB1}^2 \cdot A_{CB2}^2}{\left(A_{CB1}^2 + A_{CB2}^2\right)}} \quad \text{for two center bypass areas} \tag{3.47}$$

For as long as no center bypass is fully closed ($A_{CBi} \neq 0\ mm^2$) and there is oil flowing to at least one actuator (cylinder C or motor M) any additional pilot actuation will lead to a change in NC pressure p_{NC} and with that to a change in pump flow.

The basic principle of the machine control behavior (i.e., the load dependency) is comparable with the open center throttle control (described above), only with the difference that the variable displacement pump leads to significantly better energy efficiency. The priority in the parallel movement is also comparable with open center throttle control and is handled by the sequence of the functions inside the control block.

A negative control is mainly used in mini and crawler excavators, typically using hydraulic pilot control units (see Figure 3.86) for actuation (Figure 3.112).

3.4.4.3 **Positive Control (PC) and Electronic Positive Control (EPC):** A second common approach to improve the energy efficiency of open center throttle-controlled hydraulic systems, while keeping its core controllability features, is to actively set the swivel angle of the variable displacement pump with the pilot actuation (PA) of the driver. This is called positive control (PC) or positive flow control.

In a 1-loop OC, NC, or PC system, typically all actuators (cylinders and motors) are designed that they can take the full pump flow. In a hydraulic positive control system (PC), as shown in Figure 3.113, the highest pilot pressure of the pilot actuation (PA) is forwarded to the pump control. Like in all hydraulic pilot actuated control systems, when a hydraulic pilot control unit of one or more machine functions is activated, the corresponding pilot pressure signals p_{St} (a_1, b_1, etc.) are applied to the valve spools in the main control valve (MCV). In a PC system, additionally alternating check valves (ACV) determine the highest pilot actuation signal and use this signal to actively swivel out the pump.

If the driver commands more than one machine function at the same time, the highest pilot actuation pressure is forwarded to the pump. As a consequence, there is no change in pump flow Q_P if a second function is actuated with a lower pilot actuation pressure as the first. On the right side in Figure 3.113 the spool stroke s_i and the pump flow Q_P are shown with the pilot actuation pressure p_{St}. Just like negative control (NC), also positive control (PC) usually has a minimal pump flow Q_{min}. Part of the reason is identical to NC (cooling, heating, and filtering), but it also improves the fine

FIGURE 3.113 Hydraulic concept of positive control (PC) system incl. pump swivel characteristics.

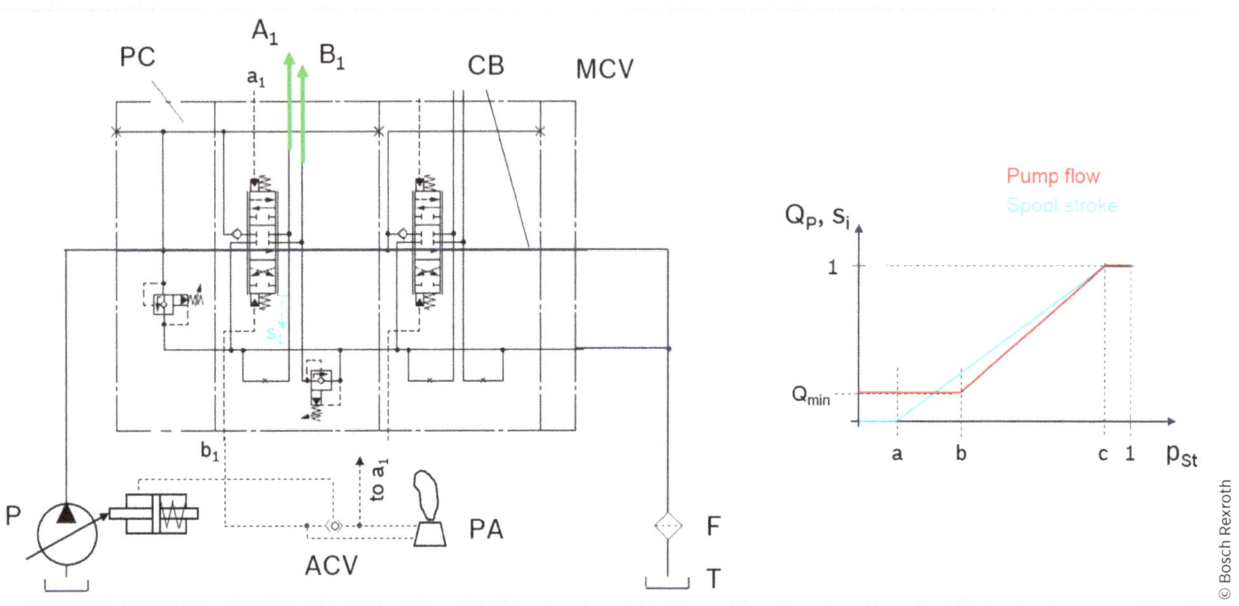

controllability in parallel movement. As long as the pilot pressure p_{St} does not exceed the pressure point b, the oil flow to the actuator is only dependent on the spool displacement/stroke and not on the pump swash angle and it's dynamic, making it easier to control for the driver/operator.

Positive control (PC) systems very often are 2-loop systems (meaning a system with two pumps for the hydraulic system, see Section 3.4.4.9) with 6–8 actuators. The corresponding alternating check valves logic of the hydraulic pilot actuation becomes complex, especially for excavators, where typically the machine function to pump mapping is changed when tracking and working simultaneously. This becomes significantly easier with an electronic positive control system (EPC). Here the hydraulic pilot control units (PA) are either replaced by electronic joysticks, see **Figure 3.114**, or the hydraulic pilot pressures are measured with individual electronic pressure sensors. When using electronic joysticks instead of hydraulic pilot control units, additional solenoid valves (i.e., pressure reducing valves) are needed to control the spools in the main control block. Figure 3.114 shows them integrated into the spool caps in the main control valve (MCV).

This electronic signal is then used in an electronic control unit (ECU) to calculate the required pump flow and with the help of an electronic flow controller on the pump, the oil flow is set. In an EPC system, the software on the ECU replaces the alternating check valves logic of the hydraulic PC system, with the additional benefit, that the logic in software can be adapted in certain situations, i.e., based on the engine speed. Additionally, an EPC system can adapt the pump swivel dynamics concerning the operated machine function (i.e., on a crawler excavator the bucket requires a much higher dynamic compared to the boom or slew), something that is very difficult to achieve with a hydraulic positive control (PC). With the help of pump pressure sensors

FIGURE 3.114 2-loop electronic positive control (EPC) system with an electronic pilot actuation.

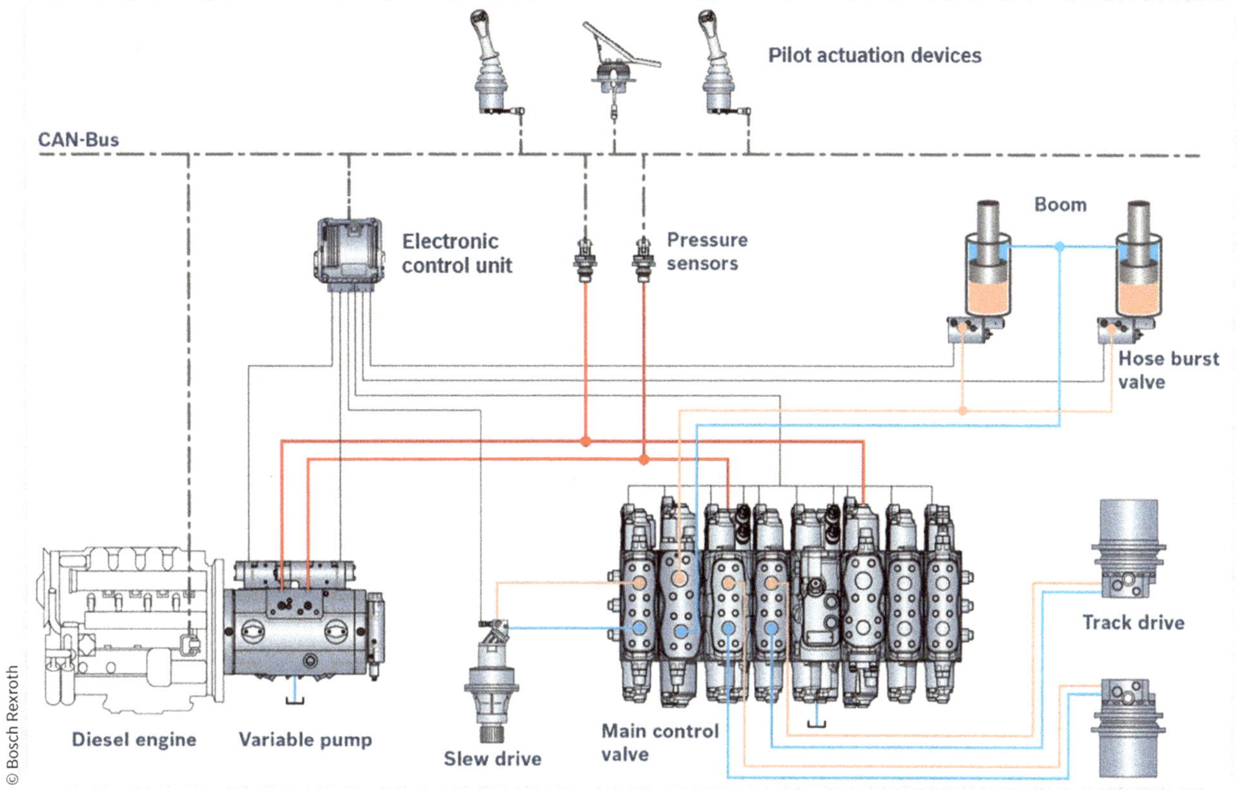

Pilot actuation devices
CAN-Bus
Electronic control unit
Pressure sensors
Boom
Hose burst valve
Track drive
Diesel engine
Variable pump
Slew drive
Main control valve
© Bosch Rexroth

in each pump loop, it is also possible to do the load limiting control (LR, a function which prevents a total engine stop because of too high load conditions) of the diesel engine with the ECU by limiting the pump swash angle based on the current pump pressure.

Positive control (PC) and electronic positive control (EPC) is mainly used in mining and crawler excavators, but also in large drill and pile rigs (Figure 3.115).

FIGURE 3.115 Typical machines that use positive control (PC) and electronic positive control (EPC) systems.

© Bosch Rexroth

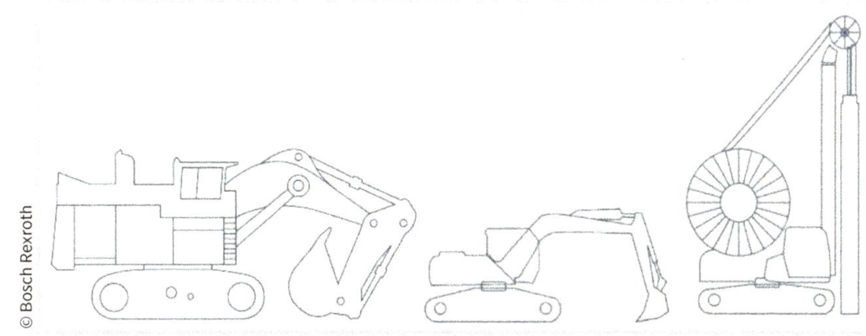

3.4.4.4 **Virtual Bleed Off (VBO):** While negative (NC) and positive control (PC and EPC) both improve the energy efficiency (in partial actuation) compared to the open center throttle control, they still have a center bypass (CB) where oil flows to tank in certain machine conditions (see Figures 3.111 and 3.113). Virtual bleed off (VBO) improves the system efficiency by removing the center bypass losses by using a closed center valve and placing the center bypass into the software and so making it virtual. Like in the EPC system, the driver command is translated into an electronic signal, either by using electronic joysticks (see Figure 3.88) or pressure sensors with the hydraulic pilot control unit. For ease of explanation, the former will be assumed in the following description. As the name virtual bleed off (VBO) indicates, the concept of this system approach is to eliminate the center bypass (CB) flow (also called bleed off flow) and only simulate it virtually in the software on the electronic control unit (ECU).

To be able to do the virtual bleed off (VBO) calculation the center bypass areas of each control spool are available in the ECU software code as a mathematical parameter/table. As shown in Figure 3.116, when the operator moves the joystick, the software will calculate the pressure drop that would result from the virtual pump oil Q_{Pv} flowing through the virtual center bypass area A_{CBv} of the actuated machine function. In the first step, it is assumed that no pump oil flows to the actuator. The calculated pressure drop is the virtual pressure p_{Pv} and is set as the target pressure p_P for the real electronic pressure-controlled pump. As soon as the real pump pressure is high enough that some oil will start to flow to the actuator, the real variable displacement pump will swash out (to a higher displacement) in the same manner. The swivel angle sensor on the pump detects this. This swivel angle information α, which corresponds to the displacement of the variable pump, is used in the software on the ECU to calculate the new virtual pump pressure p_{Pv} using the flow dividing equation. As some oil Q_P (α) is flowing to the actuator, only a reduced oil flow Q_{CBv} would pass through the

FIGURE 3.116 Virtual bleed off (VBO) concept.

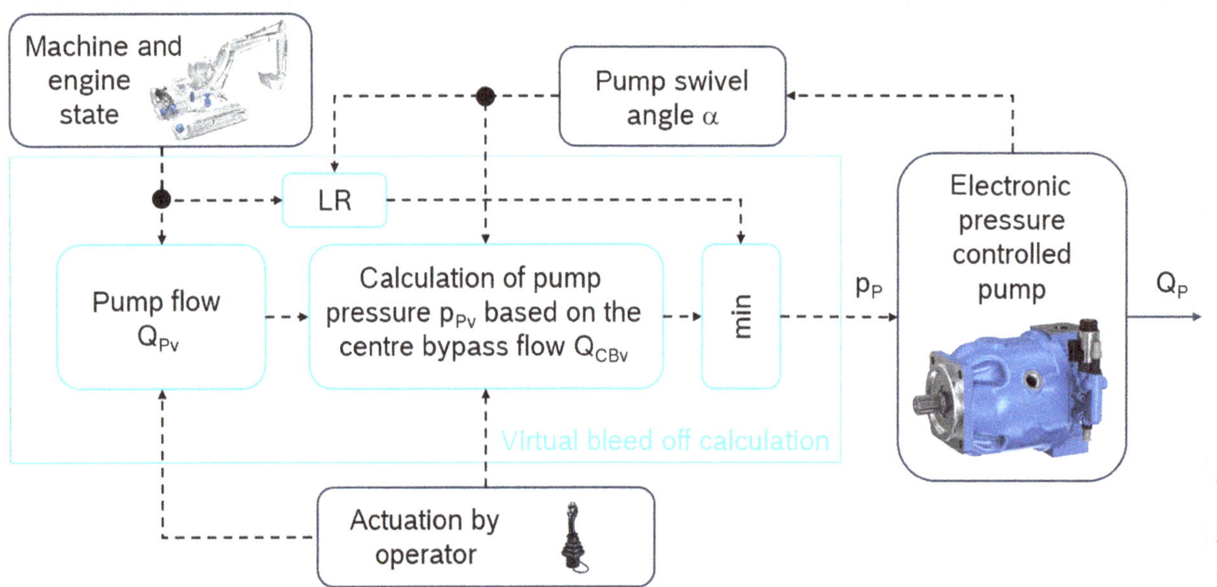

center bypass channel, with $Q_{CBv} = Q_{Pv} - Q_p(\alpha)$. This virtual center bypass flow Q_{CBv} is again used to calculate the new virtual pump pressure p_{Pv} and with that the pump target pressure p_P.

Depending on the virtual pump model inside the software, the system behavior of the VBO system will either be similar to an open center throttle control or positive control (PC), but in both cases without the center bypass energy or oil flow losses [57].

As both the pump pressure (by the pump controller) and the pump flow (by the swash angle sensor) are known at all times, a load limiting control (LR) can easily be integrated into the VBO software. Based on the available power P from the prime mover and the current swash angle α the maximum pressure p_{Pmax} is limited.

$$p_{Pmax} = \frac{P}{Q_P(\alpha)} \tag{3.48}$$

Apart from the elimination of the center bypass losses compared to throttle control, NC, PC and EPC, and the easy inclusion of an LLC (like in EPC), there are additional benefits. As the pump is pressure controlled the maximum pump pressure can be limited by the software flexibly (i.e., with a lower value for auxiliary functions) and loss-free. With a pressure controlled pump it is easy to adapt the pressure increase rate of the system, also with a specific machine function or based on a machine state (mode selection or engine speed). This greatly improves the smoothness of the machine operation.

When accelerating a high load, i.e., when rotating the upper structure of an excavator, the actuator (motor or cylinder) will not follow the command immediately. Figure 3.117 shows such a situation. The driver wants to accelerate the motor as fast as possible (driver command/set value), but due to the high inertia and the maximum

FIGURE 3.117 Acceleration of a heavy load.

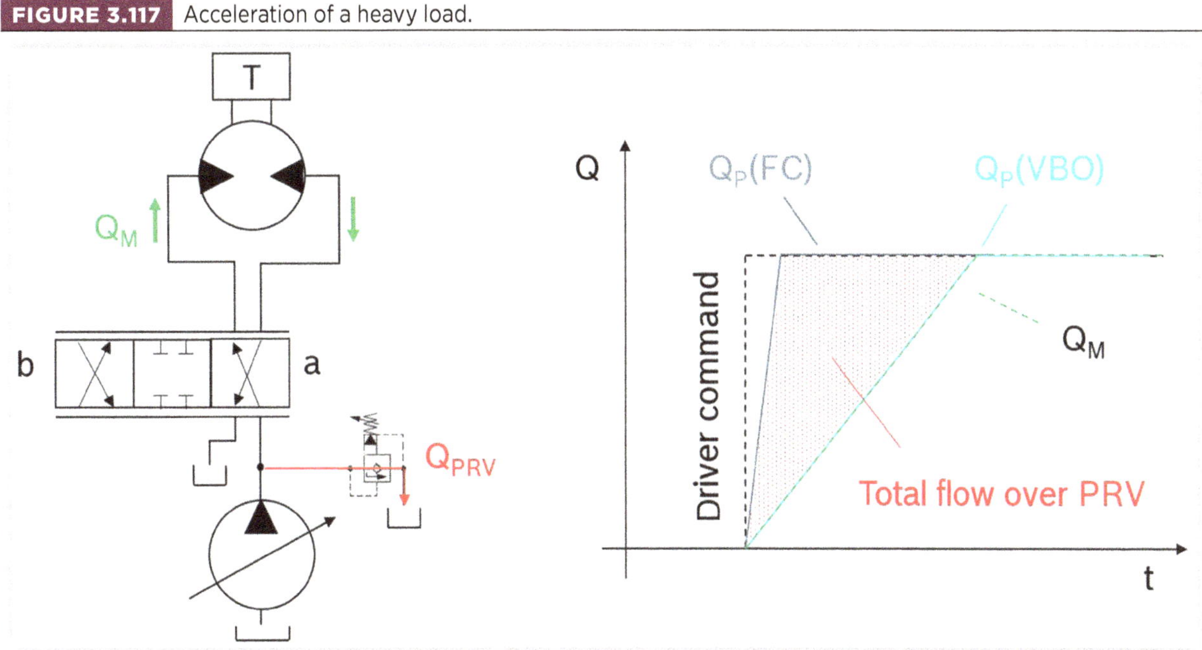

FIGURE 3.118 Machine that uses virtual bleed off (VBO).

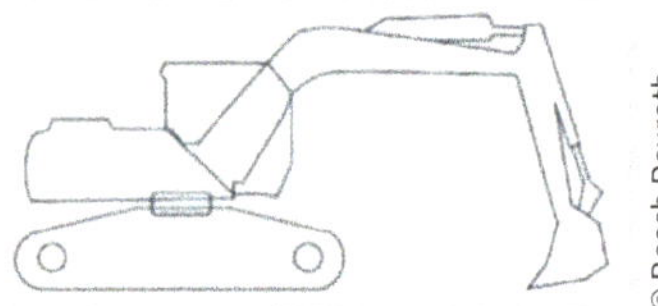

© Bosch Rexroth

allowable system pressure, the acceleration is limited. The flow Q_M represents the flow under this acceleration. In a flow-controlled system, i.e., a positive control system, the pump supplies the maximum flow Q_P (FC), based on the driver command, after a short ramp-up. However, as the flow cannot go through the motor the pump pressure rises to the maximum system pressure set by the primary pressure relief valve (prim. PRV) and all excess oil Q_{PRV} flows through this pressure relief valve (PRV) to the tank. In this case, the pump supplies more oil flow as needed, with $Q_P(FC) = Q_M + Q_{PRV}$. In a pressure-regulated system, such as VBO, the pump pressure will not exceed the maximum pressure by principle, so that the pump only supplies the oil flow that goes to the motor, with $Q_P(VBO) = Q_M$, saving energy.

VBO is mainly used in crawler excavators (Figure 3.118).

3.4.4.5 **Classic Load Sensing:** Load-sensing systems (LS systems) with closed-center main control valves (MCV) have been developed to eliminate the dependency of the open center systems (OC, NC, and PC) on the load pressure and to reduce the power loss in partial actuation. This resulted in efficient systems that can be operated more predictable and therefore for certain applications more easily. The efficiency improvement is achieved by the combination of a pump that is regulated by the load pressure and a closed center valve with a pressure compensator. The current load pressure is constantly forwarded to the pump and used for the pump controller and the pressure compensators in the main control valve. This use of the load pressure is called load sensing (LS). The load pressure signal is generally also referred to as LS signal. For the correct behavior, it is important to always record and forward the highest load pressure signal. For that purpose, the pressure is recorded by the relevant working port using bores in the directional valve spools and forwarded to the pump controller via alternating check valves (ACV), also called shuttle valves (SV). The ACV guarantee that in parallel operation, always the highest load pressure is forwarded to the pump controller (see Figure 3.119).

The pump controller has been designed so that the pump pressure p_P always exceeds the load sensing pressure p_{LS} by a given value (e.g., 25 bar (362 psi)) often called Δp_{LS}. The pump controller always tries to keep the pressure differential Δp_{LS} constant, in consequence, changes in the load pressure also result in changes in the pump pressure.

The LS flow controller valve (3), a flow-controlled valve that bleeds off the LS pressure to the tank with a constant flow, ensures that the pump pressure is reduced once the

FIGURE 3.119 Pump with the load pressure signal (LS).

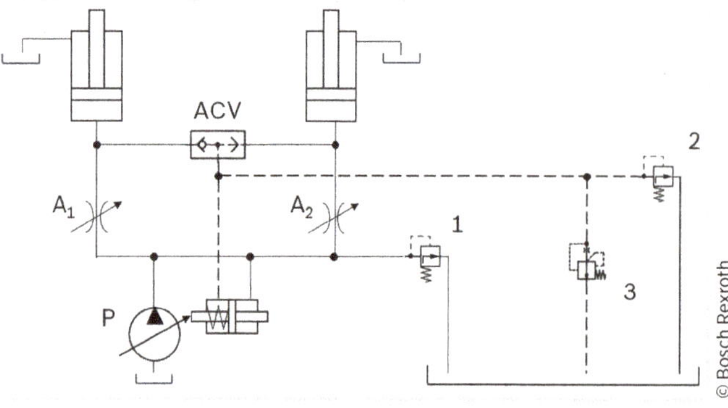

© Bosch Rexroth

load pressure (LS) reduces. Typically, the flow through the LS flow controller valve is below 1 L/min. To prevent the pump pressure from exceeding a given value, also in an LS system a pressure relief valve (PRV, 1) in the pump line can be used. A more energy-efficient way is instead to limit the LS pressure with an LS pressure relief valve (PRV, 2), as shown in Figure 3.119. In the latter case, only a small oil flow goes through the LS pressure relief valve (PRV) to the tank and the pump swivels back when the maximum system pressure is reached [58].

Depending on the type of closed center valve combined with the load sensing (LS) pump, different LS systems architecture results. The most common combination is a load sensing variable displacement pump controller with a closed center valve and a traditional LS pressure compensator (see Figure 3.102).

In a classic closed center LS valve, as shown in Figure 3.120, the load sensing connection holes in the control spool (2) of the valve and the LS shuttle valve (8) make sure that the highest load pressure of all active functions is forwarded to the pump controller. In parallel operation, the pump pressure can be higher than the individual load pressure

FIGURE 3.120 Cross-section and hydraulic scheme of a classic closed center LS valve [59].

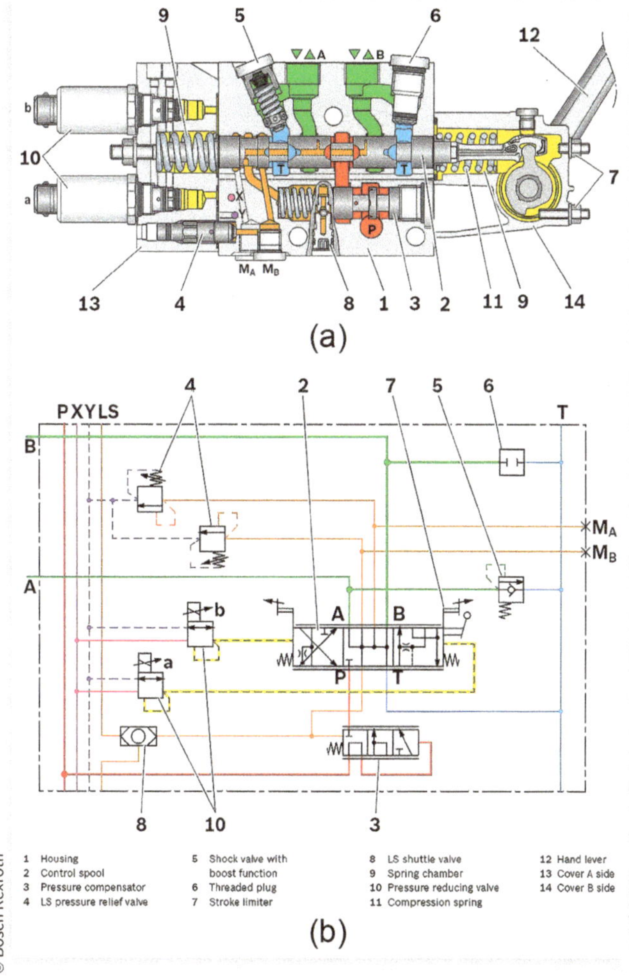

FIGURE 3.121 Common mobile machines that use classic load-sensing (LS) systems.

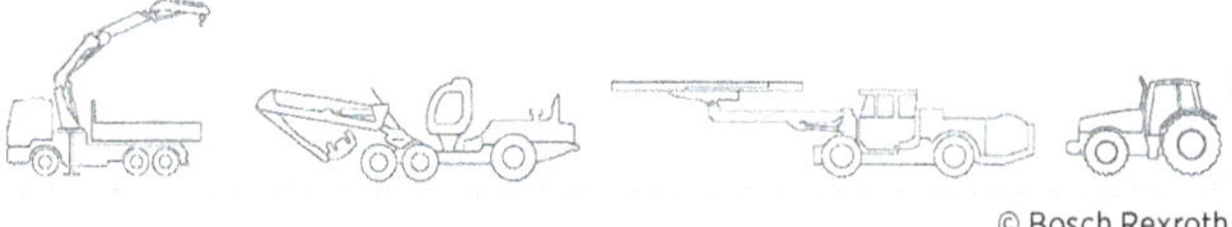

© Bosch Rexroth

p_{Li} plus the Δp_{LS}. In that case, the pressure compensator (3) ensures that Δp over the meter in the area of the control spool is kept constant, as described in Section 3.4.3.2.

Figure 3.120 shows a cross-section through a classic LS valve with many additional features. This closed center valve can limit the spool stroke (7) and with that, the maximum flow to the actuator. In individual cases in which certain actuators must not work with the maximum system pressure, the actuator pressure must be limited. You can do so using the secondary pressure relief valves (5), as described in the Section 3.4.4.11, but when the actuator pressure is reached, the entire actuator oil flow will then, however, be discharged to the tank (and can lead to a high power or oil flow loss). In an LS system, it is possible to limit the LS pressure of each machine function individually, by limiting the pressure in the spring chamber of the pressure compensator with an LS pressure relief valve (4). If the set maximum actuator pressure has been reached, the pressure in the pressure compensator spring chamber will stop increasing and the pressure compensator (3) closes. The oil flow over the LS relief valve (4) is very small, as it is only a pilot oil flow. This makes LS systems very attractive for machines that have different maximum pressures for their actuators.

If the requested actuator flow by the directional valves is greater than the maximum displacement of the pump at a given time, the system is undersaturated. In this case, the pump is at its maximum swivel angle or, if the system has power regulation, at the point of maximum flow based on the available power of the engine. As the oil preferably flows to the actuator with the lower load pressure the required system pressure for the actuator with the highest load pressure can no longer be maintained. For the actuator with the highest load pressure, this means that it only receives the remaining oil flow, so that it drives slower, or it can even stop completely if the undersaturation is bigger than the corresponding flow request of the highest load. When operating a machine, this less advantageous behavior may be annoying. Therefore in a load-sensing system, the maximum oil flow of the pump and the available power are typically sufficient for all common simultaneous movements.

Typical applications that use LS for the implement control are machines that need to precisely control the movement of a cylinder or motor with a high repeat accuracy and especially without interaction during parallel operation (**Figure 3.121**). This is i.e. the case for knuckle-boom cranes, forestry equipment like the harvester, underground drills but also tractors.

3.4.4.6 **Flow Sharing (LUDV):** The less advantageous LS behavior in under-saturation described in the chapter above, where the actuator with the highest load will reduce its speed, is not ideal for every application. Especially machines that are designed in such a way that nearly all actuators can use the maximum pump flow by themselves, like excavators or wheel loaders, become very difficult to handle. To overcome this problem, flow sharing or LUDV was invented. LUVD, from the German term *Lastdruck Unabhängige Durchfluss Verteilung* (load pressure independent flow sharing), has many

FIGURE 3.122 Cross-section and hydraulic scheme of a closed center flow sharing (LUDV) valve [60].

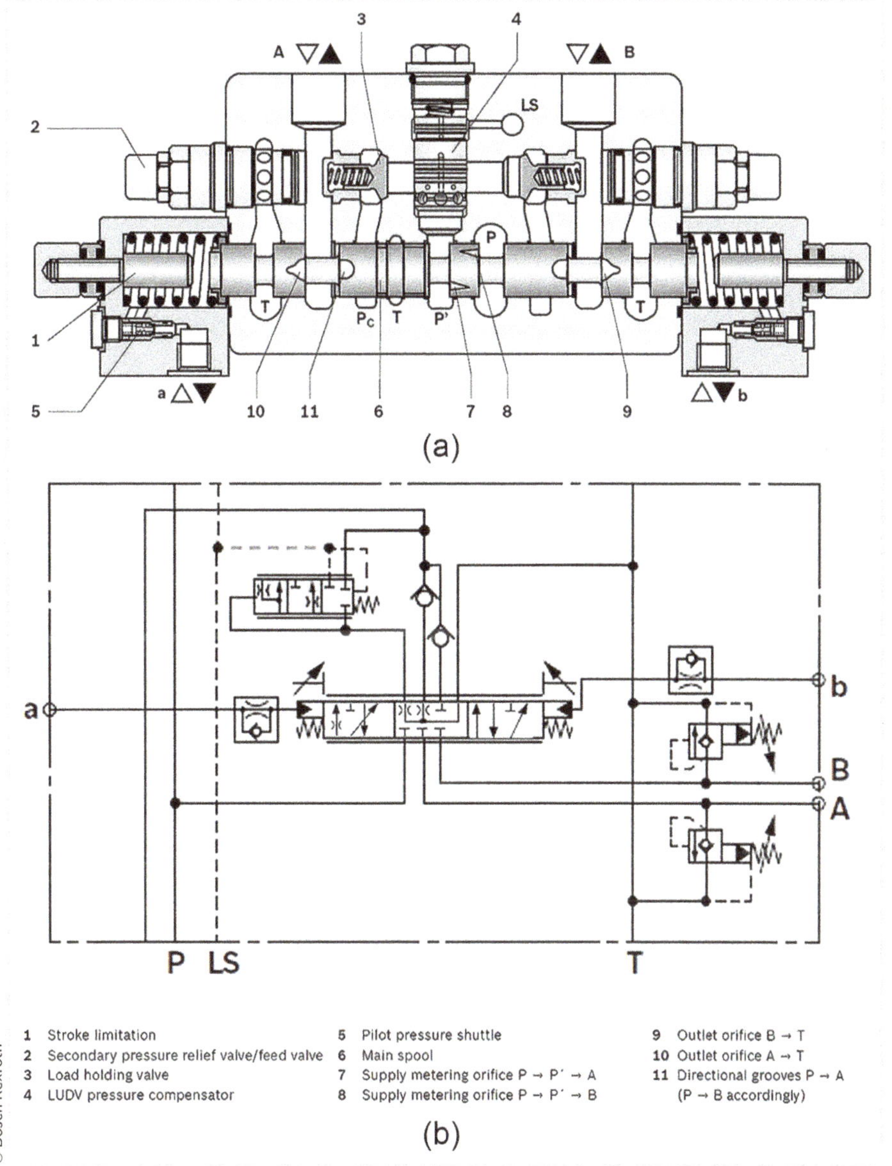

similarities with the LS system. In both, the variable displacement pump operates as a hydraulic-mechanical pressure regulated closed-loop control that ensures the supply pressure exceeds the highest load pressure by a fixed excess pressure Δp_{LS}.

The difference between classic LS and flow sharing (LUDV) is the pressure compensators. A flow sharing (LUDV) system uses flow sharing pressure compensators instead of traditional LS compensators. Figure 3.122 shows a cross-section of a closed center flow sharing (LUDV) valve and its corresponding hydraulic scheme. In the neutral position of the main spool, when no pilot control pressure is at the ports a or b, the connection from the pump channel

P to the P′ channel (7 and 8) is blocked by the main spool (6). The load holding valves (3) and the flow sharing (LUDV) pressure compensator (4) are closed. The actuator ports are blocked by the main spool (6) overlap in the housing. The flow sharing (LUDV) pressure compensator (4) consists of the main spool and a compression spring defining a stable initial position. The main spool (6) is proportionally moved to the right against the spring force by the applied pilot control pressure of the pilot control unit in the control cover (i.e., at a). The supply metering orifice (7) of the main spool opens the connection from the pump port P to the channel P′. The resulting pressure in this chamber opens the pressure compensator (4) and is applied to the load holding valves (3). The actuator pressure p_C of port A keeps the left load holding valve (3) closed via the control notches in the main spool (11). When the value of P′ exceeds that of p_C, the load holding check valve (3) is opened. The connection from the pump to the actuator is established and oil flows to the actuator. Downstream the oil from the other side of the actuator flows from B via the meter out or outlet orifice (9) back to the tank. The secondary pressure relief valves (2) remain closed as long as the pressure in the actuator port remains below their pressure setting. The main poppet of the combined pressure relief/feed valve (2) in the supply (side A) opens in the event of cavitation in the actuator port and enables feed-in from the tank channel [60].

Apart from the main spool and the flow sharing pressure compensator a standard flow sharing (LUDV) system also typically contains one primary (1) and a few secondary pressure relief valves (3) as shown in Figure 3.123. They have the same task as in the other hydraulics systems. What is specific to the flow sharing (LUDV) system are the load sensing pressure relief valve (5), the load sensing flow controller (4) and the load sensing shuttle (6) in the pump connection. The LS pressure relief valve allows an energy-efficient maximum pressure setting, as it limits the swiveling pressure so that the pump never supplies a higher pressure. The primary pressure relief valve (1) is only needed in case of pump malfunctioning, in normal operation, the LS pressure

FIGURE 3.123 Common flow sharing (LUDV) system elements [60].

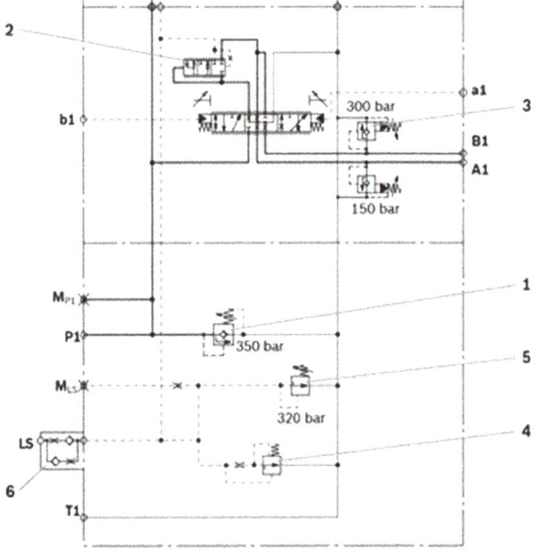

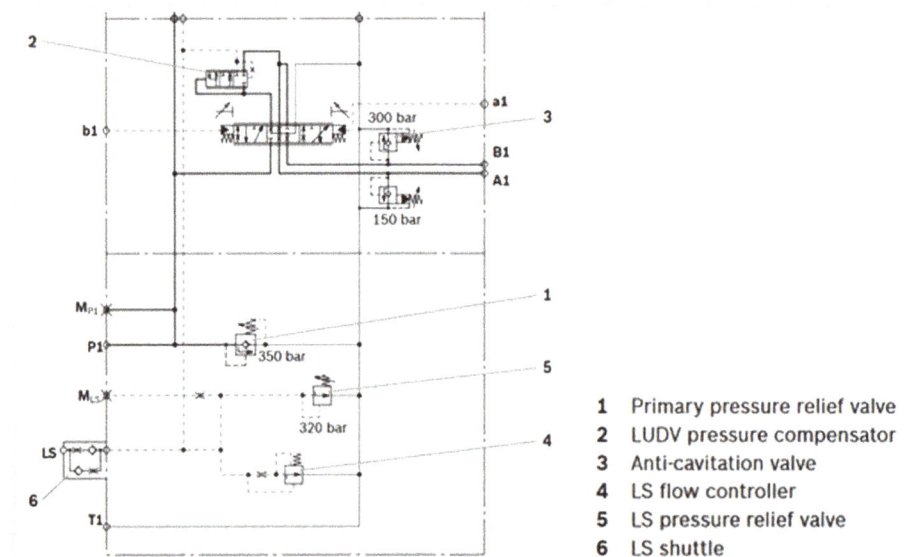

1 Primary pressure relief valve
2 LUDV pressure compensator
3 Anti-cavitation valve
4 LS flow controller
5 LS pressure relief valve
6 LS shuttle

FIGURE 3.124 Common mobile machines that use flow sharing (LUDV).

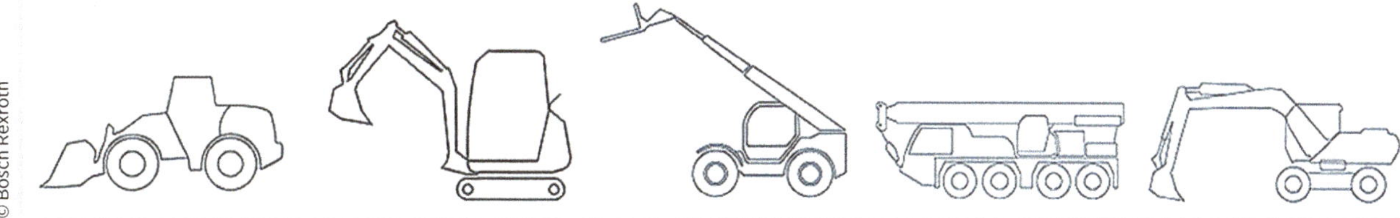

relief ensures the maximum system pressure. The LS flow controller (4) constantly releases oil from the LS line, making sure that if no axis is actuated the LS pressure is zero. The LS shuttle (6), a combination of two orifices with check valves, is used in the LS line that transfers the LS signal to the pump. This valve enables hydraulic damping to adapt the system dynamics to the machine needs. In certain situations, the driver will request more oil flow than the pump can supply. The undersupply will lead to a reduced pressure differential Δp_{LS} at all valve openings and that results in a reduced flow at all the actuators. The closing of all active pressure compensators simultaneously prevents the standstill of any actuator (in comparison to the classic load sensing system, where only the speed of the highest load is reduced). As the reduced pressure differential Δp_{LS} is the same again at all valve openings, the reduced oil flow rates to the actuators are in the same ratio as well. This is why the system is called flow sharing, as all actuators receive their share of the pump flow following the pilot actuation request [60].

Especially machines that are used as tool carriers, i.e., wheeled excavators, very often use flow sharing (LUDV), as the auxiliary functions are easy to operate in parallel to the main machine functions, without any interference based on the load pressures.

Flow sharing (LUDV) systems can be found today in many applications as in mini, midi and wheeled excavators, mobile cranes, wheel loaders or telehandlers (Figure 3.124).

3.4.4.7 **Electronic Load Sensing (eLS):** Both the traditional load sensing (classic LS) and the flow sharing (LUDV) system connect the maximum load pressure to control the pump. The pump controller adapts the load pressure plus a differential pressure Δp_{LS}, this keeps system losses low. The connection between the valve and the pump contains hydraulic damping elements to optimize the system behavior. The disadvantage of this setup is that these elements are fixed and need to work under all conditions the machine will be operated in.

In an electronic load sensing (eLS) system, as shown in Figure 3.125, the hydraulic LS line connection is replaced by a pressure sensor s_1 at the load sensing port of the main control valve (MCV). With the help of an electronic control unit (ECU), the electro-hydraulic pressure controller of the pump is set.

It is now possible to filter the LS pressure signal and stabilize the pump pressure setting. The filter and dampening parameters can be adapted based on various machines or system states, i.e., also as set inputs by the driver/operator on a human-machine interface (HMI) display. The possible situational adaptions the system can offer is increased, when combined with an electronic pilot actuation (PA).

FIGURE 3.125 Electronic load sensing (eLS) system for a tractor [62].

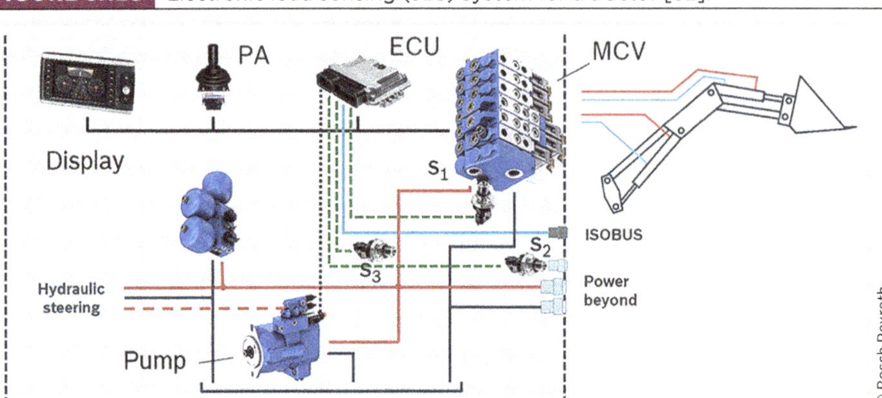

Electronic load sensing (eLS) is still relatively new and is not yet widely spread on mobile machines.

3.4.4.8 **Electrohydraulic Flow Matching (EFM) or Electrohydraulic Flow-on-Demand System:** In both load sensing systems, classic LS and flow sharing (LUDV), the main spool first needs to move to open a connection from the pump channel to the LS chamber to generate a load signal and then use a hydraulic hose or pipe to transport the LS pressure signal from the main control block to the pump regulator. This leads to a certain response time of the movement and this response dynamic is dependent on the oil temperature. In an electrohydraulic flow matching (EFM) system, also called Electrohydraulic Flow-on-Demand System [61], the pressure controlled variable displacement pump of a classic LS or LUVD system is replaced with an electronically flow-controlled pump that supplies the required flow at the same time as the valve opens. To do this, the driver demand needs to be known, either with the use of electronic joysticks or by using a hydraulic pilot actuation unit and pilot pressure sensors. The driver demands are translated into a flow request inside the electronic control unit (ECU) and given as a command to an electro-hydraulic flow controller of the variable displacement pump (see Figure 3.126).

As the pump in an EFM system does not operate with the load sensing pressure p_{LS}, but with an electro proportional variable displacement pump in an open control loop, it no longer responds to changes in load pressure and, instead, operates independently without interaction with the pressure compensators. Furthermore, the pump and valve are actuated synchronously and therefore EFM eliminates delays between the pilot actuation input and the LS signal arriving at the pump. This improves the system response and increases stability for disturbance variables. Both aspects making the working hydraulics more agile and less prone to oscillations. The electrohydraulic flow matching (EFM) concept is a pump control strategy similar to an electronic positive control (EPC) but combined with a closed center valve with a pressure compensator [63, 64]. This can be a classic LS compensator (see Figure 3.102) or a flow sharing (LUDV) compensator (see Figure 3.103).

EFM systems combined with the main control valve using a classic LS pressure compensator are not so common (Figure 3.127). This system shows all the

FIGURE 3.126 Electrohydraulic flow matching (EFM) setup.

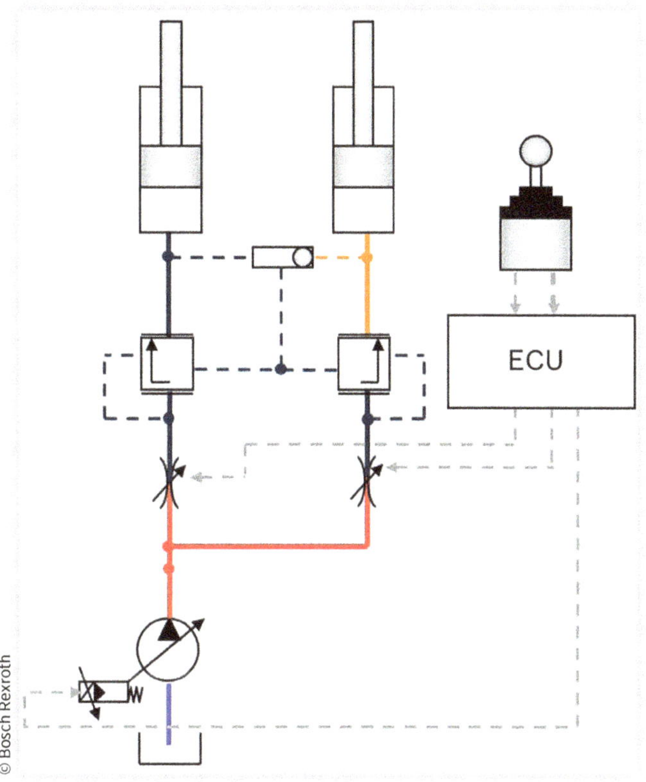

FIGURE 3.127 Common mobile machines that use electrohydraulic flow matching (EFM).

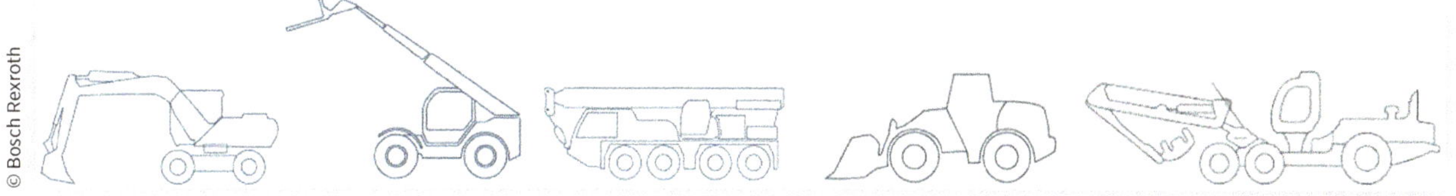

above-described benefits but requires a very precise pump control (i.e., a closed-loop flow control). When using a classic LS pressure compensator too much pump flow leads to an unwanted pump pressure increase as the classic LS pressure compensators meter a defined amount of oil to the actuator, based on the meter-in orifice [65]. Too little pump flow will lead to a reduced speed of the actuator with the highest load. It seems that the concept of electronic load sensing (eLS or ELS) is the more common solution when electronifying classic LS valve systems.

The case is different for EFM combined with the main control valve using a flow sharing (LUDV) pressure compensator. The flow sharing (LUDV) pressure compensators share the supplied pump flow all evenly. Therefore, the system can handle a flow shortage or surplus perfectly. The EFM pump control concept does not rely on a set pressure surplus to the load pressure so it reduces the pump pressure needed compared to traditional flow sharing (LUDV) systems, reducing the Δp over spools in partial operation, saving energy.

EFM LUDV systems are used in machines that appreciate the flow sharing behavior of a LUDV system but also benefit from the increased dynamic pump behavior. Typical applications are mobile cranes, wheeled and industry excavators, forestry cranes [66] or telehandlers.

3.4.4.9 **Multi-Loop Systems:** The above chapter describes different working hydraulic system concepts as 1-loop solutions. In the hydraulics, loops are defined by the amount/number of pumps that are used to operate/supply the main functions of a machine. For example, an excavator that uses two pumps combined for the implement and crawler control, one pump for the pilot oil supply and one pump for the auxiliary functions is using a 2-loop system. There are different reasons why a machine has more than one loop. The most common reasons are function separation, energy efficiency, available component sizes, and cost.

When working functions of a machine are often used at the same time but have significantly different pressure levels, from an energy-efficient point of view it makes sense to have them both run on different pumps. For some functions of a machine, i.e., winches or the rotary movement of an upper structure, it is important to have limited interaction between functions. This can be achieved with an LS System (see Sections 3.4.4.5 to 3.4.4.8) but this is not always appropriate for the other functions. Therefore, on bigger excavators, some machines have a closed-loop only for the rotational slew drive, so it is completely independent of the other machine functions, such as the control of the boom, bucket and arm cylinder. Lifting the heavy load (e.g., a sewage pipe on a rope) while slewing will not cause any oscillation of the load.

In many cases, there are possibilities to combine the oil flow of the different loops under certain conditions. Usually, this is called summation.

Figures 3.128 and 3.129 show the two most common summation strategies. In the case of a pump line summation (Figure 3.128), the pump flow from one pump that is not needed in its loop is added to the other pump line. To do this, a valve (sum) is activated and makes the connection. In case the "a" side of the sum valve is activated, only oil from pump P2 can flow to the P1 loop. In case the two-loop system is an LS system,

FIGURE 3.128 2-loop with pump line summation.

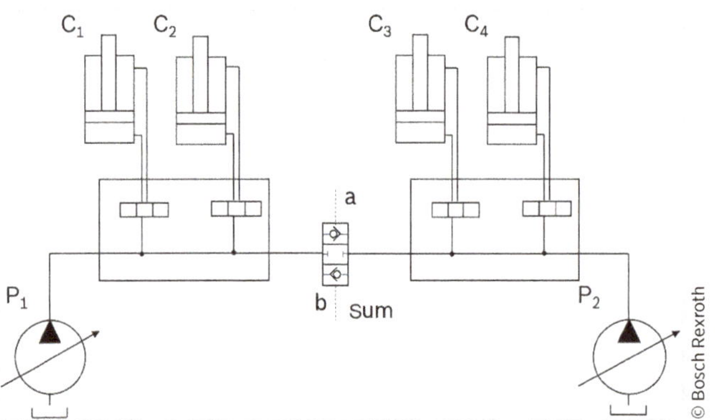

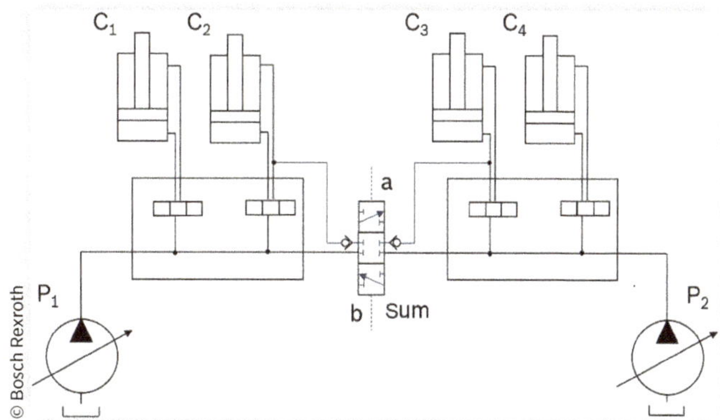

FIGURE 3.129 2-loop with cylinder summation.

then also the LS pressure needs to be connected between the two pumps [67, 68]. Typically the summation valve (sum) is either activated in accordance to the main spool actuations or based on the requested power in each loop. The direct cylinder summation (Figure 3.129) adds the flow from one loop not into the other pump line but directly to a cylinder head or rod side. This is common for machines where only very few movements need the flow from two pumps.

If the machine control and its energy efficiency benefit from full flexibility it is also possible to have a system where each function can be connected either to pump 1, to pump 2 or to both. Figure 3.130 shows such a solution including a cross-section of a valve that allows this flexibility.

Different original equipment suppliers (OEMs) will use different strategies for what hydraulic system architecture and the number of loops they use for their machines. Figure 3.131 shows a simplification of the common systems used in the different excavator machine sizes and the amount/number of pump loops. The smallest machines

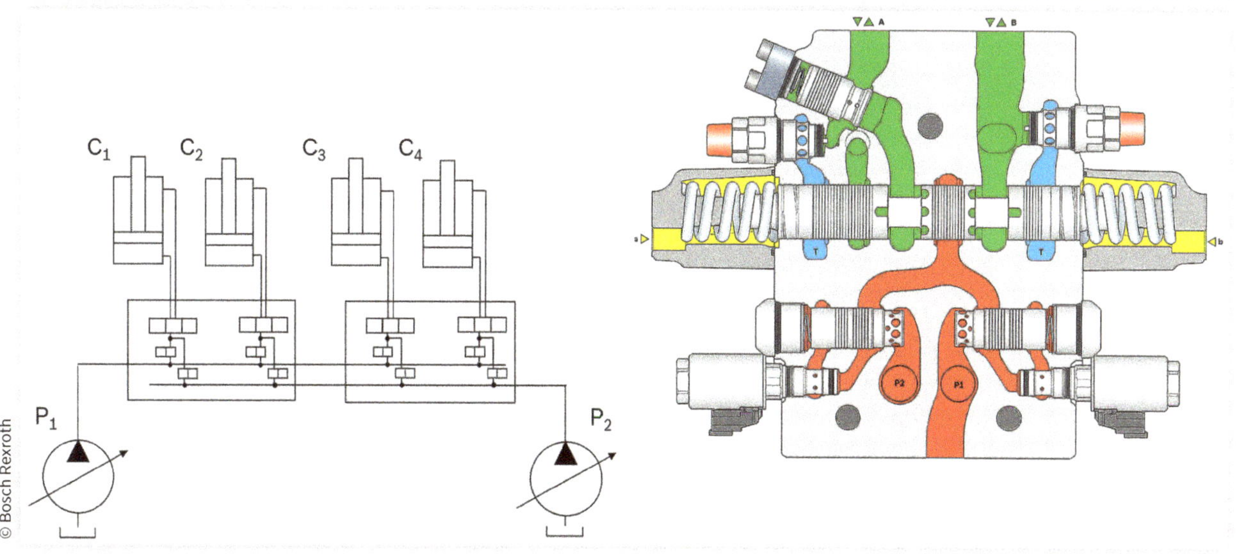

FIGURE 3.130 Flexible 2-loop system with corresponding valve cross-section [69].

FIGURE 3.131 Amount of loops commonly used in the excavator application.

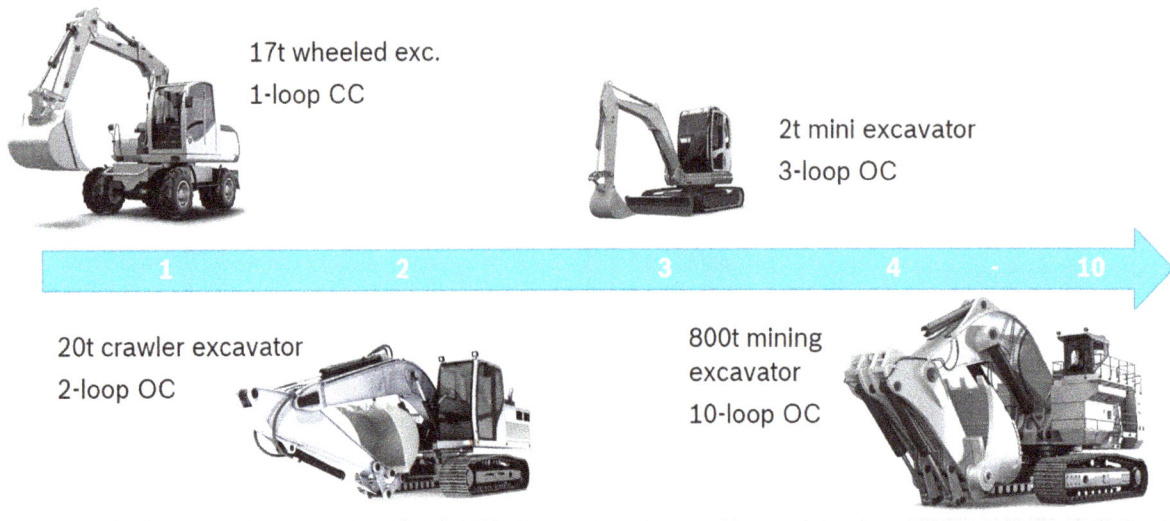

have often a 3-loop throttle control. This offers the best compromise on cost and performance for this machine size. The wheeled excavators are often controlled with a 1-loop flow sharing (LUDV) closed-center system as this offers from the best compromise between cost and a good energy efficiency paired with perfect auxiliary control, as wheeled excavators are flexible tool carrier machines. The 20–40 ton crawler excavators are used in quarries and on big constructions. Here a 2-loop open center (typically NC or PC) system offers the best compromise in energy efficiency, controllability, and cost. For bigger mining machines often 4–10 pump loops are used to achieve the perfect combination of energy efficiency, robustness, availability, and cost.

3.4.4.10 **Regeneration and Differential Connection:** Mobile machines work in many conditions on the limit of their power and/or flow. Still, the operator would like to have a higher speed of the machine movement. One way to increase the available flow for a single consumer is to include a special control that allows re-using the meter-out flow for the meter-in. The most common versions of this concept to re-use the oil flow directly are the regeneration and the differential connection.

The maximum speed of a cylinder is typically limited by the meter in flow. When lifting a load, the pump oil flow is the limiting factor and when lowering load cavitation needs to be prevented in the cylinder, so without anti-cavitation valves again the speed is limited by the pump flow. The concept of regeneration aims to use the potential energy of a cylinder to increase the speed of the movement. Figure 3.132 shows a regeneration solution that is often incorporated inside the main spool of the valve section that controls the cylinder. In the shown example a load M is pulling in the same direction as the cylinder movement s as an aiding load. In consequence to control the speed of the cylinder the oil flow on the meter-out side needs to be throttled. With that, the pressure in the rod side of the cylinder p_A is higher than the pressure in the head side p_B.

FIGURE 3.132 Common regeneration for a PC system, i.e., for the arm cylinder of an excavator.

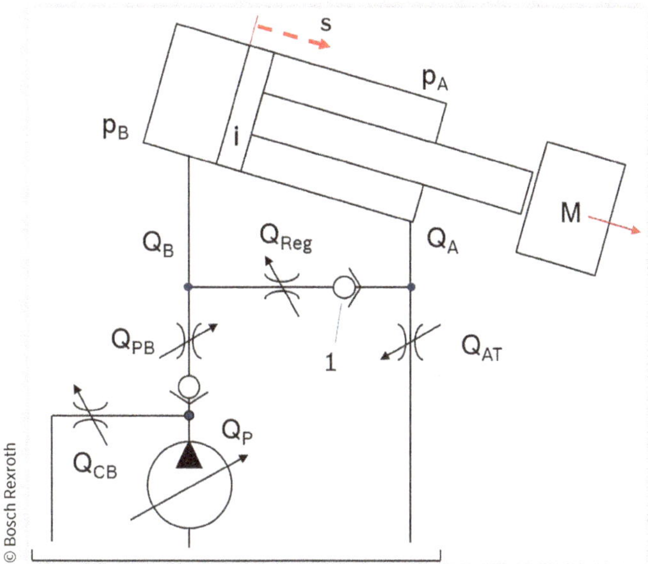

An additional metering orifice A_{Reg} connects the head and the rot side of the differential cylinder. Now based on the pressures the flow from the rod side Q_A is divided between a tank flow Q_{AT} and a regeneration flow Q_{Reg}. This regeneration flow is now added to the pump flow Q_{PB}, so that the meter-in flow Q_B is higher and with that the maximum cylinder speed without cavitation. The regeneration flow is dependent on the load M, the machine requirements and the cylinder ratio i and typically can reach about 30–50% (for i = 2) of Q_P. To prevent the loss of force on the cylinder in case the load direction is in the opposite direction of the movement a check valve (1) is added.

A second common way to increase the cylinder speed for a given pump oil flow is the use of a differential connection.

The differential solution, shown in Figure 3.133, eliminates the meter-out tank connection so that all meter-out flow goes through the regeneration to the meter-in again. By this, the cylinder speed is increased following the cylinder ratio i independent of the load.

$$Q_B = \frac{Q_{PB}}{\left(1 - \frac{1}{i}\right)} \tag{3.49}$$

with the differential cylinder area ratio $i = \frac{A_B}{A_A}$.

The downside of the differential connection is the increase of the pressure p_B when lifting a load (in the same ratio as the flow is increased), limiting the maximum performance of the machine. For that reason, in most machines, that use the differential mode, it can be deactivated.

FIGURE 3.133 Differential connection for a PC system.

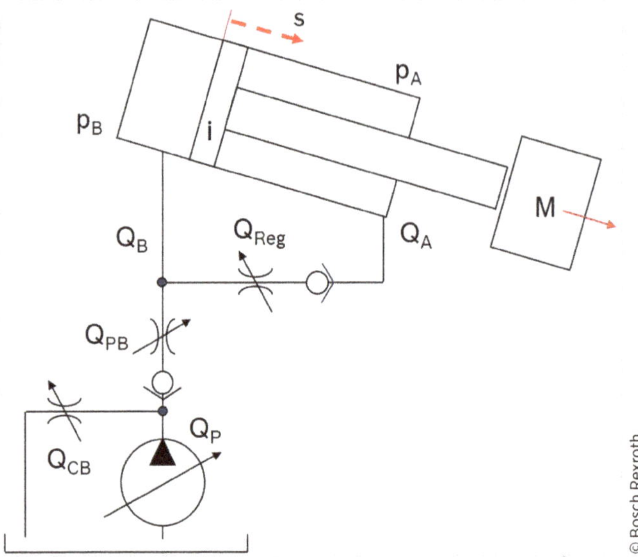

Figure 3.134 shows two typical solutions common for crawler excavators for the arm cylinder. As long as the pressure in the cylinder head side p_B is below a certain value, the connection from the rod side to the tank is closed with help of an additional valve (2) or (3). In this closed condition, it behaves as a differential connection with the flow benefits. As soon as the pressure p_B rises above a defined (application-specific) value the connection to the tank is gradually opened. This is either done by a hydraulic piloted valve (2) or with an electro-hydraulic valve (3) that uses a pressure sensor as a trigger. As a result, the differential mode is turned into a common regeneration with

FIGURE 3.134 Regeneration valve with a braking valve or full differential mode for a PC system.

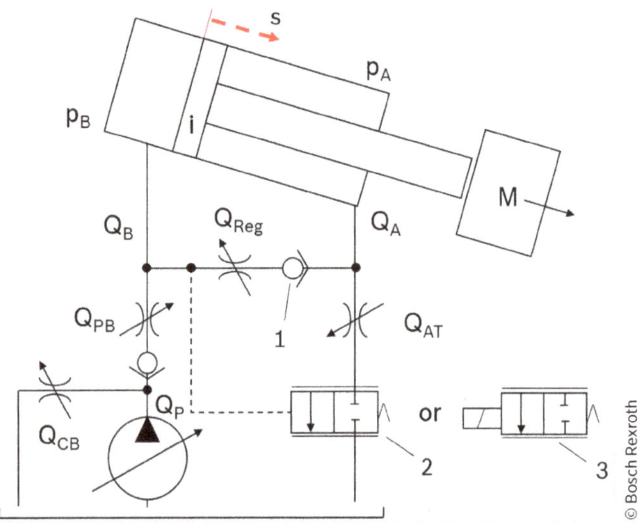

no negative impact on the lifting (or in case of the excavator's arm the digging) performance. The advantage of this third, combined solution mode is, that both modes, the differential as well as common mode could be used dependent on the best hydraulic condition of both.

3.4.4.11 **Pressure Relief Valve (PRV):** Figure 3.107 shows a primary pressure relief valve (prim. PRV) in the inlet section and additionally a secondary pressure relief valve (sec. PRV) in port B. These valves protect the port, the connected hoses/pipes and the pump/cylinder/motor from excessive/overpressure. **Figure 3.135** shows a cross-section through a pilot-operated pressure relief valve (PRV).

The pressure relief valve is closed in the initial position. The pressure in the main port B acts on the spool (1). Simultaneously, the pressure is applied to the spring-loaded side of the spool (1) and the pilot poppet (6) via the nozzle (2). If the pressure in main port B exceeds the value set at the spring (5), the pilot poppet (6) opens. Hydraulic fluid flows from the spring-loaded side of the spool (1), via a nozzle (3) and channel (7), into main port T (tank). The resulting pressure drop moves the spool (1) and thus opens the connection from main port A to T while maintaining the pressure set at the spring (5). The pilot oil return is implemented internally via the channel (7) into main port T [70].

The anti-cavitation function makes up for lacking hydraulic fluid volumes caused, e.g., in the case of leading/aiding loads. If the pressure at main port B is lower than the one at main port T, the spool (1) will be lifted out of its seat. Hydraulic fluid flows from the main port T to main port B. For this to work, the pressure in the tank line needs to be high enough to overcome the pressure drop of the housing and the relief valve. Therefore, normally a preloaded check-valve (set at 3–7 bar (44–102 psi)) is installed between the tank port of the main control valve and the oil reservoir (tank), so that enough oil can flow through the anti-cavitation function to prevent cavitation under all circumstances.

Pressure relief valves are used in nearly all hydraulic systems because of safety against excessive pressure. Like in Figure 3.107, the primary relief valve (prim. PRV) protects the system from the pressure generated by the pump and is common in flow controlled pump systems like a throttle or positive control. Secondary pressure relief valves (sec. PRV) are used for functions that can't take the maximum pump pressure

FIGURE 3.135 Scheme and cross-section of a pilot-operated pressure relief valve (PRV) [70].

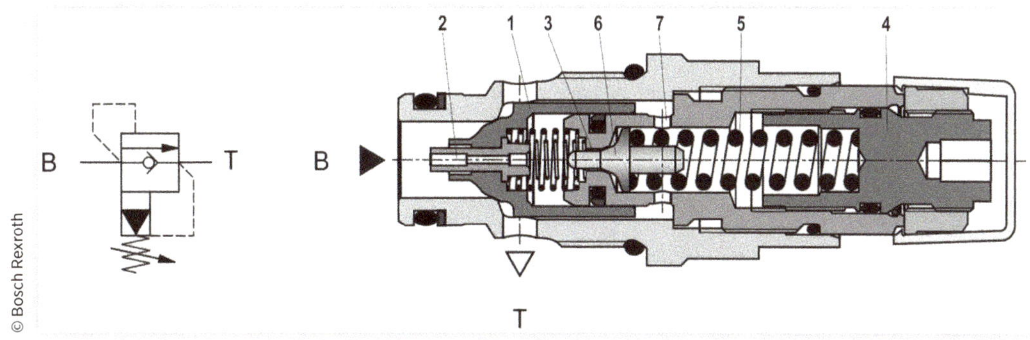

and in all applications where an external load can lead to a bursting pressure for the hydraulic system.

3.4.5 Comparison between the Systems

For a mobile working machine, the hydraulic system is always a compromise between energy efficiency, load dependency, its behavior during flow or power shortage, system stiffness, flow and pressure accuracy, control complexity, expandability, size, and cost. As described in the above chapters, nowadays the most common mobile hydraulic system architectures have different functional targets based on the main application it is/was designed for. To decide what hydraulic system is the best for a new mobile working machine or when upgrading an existing system, the following graphs explain the difference in system behavior in relevant conditions. The p-Q diagrams show the pressure p and the flow Q for two actuators and the pump in three conditions including losses. Figure 3.136 explains the meaning of the colors.

The load sensitivity of the different solutions is shown in Figure 3.140 and the change in behavior based on the prime mover speed in Figure 3.141.

3.4.5.1 Pressure p and Oil Flow Q in the Different Machine States:

The following figures describe the behavior of the most common 1-loop systems in three situations. Figure 3.137 shows the flows and pressures when two actuators (with the pressures p_A and $p_B < p_{max}$) are partially activated. The valve sections A and B are parallel connected in the main control valve (MCV), and the spool opening areas A_{PA} und A_{PB} (see Figure 3.96) are adequate for the oil flows Q_A respectively Q_B. This flow is less than the maximum pump flow and there is no power limitation. This is the typical state for machines during normal operation, i.e., while fine grading (excavators), positioning loads (cranes) and working with auxiliaries (tractors).

In this condition virtual bleed off (VBO) and a flow sharing system with an electrohydraulic flow matching (EFM) pump control are the most efficient. In both cases, the pump only supplies the exact wanted to flow and the pump pressure is only slightly higher (typically 10 bar (145 psi) caused by the pressure drop Δp over the spool opening area A_{PA}) than the highest load pressure p_B. Compared to these two system solutions the classic LS and the pure flow sharing system (LUDV) have a slightly higher pump pressure, caused by the LS pump regulator setup in combination with the pressure compensators.

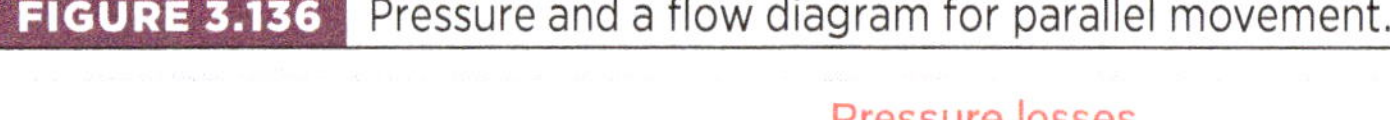
FIGURE 3.136 Pressure and a flow diagram for parallel movement.

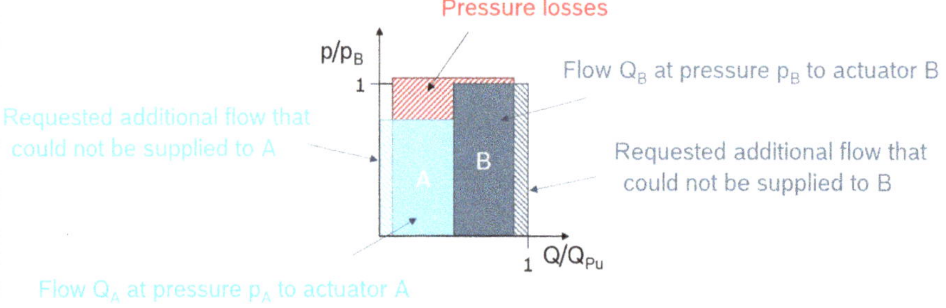

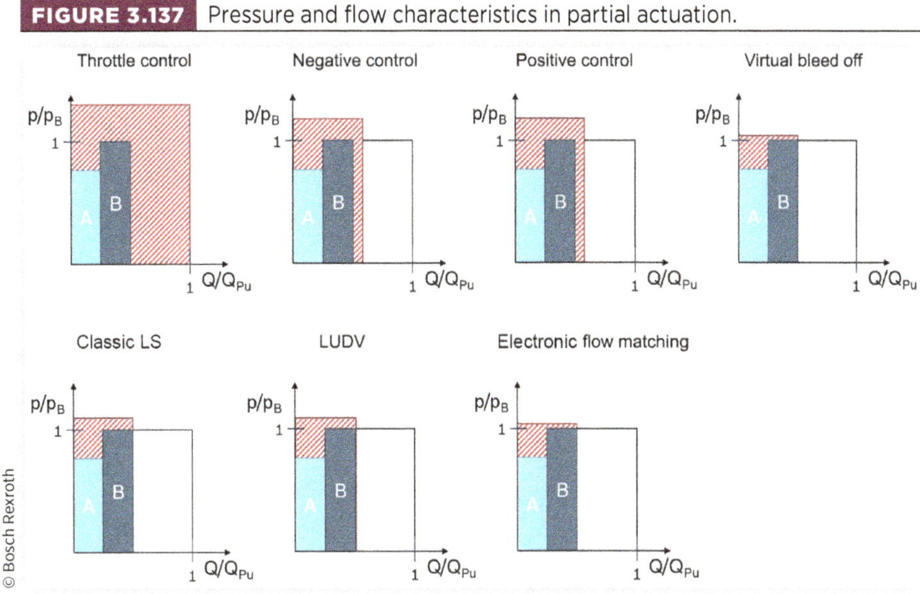

FIGURE 3.137 Pressure and flow characteristics in partial actuation.

Both negative and positive control has an even higher pump pressure and adds some pump flow goes through the neutral gallery to tank due to the open center valve design. In a throttle control system, the pump displacement is constant and therefore the unused pump flow goes to the tank, causing the highest losses.

Figure 3.138 shows the situation when both actuators together request exactly the maximum pump flow Q_{Pu} (no oil flow under saturation) but the system is not power limited. This is a situation that occurs on machines that have a bigger available power

FIGURE 3.138 Pressure and flow characteristics for movement with full pump flow but no power limitation.

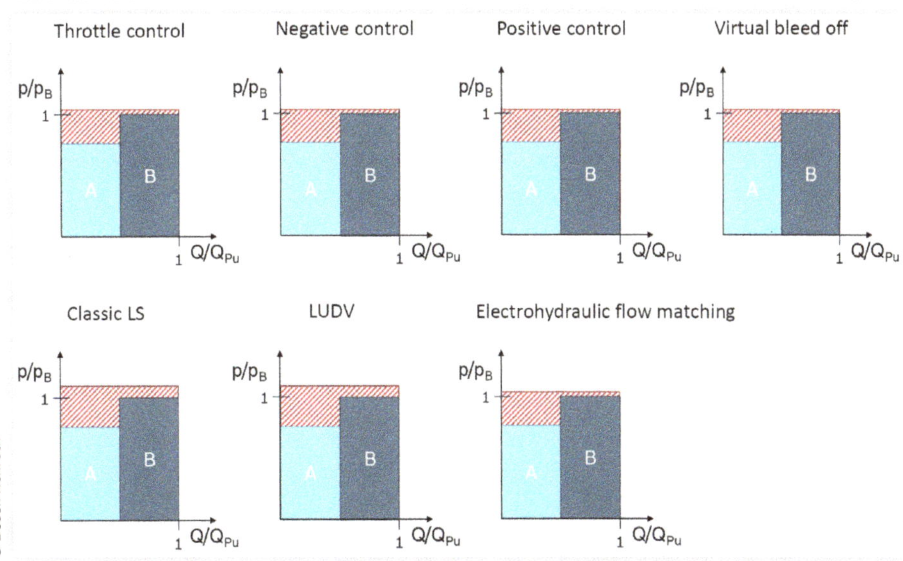

FIGURE 3.139 Pressure and flow characteristics with power limited pump flow.

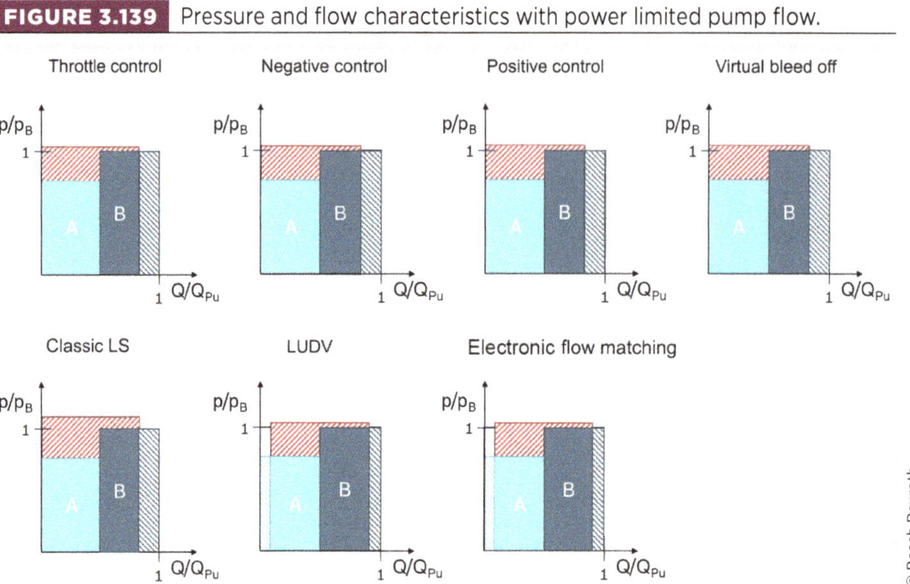

than used for the working hydraulic functions, typically because the prime mover is sized for driving, i.e., on knuckle boom cranes.

Under this condition for most systems the pump pressure is only slightly higher (typically 10 bar (145 psi) caused by the pressure drop Δp over the spool opening area A_{PA}) than the highest load pressure p_B. Only the classic LS and the pure flow sharing system (LUDV) have a slightly higher pump pressure, caused by the LS pump regulator setup in combination with the pressure compensators. The hydraulic systems in Figure 3.138 need significant different pilot actuation to achieve the shown state. Especially for throttle control, this is not a typical condition in reality.

Figure 3.139 shows the situation when the actuators request the maximum pump flow ($Q_A + Q_B = Q_{Pu}$) but due to a power limitation, the maximum available pump flow is less. This is the situation most machines are in when executing heavy work, such as the excavator while digging and loading, the wheel loader during truck loading and the backhoe loader during trenching.

With power limited pump flow, the oil flow to the actuators is less than requested. The way this oil is shared between the two actuators A and B depends on the system architecture. In the open center systems (throttle control (OC), NC and PC) (OC), VBO and the classic LS system the oil flow to the higher load (in this case actuator B) is reduced. With flow sharing (LUDV) and electrohydraulic flow matching (EFM), the undersaturation is evenly shared between the actuators, so that both the actuators A and B receive less oil with their oil request. Additionally, due to the undersaturation, the losses in the flow sharing (LUDV) system caused by the LS pump regulator setup in combination with the flow sharing pressure compensators are reduced or even eliminated, dependent on the grade of undersaturation. For a single loop in a parallel movement with undersaturation, the flow sharing (LUDV) and the electrohydraulic flow matching (EFM) system is the most energy-efficient.

Also in Figure 3.139, the hydraulic systems need significant different pilot actuation to achieve the shown states.

3.4.5.2 Load Dependency of the Flow: The start of the movement and the flow increase in some of the described standard hydraulic implement solutions is load dependent. **Figure 3.140** shows the oil flow Q to the actuator for three different load situations (represented by the load pressure p_L) with the pilot actuation i.

When increasing a load of an actuator the start of move-in an LS system (including flow sharing and electrohydraulic flow matching (EFM)) is not changed, hence the name load sensing. As visible in **Figure 3.140** this is not the same as the other shown systems. VBO and positive control (PC) have a slight load feeling, so that the start of movement changes noticeable, but not to a point that the resolution for the rest of the movement is significantly reduced. This load feeling is even stronger for negative control (NC). The biggest influence of the load can be seen on throttle control and, depending on the spool design, this can lead to a noticeable reduction in controllability of the actuator speed under heavy loads.

3.4.5.3 Reaction to Change of Speed of the Prime Mover: One additional point where the behavior of the described mobile hydraulic systems differs significantly is the reaction of the machine behavior when the prime mover's speed is changed. In some applications machines are used at different engine speeds, like wheel loaders, sometimes based on the allowed noise level (such as excavators) and in other applications the engine speed is varied to change the machine reaction compared to the pilot activation. **Figure 3.141** shows what happens if the prime mover speed is reduced from its standard/rated speed n_r down to zero.

The red line shows the maximum oil flow a pump can supply for the given prime mover speed. In this example, the oil flow for all system architectures is set to a reference flow Q_r (with $Q_r < Q_{Pmax}$) at the rated speed n_r, which, i.e., could be 1,800 1/min for a diesel engine. When reducing the speed of the engine ($n > n_s$), in a classic LS or flow sharing (LUDV) system the oil flow Q to the cylinder/motor is not changing. This is caused by the system concept, where the pump tries to supply as much oil as needed to keep a given Δp over the load pressure and swivels out further if the engine speed is reduced. In an electrohydraulic flow matching (EFM) system there is also no change in flow to the cylinder, but in this case, the software compensates the reduced engine speed

FIGURE 3.140 Load dependency of different hydraulic systems.

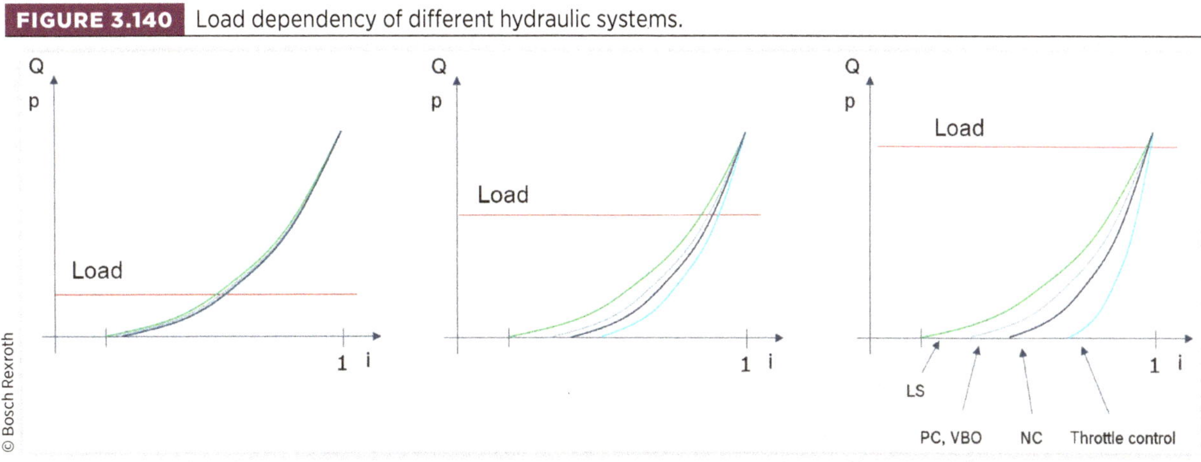

FIGURE 3.141 Actuator flow about the engine speed.

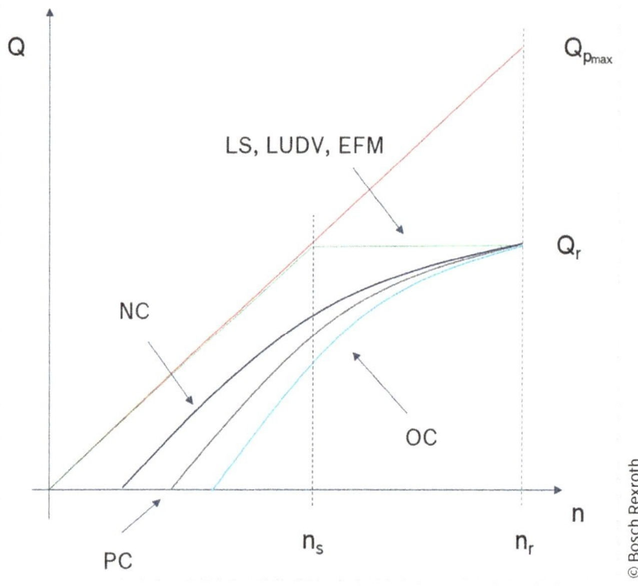

with a bigger swash angle command to the variable displacement pump. However, as soon as the engine speed reaches the saturation speed n_s (n_s is a function of Q_r) it is not possible to increase the swash angle of the pump anymore, as it is at its maximum. In comparison in throttle control (OC), NC, and PC in partial pilot actuation (Q_r) the flow to the actuator is more influenced by a reduction in engine speed, especially in throttle control system with constant displacement pumps.

Both behaviors have their benefits. The limited engine speed influence on the oil flow of the LS systems makes working with changing engine speeds easier. On the other hand, the significant influence of the prime mover speed on the actuator flow with the open center systems (throttle control (OC), NC and PC) allows for a higher fine control resolution at reduced speed.

References

1. Thiebes, P. and Geimer, M., "Energiespeicher für mobile Arbeitsmaschinen mit Hybridantrieben (Energy Storages for Mobile Working Machines with Hybrid Drives)," in: *1st VDI Conference "Getriebe in mobilen Arbeitsmaschinen" (Gear Boxes in Mobile Working Machines)*, Friedrichshafen, June 7-8, 2011 (VDI-Wissensforum, Düsseldorf, 2011), ISBN:978-3-942980-03-6.
2. Regulation (EU) 2016/1628 of the European Parliament and the Council of 14 September 2016 on Requirements Relating to Gaseous and Particulate Pollutant Emission Limits and Type-Approval for Internal Combustion Engines for Non-Road Mobile Machinery, L 252/53, September 16, 2016.
3. Pohlandt, C., "Intelligentes Gesamtmaschinenmanagement für elektrische Antriebssysteme (Intelligent Machine Management for Electric Drive Systems)," Dissertation, Karlsruher Schriftenreihe Fahrzeugsystemtechnik, Vol. 60 (Karlsruhe: KIT Scientific Publishing, 2018), 33.

4. Renius, K.Th. and Resch, R., "Continuously Variable Transmissions," ASAE Distinguished Lecture Series No. 29, *ASAE-ATEC Meeting Louisville*, St. Josef, MI, February 14-16, 2005 (ASAE Publishing, 2005).

5. ISO 1219-1:2012, "Fluid Power Systems and Components - Graphic Symbols and Circuit Diagrams - Part 1: Graphic Symbols for Conventional Use and Data-Processing Applications," Revision 2012.

6. Renius, K.Th., "Grundkonzepte der Stufengetriebe moderner Ackerschlepper (Basic Concepts of Step Transmissions for Modern Tractors)," *Grundlagen Landtechnik* 18, no. 3 (1968): 97-106.

7. Geimer, M., Renius, K.Th., and Stirnimann, R., "Motoren und Getriebe in Traktoren (Motors and Transmissions in Tractors)," in: Yearbook Agricultural Engineering, 2018, https://doi.org/10.24355/dbbs.084-201901211131-0, last view June 15, 2019.

8. Kohmäscher, T., "Modellbildung, Analyse und Auslegung hydrostatischer Antriebsstrangkonzepte (Modeling, Analysis and Design of Hydrostatic Drive Line Concepts)," Dissertation, RWTH Aachen University, Aachen, 2008.

9. Murrenhoff, H. and Eckstein, L., *Fluidtechnik für mobile Anwendungen (Fluid Power for Mobile Applications)*, 6th revised edn. (Aachen: Shaker Publishing, 2014), ISBN:978-3-8440-2919-2.

10. Looman, J., *Zahnradgetriebe, Grundlagen, Konstruktionen, Anwendungen in Fahrzeugen (Gear Drives, Basics, Designs, Applications in Vehicles)* (Berlin: Springer, 1996).

11. Kohmäscher, T. and Murrenhoff, H., "Advanced Modeling of Hydro-Mechanical Power Split Transmissions," *6th PhD Symposium 2010*, West Lafayette, 2010.

12. Ramm, M., "Systematische Entwicklung und Analyse stufenlos verstellbarer Getriebe mit innerer Leistungsverzweigung für mobile Arbeitsmaschinen (Systematic Design and Analysis of Infinitely Variable Transmissions with Internal Power Split for Mobile Working Machines)," Dissertation, RWTH Aachen University, Aachen, 2015.

13. Gelmini, M., "Development of a Simulation Tool for Thermal Clutch Behavior Analysis," Dissertation, UNI-TN Trento University, 2015.

14. Zagrodzki, P., "Numerical Analysis of Temperature Fields and Thermal Stresses in the Friction Discs of a Multidisc Wet Clutch," *Wear* 101, no. 3 (1985): 255-271.

15. Payvar, P., "Laminar Heat Transfer in the Oil Groove of a Wet Clutch," *International Journal of Heat and Mass Transfer* 34, no.7 (1991): 1791-1798.

16. Jianzhong, C. et al., "Numerical Investigation on Transient Thermal Behavior of Multidisc Friction Pairs in Hydro-Viscous Drive," *Applied Thermal Engineering* 67 (2014): 409-422.

17. Jen, T.-C. and Nemecek, D.J., "Thermal Analysis of a Wet Disc Clutch Subjected to a Constant Energy Engagement," *International Journal of Heat and Mass Transfer* 51 (2008): 1757-1769.

18. Shoaib, I. et al., "Mathematical Model and Experimental Evaluation of Drag Torque in Disengaged Wet Clutches," *ISRN Tribology* 2013 (2013): 1-16.

19. Lechner, G. and Naunheimer, H., *Fahrzeuggetriebe, Grundlagen, Auswahl, Auslegung und Konstruktion (Vehicle Transmissions, Basics, Selection, Design and Construction)* (Berlin: Springer, 1994).

20. Ivantysyn, J. and Ivantysynova, M., *Hydrostatic Pumps and Motors: Principles, Design, Performance, Modelling, Analysis, Control and Testing*, 1st English edn. (abi-Verlag, 2001).

21. Murrenhoff, H., *Grundlagen der Fluidtechnik, Teil 1: Hydraulik (Basics of Fluid Power, Part 1: Hydraulics)* (Aachen: Shaker Verlag GmbH, 2016), ISBN:978-3-8840-4816-2.

22. Danfoss Power Solutions, "Basic Information - H1 Axial Piston Pumps - Single and Tandem," BC00000057en-000703, November 2019.

23. Danfoss Power Solutions, "Technical Information - H1 Bent Axis Motors - Size 060/080/110/160/210/250", BC00000043en-US1108, March 2018.

24. Kohmäscher, T., Rahmfeld, R., Murrenhoff, H., and Skirde, E., "Improved Loss Modeling of Hydrostatic Units – Requirements for Precise Simulation of Mobile Working Machine

Drivelines," *Proceedings of IMECE07, 2007 ASME International Mechanical Engineering Congress and Exposition*, Seattle, WA, November 11-15, 2007.

25. Rashid, M.H., *Power Electronics Handbook: Devices, Circuits and Applications* (Amsterdam: Elsevier, 2010).
26. Kabziński, J., *Advanced Control of Electrical Drives and Power Electronic Converters* (Berlin: Springer, 2016).
27. Geyer, T., *Model Predictive Control of High Power Converters and Industrial Drives* (New York: John Wiley & Sons, 2017).
28. Leonhard, W., *Control of Electrical Drives* (Berlin: Springer Science & Business Media, 2001).
29. De Doncker, R., Pulle, D.W.J., and Veltman, A., *Advanced Electrical Drives: Analysis, Modeling, Control* (Berlin: Springer Science & Business Media, 2010).
30. Dana Corporation, "Mantenance and Service Manual T16000 Powershift Transmission," 1990. Dana Power Shift Technical Literature and Manuals.
31. Goering, C.E., *Off-Road Vehicle Engineering Principles* (St. Joseph, MI: ASAE, 2006), 255-349, ISBN 1-892769-26-3.
32. Geimer, M. and Pohlandt, C., *Grundlagen mobiler Arbeitsmaschinen (Basics of Mobile Working Machines)*, Karlsruher Schriftenreihe Fahrzeugsystemtechnik, Vol. 22 (Karlsruhe: KIT Scientific Publishing, 2014), ISBN:978-3-7315-0188-6.
33. Kress, J.H., "Hydrostatic Power-Splitting Transmission for Wheeled Vehicles - Classification and Theory of Operation, Society of Automotive Engineers," SAE Technical Paper 680549, 1968, https://doi.org/10.4271/680549.
34. Jacobson, E., Wright, J., and Kohmäscher, T., "Hydro-Mechanical Power Split Transmissions (HMT) - Superior Technology to Solve the Conflict: Tier 4 vs. Machine Performance," *IFPE 2011*, Las Vegas, NV, 2011.
35. Göllner, W. et al., "Das Doppeljoch - ein innovativer Ansatz für neue hocheffiziente Leistungsverzweigungsgetriebe basierend auf Großwinkel-Schrägachsenmaschinen (The Dual Yoke - An Innovative Approach for New Highly Efficient Power-Split Gears Based on Large Angle Bent Axis Machines)," *5th Colloquium Mobile Hydraulics*, Karlsruhe, October 16-17, 2008, in: *Conference Proceedings*, 54-68.
36. Mistry, S.I. and Sparks, G.E., "Infinitely Variable Transmission (IVT) of John Deere 7000 TEN Series Tractors," *ASME International Mechanical Engineering Congress*, New Orleans, November 17–22, 2002, Paper IMECE2002-39347 (New York: AMSE).
37. Renius, K.Th., "Traktoren 2006," *LANDTECHNIK* 61, no. 4 (2006): 186-187.
38. Bosch Rexroth, *Hydraulik in mobilen Arbeitsmaschinen (Hydraulics in Mobile Working Machines)*, 2nd ed. (Ditzingen: Omegon Fachliteratur, 2001) ISBN:3-933698-15-4.
39. Bosch Rexroth Hydraulik Trainer, Vol. 1, RD 00290/10.1991, Grundlagen und Komponenten der Fluidtechnik (Basics and Components of Fluid Power), ISBN:3-8023-0266-4.
40. Bosch Rexroth Hydraulik Trainer, Vol. 3, RD 00281/10.88, Projektierung und Konstruktion von Hydroanlagen (Project Planning and Construction of Hydraulic Systems), ISBN:3-8023-0266-4.
41. Geerling, G., *Entwicklung und Untersuchung neuer Konzepte elektrohydraulischer Antriebe von Flugzeug-Landeklappensystemen (Development and Investigation of New Concepts for Electro-Hydraulic Drives for Aircraft Landing Flap Systems)*, VDI-Fortschrittsbericht Reihe 12, Nr. 538 (Düsseldorf: VDI-Verl., 2003), ISBN:3-18-353812-1.
42. Bosch Rexroth L8_L1… (ED-LV) Product Documentation RE 18301-08/02.2016.
43. Bosch Rexroth SP08 Product Documentation RE 64139/07.2015.
44. Bosch Rexroth 4THH5 Product Documentation RE 64555/04.2006.

45. Bosch Rexroth 4THE5 Product Documentation RE 29881/09.2010.
46. Bosch Rexroth A11 Product Documentation RE 92500/10.2009.
47. Bosch Rexroth M8 Product Documentation RE 64294/03.2017.
48. Bosch Rexroth RS12 Product Documentation, Bauma, 2016.
49. Bosch Rexroth Project Manual Mobile Hydraulics LUDV RE 09972/03.2011.
50. Axin, M., *Mobile Working Hydraulic System Dynamics* (Berlin: Linköping University Electronic Press, 2015), ISBN:978-91-7685-971-1.
51. Gro, S., *Concepts of Hydraulic Circuit Design Integrating the Combustion Engine* (Shaker Verlag Gmbh, 2014), ISBN:978-3-8440-2660-3.
52. Vukovic, M., *Hydraulic Hybrid Systems for Excavators* (Aachen: Shaker, 2017), ISBN:978-3-8440-5312-8.
53. Heemskerk, E., Burgis, S., and Hörner, F., "A Multi-Pressure Network System for Mobile Applications," *16th Scandinavian International Conference on Fluid Power (SICFP)*, Tampere, Finland, May 22-24, 2019.
54. Linjama, M., "Digital Fluid Power – State of the Art," *SICFP 2011*, Tampere, Finland, 2011.
55. Heemskerk, E., Bonefeld, R., and Buschmann, H., "Control of a Semi-Binary Hydraulic Four-Chamber Cylinder," *14th Scandinavian International Conference on Fluid Power (SICFP)*, Tampere, Finland, May 20-22, 2015.
56. Bosch Rexroth SM12 Product Documentation RE 64122/07.2016.
57. Yamaji, K., "Method for controlling variable displacement pump," European Patent EP2703652A1, 2014.
58. Bosch Rexroth Project Manual Mobile Hydraulics Load Sensing RE 09970/03.211.
59. Bosch Rexroth M4-12 Product Documentation RE 64276/02.2019.
60. Bosch Rexroth M7-20 Product Documentation RE 64293/06.2013.
61. Scherer, M., Geimer, M., and Weiß, B., "Contribution on Control Strategies of Flow-on-Demand Hydraulic Circuits," *13th Scandinavian International Conference on Fluid Power (SICFP)*, Linköping, Sweden, June 3-5, 2013.
62. Bosch Rexroth Video E-Load-Sensing, https://www.youtube.com/watch?v=HBEOBdl_Mr0.
63. Latour, Ch., "Electrohydraulic Flow Matching: The Next Generation of Load-Sensing Controls," O+P, 2007.
64. Scherer, M., "Beitrag zur Effizienzsteigerung mobiler Arbeitsmaschinen: Entwicklung einer elektrohydraulischen Bedarfsstromsteuerung mit aufgeprägtem Volumenstrom (Contribution to Increasing the Efficiency of Mobile Machines: Development of an Electrohydraulic Flow-on-Demand Control System with Impressed Flow Rate)," Dissertation, Karlsruhe Institute of Technology (KIT), 2015 (KIT Scientific Publishing).
65. Djurovic, M., *Energiesparende Antriebssysteme für die Arbeitshydraulik mobiler Arbeitsmaschinen "Elektrohydraulisches Flow Matching" (Energy-Saving Drives for the Function Drives of Mobile Working Machines "Electrohydraulic Flow Matching")* (Aachen: Shaker, 2007), ISBN:978-3-8322-6361-4.
66. Scherer, M., Geimer, M., and Weiß, B., "Forestry Crane with Electrohydraulic Flow-on-Demand System," *71st Conference "Land. Technik – AgEng"*, 2013, in: *Conference Proceedings* (Düsseldorf, VDI-Verlag), 175-182.
67. Finzel, R., *Elektrohydraulische Steuerungssysteme für mobile Arbeitsmaschinen (Electro-Hydraulic Controls for Mobile Working Machines)* (Aachen: Shaker, 2011), ISBN:978-3-8322-9786-2.
68. Weickert, Th., "States of Independence," iVT International Off-Highway, 2016.
69. Bosch Rexroth RCS30 Product Documentation RD 64155/02.2019.
70. Bosch Rexroth MHDBN Product Documentation RE 64602/02.2016.

CHAPTER 4

Machine Control Concepts

Detlef Hawlitzek

Life today cannot be imagined without electronics in motor vehicles and mobile working machines. Many necessary and convenient functions could not be implemented without electronic systems. The requirements on mobile working machines, e.g., construction machines, municipal vehicles, or agricultural machines, have changed very much over the last years. The demand for continuously increasing productivity while at the same time increasing the driving comfort by means of state-of-the-art display and operating systems keeps facing the manufacturers with new challenges. As compared to electronics in consumer goods or in "normal" industrial use, e.g., in packaging machines or conveying systems, there are considerably higher requirements regarding use in mobile applications, to ensure sufficient operating reliability in all work situations.

4.1 Requirements and Trends

Mobile working machines and plants are often tailored to specific applications and thus provide optimized functions. Since they can therefore not easily be replaced by other machines, high availability and reliability are of utmost priority. The integration of system diagnosis (in hardware and software) which enables the operating staff to quickly localize faults minimizes downtime. This can be supported by the use of dialogue units (HMI = Human Machine Interface). These dialogue modules provide the machine operator with clear information and messages adapted to the operating situation, either as text or as graphics. Incorrect operation of the machine is also prevented this way. For the machine operator the dialogue units have to be clearly visible and easily accessible, so the best possible mounting locations on the machine have to be selected. The requirements for a clear, ergonomic design permanently withstanding the hard conditions in field applications are increasing.

4.1.1 Electronics and Hydraulics

Increasingly harder regulations on emission such as TIER 4 (final) have made the demand for energy efficient solutions grow worldwide. In these solutions, the electronics in conjunction with work function and traction drive control plays an important role. The interaction of these components helps reduce diesel consumption and thus stack emissions of mobile working machines. Energy efficient hydraulics with high user-friendliness will also contribute its share. That means that the further electrification of hydraulics as well as the integration of hydraulics in networked systems of mobile working machines will be a challenge in the future. Besides, the increasing use of electric drives in mobile working machines as a replacement for hydraulic functions makes the use of intelligent efficient control units necessary. This trend continues through hybrid vehicles to completely electrically driven vehicles, such as compact sweeper vehicles.

4.1.2 Safety

Almost all mobile working machines have functions that may endanger persons and material. Therefore, every manufacturer has to comply with the general regulations for a safe machine design. Given that these regulations and standards are defined for a broad range of different machines, they cannot be precisely adapted to the function of a mobile working machine. Therefore, there is an increasing number of product standards which are tailored to specific requirements. In certain applications like vehicle lifts, clearly defined product standards have been in place for a long time. The employers' liability insurance associations also often have clear requirements towards manufacturers of machines. For this reason, there is an increasing demand for certified electronic assemblies for mobile vehicles. So-called safety control units can be used in applications where components with different performance levels (Pl) to EN 13849 [4] or SIL to EN 62061 [5] are needed. The safety concept of these control units monitors all internal and external functions and reliably switches off in case of an error. For safe data transmission the CANopen safety protocol can be used in the network. Practical: transmission is carried out together with the "non-safe" data on the same bus cable; no additional bus cable is required.

4.1.3 Driver Assistance Systems

There is an increasing demand for so-called assistance systems to support safe machine operation. They can, for instance, monitor the working area of a mobile working machine

FIGURE 4.1 Time-to-flight principle.

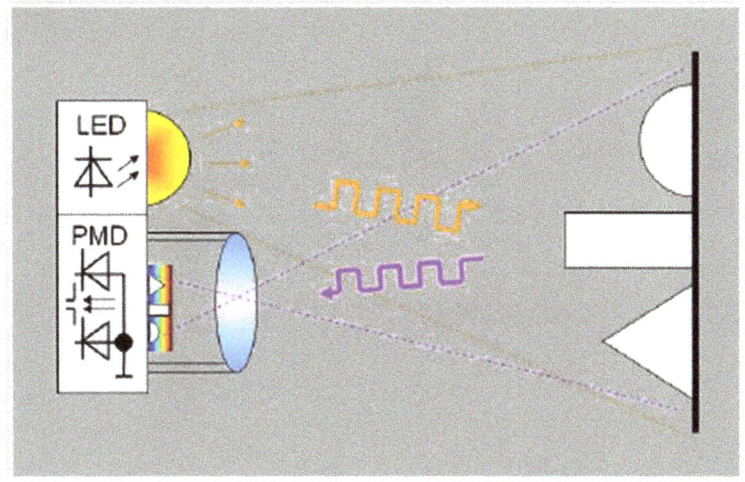

for foreign objects. Examples are 2D/3D sensors. On a photoelectric basis they detect objects in the detection range and in the result transmit size and distance of the detected objects to the downstream control unit (Figures 4.1 and 4.2).

The most important element of the 3D sensor system is a photonic mixer device (PMD) whose function is based on the time-of-flight principle. A modulated light source illuminates the detection zone with invisible infrared light. The PMD sensor that is coupled with the modulation source receives the reflected light and measures the phase shift between the transmitted and received signals. This allows to precisely determine

FIGURE 4.2 Example 3D- and 2D camera.

FIGURE 4.3 Typical applications for distance measuring in the field.

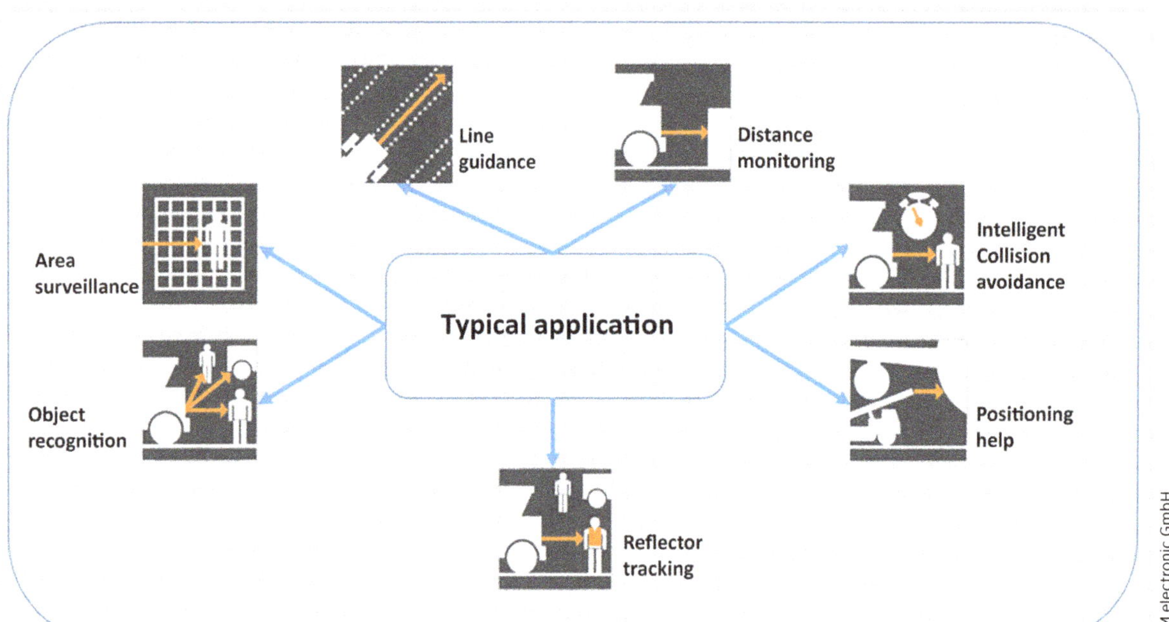

the time-of-flight of the light and thus the distance to the object. The PMD sensor has an integrated active extraneous light suppression and operates extremely reliably even in bright sunlight.

This technology is well-known in the automotive range and is also used in latest generation mobile phones.

Typical applications are collision avoidance, obstacle detection, line guidance, autonomous driving, object recognition, or area surveillance (Figure 4.3).

For similar applications without such high requirements such as requirements on precision, sensor systems on the basis of ultrasonic or radar can be used.

4.1.4 Logging of Operational Data

Another requirement: because machine manufacturers and operators want to operate their machines as efficiently as possible and want to reduce downtime to a minimum, operational data logging is becoming increasingly important. System states and machine data have to be saved for statistic evaluation and documentation. This happens via electronic memory components on the machine, with remote access to these data from any location at any time being an additional requirement. So, further technologies are needed allowing access to these data via Internet or via GSM modem. That way manufacturer as well as operator can read and monitor machine states at any time. This not only ensures the optimum operation of the machine, but also increases predictive service management.

4.1.5 Networking

Legislation's requirements concerning safety keep increasing and customers have continuously higher functional requirements to be met by machine manufacturers so the complexity of systems also increases. As a result, the demand for compact and efficient control units that can be combined depending on their functionality keeps getting stronger. Connectivity is therefore a big issue when talking about networked systems. This facilitates the use of bus systems which connect the individual control units for the purpose of data exchange. They allow a remote and intelligent arrangement of the control units in use. So, it is possible to place the control units in the machine where sensors and actuators are mounted. This has the further advantage that control units can be more compact. Another advantage is the considerable reduction of wiring which additionally helps to save weight. In addition, the components must be easy to handle for mounting and in case of service. This is achieved by the mechanical design as well as easy and clear integration during programming. Finally, the components and devices used must withstand the extreme mechanical, climatic, and electric requirements. All requirements have one goal: to produce reliable and competitive machines.

4.2 Controller and Sensors

Mobile working machines nowadays are used worldwide. Climatic and geological conditions play an important role in the construction of the machinery. On the other hand, they should not affect the functionality of a mobile working machine. The extreme mechanical stress caused by impacts and shocks as well as the use at low and high ambient temperatures require a careful selection of the electronic components. Given that the devices are often exposed to dirt, moisture, and water in the applications, high protection ratings and specially selected housing materials are necessary. Furthermore, the electrical interference affecting the entire system or individual components must also be taken into account. A wide supply voltage range and adapted protection measures allow reliable operation of the devices even in cases of large voltage fluctuations caused by the battery/generator system and high conducted interference.

4.2.1 Requirements

Modern control units and I/O extensions for mobile working machines meet the required manufacturers' specifications (**Figures 4.4-4.6**).

Extensive climatic tests take place to meet the specifications of the extended temperature range of −40°C to +80°C (−40°F to 176°F), as well as extensive mechanical tests for vibration and shock resistance. Intensive electric tests following different standards are as self-understood as certification for road traffic or TÜV approvals concerning functional safety. In addition, the machine manufacturer expects data from application-specific sensors and actuators to be processed by the control units. For this purpose, control units offer multifunctional inputs and outputs. These inputs can be configured by software as digital, analog, frequency, resistance, or ratiometric input signals, while

FIGURE 4.4 Controllers and displays.

©IFM electronic GmbH

FIGURE 4.5 I/O modules.

©IFM electronic GmbH

the outputs can be parameterized as digital, PWM (see Section 4.2.2), or analog output signals. In addition, extensive diagnostic possibilities enable the detection of excessive current, wire break or short-circuit; they can be evaluated by means of the programming software. Signalling elements indicate the function state of the control unit. This can be extremely helpful for troubleshooting.

FIGURE 4.6 Controllers.

The units provide several interfaces for networking. The CAN bus with the protocols CANopen (European Norm EN 50325-4 [1]–supported by CiA, CAN in Automation) and SAE J 1939 [11] (Society of Automotive Engineers) has successfully established itself for networking of the devices. In addition, controllers and displays provide interfaces for Ethernet or industry-specific protocols such as ISOBUS.

Since simple and efficient operating concepts are gaining more and more importance, the requirements for optical operating units are getting more complex (Figure 4.7).

FIGURE 4.7 Example displays for mobile working machines.

FIGURE 4.8 Jog-dial element.

The manufacturers of mobile working machines expect the same requirements regarding harsh environments as for the controllers. Different screen sizes and ergonomic housing concepts as well as the arrangement of operating keys and operation via touch function play an important role. Furthermore, a trouble-free readability at solar radiation, adjustable brightness and high viewing angle stability are expected. Operating concepts from the passenger vehicle sector are finding their way into mobile working machines. Jog-dial elements in combination with operating keys can be found in driver's cabins more and more often (Figure 4.8). They can enormously increase the operating comfort for the driver.

4.2.2 PWM and Dither

PWM and dither are basic functions from controllers for mobile working machines to control hydraulic valves. The abbreviation PWM stands for pulse width modulation (Figure 4.9). PWM is mainly used to trigger proportional valves (PWM valves) for mobile and robust controller applications. Also, with an additional component (accessory) for a PWM output, the pulse-width modulated output signal can be converted into an analogue output voltage.

FIGURE 4.9 PWM principle.

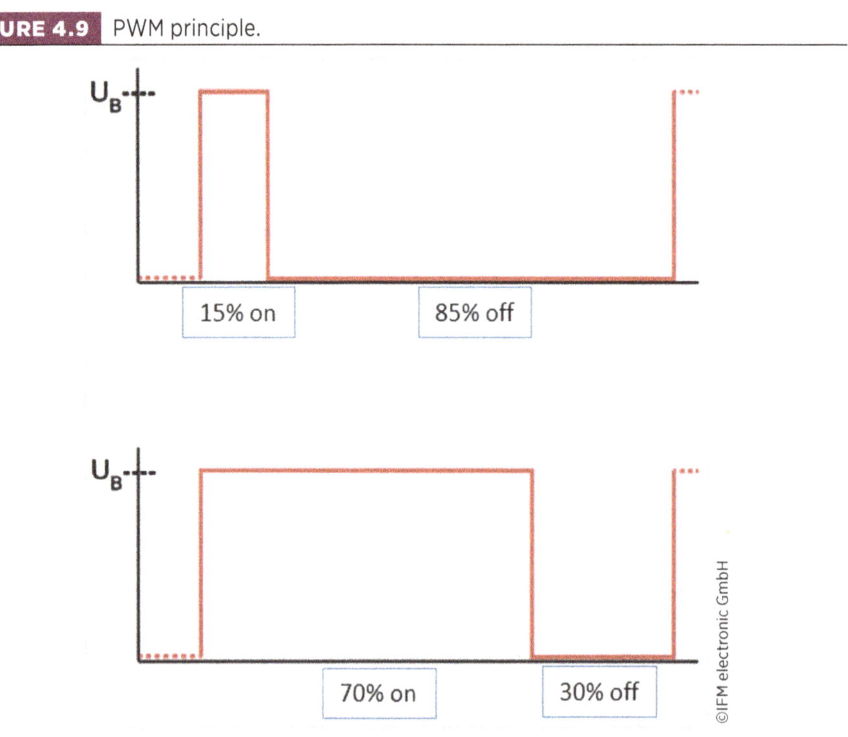

The PWM output signal is a pulsed signal between GND and supply voltage. Within a defined period (PWM frequency) the mark-to-space ratio is then varied. Depending on the mark-to-space ratio, the connected load determines the corresponding RMS current.

The PWM function of the controller is a hardware function provided by the processor. To use the integrated PWM outputs of the controller, they must be initialized in the application program and parameterized corresponding to the requested output signal

4.2.2.1 **What Does a PWM Output Do?** In general, digital outputs provide a fixed output voltage as soon as they are switched on. The value of the output voltage cannot be changed here. The PWM outputs, however, split the voltage into a quick sequence of many square wave pulse trains. The pulse duration [switched on]/pulse duration [switched off] ratio determines the effective value of the requested output voltage. This is referred to as the switch-on time in (%).

In the Figures 4.10-4.12, the current profiles are shown as a stylized straight line. In reality, the current flows to an e-function.

The profile of the PWM voltage U and the coil current I at 10% switch on time: The effective coil current Ieff is also 10%, see Figure 4.10.

The profile of the PWM voltage U and the coil current I at 50% switch-on time: The effective coil current Ieff is also 50%, see Figure 4.11.

The profile of the PWM voltage U and the coil current I at 100% switch-on time: The effective coil current Ieff is also 100%, see Figure 4.12.

FIGURE 4.10 PWM example 10%.

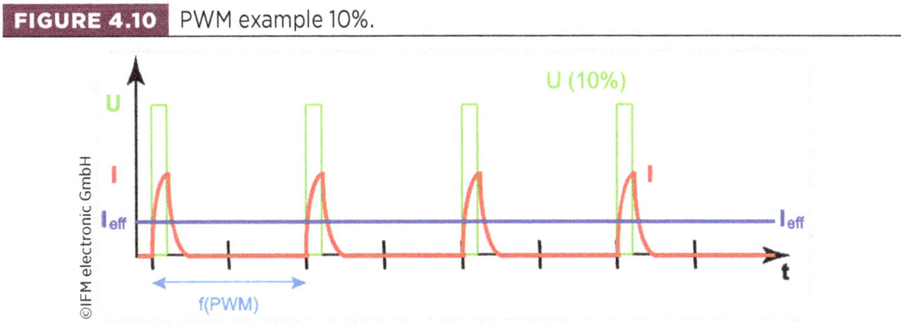

FIGURE 4.11 PWM example 50%.

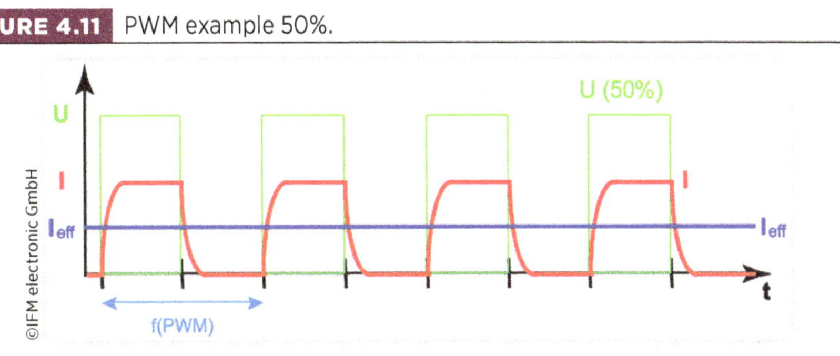

FIGURE 4.12 PWM example 100%.

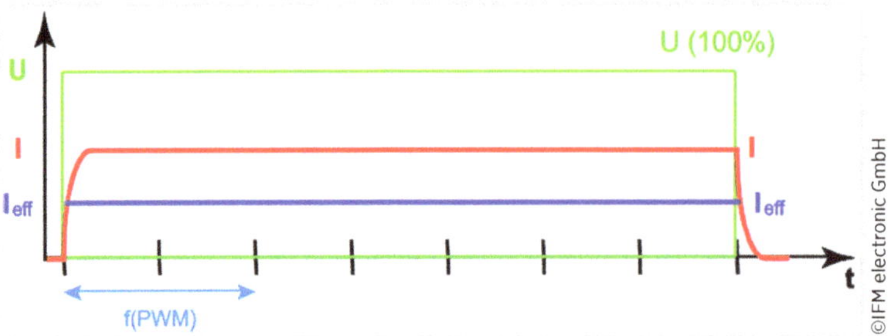

4.2.2.2 **PWM: What Is the Dither?** If a proportional hydraulic valve is controlled, its piston does not move right away and at first not proportional to the coil current. Due to this "slip stick effect," a kind of "break-away torque," the valve needs a slightly higher current at first to generate the power it needs to move the piston from its off position. The same also happens for each other change in the position of the valve piston. This effect is reflected in a jerking movement of the valve piston, especially at very low manipulating speeds.

Technology solves this problem by having the valve piston move slightly back and forth (dither). The piston is continuously vibrating and cannot "stick." Also, a small change in position is now performed without any delay, a "flying splice" so to speak.

- **Advantage**: The hydraulic cylinder controlled in that way can be moved more sensitively.
- **Disadvantage**: The valve becomes measurably hotter with dither than without because the valve coil is now working continuously.

That means that the "golden means" has to be found.

4.2.2.3 **When Is a Dither Useful?** When the PWM output provides a pulse frequency that is small enough (standard value: up to 250 Hz) so that the valve piston continuously moves at a minimum stroke, an additional dither is not required (see Figure 4.13):

At a higher PWM frequency (standard value 250 Hz up to 1 kHz) the remaining movement of the valve piston is so short or so slow that this effectively results in a standstill so that the valve piston can again get stuck in its current position (and will do so!) (see Figures 4.14 and 4.15):

FIGURE 4.13 PWM signal, no dither.

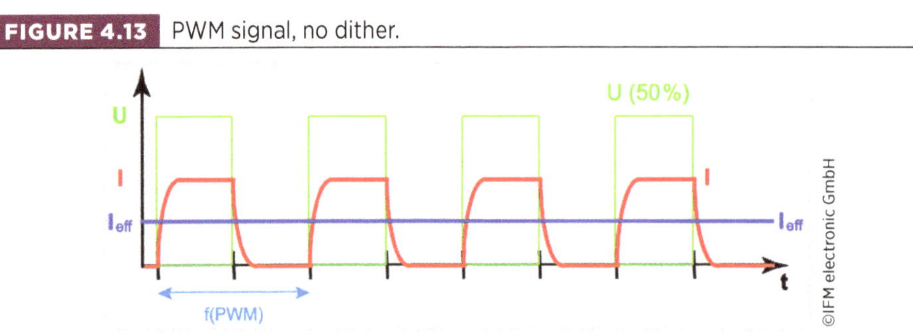

FIGURE 4.14 PWM higher frequency.

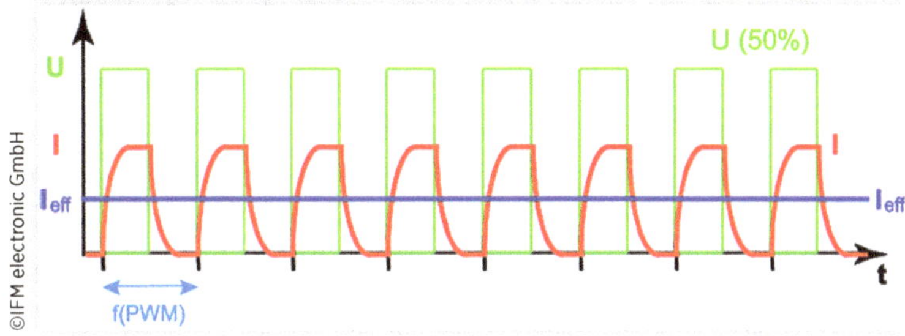

FIGURE 4.15 PWM low frequency.

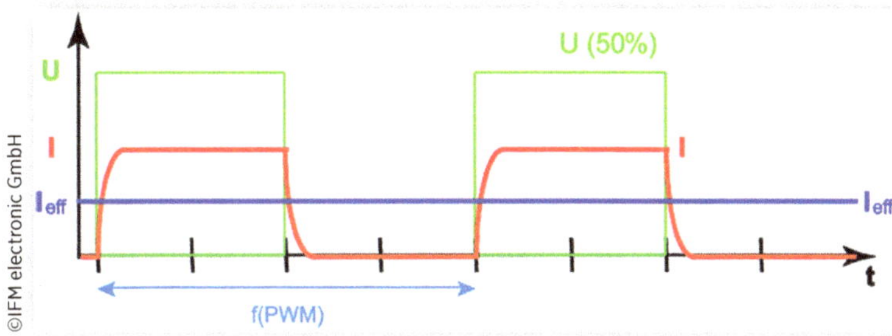

A high frequency of the PWM signal results in an almost direct current in the coil. The valve piston does not move enough any longer. With each signal change, the valve piston has to overcome the break-away torque again.

Too low frequencies of the PWM signal only allow rare, jerking movements of the valve piston. Each pulse moves the valve piston again from its off position; everytime the valve piston has to overcome the break-away torque again.

In case of a PWM switch-on time under 10% and over 90%, it is adequate and necessary to superimpose the PWM signal with a dither signal.

4.2.2.4 **Dither Frequency and Amplitude:** The mark/space ratio (the switch-on time) of the PWM output signal is switched with the dither frequency. The dither amplitude determines the difference of the switch-on times in the two dither half-waves.

The dither frequency must be an integer part of the PWM frequency. Otherwise, the hydraulic system would not work evenly but it would oscillate.

4.2.2.5 **Example Dither:** The dither frequency is 1/8 of the PWM frequency. The dither amplitude is 10%.

With the switch-on time of 50% in the figure, the actual switch-on time for 4 pulses is 60% and for the next 4 pulses it is 40% which means an average of 50% switch-on time (see **Figure 4.16**). The resulting effective coil current will be 50% of the maximum coil current.

The result is that the valve piston always oscillates around its off position to be ready to take a new position with the next signal change without having to overcome the break-away torque before.

FIGURE 4.16 Example dither.

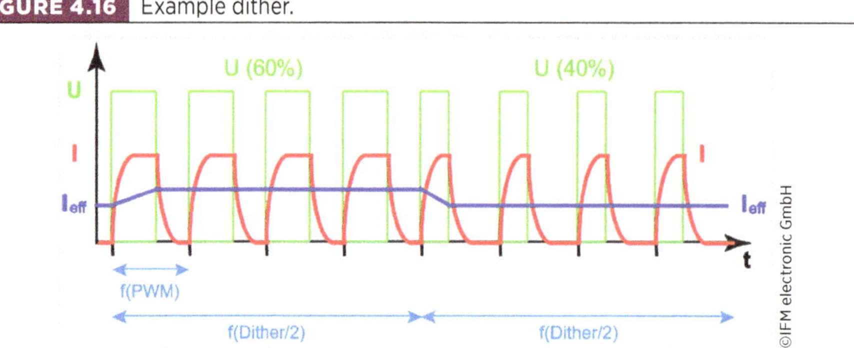

4.3 Automation Technologies

The choice of components for the different applications mainly depends on the design of the machine and installation. The following questions should be asked when constructing a system architecture:

- Where in the machine will the system components be placed? Are they protected against extreme mechanical impact, e.g., rocks falling onto the connectors? Also, extreme temperatures (e.g., hot asphalt) may influence the components.
- Which wiring is chosen (material and cable design)?
- How can an ideal network structure be implemented?
- Which operating concept is planned? Classical layout with mechanical switches and keys or operation of the machine via dialogue unit?
- Which diagnostic and service information is needed? Should this information only be available within the machine or should it be called off or transmitted via remote maintenance, if required?
- Can machine functions be combined?
- Is an extension of functions planned for the future?

Apart from these basic project planning questions, two system concepts exist in practical use.

4.3.1 Machines with a Central Controller

These are mostly machines with only a few work functions. The need for controller inputs and outputs is limited. A single central control module with application-dependent performance takes over the control of the complete machine (traction drive and work functions). All sensors and actuators implemented in the machine are connected to the central control unit which processes their signals (Figure 4.17). In general, a dialogue module is used additionally for displaying system states and diagnostic data and for entering operation requests.

The advantage of this system architecture is the compact design for machines with reduced function scope. An extension of the system is only possible with restrictions, unless taken into account beforehand.

FIGURE 4.17 Example central controller.

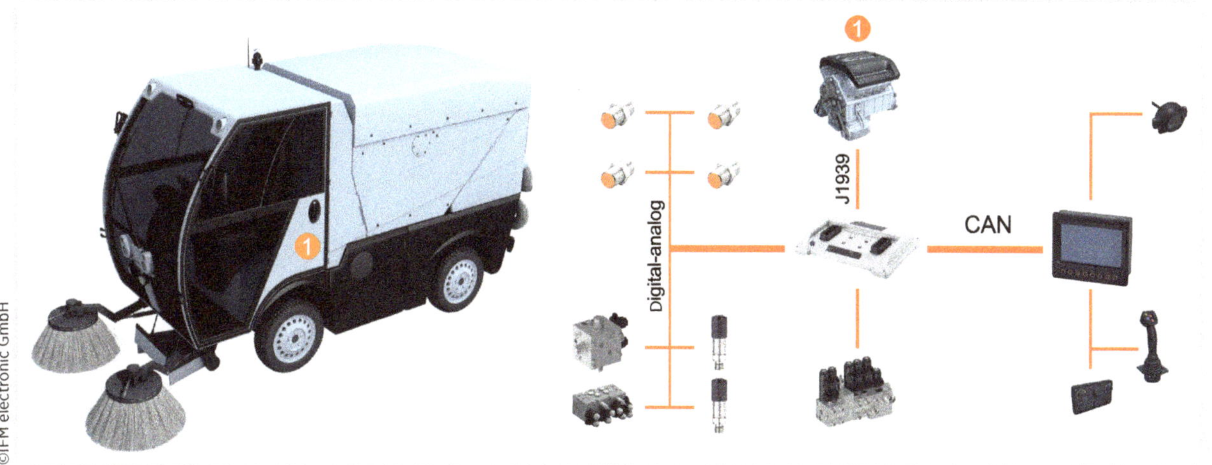

4.3.1.1 **Typical Applications:** Agricultural implements, access platforms, compact construction machines, compact sweeper vehicles, simple monitoring, and diagnostic tasks.

4.3.2 Machines with Decentralized Control Technology/Distributed Intelligence

Once the function scope of a mobile working machine increases the system architecture has to become more complex (Figure 4.18). Here it might help to divide the function of the machine into smaller logic function blocks which can be controlled by independent control units. The individual control units are interconnected for data exchange via a bus system. The most common bus system is the CAN bus. This structure can cover a high requirement of inputs and outputs, like sensors and actuators, at the same time considerably reducing wiring. Another important advantage of the decentralized system architecture is the possibility of subsequent extension without having to redesign the complete system architecture. In general, a dialogue module is used additionally for displaying system states and diagnostic data and for entering operation requests.

4.3.2.1 **Typical Applications:** Complex construction machines, drilling machines, municipal vehicles, agricultural machines.

Some guidance on which system architecture to choose will be given in the following.

4.3.2.2 **The Most Important Functional Units:** Modern mobile working machines consist of different segments. The machine manufacturer decides on the automation components suitable for mobile use for the different tasks. Depending on the requirements, they assume control, monitor, and display functions. Networking is made via the CAN bus with the CANopen protocol.

4.3.2.3 **Segment Engine:** Beside the conventional engine with differential gear and torque converter, stepless hydrostatic drives have achieved recognition over the past

FIGURE 4.18 Example decentralized controller.

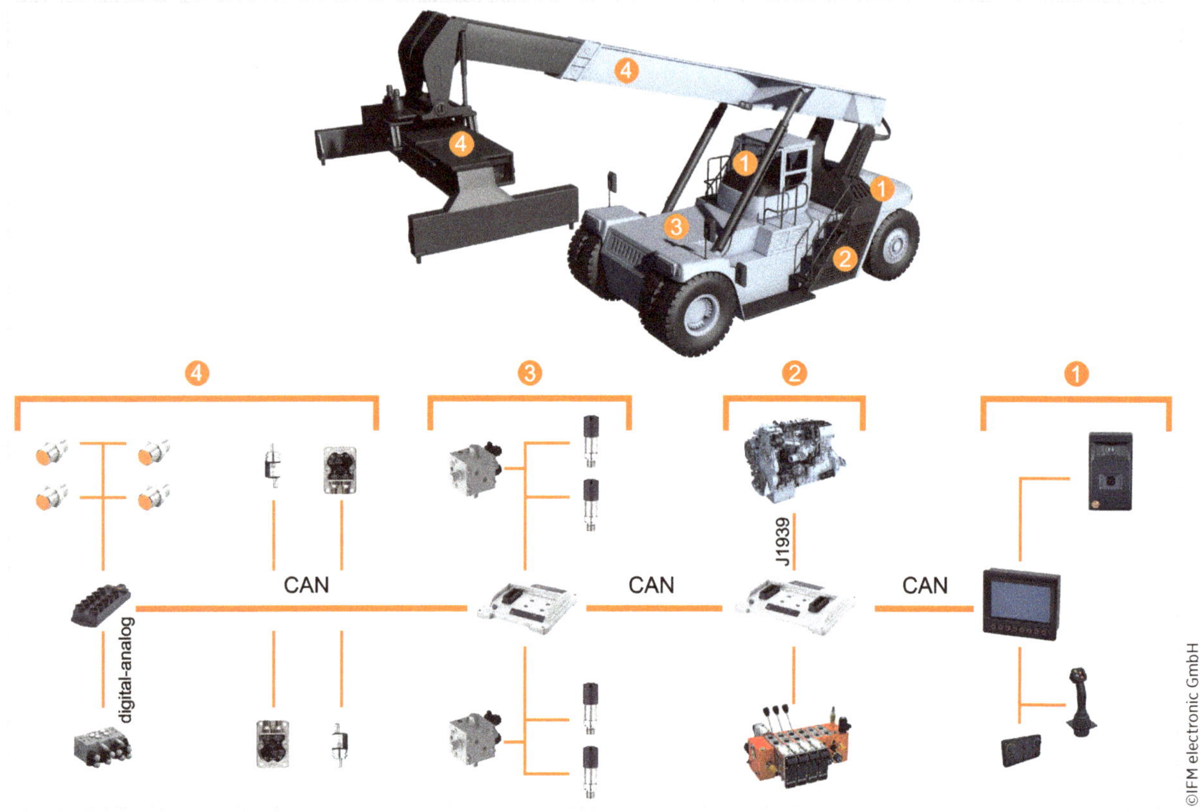

years in many mobile working machines. Main advantages are an increased productivity (stepless driving, braking, and reversing) and a considerably improved driving comfort. In addition, they provide a very high automation potential.

4.3.2.4 **Segment Motor:** Depending on the task, engines of different performance classes are used. Modern combustion engines have a CAN interface which is accessible to the user. Most of the time, it is designed to the standard SAE J 1939 [11]. It, e.g., encodes engine parameters such as operating temperature, oil pressure, torque, or rotational speed which are then processed by the control unit. In an electric motor, additional components such as inverter, charger, and DC converter are used. They also supply characteristic data to a control unit which filters and processes the data.

4.3.2.5 **Segment Machine Interface:** In mobile working machines a variety of operating elements is used for machine control. Information provided by joysticks, switches, and pushbuttons can be passed on to the control unit via CAN bus either individually or summed up. Dialogue modules (HMI) inform the operator about the operating states of the machine and are used for operator inputs. In case of disturbances, informative symbols and texts help to quickly localize and remove the fault. In addition, operating and diagnostic data can be saved permanently. In case of service, they provide significant information about the application and problems occurred.

With additional remote solutions error messages can be directly passed on to the service department or the machine manufacturer. In particular for machines in worldwide use, this allows considerable service cost savings.

4.3.2.6 **Segment Work Facilities/Vehicle Body:** This segment represents the actual work process of the machine. Proportional hydraulic valves, e.g., are controlled via the current-controlled PWM outputs of the decentralized output modules. Sensor signals, too, are detected via the I/O modules and transmitted to the control unit via CAN bus.

In small machines as well as in large and complex systems, whole functional units are controlled by intelligent control units. These freely programmable devices directly process all relevant process signals of their assigned machine units. Only relevant, preprocessed data or status messages are transmitted to other participants in the network.

4.4 Communication and Networking

4.4.1 Bus Systems in Mobile Working Machinery

Modern vehicle industry cannot be imagined without bus systems. The increasing technical requirements for more comfort, safety, and functionality require the replacement of parallel wiring with its extensive cable trees and the resulting weight by more suitable solutions. Against this background the companies Bosch and Intel in cooperation with the automobile industry developed the CAN bus (Controller Area Network) in the 1980s. CAN bus is an asynchronous serial fieldbus system via which actuators and sensors can be networked for communication. It is standardized via ISO 11898 [9], another reason for its wide distribution and use.

Standardization of the communication allows communication between the devices and thus processing and exchange of information independent of the manufacturer. Robustness and reliability of the data exchange make the CAN bus the suitable bus system in vehicles and mobile working machines.

In order to compare and clearly describe different kinds of bus systems and subject them to standardization, the so-called OSI reference model (OSI = Open System Interconnection) has proven successful (see **Figure 4.19**).

The OSI reference model consists of seven layers and describes the requirements on the bus system.

- Layer 1 (Physical Layer, bit transmission layer) just describes the physical kind of data transmission. This concerns the type of cables, connectors, signal conventions, and other details of physical requirements.
- Layer 2 (Data Link Layer, Data Saving Layer) defines the implementation of faultless and reliable data transmission and the access to the data transmission medium (layer 1).
- Layers 3 (Network Layer), 4 (Transport Layer), 5 (Session Layer) and 6 (Presentation Layer) are not used by the CAN bus.
- Layer 7 (Application Layer) is the interface between the application and the lower layers (e.g., layers 1 and 2). This is also where data input and output takes place (defined for CANopen and SAE J 1939 [11]).

FIGURE 4.19 OSI reference model.

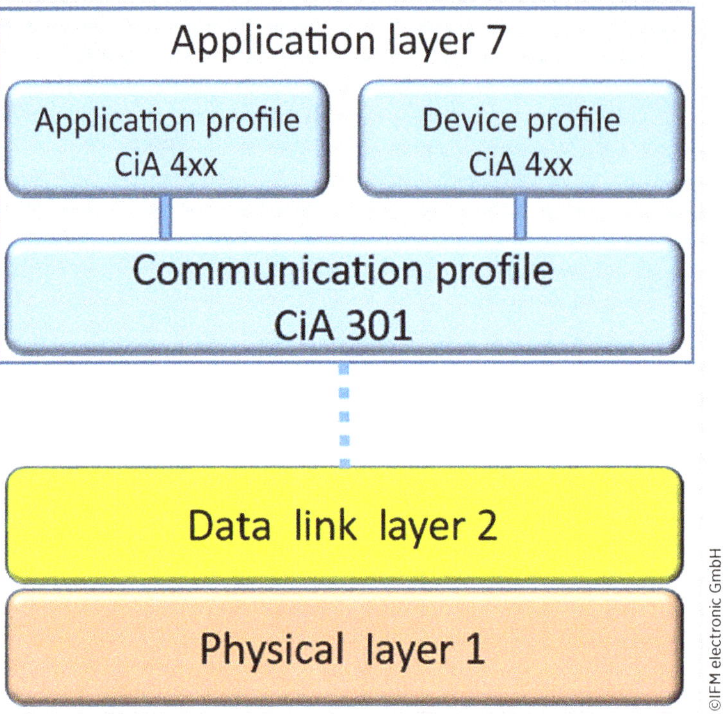

4.4.1.1 **The Physical Properties of the CAN Bus:** In the case of the CAN bus, layer 1 is a three-wire technology consisting of the signal wires CAN-high, CAN-low, and CAN-GND. As required, the bus cable is often twisted or additionally screened. Differential voltage levels ensure the optimum noise immunity. Interferences on the cable usually have the same effect on both signal wires and are suppressed by means of the proven method of difference formation (Table 4.1).

4.4.1.2 **The CAN Topology:** The network topology of the CAN bus is the linear structure (see Figures 4.20 and 4.21). The individual nodes are connected to a continuous cable via spurs that should be as short as possible. 120 Ohm resistors are located on either end of the bus cable as termination (bus termination). The bus cables should also be screened twisted pair cables (Figure 4.22).

TABLE 4.1 Voltage level (ISO 11 898).

	Recessive		Dominant	
	Min.	**Max.**	**Min.**	**Max.**
$V_{CAN\ H}$	2.0 V	3.0 V	2.75 V	4.5 V
$V_{CAN\ L}$	2.0 V	3.0 V	0.5 V	2.25 V
$V_{Diff\ Tx}$	−0.5 V	0.05 V	1.5 V	3.0 V
$V_{Diff\ Rx}$	−1.0 V	0.5 V	0.9 V	5.0 V

©IFM electronic GmbH

FIGURE 4.20 Voltage level CAN bus.

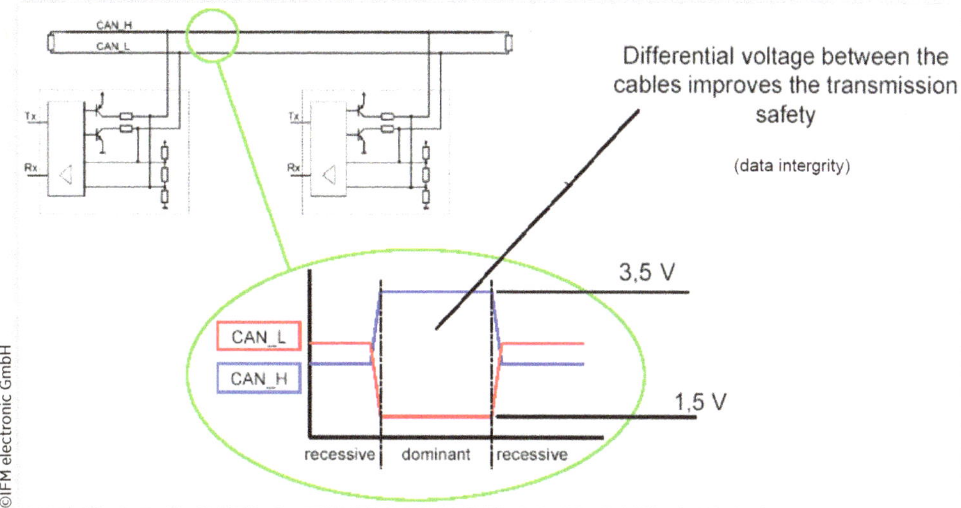

FIGURE 4.21 CAN topology.

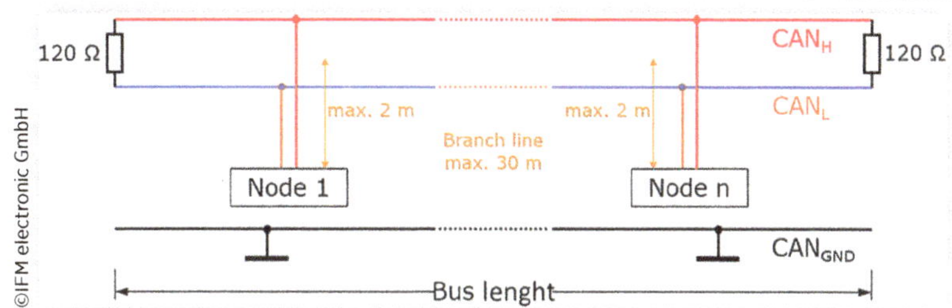

FIGURE 4.22 CAN cable structure.

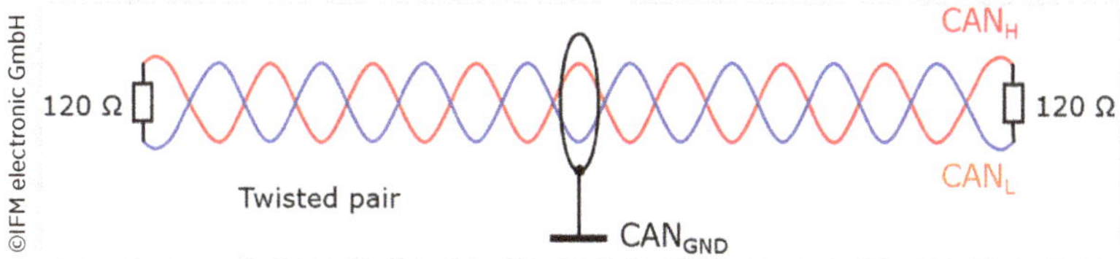

4.4.1.3 **Speed and Cable Length:** The possible data transmission rate in CAN networks is currently 1 Mbit/s (or 1 MBaud). The data speed is called baud rate. The technically possible speed mainly depends on the complete cable length used in the bus system (see Table 4.2). All values additionally depend on the quality of the cable and the quality of the bus cabling.

TABLE 4.2 Length and bit rate.

Bit rate (kbit/s)	Length (max. m)
1,000	40
500	100
250	250
125	500
100	600
50	1,200
25	3,000

©IFM electronic GmbH

4.4.1.4 **CAN Communication:** Communication can be schematically explained with the following example, see Figure 4.23.

The node (node 4) that wants to transmit a message via CAN gives the data to be transmitted and a message ID to the application layer (layer 7) which transmits the data in a suitable form to the data link layer (layer 2).

FIGURE 4.23 Example CAN communication.

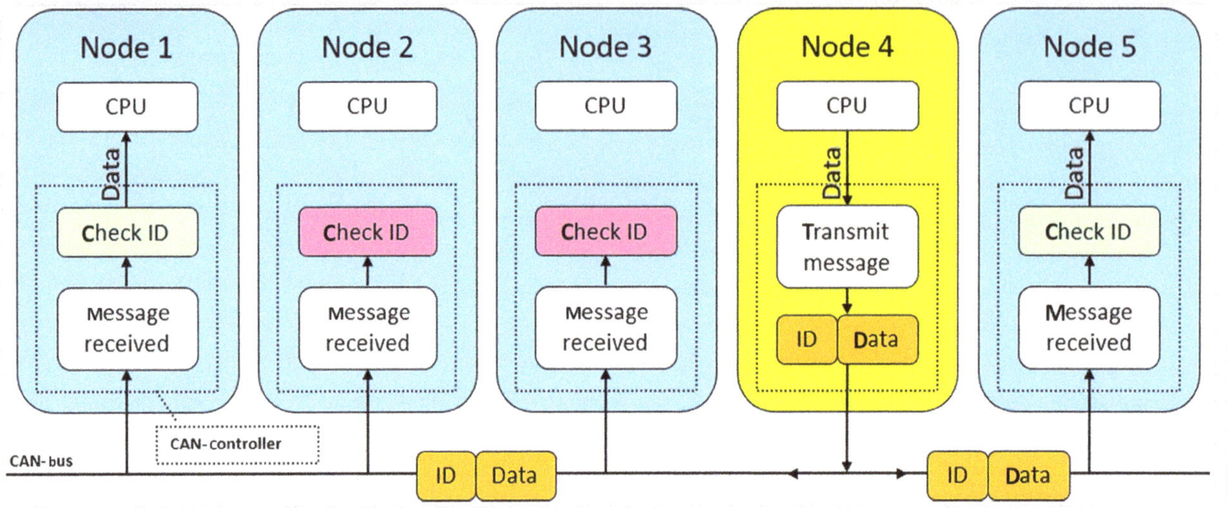

There the data are packed in a defined form/frame. This frame is then transmitted to the bit transmission layer (layer 1) and transmitted to the bus cable as a signal.

Each of the nodes connected to the CAN bus (nodes 1, 2, 3, 5) listens to the messages transmitted on the bus. Once a node detects a certain "key word" (message ID) by which it identifies the message (nodes 1 and 5), the data are transmitted to the application of this node and processed accordingly.

It is possible, of course, to have not just one but several nodes respond to a message. Important are the "key words" (message ID).

4.4.1.5 **Structure of CAN Messages:** A deeper insight into the representation of a CAN message on the CAN bus cable is given when having a closer look at the frames defined in the protocol. At first, we distinguish between:

- Data frame
- Remote frame
- Error frame

The different frame types each take on a special function in the data traffic (see Figure 4.24).

The **data frame** is used to transmit the actual information.

- **SOF**

  The SOF bit (Start of Frame) starts the CAN frame and is also used to synchronize the nodes.

- **Arbitration Field**

  The arbitration field sends the ID of the message (11 bits for base frames, 29 bits for extended frames). The ID defines the priority and the logical address of the message. The type of frame (data frame or remote frame) is distinguished by the RTR bit.

FIGURE 4.24 CAN data frame.

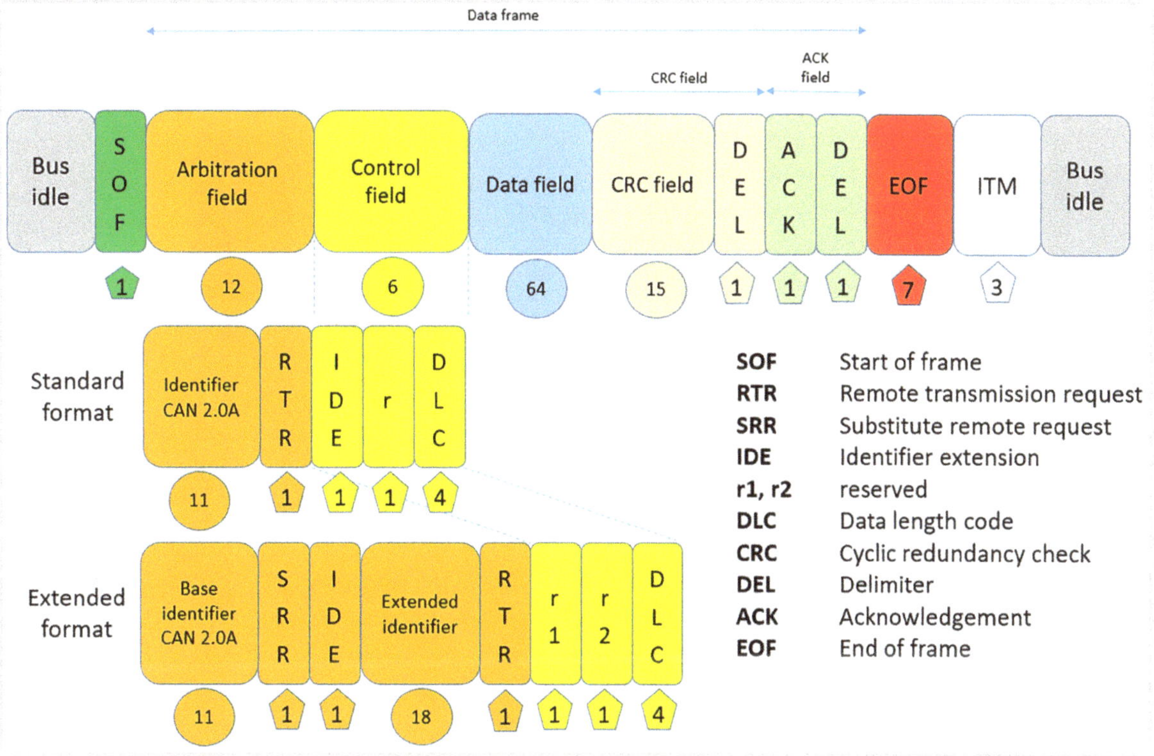

SOF	Start of frame
RTR	Remote transmission request
SRR	Substitute remote request
IDE	Identifier extension
r1, r2	reserved
DLC	Data length code
CRC	Cyclic redundancy check
DEL	Delimiter
ACK	Acknowledgement
EOF	End of frame

- **Control Field**

  In the control field, the IDE bit indicates whether the standard ID (11 bit) or the extended ID (29 bit) is used. Bit r is intended for future extensions. DLC (Data Length Code) indicates in 4 bit the number of the following data bytes.

  The IDE bit is set to send a data frame with extended ID.

- **Data Field**

  The data field contains the user data to be transmitted. Maximum 8 byte can be transmitted in one frame.

- **CRC Field**

  The content of the CRC field is a check sum entered by the data link layer with which the receiver can check the correct transmission of the message. Its length is 15 bit.

- **Acknowledgement Field (ACK)**

  Contains the feedback of other nodes upon correct reception of the message. If an error is detected, this feedback is not given and one of the nodes immediately sends an error frame.

- **EOF**

  Signifies the end of the data telegram (7 recessive bits). It is called EOF (End of Frame). The extended ID (Extended Data Frame) can handle considerably more messages (536 mill.) than the standard ID (Base Data Frame) (2,032). On the other hand, the frames are extended by 20 bit which might cause a speed loss.

FIGURE 4.25 CAN remote frame.

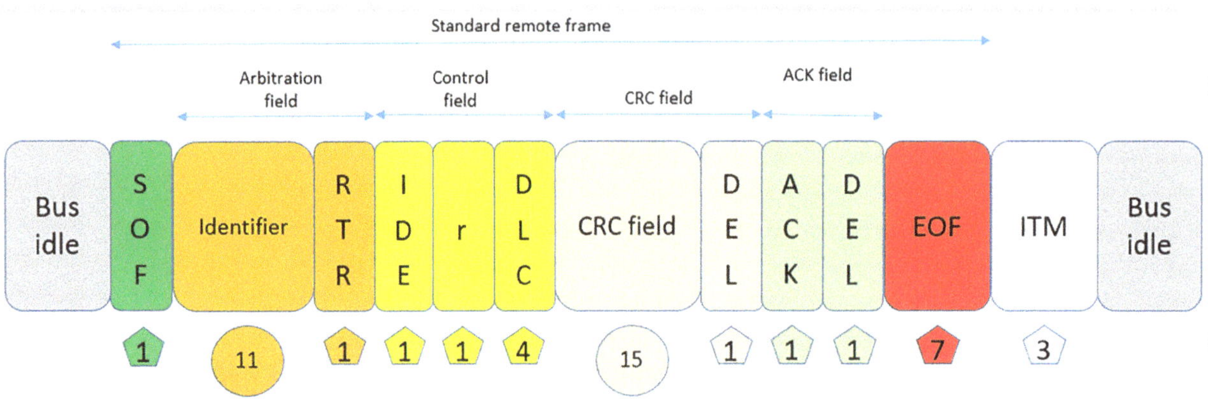

SOF	Start of frame
RTR	Remote transmission request (dom.= data frame, rec.= remote frame
IDE	Identifier extension (dom.= Standard ID, rec.= Extended ID
r	reserved
DLC	Data length code
CRC	Cyclic redundancy check
DEL	Delimiter
ACK	Acknowledgement
EOF	End of frame

When a remote frame is sent, all nodes are asked to provide data. The data requester sets the RTR bit to create a remote message. These remote frames do not contain any data field (see Figure 4.25). The addressed nodes respond to this frame by sending the requested data.

At the end of each frame three ITM bits (Intermission) are inserted to create a guaranteed gap between two sent frames.

The so-called **Error Frames** have a slightly different form (see Figure 4.26). They are used to inform all nodes when one node has noticed an error in communication. They contain an overlaying of error flags of the different nodes.

FIGURE 4.26 CAN error frame.

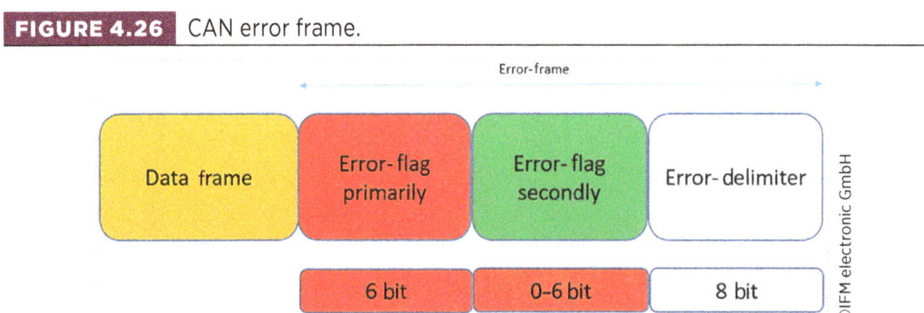

FIGURE 4.27 Transmission process.

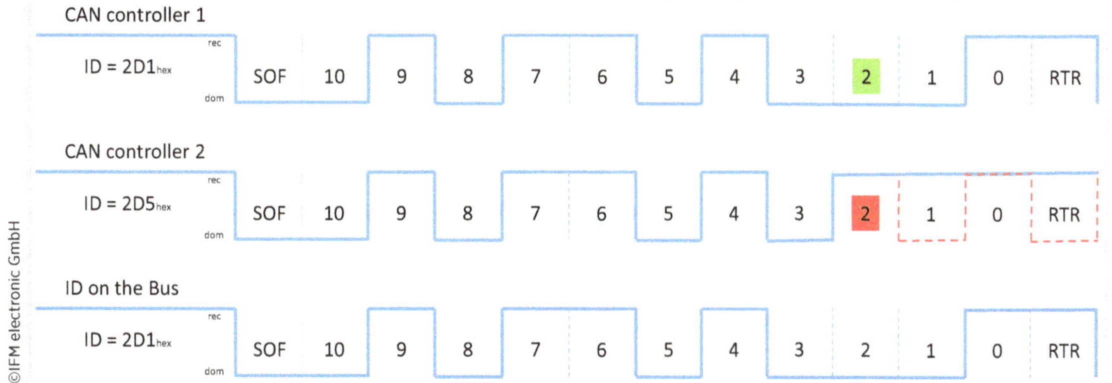

4.4.1.6 **Transmitting CAN Messages:** The transmission process is controlled by a prioritizing of messages. The information is contained in the message ID. The smaller the ID, the higher the priority of the message.

The following example will provide some explanation.

Two bus nodes are trying simultaneously to send a message over the bus. Bus node (1) sends a message with ID $2D1_{hex}$, bus node 2 with ID $2D5_{hex}$.

The messages are shown in the Figure 4.27.

Nodes Controller 1 (ID-$2D1_{hex}$) and Controller 2 (ID-$2D5_{hex}$) want to simultaneously transmit a message on the bus. The bits of the individual nodes are transmitted to the bus one after the other (bit by bit, bit 10 to bit 0) and are compared. Bits with recessive level ("1" signal) are overwritten by bits with dominant level ("0" signal). The example shows a difference between ID-$2D1_{hex}$ and ID-$2D5_{hex}$ in bit "2". The recessive level of bit "2" of ID-$2D5_{hex}$ is overwritten by the dominant level of bit "2" of ID-$2D1_{hex}$, see "ID on the Bus." ID-$2D5_{hex}$ thus loses its sending authorization (red dashed line shows the desired sending) and goes into the receiving mode (recessive bits, blue line). ID-$2D1_{hex}$ continues sending his ID and is given bus access after sending bit "0." Once the bus is free again (EOF and TM transmitted) the other nodes can try again to place their message on the bus. The process described here is known as "CSMA/CA" process (Carrier Sense Multiple Access with Collision Avoidance).

4.4.1.7 **SAE J 1939:** The SAE J 1939 protocol [11] was developed for data communication in utility vehicles (Figure 4.28). It is used to transmit diagnostic data such as motor speed, temperature and control data and commands. The information sent via the bus is grouped in parameter groups with associated numbers (PGN). The SAE J 1939

FIGURE 4.28 SAE J 1939 structure.

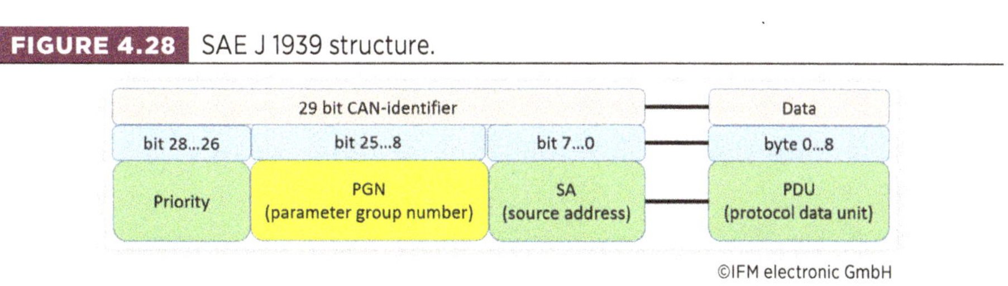

protocol uses the extended identifier to pack a large variety of information into the 29 bit ID. The data transmission speed is defined as 250 kbit/s.

The top three bits are prioritized. They are followed by the parameter group number in 18 bit and the address of the sender. In the SAE J 1939 protocol, too, each device is assigned its own node ID.

There is no clear separation between transmitted data and transport services.

The data or signals are included in the so-called parameter groups which can be identified via a unique Parameter Group Number (PGN). One PGN contains several Suspect Parameter Numbers (SPN) describing an individual date. The PGN are mostly optimized to a data length of 8 byte and are mostly transmitted cyclically as individual CAN messages. The content of the PGN and the data types, value ranges, meanings of the SPN are defined in SAE J 1939/71. In addition, there are services for assigning node numbers and for transporting data larger than 8 byte.

4.4.1.8 **"Higher" CAN Protocols CANopen and SAE J 1939 (Short Comparison):** The "higher" CAN protocols CANopen [1] and SAE J 1939 have established themselves as quasi standard in utility vehicles and mobile working machines.

Both protocols use CAN to ISO 11898-2 [9]. In CANopen the bit rates are variable between 10 kbit/s up to 1,000 kbit/s with a fixed set of bit rates defined. For SAE J 1939 the bit rate is set to 250 kbit/s for most applications, as defined. CANopen defines no cable whereas SAE J 1939 defines a screened twisted cable. In CiA 303 [2] CANopen defines the pin connection for more than 20 different connectors, but without stipulating a connector. SAE J 1939 only defines two types of connectors.

Special cases with other bit rates and varying cable definitions are defined in SAE J 1939-14 (500 kbit/s) and SAE J 1939-15 [250 kbit/s, Unshielded Twisted Pair (UTP)].

CANopen theoretically allows up to 127 nodes in the network, SAE J 1939 supports up to 254 logical nodes and limits it to 30 nodes per segment. Due to the defined bit rate the cable length is limited to 40 m (44 yards) while CANopen allows up to 5,000 m (5,468 yards) for 10 kbit/s cables.

CANopen mainly uses 11 bit CAN IDs and SAE J 1939 uses 29 bit CAN IDs while in practical use it is common to use both protocols on the same physical bus (if supported by the CAN chip).

As opposed to SAE J 1939 CANopen clearly distinguishes between data and services (Table 4.3). In CANopen data (process data, configuration data, or device ID information) are globally stored in an object directory. Each date can be addressed via a 16 bit index and an 8 bit subindex. For data transmission various services and objects are defined which differ with regard to speed, access possibility, flexibility, and maximum load capacity (>8 byte).

TABLE 4.3 Services for CANopen objects.

CANopen objects	Notice
PDO (process data object)	Real-time transmission of process data
SDO (service data objects)	Read and write access to the object directory
SYNC-Object (synchronization object)	Synchronization of network nodes
EMCY-Object (emergency object)	Error indication of a device or its periphery
NMT (network management)	Initialization and network control

In general, CANopen works based on the master-node principle. The master is the central control unit of the CANopen network.

Each bus node is assigned a node number (node ID) which is always integrated in the message ID when data are transmitted so that the bus nodes always know where the data come from. This allows a targeted addressing of bus nodes.

A CANopen device always includes a so-called EDS file (Electronic Data Sheet). It describes the interface with its parameters which are important for the user of the device. This EDS file is provided by the manufacturer of the CANopen device. With the help of the data from the EDS file the device can be easily integrated in the network.

4.4.1.9 **Ethernet:** The requirements, functionality and thus also the data amounts have continuously increased over the time. For this reason, many manufacturers use several CAN networks in their machines. Additional assistance systems such as 3D cameras, laser scanners, or radar sensors further increase the amount of data to be transmitted.

The CAN bus, however, only has a limited band width so that alternative network systems are inevitably used. Ethernet offers a standardized communication technology with the possibility to transmit larger data amounts in a short time. Existing applications so far were mainly diagnostic functions and the download and upload of software.

Assistance systems such as 3D cameras for object recognition generate a large amount of data information whose transmission can hardly be handled via the CAN bus within the requested time. Ethernet with is characteristics opens up new possibilities of data transmission in mobile working machines. In this case in particular IP-based cameras which communicate with the dialogue units directly via Ethernet are used more and more often.

Each connection and use of IoT requires open, standardized and thus cost-efficient communication technologies. That makes it possible to simply and cost-efficiently communicate with existing IT solutions. Data from a mobile working machine can thus be simply and efficiently integrated into the IT world of a company. The evaluation of the gained data can be used to optimize the machine, reduce its downtime and also reduce maintenance costs.

4.4.1.10 **ISOBUS:** Agricultural machinery cannot be imagined without electronics. Tractors, implements, and self-driving agricultural vehicles are getting more and more complex and the requirements on efficiency keep increasing. In practice, the users do not work with tractors, devices, and machines of just one manufacturer, but normally with different suppliers. To ensure communication of the different machines with each other, electronics and the required data exchange have been defined. Result of these definitions is the ISOBUS.

ISOBUS (trade name) is a fieldbus system used in agricultural machinery. This system is defined in the Standard ISO 11783 (Tractors and machinery for agriculture and forestry, Part 1-14) [8]. The Standard ISO 11783 is updated by AEM (Association of Equipment Manufacturer - North America), EAF (Agriculture Industry Electronics Foundation) and VDMA (German Mechanical and Systems Engineering Association).

The ISO 11783 was developed with the goal to combine any tractors with any implements via ISOBUS and to promote new developments in agricultural machinery such as

FIGURE 4.29 ISOBUS advantage.

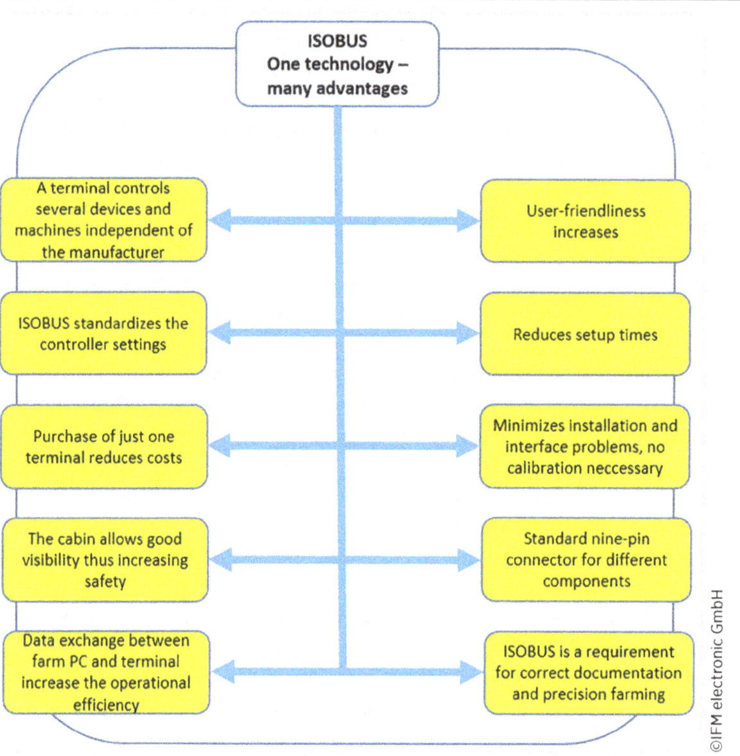

precision farming (see Figure 4.29). The standard is based on SAE J 1939 with fully dynamic addressing and adds special components such as satellite navigation from NEMA 2000. Furthermore, ISO 11783 as compared to SAE J 1939 uses two additional transport protocols: the extended TP for high data volumes and FastPacket for navigation data. It also restricts the diagnosis to the active and recently active errors (DM1-DM3).

4.4.1.11 What Does an ISOBUS System Consist Of?

A modern ISOBUS system consists of different components including tractor, terminal, and device (Figure 4.30). It always depends on the capabilities of terminal and device and, last but not least, on the equipment used. Functionalities have been defined to increase clarity.

- **UT - Universal Terminal**

  The possibility to operate a device on any terminal or the possibility to use one terminal for operating different devices.
- **AUX-N - Auxiliary Control**

  Additional operating elements intended to facilitate the operation of complex devices, such as a joystick; or, on the unit side, the possibility to trigger functions via an additional operating element.
- **TC-BAS - Task Controller basic (totals)**

  Documents total values, e.g., values of the completed work. The device provides the values. Data exchange between acreage index and the task controller takes place

FIGURE 4.30 Example ISOBUS in a tractor.

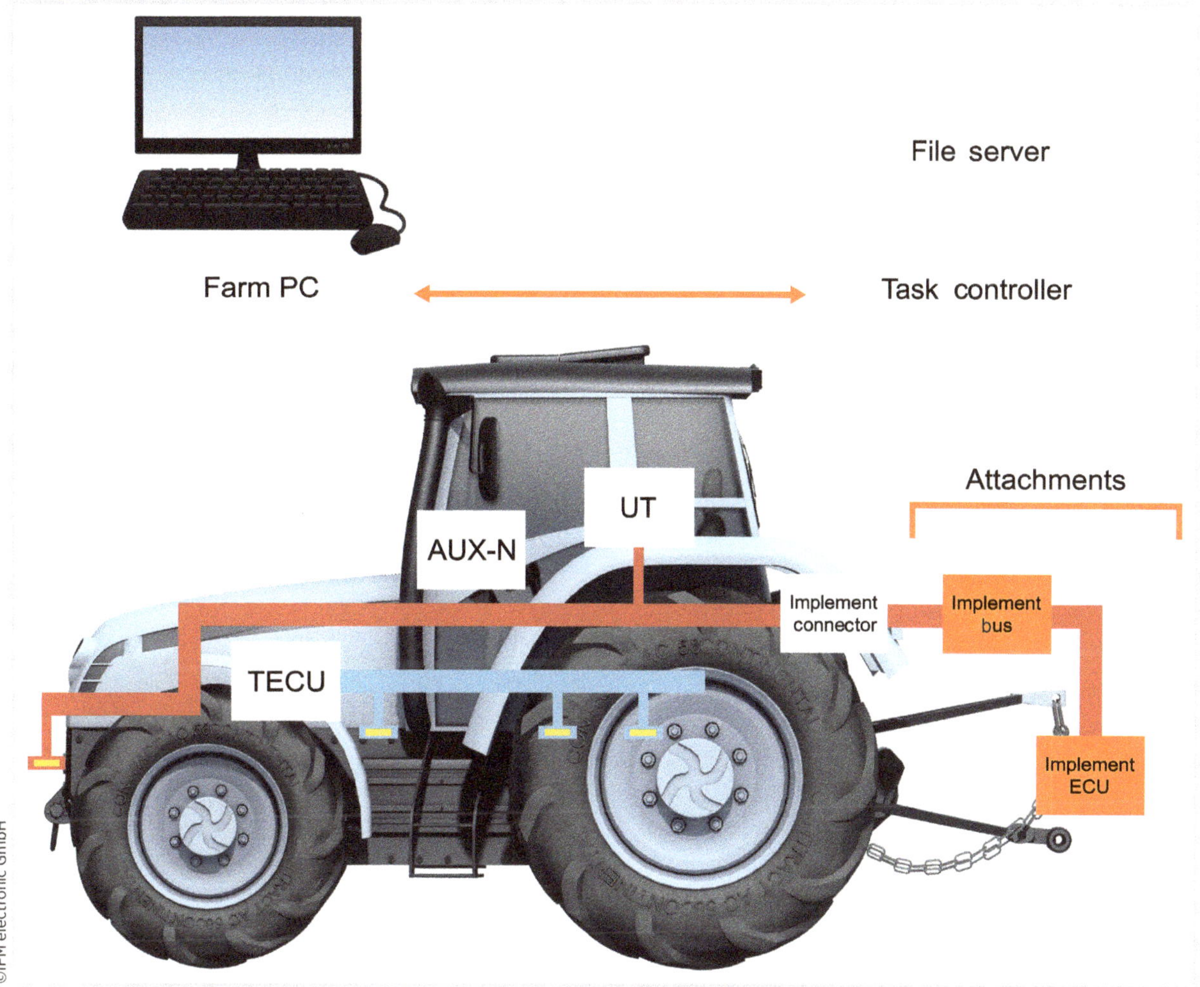

via the ISO XML file format. Orders can easily be imported in the task controller and/or the finished documentation can be exported.

- **TC-GEO - Task Controller geo-based (variables)**

  Offers the additional possibility to collect location-based data or to plan orders location-based, via application maps.

- **TC-SC Task Controller Section Control**

  Handles the automatic switching of part widths, like pesticide sprayers, seeders, and manure distributors depending on the GPS position and the required degree of overlap.

- **TECU - Basic Tractor ECU**

  The tractor ECU is the "job calculator" of the tractor. Central information such as speed or PTO shaft speed are provided. In addition, a device socket at the tractor rear and a terminal socket in the cabin are required.

- **TIM - Tractor-Implement Management (in development)**

  While in the case of TECU communication is one-directional, i.e., the tractor provides certain data, TIM offers the possibility of bidirectional communication. By means of the tractor implement management system (TIM) an implement can automatically control certain tractor functions-e.g., driving speed or control valves of the tractor. The unit self-optimizes its operation, the driver gets less tired, and the complete system is more productive.
- **LOG - Acquisition of device values independent of the order (in development)**

  Describes the acquisition of device values (tractor, implement, etc.) which can be collected independent of the order. These values can be total values such as total area or total harvest or all other measurement values that a unit can send. LOG data can be exported as ISO XML data (such as task controller data). This function can be used for products such as Telematik loggers.
- **ISB - ISOBUS Shortcut Button (in development)**

  The ISB allows deactivation of unit functions that were activated via an ISOBUS terminal. This is necessary when the unit in question is not in the foreground, maybe because several units are operated via a single ISOBUS terminal. The exact functions that can be deactivated by the ISB vary and have to be defined by the manufacturer.

4.4.2 Telemetry and Remote Service

Manufacturers and operators of mobile working machines have one common goal: The machines should operate with the highest possible availability, efficiency, and reliability. This applies, of course, for worldwide use of the machines.

More and more complex machines with remote control units, assistance systems, and "more intelligent" sensors also ask for more complex service processes of manufacturers and operators. Apart from the acquisition of machine and process data, monitoring of the electronic components has a high priority. Remote access solutions or even cloud solutions for machines are gaining importance.

The use of sensors and actuators with special features are of particular importance.

4.4.2.1 **IO-Link:** IO-Link is a global standardized I/O technology and complies with the international standard IEC 61131-9 (Part-9: Single-drop digital communication interface for small sensors and actuators SDCI) [6].

To better benefit from the performance of modern sensors and actuators thus being able to operate machines and systems more productively, consistent communication down to the lowest field level is required. Leading automation manufacturers have created a standard with IO-Link that now solves the problem of the "last mile."

In all industrial areas machines and systems are continuously optimized to improve their productivity across the whole life cycle. Besides a reduction of the total cost of ownership (TCO) the increase of the output and uptime contribute to an increase in competitiveness. Due to the more flexible adaptation to different requirements and ambient temperatures and faster retrofitting in spite of comprehensive configuration data the machines are becoming more and more complex. The more functionality they provide, the more information has to be exchanged across all company levels. Against this background the demand for communication with sensors and actuators is also increasing since they constitute a direct process interface.

It is not possible to exchange any other data than the actual process value via the standard interfaces used so far on the sensor/actuator level. Sensor and actuators provide more and more complex functions with integrated intelligence. Simple switching status or measured value interfaces restrict communication and lead to information congestion. For an overall consideration of a machine with respect to optimization it is, however, necessary to network all levels in a very transparent way.

It must be possible to map each component with its necessary information depth in the complete system pool. In detail this means for sensors and actuators that not only process data has to be exchanged but also parameter and diagnostic data.

In existing sensor/actuator installations a specialized and therefore expensive module had to be used for each signal type. The heterogeneous wiring with screened cables causes costs and still analog signals were usually transmitted in poor quality. Networking of mechatronic units that are used more and more frequently in mechanical engineering proved to be quite complex due to the manifold interfaces. During each machine retrofit, all relevant parameters had to be set manually or via a separate tool directly on each sensor or actuator which led to inconsistent data storage and long downtimes. Besides the complex saving and documentation mechanisms the fear of manipulation remained because the parameters were often directly stored in the sensor or actuator without any additional backup.

Since there was no continuous communication with the superposed levels, the diagnostic data of the sensors and actuators was not available in the engineering tool. Due to their exposed position in the process more errors occur with these components than, e.g., with I/O devices, drives, or controllers. If, however, diagnostic information is missing, troubleshooting and error elimination is often difficult and consumes a lot of time. Moreover, no preventive maintenance can be made with the aim to increase machine uptime. To eliminate the described weaknesses leading automation manufacturers have defined an open interface between sensors and actuators as well as I/O modules, namely IO-Link. Taking into consideration the current standard of I/O networking via a point-to-point connection a communication channel for continuous transmission of process, parameter and diagnostic data has been created.

An IO-Link system consists of the following components (see **Figure 4.31**):

- IO-Link master
- IO-Link device (e.g., sensors, valves, motor starters)
- Unscreened 3, 4, or 5-wire standard cables
- Engineering tool for projection and parameter setting of IO-Link

The IO-Link master establishes the connection between the IO-Link devices (sensors, actuators) and the automation system. As part of a peripheral system the IO-Link master is either installed in the control cabinet or as remote I/O with protection rating IP65/67 directly in the field.

The IO-Link master communicates with the control unit via various fieldbuses or product-specific backplane buses. An IO-Link master can have several IO-Link ports (channels). It is possible to connect an IO-Link device to each port.

FIGURE 4.31 Example IO-Link structure.

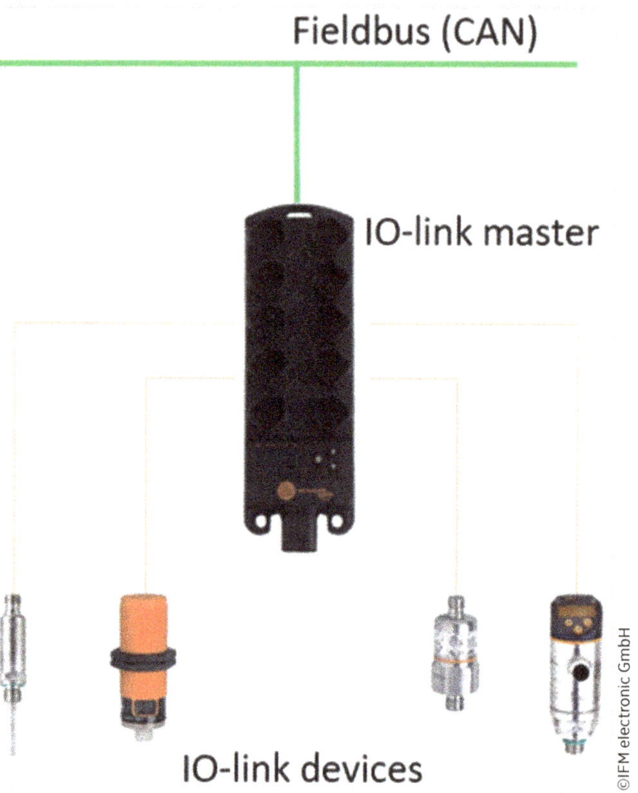

IO-Link is a point-to-point interface for the connection of any sensors and actuators to a control system. In combination with other bus systems and network levels, it enables for the first time a consistent vertical communication from the process control level down to each individual sensor.

IO-Link is not another bus system, but an add-on for the connection of the "last meter" to the sensor or actuator. In contrast to conventional fieldbus systems, there is no bus wiring but parallel wiring. IO-Link is suitable for binary and analog sensors. The feature of IO-Link sensors is the extension of the switching status channel by a communication mode. It is connected to the same pin as the switching output of conventional sensors (pin 4 for M12 connectors). The signal level is at standardized 24 V DC. So, all existing connection cables can still be used for IO-Link units. Sensors with IO-Link functionality are 100% backwards compatible to standard sensors.

4.4.2.2 IO-Link Sensors Command Three Types of Communication:

- Switching status (operating principle like conventional sensors).
- Cyclic process data (e.g., digital transmission of measured values).
- Transmission of device parameters and diagnostic data (e.g., "faulty electronics of the sensor", or "device temperature too high").

The connection of the "last meter" is done via an unscreened 3-wire M12 standard cable up to a length of 20 m (22 yards). Process data like measured values is transmitted digitally, thus reducing D/A and A/D conversion losses to a minimum. This ensures, e.g., that the controller can process the high accuracy of a pressure measuring cell without any loss. In the communication mode, e.g., 16 bits of process data are transmitted per cycle. With a baud rate of 38.4 kbaud the cycle time is 2.3 ms. Per cycle, up to 32 bytes of process data can be transmitted in both directions (input/output).

4.4.2.3 **Example Conversion and EMC Influence:** For the transfer of the analog signal to the controller several conversions from analog to digital and from digital to analog are necessary (Figure 4.32). Each of these conversions affects the quality of the process value transferred. In addition, there are electromagnetic influences, in particular if laying of the cable and connection of the screen are not handled professionally.

IO-Link requires only one conversion (Figure 4.33). The analog signal from the element detecting the measurand is converted into a digital signal. Then the signal processing and transfer are digital. This avoids losses due to conversion. Another advantage is that EMC influences due to the digital transfer can no longer influence the quality of the process value.

The following Table 4.4 shows the cost advantages of IO-Links, Table 4.5 shows the commissioning advantages, Table 4.6 the quality advantages and Table 4.7 the maintenance advantages.

Parameter setting of sensors and actuators can be done centrally "by the press of a button" or even automatically. Complex parameter setting of each sensor on site, e.g., if recipes or tools are changed, is a matter of the past. Data storage in the IO-Link master enables a simple and reliable replacement of units. The IO-Link master immediately detects when a faulty unit is replaced by a new identical one and writes the saved

FIGURE 4.32 IO-Link conversion.

FIGURE 4.33 Analog conversion.

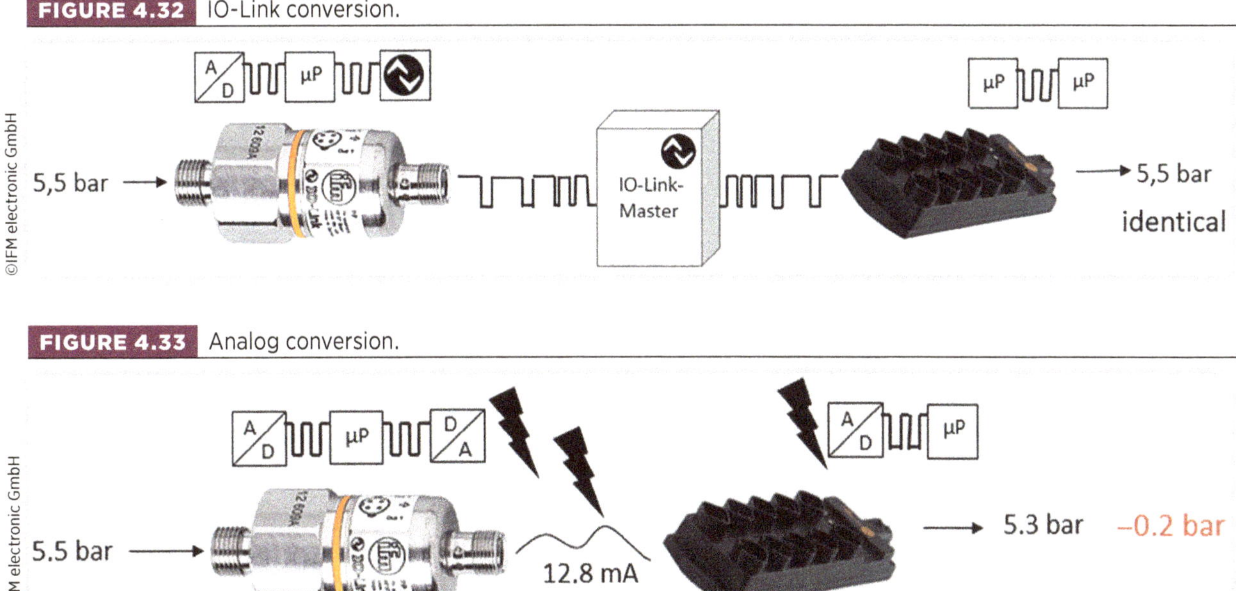

TABLE 4.4 Cost advantages of IO-Links.

Manufacturer-independent standard: IO-Link devices from other manufactures can be connected in the same way.
Twenty meters of cable are no problem
Only one type of port for all kinds of terminal equipment > cost savings for storage
Not only sensors, but also actuators can be connected. One fieldbus node on the machine for sensors, valve islands, frequency converters, e.g., > cost savings for the hardware
Binary inputs, e.g., feedbacks from valves, can be collected locally on input modules and transmitted via an IO-Link port.
Simultaneous availability of binary and analog signals.
Transfer of several signals via one cable (e.g., pressure sensors: two process values and two switch points).
Cost savings thanks to prewired cables and reduced storage space requirements.
Existing infrastructure can be used; no need to exchange existing cables.

TABLE 4.5 Commissioning advantages.

Prevention and discovery of wiring faults
Mounting location can be optimized, remote parameter setting from the workplace
Saving of the settings and providing them to the end user in electronic format. No labelling of the devices since in the settings are available in electronic format
Parameter setting of all devices with only a few mouse clicks
Commissioning with tools from the master manufacturer, e.g., ifm electronic LR Device
Point-to-point parameter setting

TABLE 4.6 Quality advantages.

Availability of the software and hardware status in the system documentation
Identification of all IO-Link devices in the system by means of serial number
Higher process data precision thanks to digital transfer. This reduces D/A and D/A conversion losses to a minimum. It is ensured, that the controller can process the high accuracy of a pressure measuring cell almost without any loss.
Signal is not influenced by EMC as in the case of analog signals
Laying the cables is uncritical; even close to FC motor cables
Logging of parameter changes: MES systems log how a parameter was changed and who did it.
Parameter storage in the database of MES
Recipe changeover inclusion of sensor parameters in orders. The parameter of the sensor will be set according to the product requirements.
Parameter setting on the device can be locked by remote access (parameter S-Loc). Nevertheless, all parameter can be read on the device.

TABLE 4.7 Maintenance advantages.

Parameters are stored automatically
Usability: Inclusion of the terminal equipment in the HMI; will be created according to the plant standard. Operators and maintenance staff work in the familiar environment.
HMI and MES can record trend curves; recorded process values make troubleshooting easier.
Preventive maintenance is supported by diagnostic information of the IO-Link device
Remote access to the IO-Link device via the PLC. The machine manufacturer can access the complete information, even about the individual sensors, via the internet.
Easy replacement of the device and damaged cables without further wiring.

parameters of the replaced unit into the new device. An alternative for parameterization via the controller would be a software solution. It allows parameter setting of IO-Link sensors on the PC via a USB adapter. Further functions are the recording and distribution of parameter sets (duplication), the detection and evaluation of sensor values over a longer period as well as the reset of sensor parameters to factory settings. The import of sensor libraries (IODDs) allows administration of IO-Link sensors from other manufacturers.

Another option for "data backup" for IO-Link sensors is the memory plug (Figure 4.34).

This compact hardware component allows easy copying and saving of sensor parameters. When a data set is stored on the memory plug it is copied from the plug's memory to a new sensor as soon as a sensor of the same type is connected.

The memory plug can be temporarily plugged onto the sensor for simple parameter setting. As an alternative, it can remain permanently connected between sensor and connection cable without any adverse effect on the function or output signals of the sensor during operation. All of this saves time and prevents parameter setting faults during setup or replacement of units. Sensors with IO-Link ensure greatest possible transparency in the machine. In addition to central parameter setting, extensive diagnosis is a big plus for machine maintenance and fault diagnosis.

4.4.2.4 **IoT in Mobile Working Machines:** Mobile working machines today are equipped with a large variety of sensors and control units whose signals provide a very high degree of information. For reasons of costs and for lack of evaluation methods, however, in most cases only the limit values of the sensor signals are monitored as protection against critical operating states; the acquired data are not analyzed or used to forecast the machine state. All available sensor and operating data of a mobile working machine should be used in a more intelligent way. A modern and future-proof machine concept therefore imposes different requirements on machine manufacturers:

- Minimizing service and maintenance work
- Optimizing machine utilization
- Diagnosis of process and machine data
- Fleet management

FIGURE 4.34 IO-Link memory plug.

FIGURE 4.35 Example cloud solution.

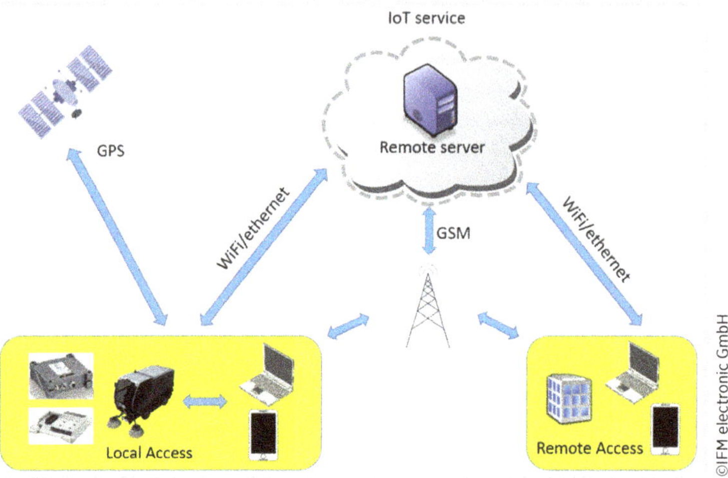

In this context the term IoT (Internet of Things) keeps coming up. Which conditions have to be met on a mobile working machine to implement IoT?

One of the most important prerequisites is the use of a suitable gateway as interface between the electronic components in a mobile working machine and the world of communication.

The principle overall structure is as shown in Figure 4.35.

4.4.2.5 Main Tasks of This Gateway Can Be:

- Connection to machine interfaces such as Ethernet and CAN
- Wireless connection to the communication world via mobile and Wifi
- Position detection via GPS
- Data logging of process and machine data
- Remote access of control units

So how can the acquired data be transmitted? In a mobile working machine main emphasis is placed on wired solutions. This is supported by the use of wiring systems and sensors with, e.g., IO-Link. If you want to transfer data from the machine, however, solutions with wireless connectivity are required. For short distances the range of Bluetooth, NFC, RFID, or Wifi is sufficient. For wider coverage also reaching remote areas, a mobile network is the preferred transmission medium.

Processing, treatment, and administration of the transmitted data are handled via software (Table 4.8). This also includes the so-called cloud solution. The acquired data is managed via IT infrastructure (computer performance, memory, software) over the internet (Table 4.9).

The features shown in Figure 4.36 can be found in conjunction with a cloud solution.

TABLE 4.8 Server side.

Function	Notice
Analysis	Standardization and analysis of the stored data on request
Storage	Storage and archiving of the transmitted data
Processing	Processing of the received data for visualization
Security	Ensuring the safe connection from and to the machine
API/apps	Access to data via customer-specific software solutions

TABLE 4.9 Operator side.

Function	Notice
Charting	Graphical representation of the individual machine data
Dashboard	Overview of the stored data
Parameterization	Tool for parameterization of machine data
Administration	Administration tool for registered machines and users
Asset list	List of all implemented machines
Geo position	Localization and tracking of machines
Notifications	Analysis of received data with regard to exceeded target value
Diagnoses	Plausibility check of the data
Debugging	Direct access application software

FIGURE 4.36 Features cloud solution.

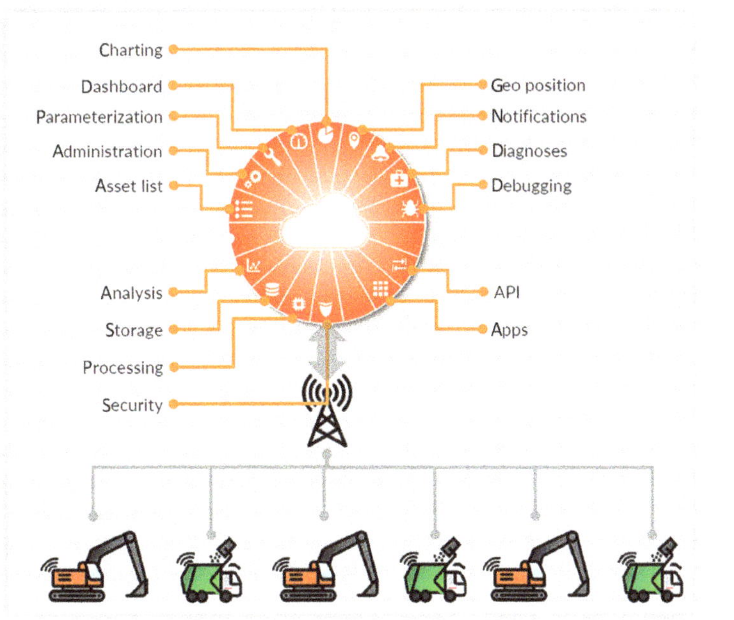

4.5 Functional Safety Aspects in Mobile Working Machines

Functional safety is the part of the safety of a system that depends on the correction function of the safety-related (sub)systems and external equipment for risk reduction. Functional safety does not include electric safety, fire protection, radiation protection. Since safety can also be achieved by stopping the intended function and by going into a safe state, we talk about safety integrity of the system.

Increasing complexity of electronic and in particular of programmable systems also increases the diversity of possible errors. So the series of standards IEC 61508 *Functional Safety of Electrical/Electronic/Programmable Electronic Safety-Related Systems* [7] demands the use of diverse methods to prevent systematic errors (errors during specification, implementation, etc. of the system) and to reliably control failures and disturbances (often caused by physical phenomenon or operating errors). See **Figures 4.37-4.41**.

FIGURE 4.37 Examples for general system influences that might present risk.

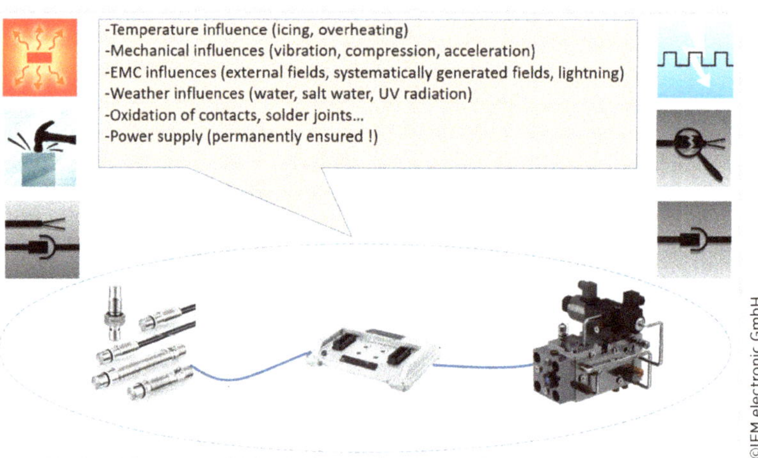

-Temperature influence (icing, overheating)
-Mechanical influences (vibration, compression, acceleration)
-EMC influences (external fields, systematically generated fields, lightning)
-Weather influences (water, salt water, UV radiation)
-Oxidation of contacts, solder joints...
-Power supply (permanently ensured !)
©IFM electronic GmbH

FIGURE 4.38 Examples for possible system influences on signal acquisition.

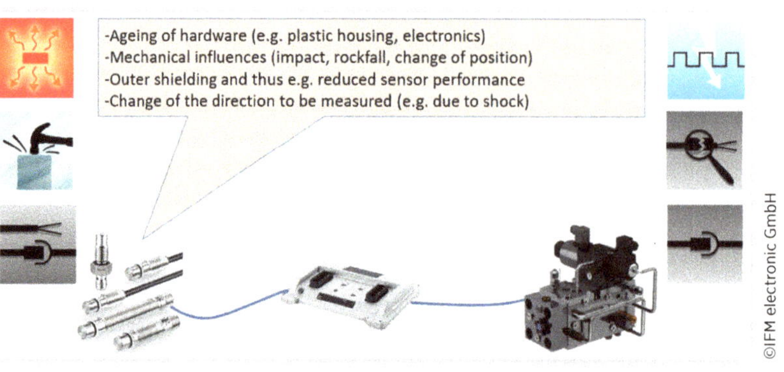

-Ageing of hardware (e.g. plastic housing, electronics)
-Mechanical influences (impact, rockfall, change of position)
-Outer shielding and thus e.g. reduced sensor performance
-Change of the direction to be measured (e.g. due to shock)
©IFM electronic GmbH

FIGURE 4.39 Example for possible system influences on signal transmission.

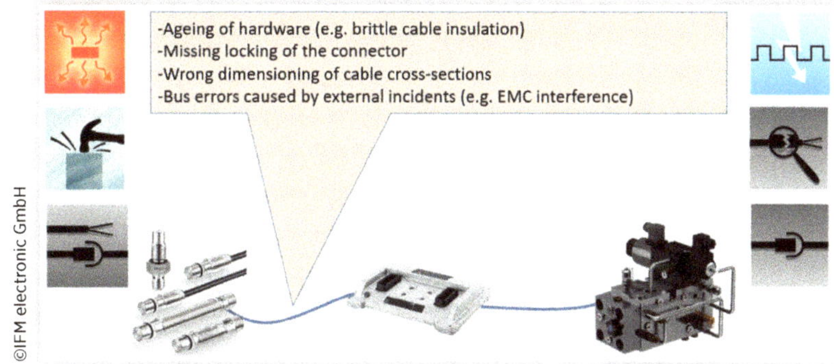

-Ageing of hardware (e.g. brittle cable insulation)
-Missing locking of the connector
-Wrong dimensioning of cable cross-sections
-Bus errors caused by external incidents (e.g. EMC interference)
©IFM electronic GmbH

FIGURE 4.40 Examples for possible system influences on control units.

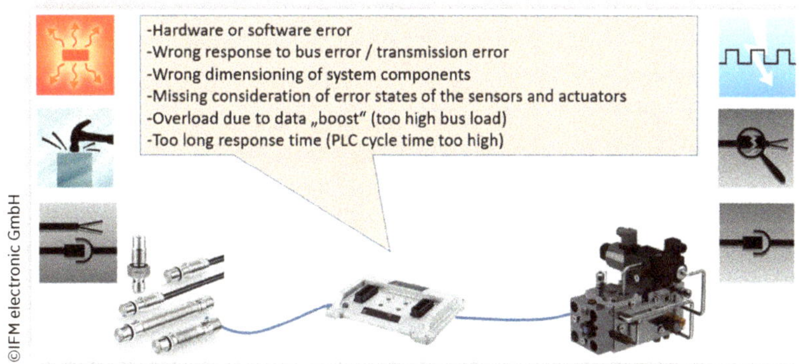

-Hardware or software error
-Wrong response to bus error / transmission error
-Wrong dimensioning of system components
-Missing consideration of error states of the sensors and actuators
-Overload due to data „boost" (too high bus load)
-Too long response time (PLC cycle time too high)
©IFM electronic GmbH

FIGURE 4.41 Examples for possible system influences of the actuators.

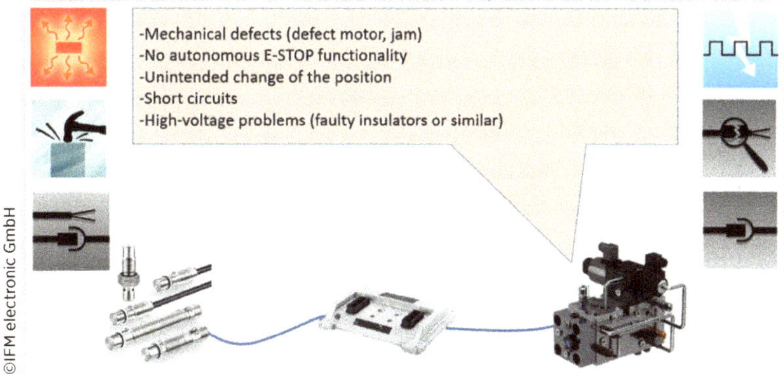

-Mechanical defects (defect motor, jam)
-No autonomous E-STOP functionality
-Unintended change of the position
-Short circuits
-High-voltage problems (faulty insulators or similar)
©IFM electronic GmbH

Almost all mobile working machines have functions that may endanger persons and material. Therefore, every manufacturer has to comply with the general regulations for a safe machine design. Given that these regulations and standards are defined for a broad range of different machines, they cannot be precisely adapted to the function of a mobile working machine. Therefore, there is an increasing number of product standards which are tailored to specific requirements.

In certain applications, e.g., vehicle lifts, there have been clearly defined product standards for a long time. The employers' liability insurance associations also often have clear requirements towards manufacturers of machines. For this reason, there is an increasing demand for certified electronic assemblies for mobile working machines.

In general, we distinguish between three different types of standards:

- **Type A standards (basic safety standards)**

  Basis safety standards contain general and basic information on the composition of a machine which will remain effective over the complete life cycle of the machine. Examples are the IEC 61508 (Functional Safety of Electrical/Electronic/ Programmable Electronic Systems) or the EN 12100-1 (Hazard Analysis). Since there is no zero risk, the aim is to achieve an acceptable residual risk.
- **Type B standards (safety group standards)**

  Contents of the safety group standards are designed for certain industries such as mechanical engineering. They describe safety-related equipment that might be used in different machines. An example is the EN ISO 13849-1 (Safety-Related Parts of Control Systems). The scope is not limited to electronic systems, but also covers the fields of hydraulics, mechanics and electrics.
- **Type C standards (safety product standards)**

  Safety product standards describe defined requirements and protective measures for certain machines or machine groups like elevating work platforms, mobile cranes, or agricultural machinery. Basic safety standards and safety group standards are also taken into consideration.

4.5.1 The Standards

Basis for all further considerations is the basic standard IEC 61508 [IEC61508]. For the design of the control systems two standards are available for the user (Figure 4.42).

EN 62061 [5] describes the design of electrical and electronic control systems for machines.

EN 13849-1 [4] describes the design of safety-related parts of a controller irrespective of technology and energy used (electric, pneumatic, hydraulic, mechanical, etc.).

Classification is made either in the Safety Integrity Level (SIL 1-3 in EN 62061) (Figure 4.43) or in the Performance Level (PL a-e in EN ISO 13849-1).

FIGURE 4.42 Overview standards.

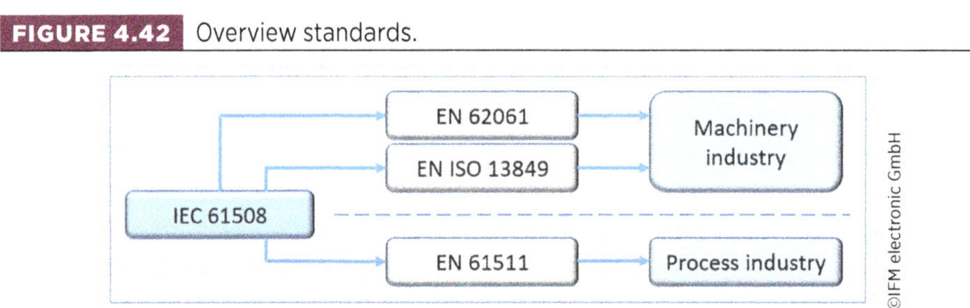

FIGURE 4.43 Safety standards.

Technology	EN ISO 13849	EN 62061
Non electrical, hydraulic valves	Covered	Not covered
Electromechanics, e. g. relays, non complex electronics	All architectures and up to PL = e	All architectures and up to SIL 3
Complex electronics, e. g. programmable device	All architectures and up to PL = e	Up to SIL 3 (when designed according to IEC 61508)
SRESW: safety related embedded software	Up to PL = e (PL e without diversity according To IEC 61508, clause 7)	Design according to IEC 61508-3
SRASW: safety related application software	Up to PL = e	Up to SIL 3
Combination of different technologies	Restrictions as above	Restriction as above (non electrical parts according to EN ISO 13849-1)

FIGURE 4.44 Type of faults.

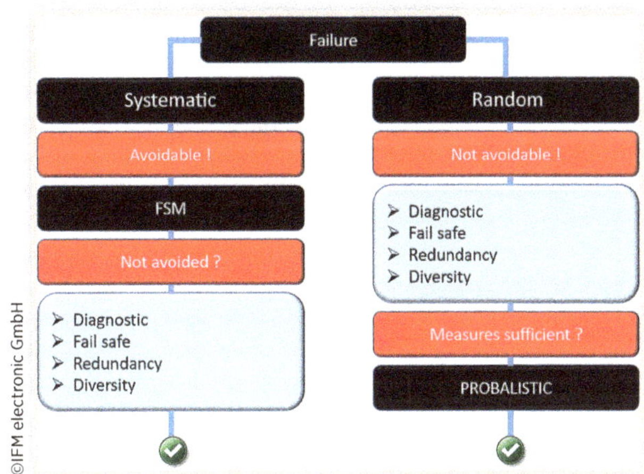

The main goal of functional safety is to have control over specific kind of failures. There are typically two kinds of failures to be covered in the context of functional safety (Figure 4.44).

Random failures usually have their roots in the quality of specific parts which are used for building a safety function. Parameters, such as MTTFd, B10d and failure rate λ (lamda) have to be considered for determining the PFH (probability of failures per HOUR) or PFD (probability of failures on DEMAND).

Systematic failures are (more or less) 100% human nature:

- Insufficient specification
- Insufficient design and implementation
- Failures during installation and commissioning
- Changes after commissioning
- Failures during operation and maintenance

Hence the management of functional safety (FSM) becomes of utmost importance in the context of systematic failures.

4.5.2 Safety Integrity

The EC Machinery Directive 2006/42/EC [3] stipulates that machinery should not present a risk.

The Machinery Directive stipulates a uniform protection level for accident prevention for machines when placed on the market within the European economic area. In Germany the implementation in national law is reached by the Machinery Directive (ninth product safety provision). It is no standard or recommended action but an act that has to be adhered to. The machine manufacturer has to carry out a risk assessment according to EN ISO 12100 [10]. Since there is no zero risk, the aim is to achieve an acceptable residual risk. If safety is dependent on control systems, these must be designed so as to minimize malfunction. For the design of the control systems two standards are available for the user:

- EN 62061 describes the design of electrical and electronic control systems for machines.
- EN 13849-1 describes the design of safety-related parts of a controller irrespective of technology and energy used (electric, pneumatic, hydraulic, mechanical, etc.).

Classification is made either in the Safety Integrity Level (SIL 1-3 in EN 62061) or in the Performance Level (PL a-e in EN ISO 13849-1).

4.5.3 Right Assessment and Consistent Implementation

The following guideline shows how to reach a target systematically, from risk assessment to final implementation and validation of the achieved safety level. Since these requirements are rather complex, it makes sense to split them into clearly defined task packages/task steps:

Step 1: Definition of the characteristics of the machine

- Which functions are to be fulfilled by the machine?
- Who shall use the machine?
- What is the intended use taking into account possible faults or misuse?

Step 2: Risk analysis

- Without any protective measures a risk will lead to harm.
- The designer has to divide the total function of the machine into sub-functions and assess the risk of every individual sub-functions as follows:
    - Which risk do the individual sub-functions provide?
    - What are the environmental conditions and the conditions of use of the sub-functions?
- Which events can cause damage?

Step 3: Risk assessment

- For each hazardous situation the risk has to be assessed:
  - Extend of injury (light, serious, lethal)
  - Probability of occurrence depending on:
  - Number of people concerned
  - Frequency of the event
  - Means of escape of the people concerned
  - Possibility of prevention or limitation
- If during the risk assessment it is found that a sub-function is too dangerous, measures how to minimize the risk have to be defined

Step 4: Determine the safety functions

- Appropriate safety functions are allocated to the risks

Step 5: Definition of the subsystem

- Definition of the subsystems (controller architectures) to EN 13849-1/EN62061.
- Definition of the subsystem architecture for every safety-related control function

Step 6: Definition of the characteristics values

- PL?, SIL?, PFH_D = ? to EN 13849-1/EN62061 for every subsystems

Step 7: Verification

- Determination of the performance level (PL) reached or the safety integrity level (SIL) reached from the characteristic values according to the standard used.
- Does the safety level reached correspond to the value defined in the risk analysis?

Step 8: Validation

- During validation, it has to be proven that the theoretically defined measures for risk minimalization have really been implemented.
- In general, this is made by practical tests on the machine.
- Are the safety-related output signals generated correctly?
- Does the behavior in case of a fault correspond to the circuit category?
- Simulation of all dangerous faults.
- Are the dimensions of the production equipment sufficient?

4.5.3.1 **Safety Integrity Level (SIL):** Hazard and risk analysis are to identify dangerous states of a machine. As a result of this analysis, safety functions have to be defined and implemented. The set target has to be to prevent danger or reduce its effects. The so-called risk graph (Figure 4.45) can be used to define the required safety integrity of a system with regard to the risk factor. The higher the level of safety integrity, the higher the probability that the system meets the safety functions.

4.5.3.2 **PFH (Probability of Dangerous Failures per Hour):** PFH is the probability of a dangerous failure per hour. This value can be seen as the unit of the PL (Performance Level) or SIL (Safety Integrity Level). The lower this value gets by using the appropriate parts, devices or subsystems the lower the probability of a hazardous

FIGURE 4.45 Safety Integrity Level according to EN 62061.

SILcl: Safety Integrity Level according to EN 62061
cl: claim limit

effect	severity S	CLASS: K = F + W + P				
		3-4	5-7	8-10	11-13	14-15
Death, loss of arm or eye	4	SIL 2	SIL 2	SIL 2	SIL 3	SIL 3
Irreversible, loss of finger	3	other measures		SIL 1	SIL 2	SIL 3
Reversible through medical treatment	2			other measures	SIL 1	SIL 2
Reversible through first aid	1				other measures	SIL 1

Duration and /or frequency of exposure F	
Less than 1 h	5
1h to 1d	5
1d to 2 w	4
2w to 1 a	3
greater than 1 a	2

Probability of event to happen W	
often	5
likely	4
possible	3
seldom	2
negligible	1

Probability of event to be avoided P	
impossible	5
possible	3
likely	1

TABLE 4.10 SIL with the particular failure limits.

SIL	Probability of dangerous failures per hour
4	$\geq 10^{-9} \ldots > 10^{-8}$
3	$\geq 10^{-8} \ldots > 10^{-7}$
2	$\geq 10^{-7} \ldots > 10^{-6}$
1	$\geq 10^{-6} \ldots > 10^{-5}$

situation at the machinery. The maximum possible PFH of the total system is the sum of the PFH of each subsystem it consists of. For each safety-impairing function a Safety Integrity Level (SIL) is defined in Table 4.10. Levels 1-4 are specified and defined by the probability of dangerous failures per hour (Probability of dangerous Failures per Hour: PFH). SIL 1 is the lowest and SIL 4 is the highest level.

4.5.3.3 **Safety Functions and Safety Performance:** Similar to the IEC 61508, the new EN ISO 13849 is used to define and implement safety functions for dangerous machine states identified in the hazard and risk analysis. By means of the risk graph the required performance level of a safety-related system is defined with regard to its risk factor.

The higher the performance level reached, the higher the probability that a system meets the safety functions.

FIGURE 4.46 Performance level according to EN 13849.

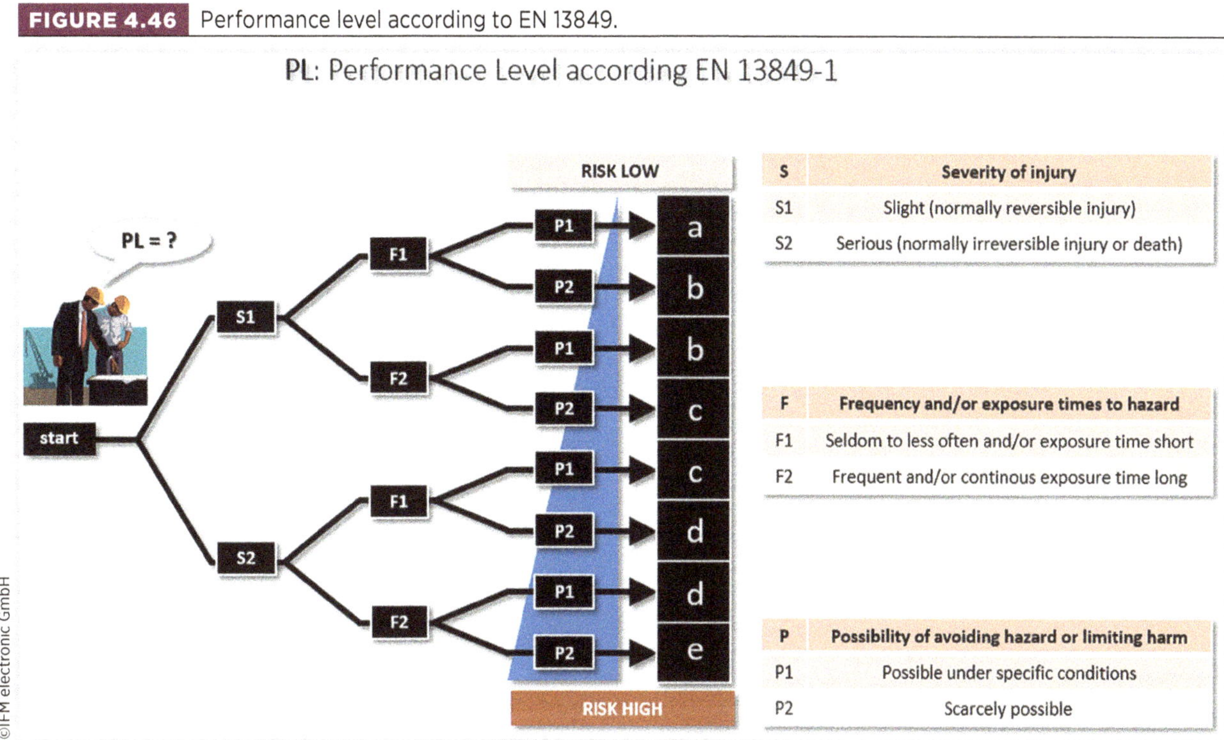

TABLE 4.11 PL with the relevant failure limits.

PL	Probability of dangerous failures per hour
a	$>10^{-5} \ldots <10^{-4}$
b	$>3 \times 10^{-6} \ldots <10^{-5}$
c	$>10^{-6} \ldots <3 \times 10^{-4}$
d	$>10^{-7} \ldots <10^{-8}$
e	$>10^{-8} \ldots. <10^{-7}$

4.5.3.4 **Performance Level (PL):** Based on hazard and risk analysis a so-called PL (Performance Level) is defined for a safety-impairing function (Figure 4.46). We distinguish between five performance levels - PL a to PL e (Table 4.11). They are defined referred to the probability of dangerous failures per hour. PL a is the lowest and PL e is the highest level. That means that a function or part of a machine assessed with PL d is potentially less safe than one assessed with PL a.

4.5.3.5 **Designated Architectures:** Furthermore, EN ISO 13849 and EN 62061 provide a group of predefined architectures and design concepts concerning the control and processing structure of each work function.

The so-called **designated architectures** are subdivided into five categories (B, 1, 2, 3, and 4). See Figures 4.47-4.50. These categories describe which diagnostic and redundant systems have to be available in a system thus increasing system stability (Table 4.12) [19].

FIGURE 4.47 Example category B and 1.

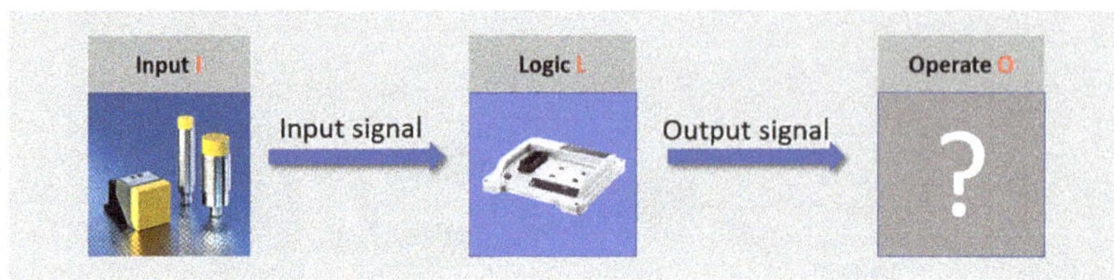

FIGURE 4.48 Example category 2.

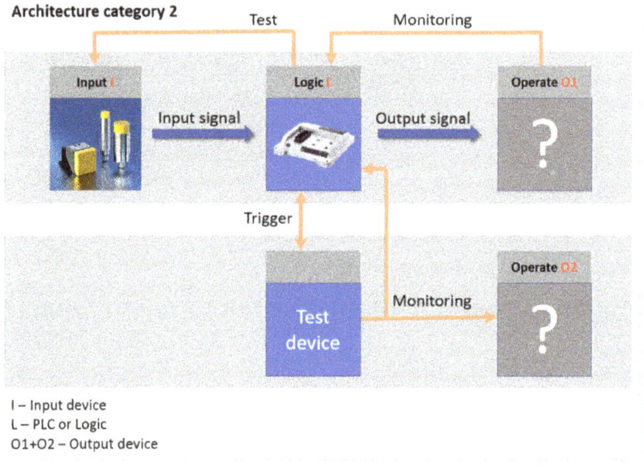

Category B, e.g., describes a simple one-channel architecture without specific test or diagnostic function. Category 2 includes test and diagnostic functions, and category 3 represents a two-channel system.

In order to use control units in machines safely and reliably, they have to meet the requirements described above. Independent of the standards, five important parameters are to be taken into particular consideration for assessing electric or electronic safety systems.

- **Designated architectures** (one-channel, two-channel, with diagnostics, without diagnostics ...)
- **DC (Diagnostic Coverage)** (probability of failure detection by means of tests)
- **PFH (Probability of dangerous failures per hour)**
- **MTTFd (Mean Time to Failure)** (number of failures per time unit or time to the first dangerous failure), manufacturer indication
- **CCF (Common Cause Failure CCF)** (influencing factors that have effects on several systems simultaneously)

FIGURE 4.49 Example category 3.

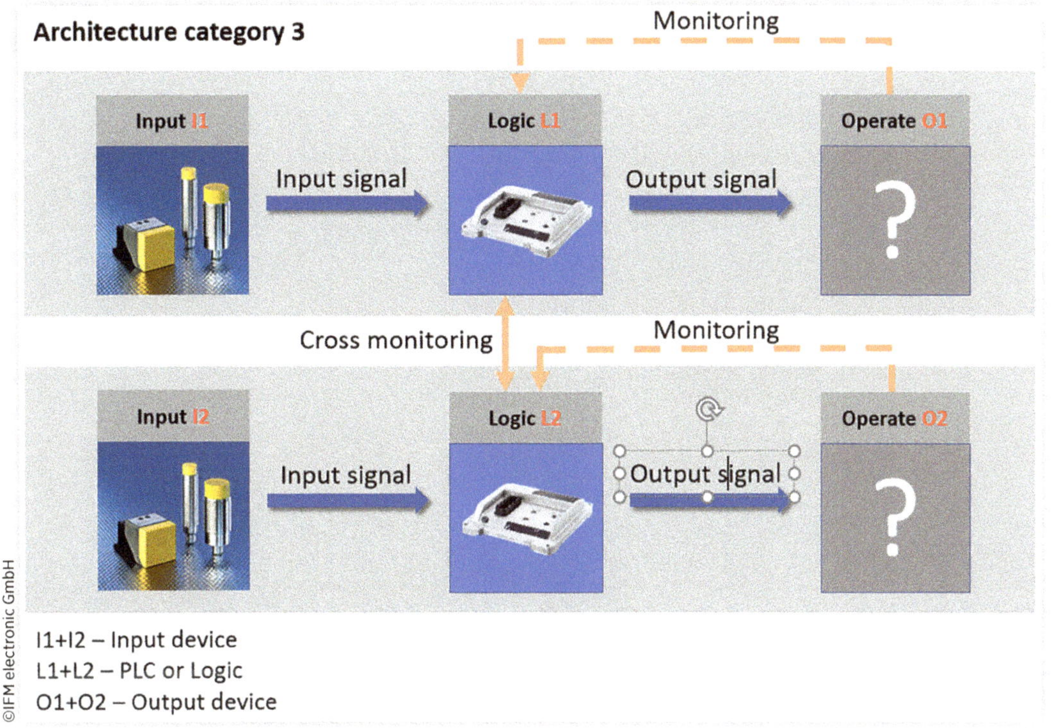

FIGURE 4.50 Example category 4.

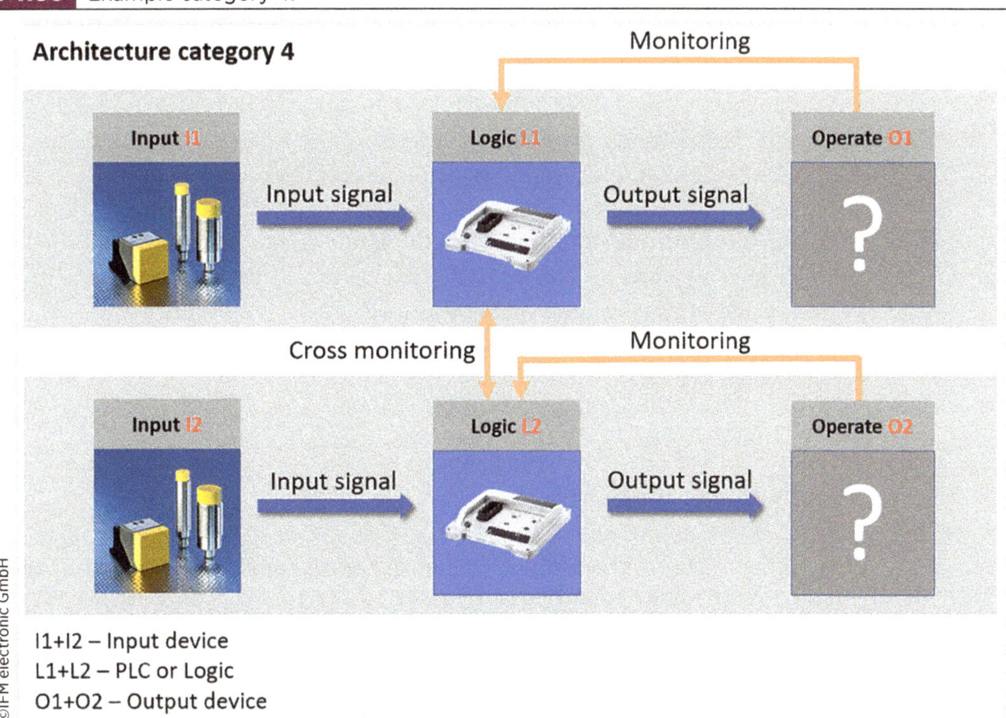

TABLE 4.12 Measures for increasing system stability.

Measures
Redundant sensors
Redundant control units/hardware
Redundant actuators
Diagnostic functions/monitoring
Mix of different technologies on different channels (e.g., brake function ABS and mechanical hand brake, mechanical and digital sensors)
Intelligent monitoring function
Autonomous E-stop function

FIGURE 4.51 PL bar diagram.

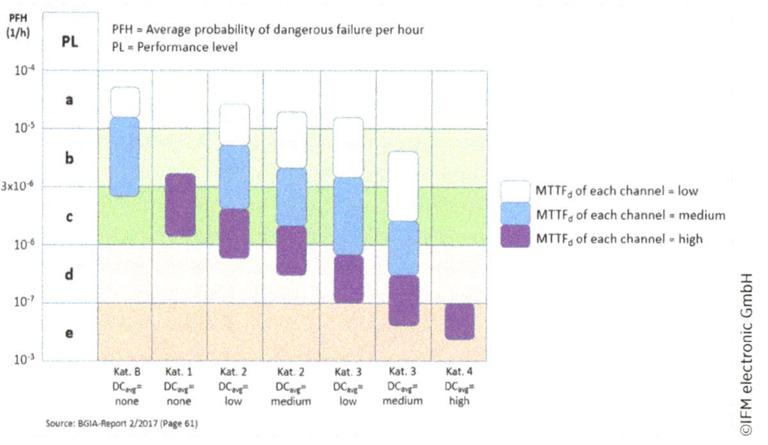

The interaction of three of these four influencing factors or parameters is shown in the so-called PL bar diagram in Figure 4.51 (see also BGIA report 2/2017, Functional safety of machine controls - Application of EN ISO 13849).

4.5.3.6 **Validation:** During validation, it has to be proven that the theoretically defined measures for risk minimization have really been implemented. In general, this is made by practical tests on the machine.

Note: All product pictures: ifm electronic gmbh, Friedrichstrasse 1, 45128 Essen, Germany.

References

1. DIN EN 50325-4:2003-07, Industrial Communications Subsystem Based on ISO 11898 (CAN) for Controller-Device Interfaces – Part 4: CANopen, dated July 2003, doi:https://doi.org/10.31030/9493503.
2. CiA 303-1 Version 1.9.0, Recommendation – Part 1: Cabeling and Connector Pin Assignment, dated September 18, 2017, https://www.can-cia.org/groups/specifications/.

3. Directive 2006/42/EC of the European Parliament and of the Council of 17 May 2006 on Machinery, and Amending Directive 95/16/EC (Recast), L 157/24, dated June 9, 2006.
4. DIN EN ISO 13849-1:2016-06, Safety of Machinery - Safety-Related Parts of Control Systems - Part 1: General Principles for Design, dated June 2016.
5. DIN EN 61061_2013-09, Safety of Machinery - Functional Safety of Safety-Related Electrical, Electronic and Programmable Electronic Control Systems, dated September 2013.
6. EN 61131-9, Programmable Controllers - Part 9: Sigle-Drop Digital Communication Interface for Small Sensors and Actuators, dated February 2015.
7. EN 61508-1:2011-02, Functional Safety of Electrical/Electronic/Programmable Electronic Safety-Related Systems - Part 1: General Requirements, dated February 2011.
8. ISO 11783:2017, Tractors and Machinery for Agriculture and Forestry - Serial Control and Communications Data Network, Internationoal Organization for Standardization, dated December 2017.
9. ISO 11898-1:2015, Road Vehicles - Controller Area Network (CAN) - Part 1 to Part 6, Internationoal Organization for Standardization, https://www.iso.org, dated December 2015.
10. EN ISO 12100:2011-03, Safety of Machinery - Gernal Principles for Design - Risk Assessment and Risk Reduction, dated March 2011.
11. SAE J1939, Recommended Practice for a Serial Control and Communications Vehicle Network, Society of Automotive Engineers (SAE), dated June 1, 2012.
12. System Manual, "Know-How ecomatmobile," ifm electronic gmbh.
13. "Training Manual 2 - Ecomat Mobile, Introduction to Working with CAN," ifm electronic gmbh, Essen, Germany.
14. CiA, "CAN in Automation," https://www.can-cia.org/can-knowledge/.
15. AEF (Agricultural Electronics Foundation), "ISOBUS in Funktionalities," Hand Fan, https://www.aef-online.org/the-aef/isobus.html.
16. CCI, "Compentence Center ISOBUS," https://www.cc-isobus.com/.
17. German Brochure, IO-Link Systembeschreibung (IO-Link System Description), IO-Link Firmengemeinschaft, c/o PROFIBUS Nutzerorganisation, www.io-link.com, Leaflet "IO-Link Systembeschreibung 2016."
18. IO-Link - We Connect You, ifm electronic gmbh, Catalogue March 2018.
19. German IFA Report 02/2017 - Funktionale Sicherheit von Maschinensteuerungen - Anwendung der DIN EN ISO 13849 (Fuctional Safety of Machine Controls - Application of DIN EN ISO 13849), 41-63.
20. Lötte, M., "Presentation Safety Technology - A Brief Guide through Functional Safety…," ifm electronic gmbh, Training Department.
21. Lötte, M., "Presentation Safety Directives & Standards," ifm electronic gmbh, Training Department.
22. German Brochure, Sicherheitstechnik - Sensoren und Systemlösungen von ifm (Safety Engineering - Sensors and System Solutions of ifm), ifm electronic, Broschüre March 2016.

CHAPTER 5

Innovative Drive Concepts

Prof. Dr. Naseem Daher, Prof. Dr.-Ing. Marcus Geimer, and Roman Ivantysyn

This chapter is designed to show which current innovative drive technologies are suited to be the future for mobile working machines. Therefore, first, an "innovation" has to be defined. Joseph Alois Schumpeter (1883–1950) is one of the fathers of innovation research and economic theories. He published his ideas in *The Theory of Economic Development* [1]. He distinguishes between an "invention" and an "innovation." The first phase of an innovation process is the generation of new ideas and called "invention." The realization and demonstration of the functioning of this idea in a second phase are called "innovation." The last and most important phase for commercialization is "diffusion." Innovation finds a widespread application here [2, 3].

The drive concepts presented in this chapter are researched and tested on demonstrators but not used in widespread applications. Therefore, the first two phases of the innovation process are completed and it is expected that the technologies will also be implemented in the future. Consequently, the concepts presented here are named "innovative."

Currently, there are several promising and innovative technologies being researched, and some have already been implemented today. Innovative machine concepts can be divided into three major classifications:

1. Efficiency improvements (process and machine)
2. Ease of use/handling improvements
3. Productivity improvements

Examples of innovative process efficiency improvements are fleet management, intelligently networked components, and process automatization. Currently, there is a strong push by governments and industry to improve in these areas such as the Industry 4.0 developments, whereby intelligent processes, such as communication between machines and cloud-controlled fleet management, will be incorporated in future mobile machines as well.

Machine efficiency entails improvements in engine technology, electrification, advanced power-train technology, system control, such as power management, and innovative components.

Ease of use and handling improvements aim to support machine operators to perform their duties more efficiently and with less effort. The trend is for machines to have additional sensors installed and to feature smarter components, which will inform the operator of malfunctions and potential failures and provide guidance on how to react properly. Under the condition-monitoring umbrella points (1) and (2) will be merged to form a smarter and more efficient machine, which will feature a dynamic maintenance schedule for when parts need to be replaced due to wear and tear, and adapt accordingly before failures occur.

Nowadays, productivity is the most important value when purchasing a new machine. The responsible person on a construction site, on a farm, or in a logistic chain needs to fulfill time-sensitive tasks and therefore, needs definable productivity of a machine. Mobile machines are often the centerpiece of a construction site or a mine, and an increase in these machines' productivity will generate higher overall productivity of the entire site.

While there are many areas of improvement in all of the aforementioned areas, in addition to new lightweight materials and digital enhancements such as fleet control, this chapter will focus on the machine improvements, and more specifically the hydraulics of the mobile working machines. In the authors' opinions, the future of mobile machines requires the most efficient and productive solution possible, while using the least number of components and energy conversions.

Several promising concepts are being researched at the moment, which can improve system efficiency and productivity. The efforts can be divided into evolutionary and revolutionary concepts. Evolutionary concepts gradually improve on the status quo by enhancing components and control schemes, or even introducing new principles, although still based on current technologies. The current mobile machines market is dominated by valve-controlled actuation (VCA) for the function drives. This means that a pump is used to set a flow or pressure level for the entire machine, and throttling valves are then used to control the movement of the individual actuators performing the work. These systems are described in Section 3.4.

On the other hand, revolutionary approaches disrupt the market by introducing new and different principles. In the case of hydraulics, one principle is based on throttle-less actuation that is featured in hydrostatic traction drives today, see Section 3.3. Instead of first generating more pressure than necessary then throttling it back down, the most efficient hydraulic concept is to generate only as much power with the pump as necessary. While many valve-based concepts already rely on adjustable pumps to minimize their throttling losses, it is still most efficient not to throttle at all. The authors see the throttle-less actuation, also called displacement controlled (DC) or pump controlled actuation (PCA), as one of the most innovative technologies since it leads to no unnecessary component losses, and uses the least number of components possible in the direct power path (pump, lines, and actuator). DC-technology can be separated into several fields such as open- and closed-loop configurations and electro-hydraulic actuation (EHA).

A second revolutionary approach is a hydraulic transformer. A hydraulic pump and a hydraulic motor, at least one of them with variable displacement, transfer flow and pressure in the required size to the actuator. While this technology does not require throttling losses, it does require additional components such as the transformers, which introduce additional losses. For an economic decision, multiple smaller pumps in PCA have to be compared to multiple hydrostatic transformers combined with a larger pump source. Moreover, usually more efficient and robust smaller pumps have to be compared to their larger counterparts.

Innovative drive concepts include solutions like power-split drives and hydraulic hybrids. Although power management schemes play an important role in the efficiency of drives, the primary energy source, such as a diesel engine, needs to be taken into consideration at all times, which is also true for all future innovative machine concepts. Sustainable alternative primary energy provisions for mobile working machines are discussed in Geimer and Ays's "Sustainable Energy Storages for Mobile Machines" [4] and "Methane-Fuel Cell CCS-Drive: The Emission-Free Working Machine" [5]. In summary, the combustion engine will remain the most important primary energy source for mobile machines, at least in the foreseeable future.

While the electrification of automobiles has made significant strides in the last decade, the electrification of mobile machines is state-of-the-art in special applications and can only be found in small low-power mobile machines like forklift trucks. Intensive research in this area can be observed in the last decade. Long established electrification solutions of mobile machines use batteries as the energy storage medium, in some cases along with a combustion engine as the prime mover. While electrification remains a controversial topic, many applications might require the locally emission-free operation of mobile machines in the future, especially in urban areas with strict emission standards.

Although electric rotary drives have been discussed in Section 3.2, several ongoing research activities aim to introduce electric drives in mobile working machines with a combustion engine as the prime mover. In addition to pure electric drives, also electro-mechanic power-split drives can be found in the traction drive [6, 7]. The possibility of an electric PTO and bringing together traction and function drives are advantages of these drives. In contrast to hydraulic systems, the energy density of electric systems is roughly 10 times less [8]. However, in the field of very low power demand, no hydraulic drives are found in practical applications. **Figure 5.1** shows the dominating power for electric and hydraulic drives as the authors see them today. The borders can be seen as dynamic and may shift according to future research activities.

Currently, there are two known alternatives for translational electric drives. The first one integrates the hydraulic system, i.e., the tank, the pump, the valves, and the cylinder, in the "electrically driven" cylinder and only the electric wires are connected to the machine system. These are highly integrated hydraulic systems that are defined in the market as linear electric drives or compact drives, see also Section 5.2.4. In contrast, the second architecture uses direct linear electric drives. An example under research is a cutter mower

FIGURE 5.1 Strong fields of electric and hydraulic drives.

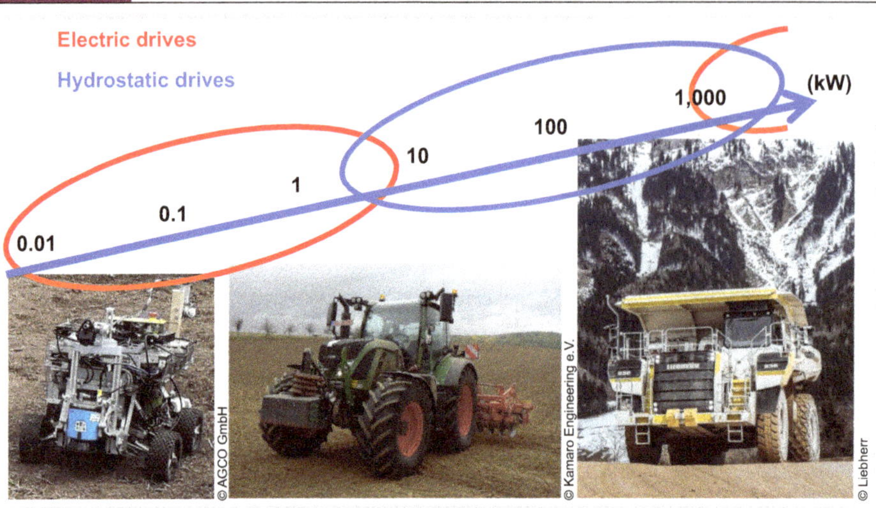

for a municipal machine that uses a battery-electric drive, which aims at reducing noise and exhaust emission. Under these conditions, the linear electric drive is expected to be more energy-efficient than the currently used hydraulic one [9].

In the last section of this chapter, innovative valve systems will be discussed. Here two innovative examples are mentioned, digital hydraulics and independent metering technology. Digital hydraulics combines fast switching valves with binary control algorithms to simplify the components used in hydraulic circuits. Simple on/off valves can control the flow of a pump to an actuator pulsing the flow just like in an electrical system.

Individual metering refers to the opportunity of separate control of in- and outflow of the hydraulic working ports. The system topology opens up for differential modes of operation of the hydraulic consumers. This technology is already being implemented in various forms by mobile machine OEMs. In combination with variable pumps, this technology is one of the most promising evolutionary system concepts.

In the authors' opinion, mobile machines need to be as light and power-dense as possible, thus giving diesel engines and purely hydraulic solutions the advantage over a wide range. In the following subchapters, hybrid systems, DC systems and concepts, as well as valve-controlled concepts, are presented.

5.1 Innovative Hybrid Drive Systems

The term "hybrid" originates from Greek and has the meaning of "mixture" in biology, is transferred into Latin, and has been integrated with today's languages. Hence, hybrids arise from a cross between different genera, species, subspecies, breeds, or breeding lines. In the technical field, a hybrid system can be seen as composed of different elements, which also can work individually.

In the field of vehicle technology, a hybrid is defined as:

a system with more than one energy source.

In this context, a "source" is a way to obtain energy for a vehicle's drive system. Hybrid passenger vehicles employ a combustion engine and an electric motor in the driveline, where the two energy storage sources are the gasoline/diesel tank and the electric battery; the battery is sized to allow the car to drive electrically for short distances. Another hybrid example is a wheel loader with a diesel engine and an additional hydraulic accumulator, which stores recapturable kinetic or potential energy for later use. A somewhat unusual example for a hybrid railway vehicle is a combination of a combustion engine and a pantograph: the energy for driving this vehicle can be taken from the combustion fuel as a first source, and the electric power supply as an alternative source.

5.1.1 Introduction

In a mobile working machine, different storage technologies are possible: electrical, mechanical, or hydraulic storages. Characterization of the different storage mediums is captured by the Ragone Plot, which plots energy density against power density, as shown in Figure 5.2. The points there represent realized storage examples. The graph can be interpreted as follows: the higher the power density, the faster the medium can be "charged." For example, a hydraulic accumulator has charge times that are in the millisecond (ms) regime. The energy density axis shows the amount of energy that can be stored. If there is a need for large amounts of energy over long periods, i.e., several hours, a high energy density storage medium is needed (e.g., a battery).

Figure 5.2 also shows that intermediate energy amounts that can be stored in hydraulic components like bladder accumulators, which are capable of storing very high

FIGURE 5.2 Ragone plot for different energy storage technologies.

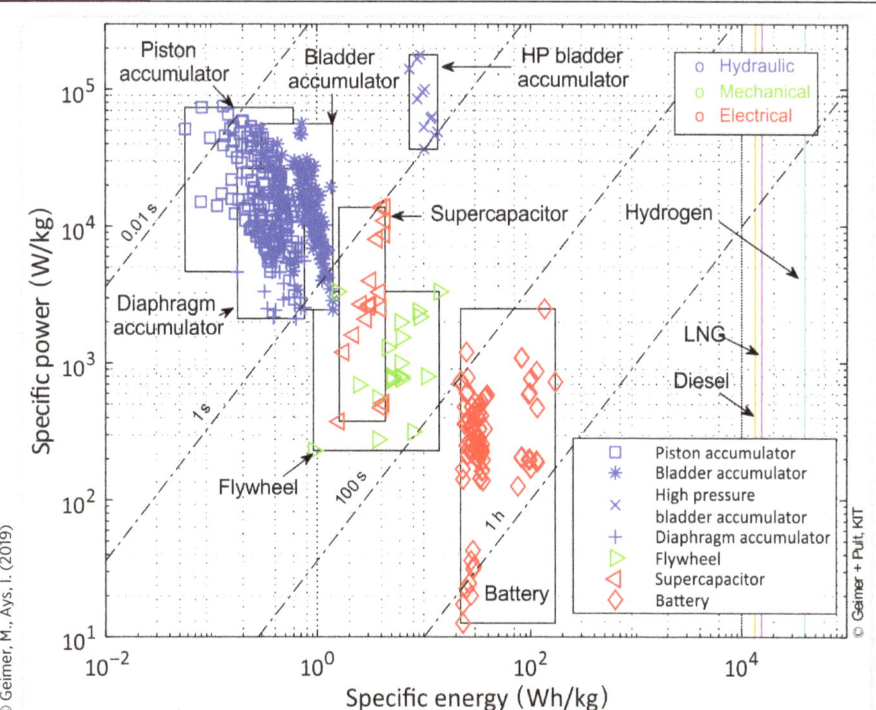

FIGURE 5.3 Basic hybrid structures: left: parallel, middle: serial, right: power-split.

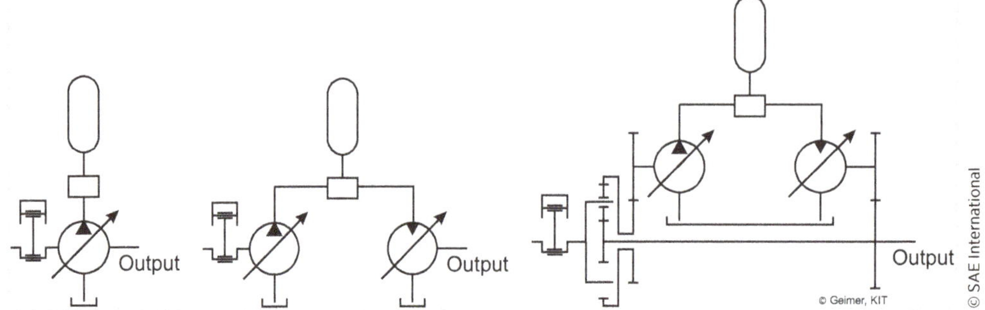

power amounts and medium energy amounts, thus puts them in competition with double-layer capacitors but not with batteries. Today, only electric energy storage mediums (storage mediums with a high energy density) are known in the market.

Different types of hybrids are known, with their basic architectures shown in Figure 5.3; from left to right: parallel, serial, and power-split hybrids. Shown is an output coupled power-split, the differences in terms of operation and efficiency are well documented [10, 11]. In the passenger car industry, different types are known and sometimes mixed with battery-electric cars and plug-in hybrids. As they are not mobile working machines, they will not be discussed here. More information can be found in Onorti [12] and Reif [13].

In a serial structure, energy from the primary storage, usually the fuel tank with a combustion engine, is transferred into another form of energy. A storage medium captures energy from the primary converter and transfers it to the actuators, the traction drive, or the working functions. Recuperation of energy into the storage medium is also possible and the combustion engine can run phlegmatized.

In parallel structures, energy from the storage system can be added to or removed from a drive, as well in the traction drive as in the function drives. Therefore, the power of the drive is usually lower than those of the primary energy converter, see Figure 5.4. The degree of hybridization α_{hybrid} can be defined by the power of the primary energy converter in proportion to the power of the secondary energy converter.

$$\alpha_{hybrid} = \frac{P_{hybrid}}{P_{primary}} \tag{5.1}$$

It can also be recognized from Figure 5.4 that the degree of hybridization of a hydraulic hybrid is usually higher than that of an electric hybrid, which is due to the higher specific weight of an electric motor-generator combination.

For the design of hybrid systems, Thiebes presents in “Hybrid drives for mobile working machines” [14] a specific method for mobile machines. The first and most important question (1) is: which goals are expected from the hybrid system? Possible goals are an increase in energy efficiency or productivity. It is important to mention here, that the chosen technology, whether electrical, mechanical, or hydraulic storage, is not imperative at this point.

FIGURE 5.4 Degree of hybridization of different hybrid structures. Modified from [16].

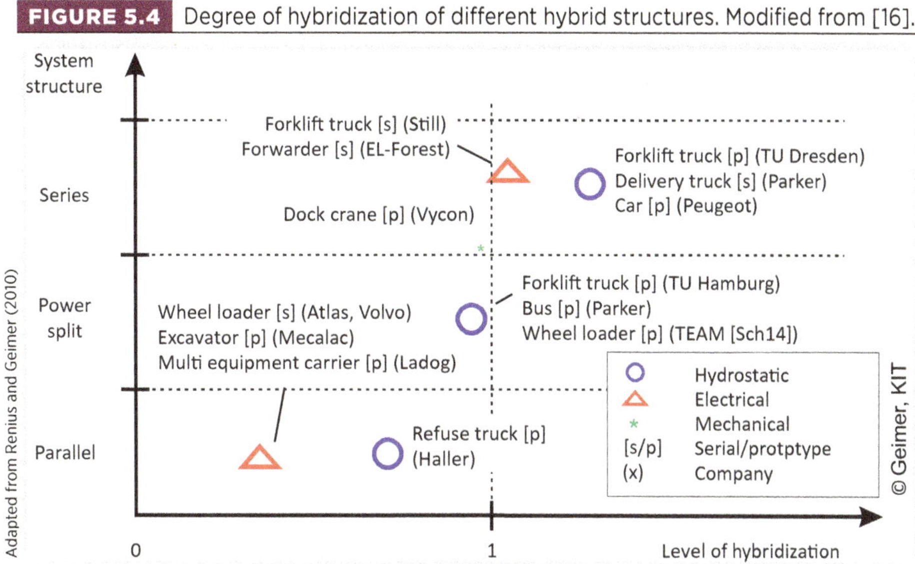

For increasing energy efficiency, several hybrid functions are known and their definitions and detailed explanations can be found in Thiebes [14] as well as other publications. Examples for hybrid functions are recuperation, regeneration, start-stop of the combustion engine, emission-free operation, phlegmatization, and rightsizing.

Later, Thiebes [14] suggests as a second step that the load cycle of the machines needs to be analyzed (2) and, as the third step, a prioritization of the functions has to be done (3). The achievement of the expected goals can be assessed as follows. The energy demand of the prioritized functions and the load cycle of the machine are the boundary conditions to design the storage of the hybrid system (4). Here the storage technology becomes essential and has to be appropriately chosen.

Hybrid variants can be generated (5) with the achieved knowledge. Thiebes [14] uses a morphological box to analyze hybrid systems and to generate the variants. Based on criteria or with the help of a simulation, the number of variants is reduced (6). The last step entails monitoring the achievements of the goals (7). Based on this information, an economical decision can be made.

Nowadays, numerous publications in the field of hybridization are available. A detailed description of a hybrid telescopic handler can be found in Nagel's "Development of an Operating Strategy for Energy Recovery in Hybrid Multi-Consumer Systems" [17] and the hybridization of a wood harvester is presented in Zu Hohenlohe's "Phlegmatisation as a Virtue of Mobile Hydraulics – The Energy Storage System of an HSM 405H2 Crane Harvester" [18]. Wheel loaders and excavators are the most studied machines for hybrid drives. A short loading cycle of a wheel loader, like shown in Filla [19], has a high potential of energy recuperation. The swing drive of an excavator also has a high potential for recuperating energy. In a 90° load cycle, e.g., the maximum turning speed is not reached; thus, the swing drive is first accelerated with maximum power and then decelerated with maximum power.

Today, several research projects, as well as established hybrid solutions in the market, can be found. Thus in this subchapter, it is not possible to show all of the existing solutions, which are presented in specialized conferences [20]. From today's solutions in real applications, it can be concluded that hybrid systems are gaining acceptance, which enables an increase in productivity. From a commercialization point of view, the current energy costs, potential energy efficiency gain, and the costs of the added components need to be considered.

5.1.2 Series-Parallel Hybrid Displacement Controlled Systems

This section discusses innovative rotary drive systems, which are mainly intended for power transmission with the capability of storing and recuperating energy during operation. Having introduced standard rotary drives in Section 3.2.2 and basic hybrid structures in Section 5.1.1, three various advanced architectures are presented here to show how DC technology can be combined with hybrid systems on mobile machines to achieve the highest possible efficiency and cost-effectiveness. Two case studies of off-highway machines are shared, as well as an on-road vehicle application to show a high-speed application.

Fuel economy standards of mobile machines are becoming more stringent, thus mobile machine manufactures are looking for ways to make their machines more efficient. One way to improve machine efficiency is to recover potential or kinetic energy. If there are not enough simultaneously acting actuators, which can use recovered energy instantaneously, the energy needs to be stored for later use. If not stored, excess energy is dissipated using the engine brake.

One innovative solution to address the above challenge is the series-parallel (S-P) hybrid system proposed in Zimmerman [21] and shown in Figure 5.5, which enables multi-actuator DC systems. In addition to the benefit of energy storage, it offers a high-pressure rail that can efficiently power multiple rotary actuators in a DC system with a single pump. The high-pressure rail also provides simultaneous control to multiple

FIGURE 5.5 Closed circuit series-parallel hybrid displacement controlled system [21].

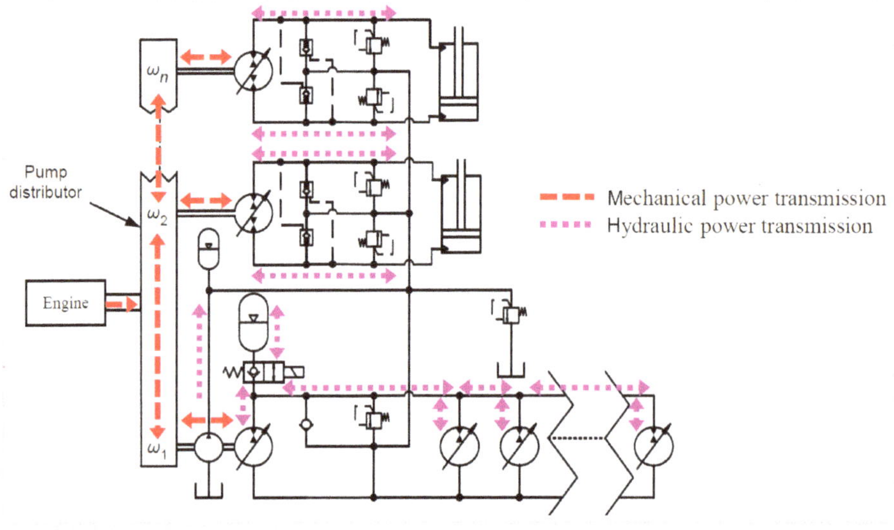

Pump distributor
ω_n
ω_2
ω_1
Engine
Mechanical power transmission
Hydraulic power transmission

linear actuators or auxiliary functions (although inefficiently), which may be low power consumers serving secondary functions on a mobile machine, without requiring additional DC servo pumps. Finally, the high-pressure rail can be used for swashplate regulation to yield faster pump adjustment dynamics or a more compact swashplate regulation system.

The proposed architecture involves the addition of a high-pressure accumulator for energy storage, adding a variable displacement pump/motor unit in place of the fixed displacement swing motor, and modifying the control law of the swing actuator to operate in secondary control mode. This system was first demonstrated in Zimmerman and Ivantysynova [22] on a 5-ton excavator employing a conservative, suboptimal supervisory power management controller dubbed as the "single-point strategy." The strategy aimed to operate the engine close to its maximum rated speed and a constant reference torque through appropriate torque control of unit 1 based on the load (pressure differential). It was shown that for an aggressive truck-loading cycle, the maximum engine power could be reduced by 50% over the standard load-sensing (LS) architecture, without sacrificing productivity while still achieving a potential fuel savings of 20% due to improved operating engine efficiency and the use of a smaller engine. For a fair energy comparison of the two systems, the cooling system, actuator relief valves, and logic valves were included in the hybrid system model as power losses but were left out of Figure 5.6 to simplify the schematic.

To further improve the proposed S-P hybrid system's efficiency, a near-optimal supervisory control was pursued in Hippalgaonkar and Ivantysynova [23]. Dynamic programming (DP) was utilized to solve for the optimal supervisory solution and serve as a benchmark for implementable control strategies. The patterns in optimal state trajectories and control input histories obtained via DP were analyzed, and a common pattern was identified for engine speed, accumulator pressure, and DC unit displacements

FIGURE 5.6 Series-parallel hybrid displacement controlled excavator system [22].

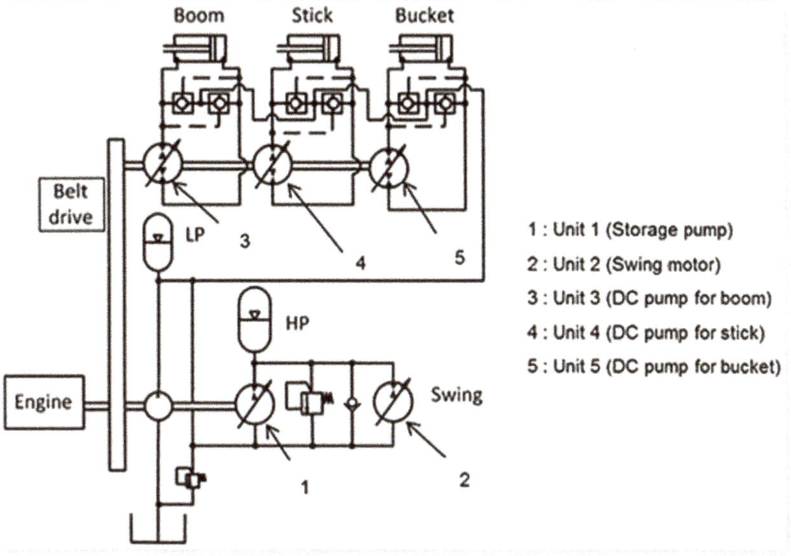

FIGURE 5.7 Controller structure on prototype hybrid excavator [24].

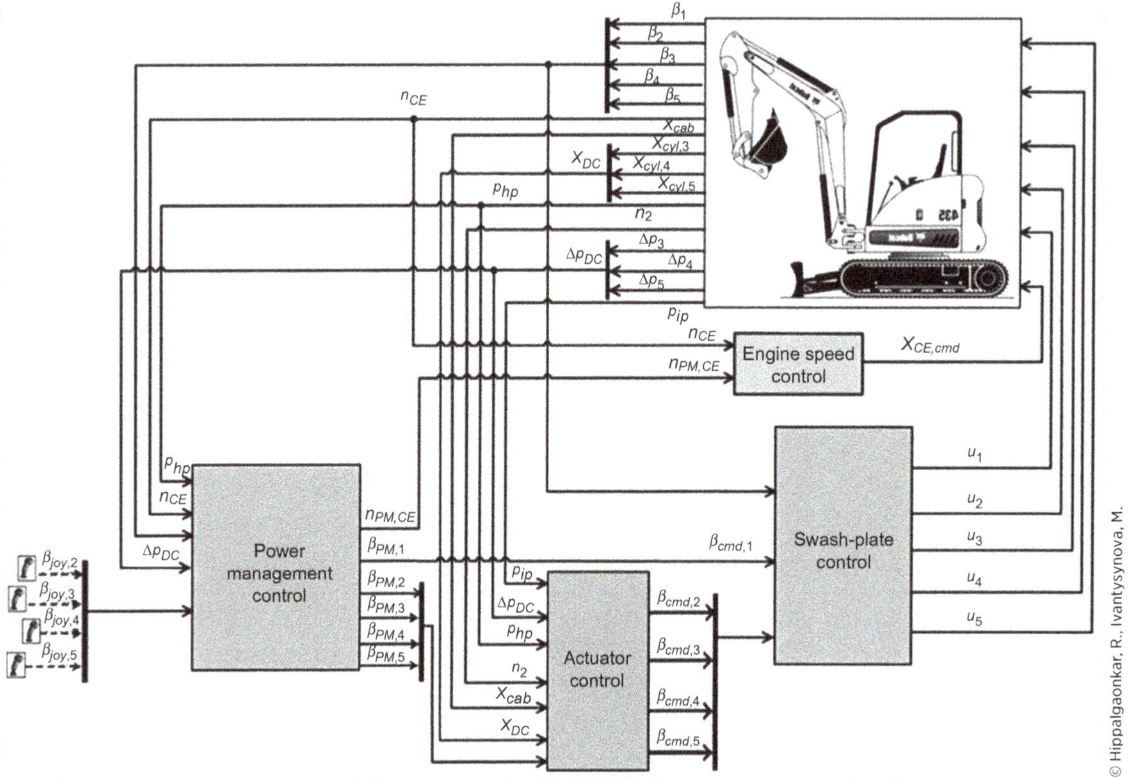

across different working cycles. All of the system's degrees-of-freedom need to be exploited to obtain near-optimal system behavior, with the engine and hydraulic units operating near their highest efficiency regions, simultaneously whenever possible. A second realizable approach, "the minimum-speed strategy," was devised and implemented on the prototype machine, which varied the commanded engine speed according to the DC actuator flow (or velocity) requirements (determined by the operator commands), and kept the engine below a prespecified power level while meeting aggressive digging cycle requirements.

The described rule-based control strategy was devised to achieve near-optimal system efficiency with the controller structure shown in Figure 5.7. The devised strategy replicated optimal system behavior with the same rule for controlling engine speed for different cycles but different rules for the primary unit of the series-hybrid swing drive for different cycles. Therefore, the practical implementation of a rule-based approach can be realized by providing a set of controllers that are selected based on the mobile machine's operating mode. It was found that downsizing the engine by 50% with the implementation of the near-optimal minimum-speed strategy results in significant fuel efficiency gains over the standard valve-controlled architectures (55%) as well as DC non-hybrid architectures (25%) in cyclical operation. The fuel efficiency improvement stems from operating the engine at a constant high speed near its rated power (single-point strategy) as shown in Figure 5.8.

FIGURE 5.8 Single-point strategy simulation: engine operation map in DC S-P hybrid excavator with 50% downsized engine in the expert truck-loading cycle [24].

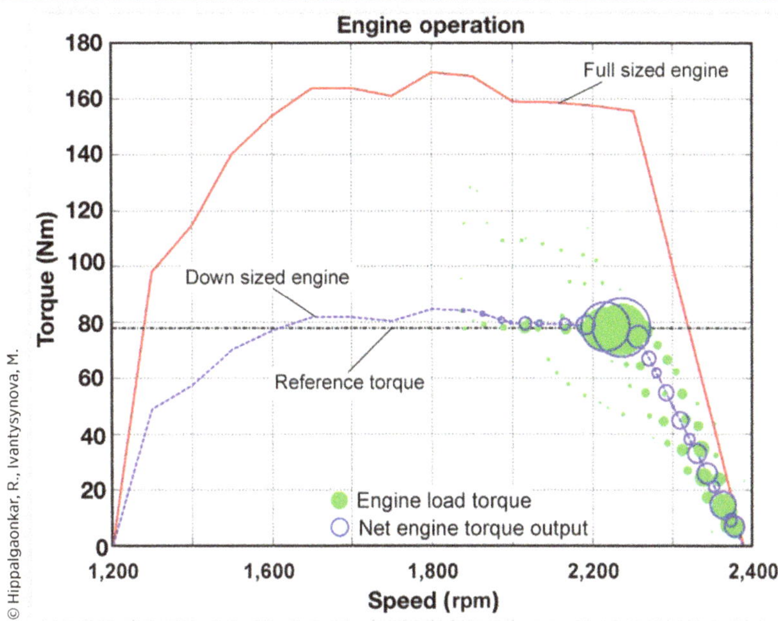

It is noted that the outlined approach, which exploits all possible system degrees-of-freedom while simultaneously keeping the engine and hydraulic units near their efficient operating regions whenever possible, is general and not specific to the DC S—P hydraulic hybrid architecture. Valve-controlled hydraulic hybrid architectures can also benefit from such a strategy, although with lower overall system efficiency and more stringent constraints on the engine speed. A similar approach was demonstrated to be quite successful in the works of Schneider, which aimed at a similar goal to use DC technology and various other innovative technologies such as a power-split hybrid drive system to operate a downsized engine of a wheel loader at a single low speed, resulting in a 22% fuel reduction over an LS system [25].

5.1.3 Coupled Displacement Controlled Actuation

DC actuation and hydraulic power-split transmissions (PST) have shown significant improvements in the fuel efficiency of mobile machinery. DC actuation eliminates metering losses in valve-controlled systems, while hydraulic PST decouples the engine and the wheels resulting in efficient engine operation. Both technologies also allow for recovering normally dissipated energy. DC actuation allows the conversion of the potential energy of a lowering load into torque that is placed back on a common drive shaft. Kinetic energy can also be captured by a PST during braking events and placing the resultant torque on the same common shaft. For certain types of mobile machines, the level of recovered energy far exceeds demand, which makes them prime candidates for hybridization.

A novel system configuration for a combined DC/PST machine was presented in Sprengel [26]. A reach stacker was chosen as the reference machine due to the large

FIGURE 5.9 Displacement controlled actuation coupled with a power-split transmission in hydraulic hybrid systems for off-highway vehicles (e.g., Reach Stacker) [26].

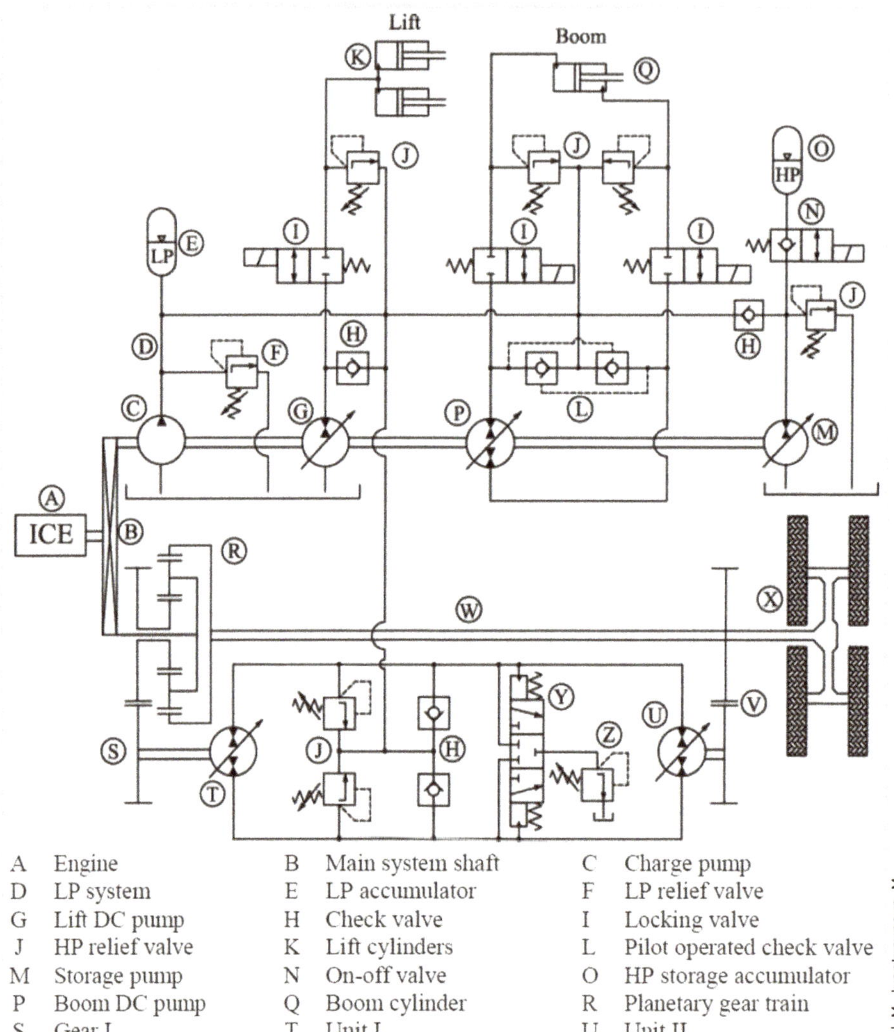

energy recovery potential since its typical working cycle can be condensed down to one of two maneuvers: unloading a truck and stacking the container or unstacking the container and transferring it to a truck. The reach stacker was modeled and simulated in both the baseline load-sensing (LS)/power-shift transmission configuration and the proposed DC/PST architecture, whose circuit is shown in Figure 5.9. Two variants of DC actuation are used: the boom cylinder (Q) uses a closed circuit that connects each of its sides directly to a pump port whereas the lift cylinders (K) use an open circuit where only the high-pressure port of pump (G) is connected with the lift cylinders. The lines connecting cylinder (K) to the tank are not shown to keep the schematic less cluttered. Both DC circuits require a low-pressure supply provided by a charge pump (C) and low-pressure accumulator (E), and it is used to supply the hydrostatic transmission. The torque converter and power-shift transmission are replaced with an output-coupled PST. Power enters the transmission from the main system shaft (B) and passes through the planetary gear train (R). The planetary gear train splits the power between the mechanical path (W) and the hydrostatic path. Power is recombined at the junction of (V) and (W). At low speeds, the entire transmission power is transmitted through the hydrostatic path. As the vehicle speed increases, a greater percentage of power transverses the mechanical path up until all of the power is mechanically transmitted at the vehicle's top speed. Vehicle speed is controlled by adjusting the displacements of units (T) and (U).

A principle concept of this circuit is that recovered energy freely flows between both subsystems. Recovered kinetic energy from the transmission can be used to power the working hydraulics, and recovered potential energy from an aiding load can be sent to the transmission. When recovered energy exceeds current system demand, storage pump (M) is used to store this excess energy in the accumulator (O). One of the energy storage system's key features is the on-off valve (N), whereby its integrated check valve in the non-energized position allows charging of the accumulator at any time. However, when no fluid is being pumped, the valve limits the pressure seen by the storage pump to that of the low-pressure system, which reduces parasitic losses of the storage pump (M) caused by exposure to the high-pressure accumulator. When stored energy is required, the valve (N) is energized to connect the accumulator to the storage pump (M), which acts as a motor that places torque back on the main system shaft.

For the two typical working cycles of a reach stacker (unloading a truck and stacking the container, or unstacking the container and transferring it to a truck), a standard driving cycle, shown in Figure 5.10, is used. This drive cycle covers a 100 m (109 yards) path and is indicative of a typical working environment.

The energy distribution of both machine architectures is illustrated using Sankey diagrams. For a clear comparison, energy consumption is normalized to the energy provided by each machine's engine. An average stacking cycle (stacked three high in the second row) with a mass of 40 tons was selected as a reference cycle. The reach stacker

FIGURE 5.10 Standard driving cycle of a reach stacker [26].

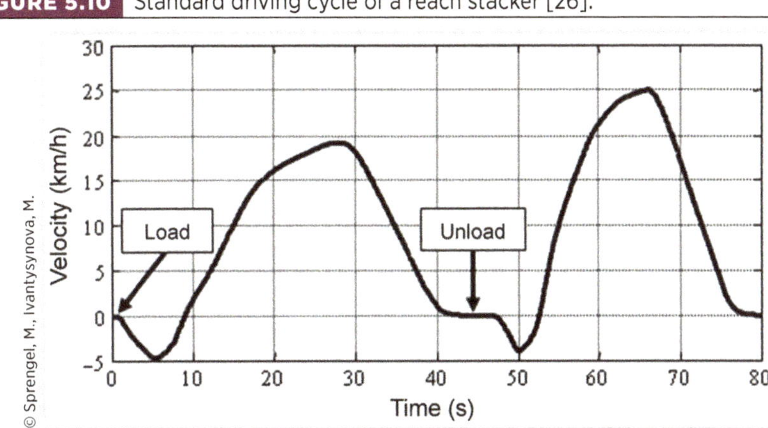

moves the container from a truck to the stack using the drive cycle in Figure 5.10 before returning to the truck.

The energy distribution of the baseline machine during the reference cycle is shown in Figure 5.11. It can be seen that all of the potential energy recovered from lowering loads is dissipated through the control valves into heat, which is inefficient and requires additional cooling energy (unmodeled). The machine's kinetic energy is also dissipated through the transmission's friction brakes since it cannot be recovered in this system architecture.

The total power consumption for the DC/PST machine decreased by a total of 28.7%, from 8.69 MJ in the baseline machine (Figure 5.11) to 6.22 MJ (Figure 5.12). This can be attributed to several factors including 0.42 MJ of energy recovered from lowering loads, and 2.24 MJ recovered through regenerative braking. These high levels of recovered energy are possible because the storage pump absorbs 2.34 MJ and releases 1.72 MJ of energy back into the system. Further efficiency improvements are attributed to the PST where 35.2% less energy is dissipated during driving. However, it is noted that lower component losses are not the only benefit of PST, but also their capacity to decouple the engine and wheel speeds to enable more efficient engine operation and allow the same amount of energy to be produced with less fuel.

While the provided Sankey diagrams provide an insight into the machine's inner workings, overall energy consumption illustrates the ultimate benefits of the proposed system. Figure 5.13 shows the simulated energy consumption during 24 cycles for both

FIGURE 5.11 Baseline machine energy distribution [26].

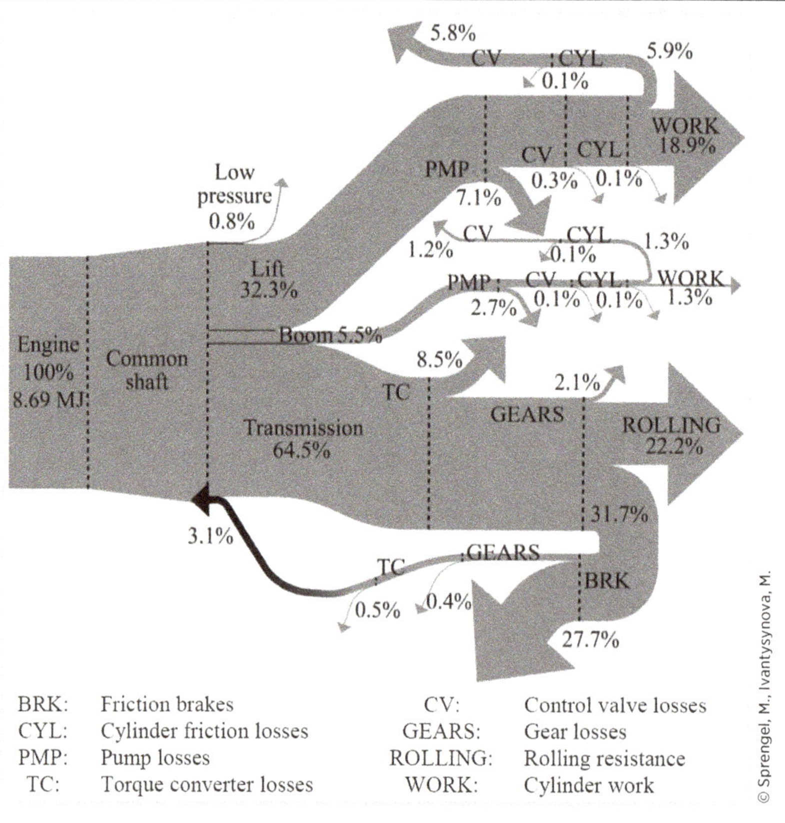

FIGURE 5.12 Proposed machine energy distribution [26].

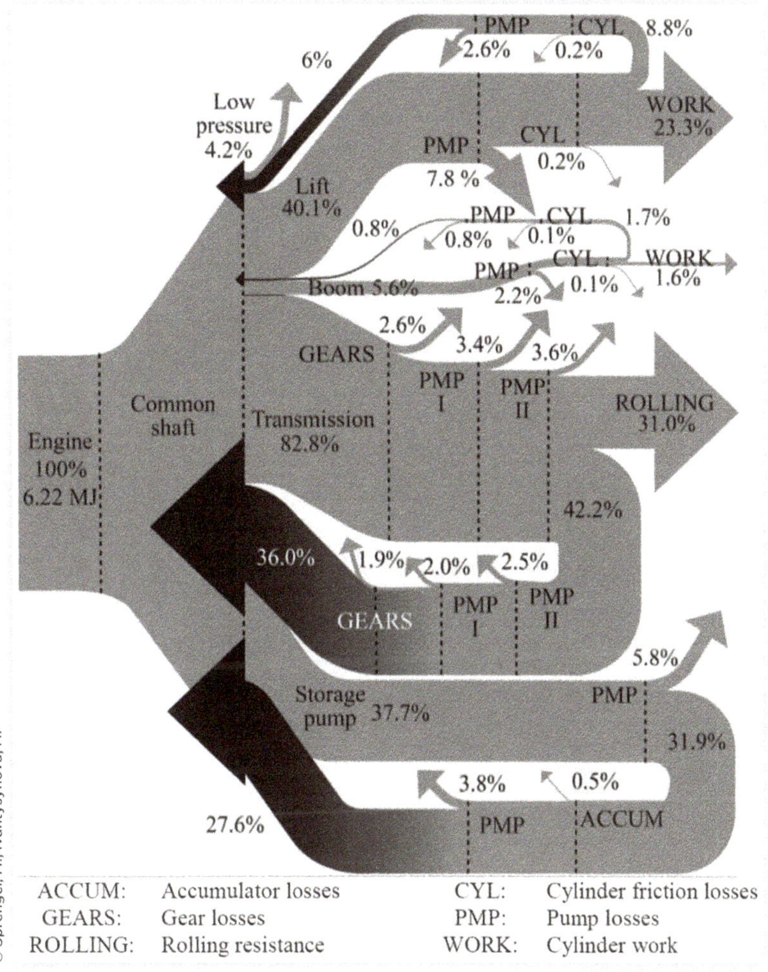

FIGURE 5.13 Simulated energy consumption of the two systems [26].

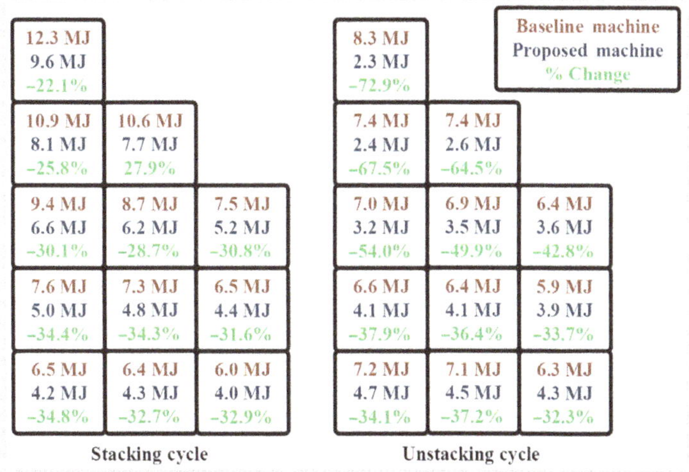

Baseline machine / Proposed machine / % Change

Stacking cycle

12.3 MJ 9.6 MJ -22.1%		
10.9 MJ 8.1 MJ -25.8%	10.6 MJ 7.7 MJ 27.9%	
9.4 MJ 6.6 MJ -30.1%	8.7 MJ 6.2 MJ -28.7%	7.5 MJ 5.2 MJ -30.8%
7.6 MJ 5.0 MJ -34.4%	7.3 MJ 4.8 MJ -34.3%	6.5 MJ 4.4 MJ -31.6%
6.5 MJ 4.2 MJ -34.8%	6.4 MJ 4.3 MJ -32.7%	6.0 MJ 4.0 MJ -32.9%

Unstacking cycle

8.3 MJ 2.3 MJ -72.9%		
7.4 MJ 2.4 MJ -67.5%	7.4 MJ 2.6 MJ -64.5%	
7.0 MJ 3.2 MJ -54.0%	6.9 MJ 3.5 MJ -49.9%	6.4 MJ 3.6 MJ -42.8%
6.6 MJ 4.1 MJ -37.9%	6.4 MJ 4.1 MJ -36.4%	5.9 MJ 3.9 MJ -33.7%
7.2 MJ 4.7 MJ -34.1%	7.1 MJ 4.5 MJ -37.2%	6.3 MJ 4.3 MJ -32.3%

machine architectures. The DC/PST machine exhibits a significant reduction in energy consumption of up to 72.9% during unstacking and by 29.7% on average during stacking. Overall cycles, energy consumption is reduced by 38.0% equating to a 2.89 MJ savings per cycle.

In brief, a mobile machine that handles large loads, such as a reach stacker, can benefit from significant energy savings by incorporating a combination of DC and PST when compared against a conventional machine. A direct comparison showed reduced energy consumption of up to 73% and an estimated saving of 200,000 L (53,000 gal) of diesel fuel over the lifetime of the machine.

5.1.4 A Blended Hydraulic Hybrid System for On-Road Vehicles

As previously mentioned, hydraulic hybrid transmissions can be grouped into three main categories: parallel, series, and power-split. While each architecture has its advantages in specific applications, they can still benefit from further improvement. The series hybrid, which is currently the most popular architecture for full hybrids such as the hydraulic hybrid bus built by Parker [27], requires over-center units that are less common and more expensive than standard units, which is especially true of the more efficient bent-axis variety. A comparison between series and parallel hybrids can be seen in **Figure 5.14**. Series hybrids often operate, inefficiently, at high pressures and low displacements due to the accumulator's state of charge. Another drawback of series hybrids is the potential for a "synthetic" or "spongy" feel which becomes more pronounced as the vehicle, and consequently the accumulator size increase. This spongy

FIGURE 5.14 Comparison of parallel and series hydraulic hybrids [27].

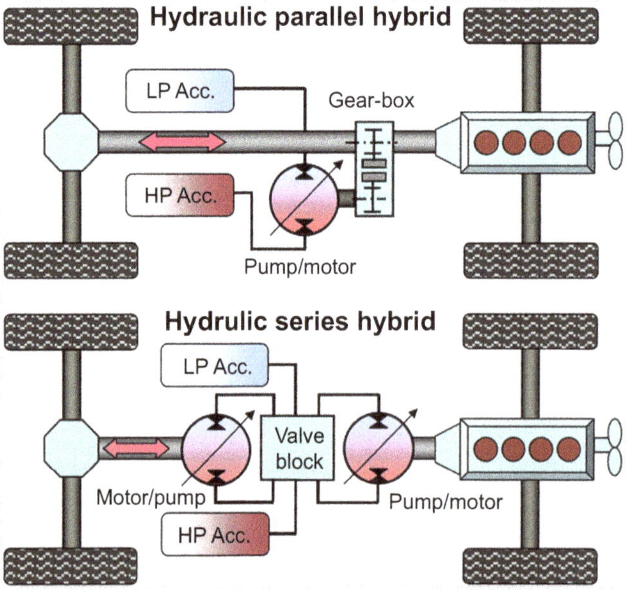

feel originates with maximum driven wheel torque and system pressure; and to increase system pressure, more flow must enter the accumulator than available. In series hybrids, a considerable delay may be felt in achieving the desired wheel torque when system pressure is below what is required, which results in a spongy feel. This inherent delay is in contrast to hydrostatic transmissions, which increase pressure virtually instantaneously, a property that forms the basis for the innovative system architecture presented in this section.

To improve the response, such as stiffness, a new hydraulic hybrid architecture, i.e., the blended hybrid was proposed in Sprengel and Ivantysynova's *Investigation and Energetic Analysis of a Novel Hydraulic Hybrid Architecture for On-Road Vehicles* [28] by blending a hydrostatic transmission with a parallel hybrid as shown in **Figure 5.15**. Given that fuel economy is of principle concern, the focus was on the overall vehicle efficiency. Simulation models were generated for a class II truck with a conventional automatic transmission, a series hybrid transmission, and the new blended hybrid transmission. Fuel consumption rates were compared using the Urban Dynamometer Driving Schedule (UDDS) driving cycle, which is a standard dynamometer test that represents city driving conditions for evaluating the fuel economy of light-duty vehicles. For a fair comparison and removing the effect of controller design on fuel consumption rates, DP was used to optimally control all three transmissions. DP is a method to determine the optimal parameterization of components and control methods, by going backward in time through a given cycle. The simulation has access to future events and can thus adjust the parameters accordingly. However, DP can only be used to find the optimal

FIGURE 5.15 Blended hybrid circuit for on-road vehicles [28].

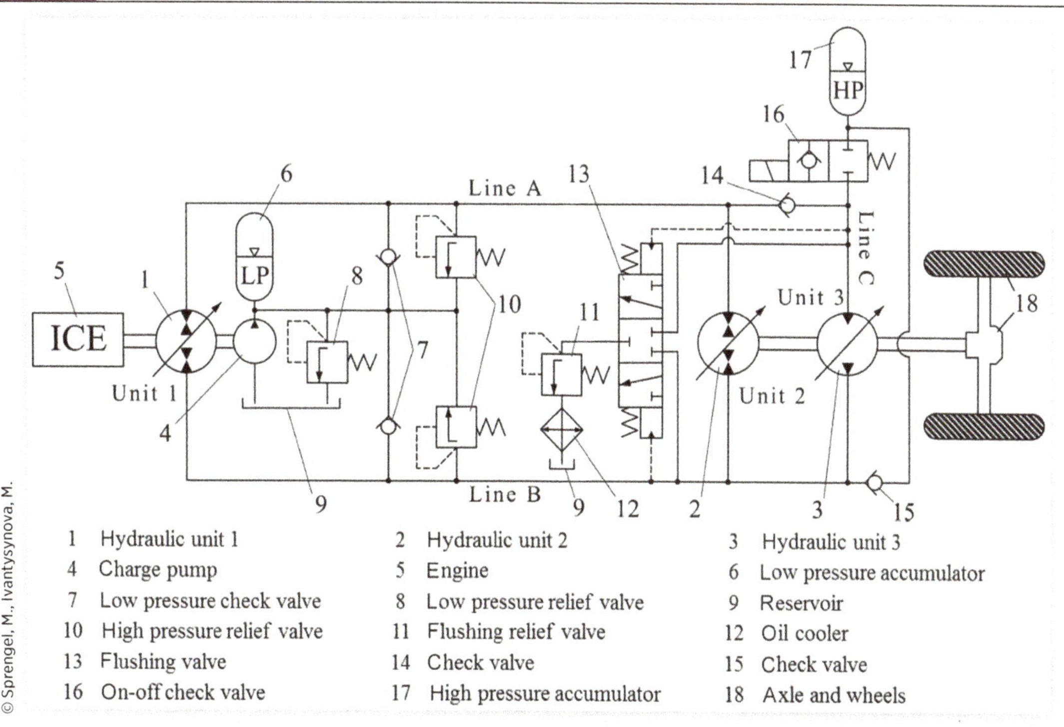

solution for known cycles and cannot be used for unknown cycles as they would occur in reality.

After analyzing the obtained simulation results, which can be found in Sprengel and Ivantysynova [28], the main differences between the series hybrid and the automatic transmission are the engine speed and throttle. Since the series hybrid decouples the engine and wheel speeds, it permits improved control over engine management to allow the engine to operate in a more efficient region. Also, the series hybrid uses the entire range of accumulator pressures allowing deep charge and discharge cycles. The series hybrid's efficiency is related to the accumulator pressure, with low pressures and high displacements being favored over high pressures and low displacements. The series hybrid generally operates at the lowest allowable engine speed with a low-to-medium throttle level, which is in contrast to the automatic transmission's operation and leads to gains in fuel economy. On the other hand, the blended hybrid's accumulator maintains a higher pressure than the series hybrid, which is due to the blended hybrid system not requiring constant operation at the accumulator's current pressure. Therefore, there is no need to increase efficiency by reducing pressure during extended periods of low torque requirement. The blended hybrid also maintains a higher minimum pressure to ensure that all of the available braking energy is captured. Unlike the series hybrid, the blended hybrid is not capable of increasing accumulator pressure while driving, which is due to a lack of a direct connection between unit 1 and the accumulator, preventing the transmission from storing excess energy from the engine in the accumulator.

The engine operation maps of the three transmission systems are shown in Figure 5.16. The engine speed is shown on the x-axis, the engine output torque is shown on the y-axis, while the colored contour lines show the specific fuel consumption. Each white dot represents a single point in time of the engine operation. It can be seen that the series hybrid's engine performs in a much narrower band close to the lower end of the engine speed spectrum, which can be explained by the decoupling of the engine from the output shaft and operating at its most efficient points. The blended hybrid does not allow for such an engine optimization and operates in a wider band.

FIGURE 5.16 Engine operation map of the automatic transmission (left), series hybrid transmission (center), and the blended hybrid transmission (right) [28].

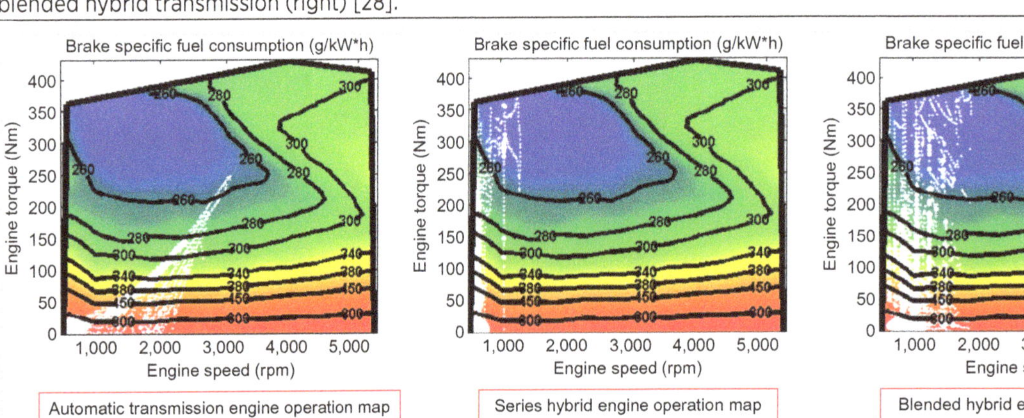

The automatic transmission achieved a fuel economy of 15.3 L/100 km (15.4 mpg); the series hybrid reached 10.5 L/100 km (22.4 mpg), which represents a 44.8% improvement over the baseline automatic transmission; and the blended hybrid achieved a fuel economy of 11.1 L/100 km (21.2 mpg), which is 37.0% better than the automatic transmission. Even though the blended hybrid system achieved a fuel economy that is 5.4% lower than the series hybrid, other considerations regarding sizing, implementable control strategies, stiffness of response, and user feel may alter the overall assessment. Fuel economy, while important, is not the sole consideration, as the blended hybrid architecture shows considerable potential in other areas over existing series hybrid configurations. The blended hybrid's stiff nature improves the transient response and the user feel to match that of traditional mechanical transmission. The blended hybrid system circumvents the need for over-center units attached to the wheels, which lowers the cost and permits the use of bent-axis units. Finally, the blended hybrid increases efficiency under certain low power operations such as driving at a constant moderate speed, during which the line pressure can be reduced below the accumulator's minimum pressure, a performance that is not possible with series hybrids.

5.2 Innovative Displacement Controlled Linear Drive Systems

5.2.1 Introduction

DC is the throttle-less control of actuators such as cylinders and rotary drives. Flow control is performed by changing the rate of the flow source, such as a pump, using either changing its speed or its volumetric displacement. It can also be referred to as PCA. While it seems intuitive to change the required flow rather than throttling it using valves, valve control is currently the state-of-the-art, especially for linear drives. To understand the fundamental differences between DC and valve control technologies, a simple schematic for a single rotary drive and a multi-drive system is shown in Figure 5.17 for valve controlled-actuation (VCA), and in Figure 5.18 for pump-controlled actuation (PCA).

FIGURE 5.17 Valve controlled actuation (VCA). Left: single rotary actuator, right: multiple actuators.

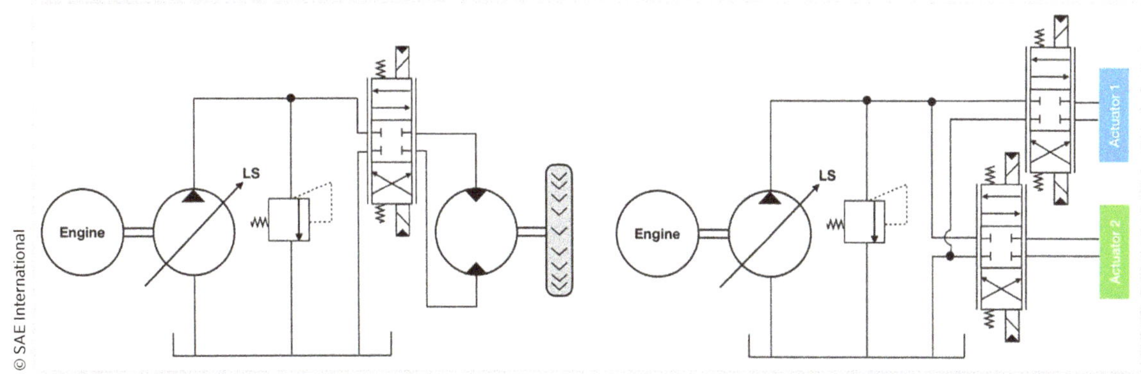

FIGURE 5.18 Pump controlled actuation (PCA). Left: single rotary actuator, right: multiple actuators.

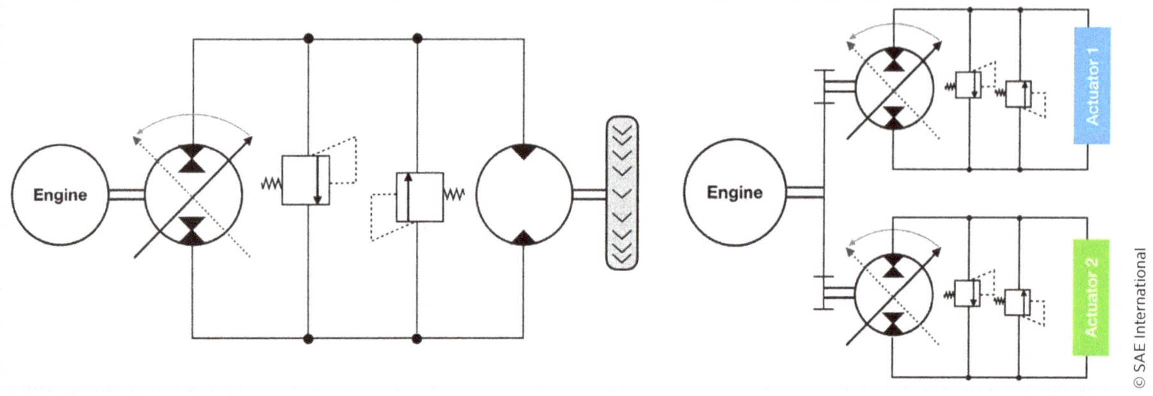

VCA systems are typically open circuit configurations, where a pump delivers flow from a tank using either a constant or adjustable pump, while a valve controls the flow and pressure the actuator sees. The remaining flow is throttled back to the tank, dissipating the excess energy into heat. While this configuration might be impractical for a single actuator, it is state-of-the-art for multi-actuator applications, in which the pump is set to the highest pressure level required by the actuators (see LS in Section 3.4), and the valves distribute the flow as required by the operator or the required task. There are numerous possibilities to realize VCA systems, ranging from standard constant pressure systems to the more advanced LS systems with pressure compensators, however, the same principle of flow throttling is applied to control the motion of the actuators.

On the other hand, PCA systems require a single flow source for each actuator. They are typically implemented in closed circuit configurations as shown in Figure 5.18, but open circuit applications and prototypes do exist. The advantages and disadvantages of open versus closed circuit solutions are covered later in this chapter. For rotary drives, it is straightforward to implement PCA, and it is already state-of-the-art for many drive systems, especially in mobile machines, with several realization options as covered in Chapter 3. For multiple actuators, a gearbox is needed to mechanically couple several pumps to the same engine, and depending on the task of each pump, they can be driven at different speeds. Alternatively, pumps with a through-shaft can be used to circumvent the need for a gearbox; however, this requires driving the pumps at the same speed and the first pump having an oversized shaft to withstand the added torque of the subsequent pumps.

To give an energetic comparison between PCA and VCA systems, pressure and flow requirements are shown in Figure 5.19 as an example. The color shaded areas represent the power requirements for each actuator, blue and green for actuators 1 and 2, respectively. The red area in the background is the power delivered by the pumps. The larger the visible red area, the larger the losses in the system. For the VCA system, it is assumed that the pump is adjusted to the required pressure level (LS). Clearly, for constant pressure or constant flow systems, the comparison would be even worse.

For single actuator applications, the difference between VCA (left) and PCA (right) systems is not very significant, as seen by the relatively small red area where the only

FIGURE 5.19 Overview of power losses. Left: VCA for single and multiple actuators. Right: PCA for single and multiple actuators.

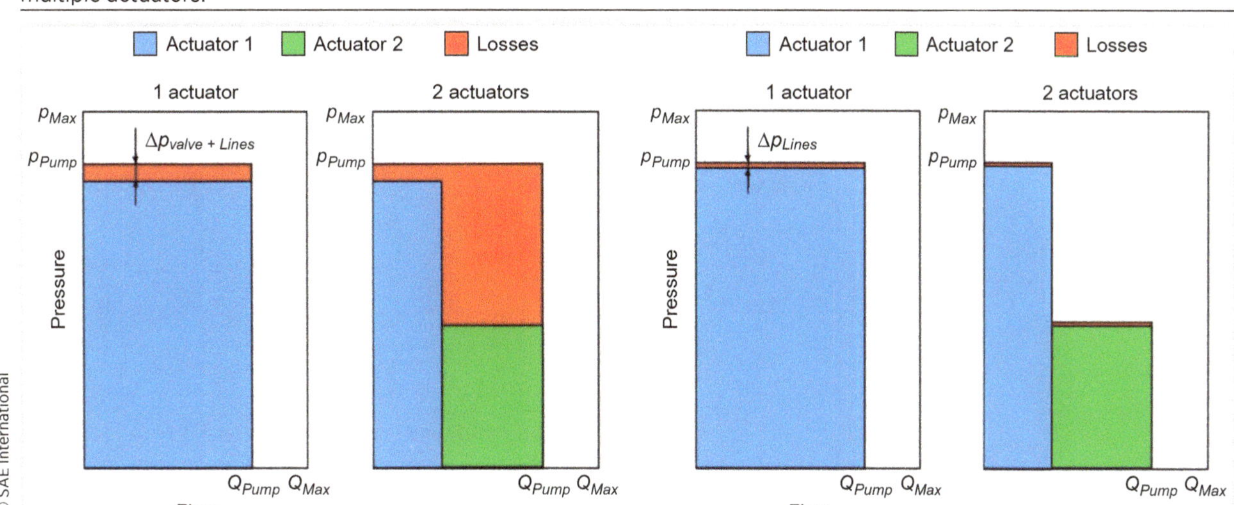

difference is generated by the pressure drop across the fully open valve. As more actuators are added, the power losses of the VCA system become more significant, which is especially true when actuators operate in different pressure ranges. This is where PCA systems show their full potential, where essentially the only losses are those of the pump and the lines.

As already mentioned, PCA systems are already widely used for single actuator rotary drives, however, as soon as linear drives are part of the system, they become virtually nonexistent in today's mobile machines. To show how linear drives can be controlled using DC-technology, it is easiest to start with a double rod cylinder as shown in Figure 5.20. In this configuration, the flow of the pump is directed into one chamber of the cylinder moving it in one direction, or by swiveling the pump over-center the flow enters the other chamber moving it in the other direction. The speed of the actuator is directly proportional to the displacement of the pump. The pressure relief valves are to protect the circuit from over-loads and are shown for completeness purposes.

FIGURE 5.20 PCA for a double rod linear actuator.

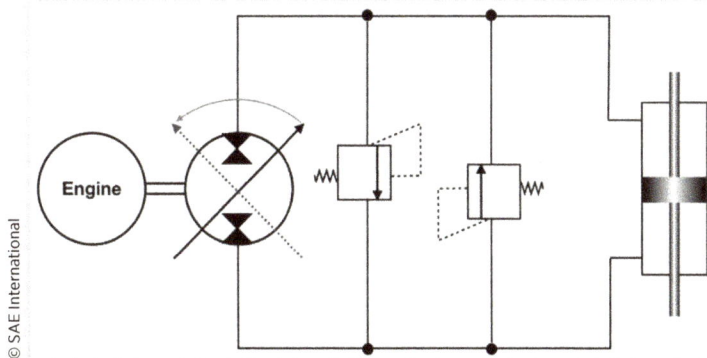

FIGURE 5.21 PCA for a single rod cylinder [29].

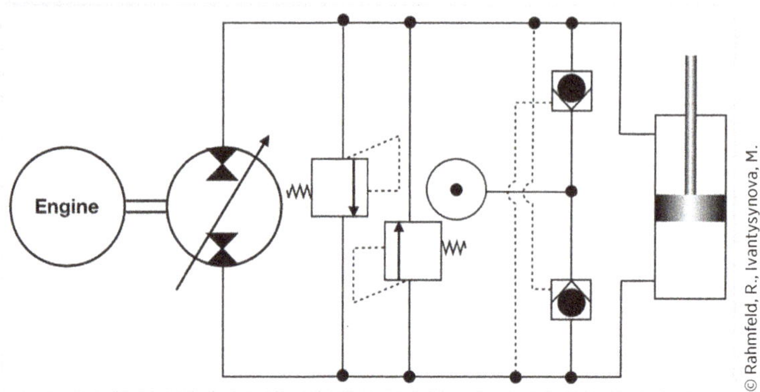

Most cylinders in today's mobile machines are differential (single rod) cylinders, meaning that the acting areas of each cylinder chamber are different. To incorporate DC technology for these systems, it is required to deal with the differential fluid volume, which is typically realized using pilot-operated check valves, as later detailed in Section 5.2.3.

The use of DC is not well-established for linear drives. Since most mobile machines use linear drives for their working functions, it is essential to control linear actuators using the pump to fully exploit the potential of DC technology. Controlling the motion of double rod cylinders is straightforward given the equal areas on both chambers.

On the other hand, single rod cylinders, also called differential cylinders, have different chamber sizes, thus the flow into the cylinder does not equal the flow out of it. This differential flow makes pump control more challenging since the pump requires the same flow in and out of it. The first work in this field of operating differential cylinders in closed circuits was reported by Lodewyks [30]. Rahmfeld, and Ivantysynova suggested a simple work-around for operating DC for differential cylinders [29]. The differential flow is diverted into a low-pressure system where it is stored in an accumulator. Then, when more flow is required by the cylinder, the flow is taken out of the accumulator. Such a system is shown in Figure 5.21. The accumulator is not shown but would be incorporated into the low-pressure source along with the charge pump. The advantage of this system is that the low-pressure system can be shared among several linear drives, which also means that an accumulator can be shared among various circuits.

The flow in and out of the low-pressure reservoir is accomplished using pilot-operated check valves. The high-pressure signal unlocks the check valve, which is connected to the low-pressure side of the cylinder. The four-quadrant operation in both directions of movement for a single rod cylinder is illustrated in Figure 5.22.

When the cylinder retracts, more flow is pushed into the pump than the cylinder is using. To compensate for this additional flow, the upper check valve, which is connected to the low-pressure line, is unlocked by the high-pressure pilot line, thus allowing the extra flow to be stored in the low-pressure accumulator. This operation is shown on the top left of Figure 5.22, where the pump is operating in motoring mode, recovering the energy of the aiding load. When the cylinder is extended, the top right side of Figure 5.22, more flow is required from the pump, than the cylinder would deliver. In this case, the

FIGURE 5.22 Examples for the four-quadrant operation for a single rod cylinder. Top left: Regeneration during aiding load retraction. Top right: resistive extraction. Bottom left: resistive retraction. Bottom right: regeneration during aided extraction.

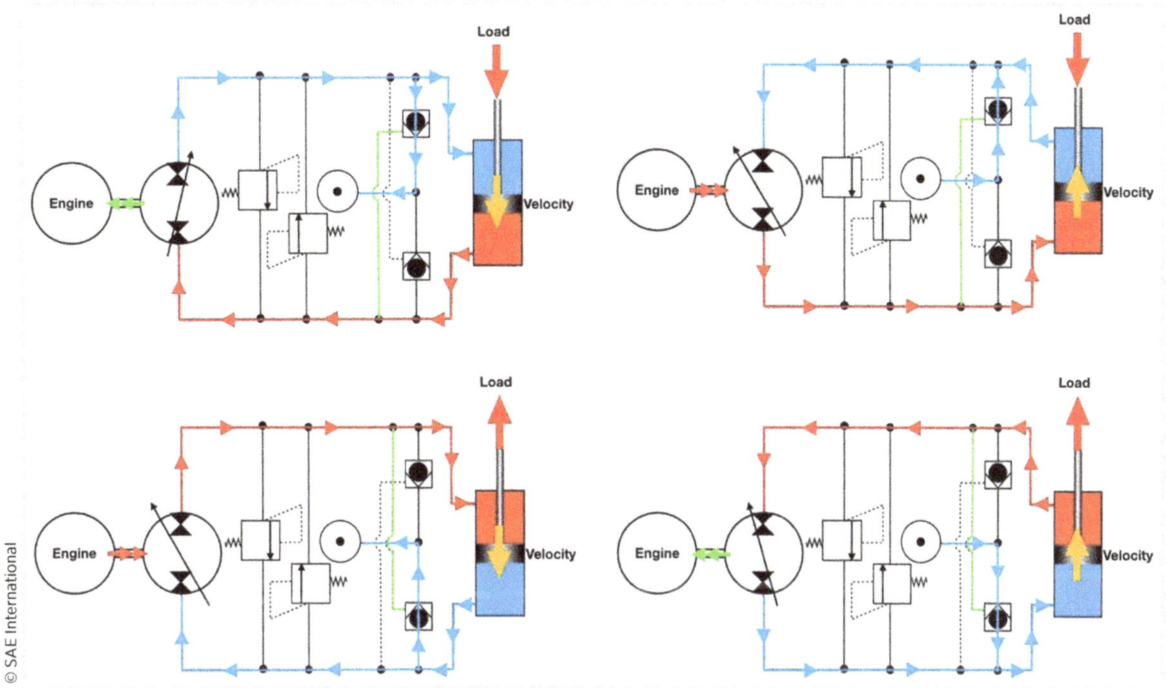

check valve opens automatically delivering the flow of the accumulator to the low-pressure side. The other two scenarios, where the load is pulling on the cylinder is shown in the bottom half of Figure 5.22. In this case, the other line is under high pressure and the second check valve is used.

This example not only shows how DC is implemented into a linear drive but also how the hydrostatic unit is capable of recovering energy. A close-up on the unit with its swashplate position is shown in **Figure 5.23**. The green arrows indicate an aiding torque that supports the engine in its task, whereas the red arrow shows the torque that is required by the engine to power the operation. If there are not enough other tasks that require torque from the engine during the recuperation phase, the engine needs to dissipate the energy using its engine brake. If enough energy can be recovered, then a hybrid unit, which is connected to an accumulator, can be installed to store the energy. At this point, it should be noted, that if a swing drive, or another comparable rotary drive, is implemented, then an additional storage module might not be necessary if the swing drive can be converted to secondary control (see Figure 5.6). In this case, the rotary drive would feature two adjustable units and an accumulator. The adjustable pump would charge the accumulator, storing the energy, while the second unit would be an adjustable motor, which would have to react to the changing pressure levels stored in the accumulator. This solution would save components, cost, and space but would result in a higher control effort.

Open circuit DC solutions can offer some advantages over closed circuit ones, especially for large flow rates. A comparison between open circuit and closed circuit schematics is

FIGURE 5.23 Four quadrant operation.

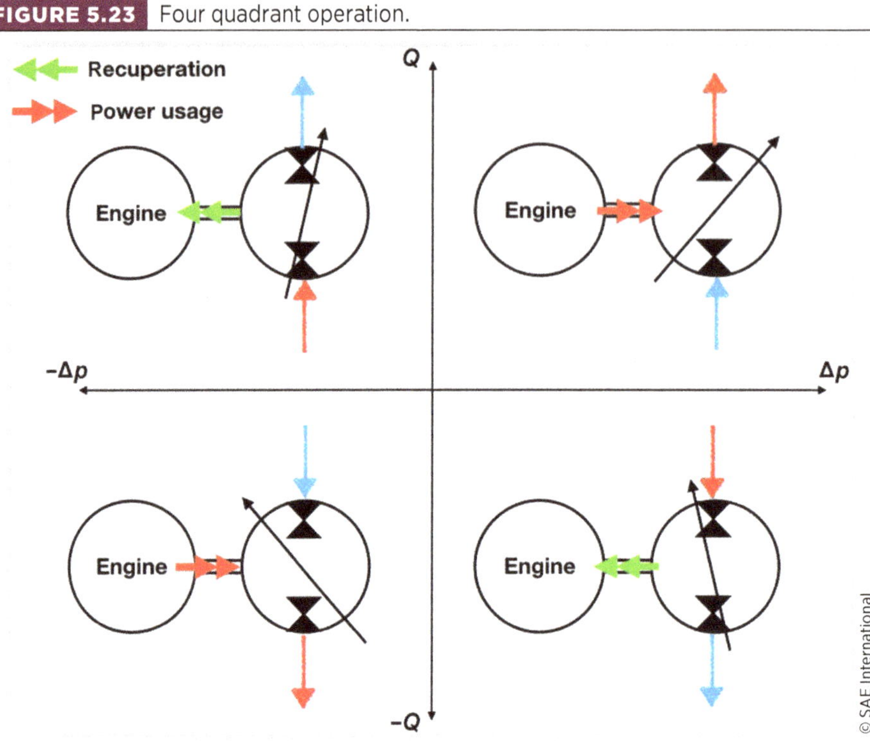

shown in Figure 5.24. For open circuit configurations, it is not necessary to store the differential volume of the single rod cylinder, which can be quite advantageous for large cylinders like the ones used in the mining industry for instance. While closed circuit solutions allow the incorporation of a float valve, which allows a faster retraction of the cylinder under aiding load, it comes at the cost of either lowering under pre-charged pressures and limiting the speed, or having to dump the differential flow into a tank and having to pump it back using the pre-charge pump. While both are not ideal solutions, they are still viable options that can enable DC to be implemented even in machines that require rapid lowering times, without having to oversize the pump. An example of such a solution can be seen in the publications of Schneider *et al.* on the "Green Wheel Loader" project, which entailed designing and building a wheel loader from the ground up to feature the newest advancements in technology, such as DC [25]. The wheel loader used a closed circuit system and required the rapid lowering of the bucket. A float valve was incorporated but it required some design effort to minimize an oversizing of the charge pump. An open circuit configuration requires no additional charge pump to install a float valve. Both the open and closed systems even allow the possibility to recover some of the energy during float valve operation if deemed feasible, as indicated by the dashed line in Figure 5.24 on the left.

The apparent disadvantage of open DC circuits is the need for switching valves that are used for the directional movement of the cylinder. While these valves can be quite cheap and designed to keep flow losses to a minimum, they result in additional components. They can also hinder the control of the actuator motion, especially when exact position control is necessary, usually around standstill with a payload. Heybroek showed in his research how such an open circuit solution could be implemented and demonstrated it on

FIGURE 5.24 Open vs. closed-loop displacement controlled circuits with float valves.

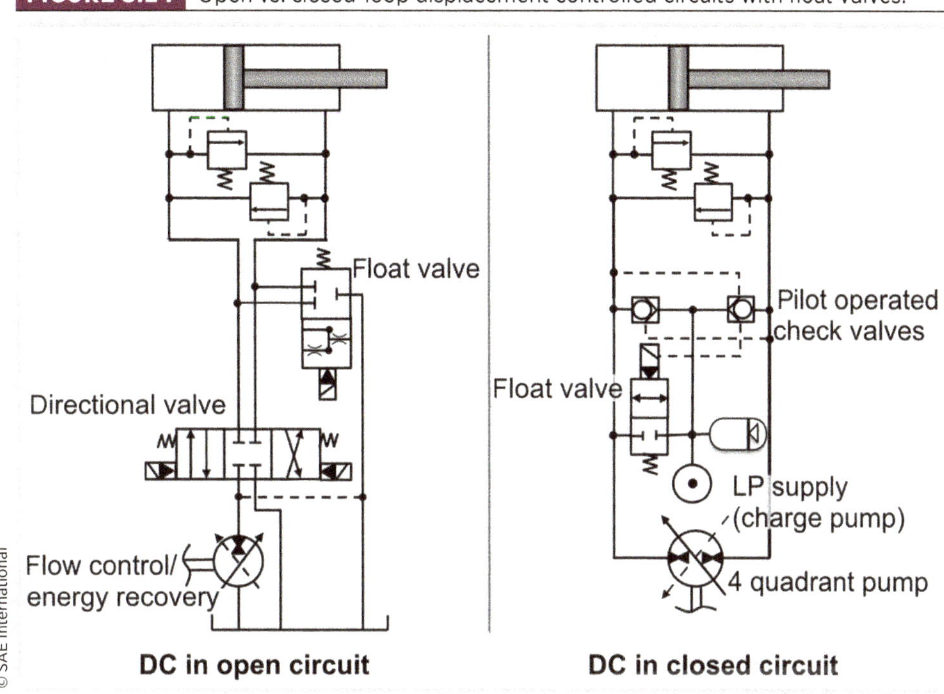

a wheel loader [31]. Another open circuit example, which shows the considerable energy saving potential that DC technology can have, especially in large scale machinery, is later shown in Section 5.2.3, where additional advantages of the open circuit are discussed. Since most valve-controlled machines operate in open circuit today, open circuit DC systems allow for easy retrofitting without the need for new components, given that flow sharing between open circuits do not require flow compensation on the low-pressure side.

In brief, the open circuit can be a feasible solution for designers, however, if exact position control around standstill is necessary, then closed circuit solutions might be the better option. A comparison between the two DC solutions is given in Figure 5.25.

FIGURE 5.25 Comparison between open and closed circuit DC solutions.

	Open circuit	Closed circuit
Energy recovery	++	++
Number of components	+	++
Control effort	+	++
Ease of flow sharing	++	+
Ease of float valves	++	+
Retrofitting of VCA	++	–
Available components on the market today	++	+

5.2.2 Closed Circuit Linear Drive Example

A simplified representation of the DC hydraulic system of an excavator is shown in Figure 5.26. The system has four DC servo pumps, where each unit is responsible for controlling one of the primary excavators working functions: boom, stick, bucket, and swing. The engine, as the prime mover, directly drives one of the servo pumps and a gear pump, which provides flow to the low-pressure charge line and the swashplate regulation systems. A belt drive transmission, which increases the nominal engine speed, drives the three remaining DC servo pumps via a pulley. Logic (on/off) valves are installed between each of the pumps and actuators to lock their positions. Note that additional valves (not shown) can be used to allow the servo pumps to control secondary actuators, which is the advanced "pump sharing" concept covered later in Section 5.3.1.

A 90° digging cycle, which includes excavating soil and transferring it onto a truck, is selected for this type of mobile machine. This cycle is one of the most common

FIGURE 5.26 Hydraulic schematic and sensors for DC excavator system [21].

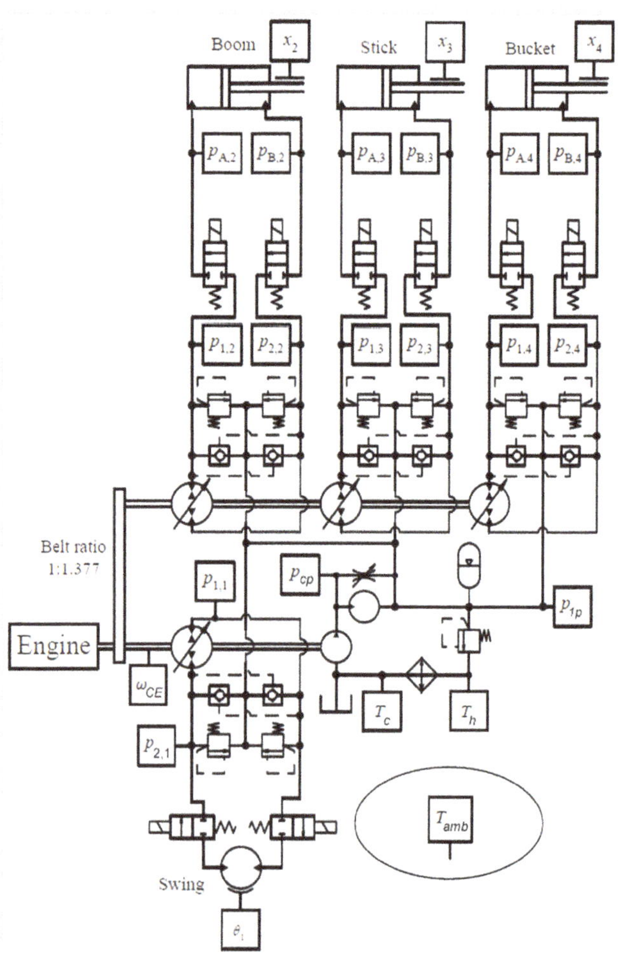

FIGURE 5.27 A picture showing the truck loading cycle performed by an "expert" operator [21].

© Zimmerman, J.

functions performed by an excavator during which all four primary working functions of the machine are routinely used, and often simultaneously. An "expert" operator performed the 90° digging cycle for both the conventional and prototype machine, as shown in Figure 5.27. Using measured data, the external loads acting on the hydraulic cylinders during the working cycle were calculated, as shown in Figure 5.28, and averaged for multiple cycles.

Each cycle is divided into four main segments: dig, lift/rotate, dump, and return. During the "dig" segment, the bucket is filled as it is pushed through the soil. This is followed by the "lift/rotate" segment during which the bucket is lifted out of the ground, and the swing is rotated to position the bucket above the truck. In the "dump" segment, the soil is released to the truck by tilting the bucket; and finally, in the "return" segment the boom is lowered, the swing is rotated, and the bucket is positioned back in the ground in preparation to recommence digging. As seen, the actuator loads are highest during

FIGURE 5.28 Measured external loads on hydraulic cylinders [21].

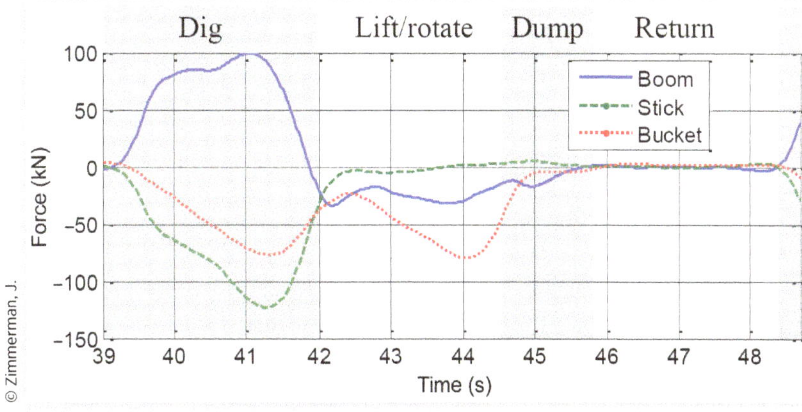

© Zimmerman, J.

FIGURE 5.29 Measured and simulated energy distribution [21].

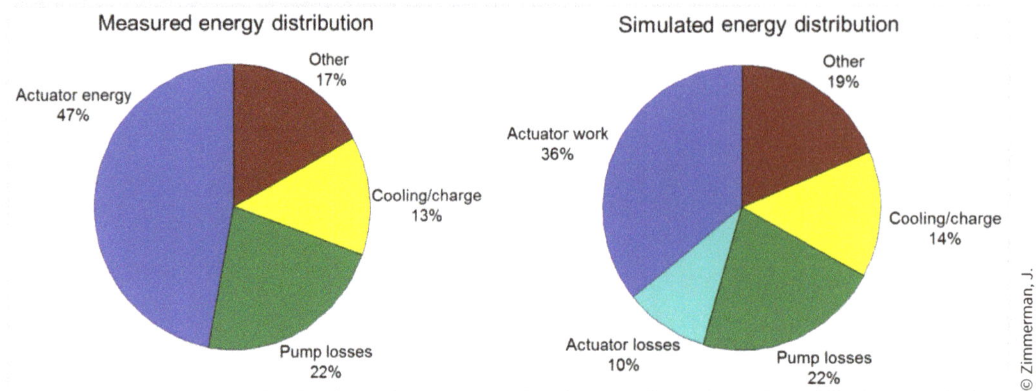

the "dig" section, as expected. Given the additional soil mass in the bucket, external loads exist during the "lift/rotate" and "dump" sections, but during the "return" section (when the machine freely rotates), the loads are zero.

The energy distribution of the hydraulic system was calculated from measured data and simulation results, which are compared in Figure 5.29. In the measurement plot, the actuator power is calculated from the measured pressures and velocities of the actuators. The losses (friction and leakage) of the actuators could not be captured with the installed sensors and as a result, they are lumped into the actuator power. In the simulation, these power losses are extracted and shown separately. The measured pump losses were calculated using the servo pump loss models along with the measured swashplate angles, pump speeds, and pump pressures. The cooling/charge loss is the total power consumption of the fixed displacement gear pump that powers these systems, as a function of the measured pump speed and pressure. Other losses include line losses (dominant), relief valve losses, and belt transmission losses.

Table 5.1 shows a comparison of energy and fuel results for 20 consecutive truck loading cycles. The first column shows the fuel consumed by the excavator as measured directly via an externally mounted tank, before and after the cycle. The second column shows the calculated results based on measured quantities. The simulated results refer to calculations based upon the outputs of the multi-body dynamic and hydraulic co-simulation model. The results show that the simulated fuel consumption is 4% higher than the measured value, and the calculated fuel consumption from measurement is 4% lower than the measured quantity. The actual shaft energy could not be measured directly and as a result, the error is unknown. In Zimmerman [21], it was shown that, for this truck loading cycle, the fuel consumption of the DC excavator

TABLE 5.1 Measured and simulated energy and fuel results (20 cycles) [21].

	Direct measurement	Calculation from measured quantities	Simulated result
Net shaft energy	NA	2.84 MJ	3.38 MJ
Total fuel	0.295 kg	0.282 kg	0.307 kg

system was 40% less than that measured on another Bobcat 435 excavator having the standard LS hydraulic system.

5.2.3 Open Circuit PCA Example

To show the vast energy-saving potential of DC, large-scale machinery was chosen as an example. Mining excavators are used to break material and transfer it to dump trucks. They are an essential tool in every open mine. They operate in short (30 s) 90° digging cycles as seen in Figure 5.30.

A typical valve-controlled hydraulic circuit of such an excavator can be seen in Figure 5.31. In this example, a 290-ton excavator is powered by two large diesel engines, which in turn power four main axial piston pumps each, and a series of smaller auxiliary pumps used for cooling and other functions. This hydraulic circuit is designed with two main goals: redundancy (robustness) and short digging times (productivity). Since the excavator forms the heart of the mine, standstills are very expensive and should be avoided at all costs. The hydraulic system is designed to continue working even if one combustion engine or one of the main axial piston pumps fail, it just continues to operate at half the speed. The main hydraulic actuators (boom, stick, and bucket) are operated in an open circuit fashion. The clam is a secondary function that opens up the back of the bucket to dump the material onto a dump truck. The distribution of the flow and the control of the actuators is performed by two open-center (OC) valve blocks. Controlling the valves essentially forms the machine intelligence, where based on their metering edges actuators are prioritized to receive most of the flow when the operator demands more power than the machine can deliver. Since the excavator is designed to work at full power and maximum digging speed, trained operators almost always

FIGURE 5.30 Typical 90° digging cycle for a mining excavator.

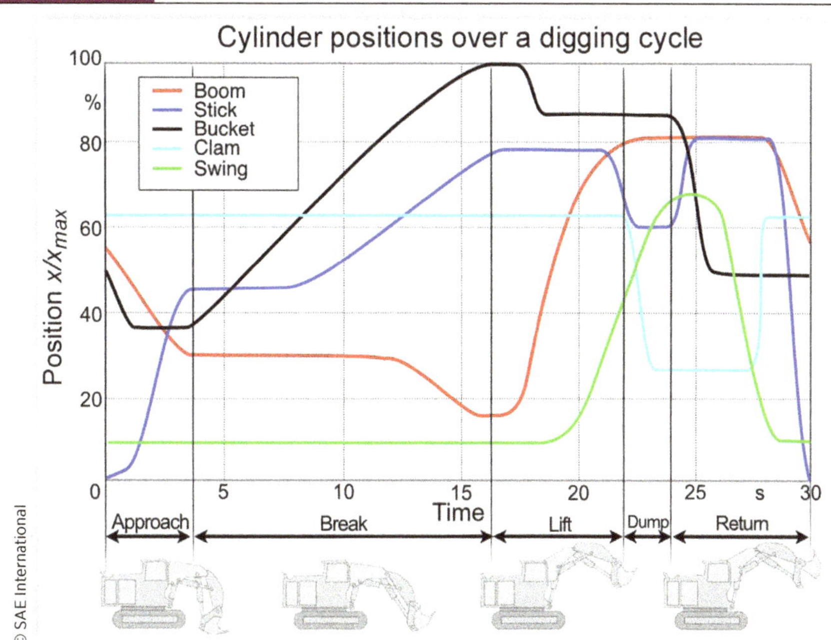

FIGURE 5.31 Case study - mining excavator. Shown is the conventional open center (OC) valve-controlled system.

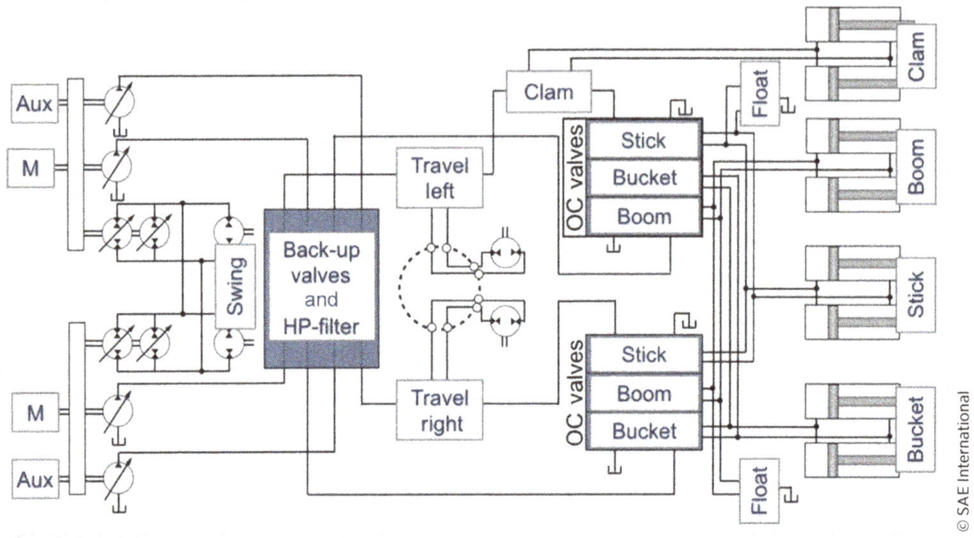

maintain the machine at its flow limits and the OC-valves distribute the power according to the machine designers' specifications. The valves play a second important function of combining the flow rates from different pumps. This flow sharing is needed to achieve minimum lifting times. The arm lowering is achieved by float valves, which simply open the actuators to the tank and lower by gravity. With an 85-ton arm weight, this task is performed in 4–5 s. At the same time that the arm is being lowered, the swing drive needs to rotate the upper carriage back to the digging site in a very rapid manner to prevent losing precious time, thus the swing drive is typically oversized and has its dedicated closed circuit DC circuit.

Such systems suffer from several downsides related to their complexity, the outdated use of OC valves that result in power losses due to flow throttling, and the intricacies involved with controlling the OC valves' metering edges to change the machine's behavior. However, robustness and lifetime have a high priority in the mining industry, thus OEMs have been hesitant to make changes to the conventional system architecture in favor of increasing energy efficiency.

A case study of the conventional valve-controlled hydraulic system was performed by Ivantysyn [32], which showed that the system has several avoidable power losses when analyzing the energy flow for a typical 30 s digging cycle, as shown in Figure 5.32. About 25% of the engine power is used to power auxiliary functions, with a large portion of this power used for cooling the fluid. About 20% of the power is lost in the pump, long hydraulic lines, and back-up valves that switch the flow in case of pump failure. More interestingly, over 30% of the energy is lost at the OC valves due to metering losses, which shows the enormous potential that a DC system could bring by eliminating these valve losses, and at the same time reducing the required cooling power since less heat is being generated through throttling.

Ivantysyn [32] has shown that introducing a novel open circuit DC system can achieve fuel savings between 28%–50%, while simultaneously preserving robustness

FIGURE 5.32 Overview of the energy usage of the conventional system.

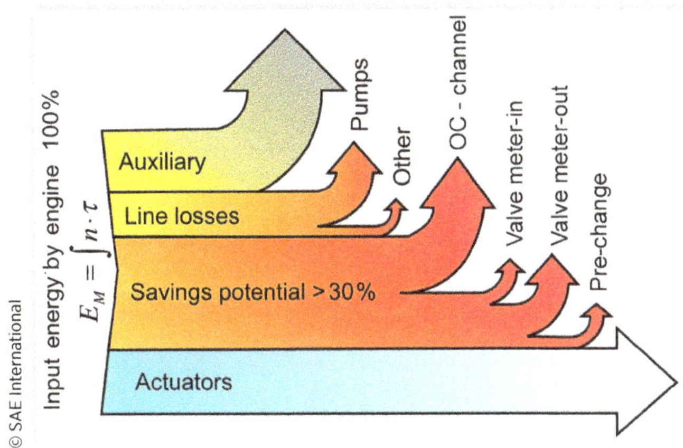

and cycle times. The proposed circuit, shown in Figure 5.33, consists of four 2-quadrant axial piston pumps (1), directional valves (2), float valves (3), flow sharing valves (4), and bypass valves (6). Unlike most DC solutions, this system does not require changes in the number and sizes of the pumps, since the baseline machine already uses flow sharing. An open circuit DC system is better suited for this application due to the required large volumetric flow rates, short lifting and lowering times, and the relative ease of pump flow sharing in open circuit designs. The peak flow rates reach 2,200 L/min (581 gpm) at lifting and over 4,000 L/min (1,057 gpm) at lowering, which can only be supported by float valves that short circuit the actuators displacement chambers to release the flow to tank. In contrast, a closed circuit system would require this flow to either be stored in massive accumulators, or large charge pumps would have to resupply this flow back to the pumps.

FIGURE 5.33 Alternative open circuit displacement controlled system.

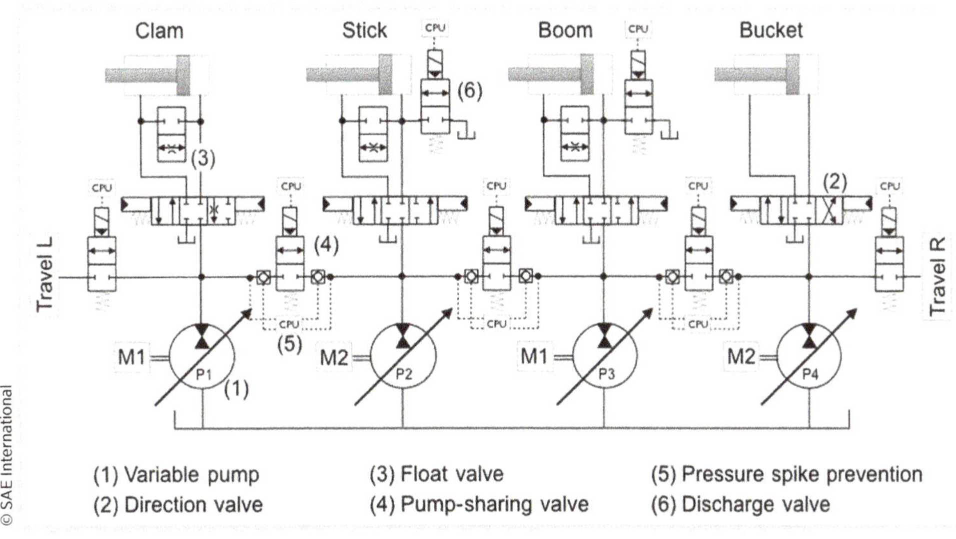

FIGURE 5.34 Example of pump sharing possibility and individual actuation.

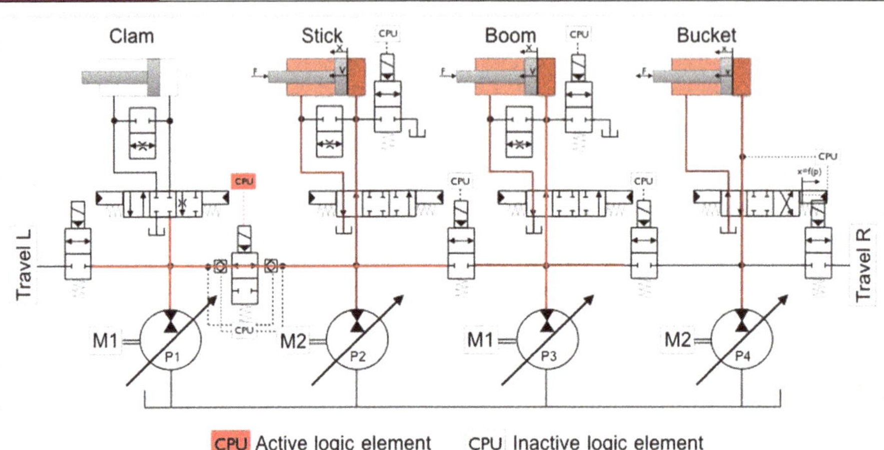

The open circuit DC system requires directional switching valves (2) since the pump can only operate in two quadrants. However, the switching valves do not perform throttling-control resulting in minor losses, as they only induce a directional change of the flow. Flow sharing switching valves (4) are used to connect one actuator to multiple pumps, thus enabling the same functionality of OC valves by distributing the flow as required without throttling, and providing a back-up function in case one pump fails (with simple additional hardware, not shown). To enable the smooth opening of the flow sharing valves without pressure spikes or drops, pilot-operated check valves are used with a hydraulic logic block (5). An example of the operation of these valves is seen in Figure 5.34, with the stick, boom, and bucket actuators being active. The stick receives flow from two pumps, as it would in the OC valve circuit during this operation.

During lowering, the proposed DC system can partially recover the potential energy stored in the arm. The float valves are engaged for the stick and the boom, however, they do not open to tank but rather to the pumps, which act as motors to recover energy. The bypass valves are only activated to achieve minimum lowering times when the operator commands full speed, and part of the flow is lost to the tank. However, if the operator does not need the full lowering speeds, it is possible to recover even more energy by guiding the entire backflow through the pumps. It was shown that the excavator does not require an energy storage medium, since all of the recovered power can be used to power the other simultaneous functions, such as the swing drive or the auxiliary drives.

In conclusion, it can be shown that DC technology can be used even in more challenging applications that require high flow rates and robust, well-established technology. New valve systems are far more complex and require new hardware while offering only a fraction of the energetic benefits. An open circuit DC system only uses existing hardware, with the only potentially needed change being a faster control valve for the pumps. In terms of durability, it is expected that this open circuit DC system would outlive its OC counterpart, since the pumps are not required to run at their corner power the entire time, and the hydraulic components are subjected to less heat due to lower overall losses.

FIGURE 5.35 Basic assembly of an electro-hydraulic compact drive [33].

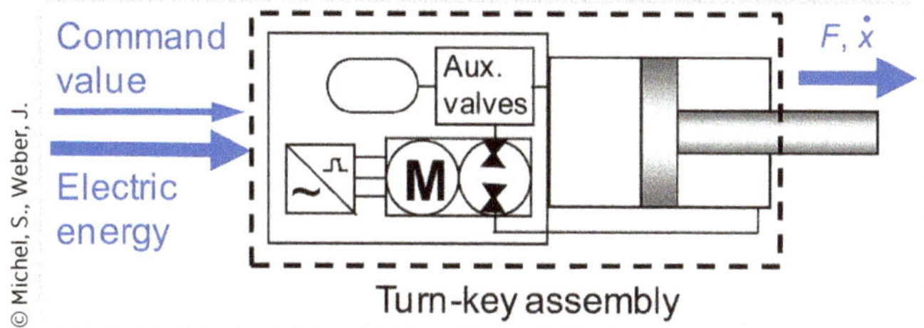

5.2.4 Electro-Hydraulic Actuation (EHA)

EHA is a well-known concept, which is well-established in aviation and stationary applications such as assembly lines. It combines an easy-to-control electrical drive with either a constant or variable hydraulic pump. The concept is very similar to the DC system shown above, with the main difference being that the shaft speed can be varied by an electrical motor quite easily. In aviation, EHA systems are used to control linear motion such as the flaps. These systems are usually sold as a compact unit with the linear actuator sharing the same housing as the electric motor and the pump, therefore they are also known as compact drives. An overview of the types of system architectures and the benefits that each inherits is given in Michel and Weber's "Electrohydraulic compact-drives for low power applications considering energy-efficiency and high inertial loads" [33].

There are multiple design possibilities for EHA drives. An overview is shown in Figure 5.36. Left in the figure are valveless options, while on the right side are options for a single hydrostatic pump. As can be seen, there are multiple options to compensate for the differential flow of the cylinder varying from the simple pump – accumulator circuits to very complex three pump drives. Which one should be used, depends on the requirements of the system.

While there is a trend that EHA may be applied to mobile machines, it would require electrification of the entire system. However, the additional electrical components such

FIGURE 5.36 Overview of design possibilities for EHA drives [33].

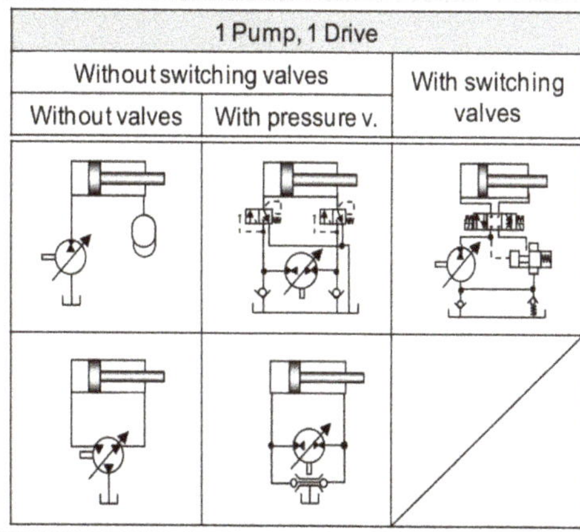

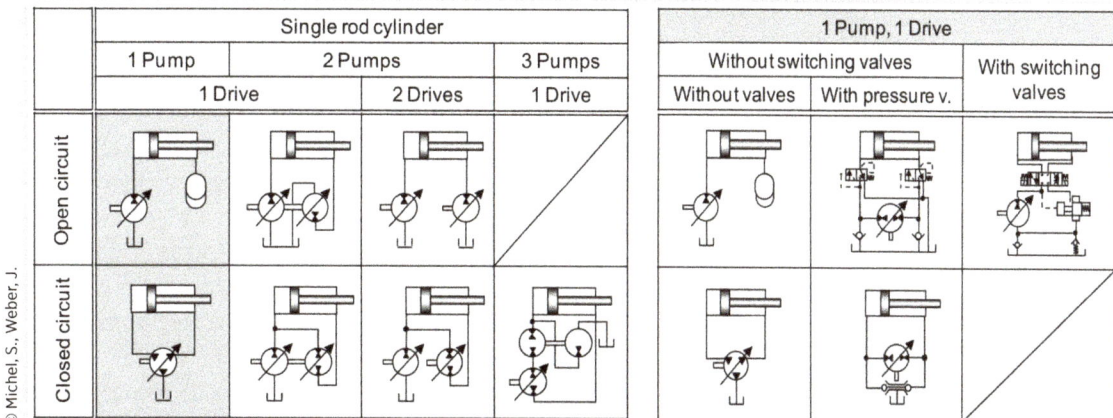

as generators, frequency controllers, and batteries can quickly outweigh the benefits that these systems might bring. Therefore, it needs to be decided on a case-by-case basis, if EHA is a better alternative to the DC systems described in the previous sections. Moreover, if the main source of power is electric and a combustion engine is not used, EHA can be a suitable and cost-effective solution that shares the benefits of the combustion engine driven DC systems.

5.2.5 Challenges for the Introduction of DC Actuation

In the previous sections, it was shown that DC technology offers a variety of advantages and potentials over valve-controlled systems. The efficiency of the system, when correctly designed, is unmatched by any other solution as the only losses that occur in DC systems are due to the pump, lines, and actuator. However, until this book was published, none of the large OEMs have introduced a machine that utilizes DC and valve-less actuation in the function drives. This chapter is meant to address the potential reasons and challenges why DC has not been utilized yet. Moreover, the authors of this chapter expect that it is just a matter of time until DC technology may disrupt the market and become the core of mobile machines in the future.

Typical arguments against the deployment of DC actuation include the increased cost and space of having more pumps, harder control effort, and slow dynamic response due to pumps being slower than valves. In the authors' opinions, these arguments are not sufficiently instantiated as will be detailed next. However, this should not mean that people have not mulled over these points before, thus the authors deem this section especially important for young engineers in the fluid power industry.

The main benefit of DC technology is derived from its throttle-less actuation. Without throttling valves, one flow source can only control one actuator. Therefore, it is true that DC technology will require multiple pumps for simultaneously acting actuators. Flow sharing between several pumps is one way to reduce the number of pumps and their required sizes, as discussed in Sections 5.2.2 and 5.3.1. Nevertheless, it is still necessary to have an individual flow source for each simultaneously working actuator. Since each pump only drives a specific actuator, it is possible to size the pump specifically for this task, and therefore the pump should be smaller than in the competing valve-controlled system. In a regular valve-controlled system, the main pump is either oversized, in case it has to be able to run all actuators at full speed simultaneously (which is in most applications, not the case), or undersized, meaning that the machine is only designed to simultaneously run, for example, two out of four actuators (which corresponds to a typical design of a machine). If more actuators are in use, then they will either move slower (pressure compensator after the control valve) or prioritized by pressure level (pressure compensator before the control valve). The designer of a DC system can face similar decisions. One can either choose the pump size to be able to run all actuators at full speed simultaneously – one pump per actuator, or undersize the pumps for flow sharing. Either way, the DC system should consist of multiple smaller pumps that roughly equal the total size of the large pump(s) of a valve-controlled system. For example, if a valve-controlled excavator has one 150 cm^3/rev pump used for the boom, bucket, and stick actuators, then the equivalent DC system would consist of three pumps that total 150 cm^3/rev (e.g., 3 × 50 cm^3/rev), which can use flow sharing when one actuator needs full flow. Smaller pumps are less costly to develop, have a higher life expectancy, and usually have high efficiency. If the system is designed from the ground up for DC actuation, then the higher efficiency can be used to reach higher peak power for overall better system efficiency. Today's approach entails retrofitting an existing machine to DC, which means that larger pumps and valves are removed and replaced with several smaller ones. Such an approach has been shown in several cases, for example, a mini excavator was retrofitted in Zimmerman and Ivantysynova [22] to fit three pumps and an accumulator in the same place where the previous larger pump and valve block was located. While it is true that three, typically oversized, pumps occupy more space than a single pump, the increase is not due to the pump's rotating kit but rather the associated parts such as the port plate, control valve, and other components. Three smaller rotating kits that match the size of a larger one can have the same weight and volume, which is explained by the scaling laws [34, 35]. Furthermore, pumps are not available in all sizes and forms, so they are usually picked to be oversized, as was done in Zimmerman and Ivantysynova [22] where a 75 cm^3/rev unit was used instead of a 65 cm^3/rev due to its unavailability on the market. For DC applications, it is possible to reduce space by designing tandem pumps that share the same housing and port plate. Advances in virtual prototyping and series production of pumps will lead to cheaper development costs, more pump manufacturers, and a larger variety of pump options. Other aspects like the potential for engine downsizing and smaller required coolers can further lead to a considerable reduction in the system size. Therefore, it can be concluded that for a properly designed DC system, it might not necessarily be true that it would require more space.

The cost of a large number of smaller pumps is economically debatable, since smaller pumps are cheaper to develop and test, last longer, and their cost can be reduced with larger production volumes and fewer overhead costs. Since the authors of this chapter are not experts in manufacturing and engineering economy, the arguments are focused on the physical aspects of cheaper development costs and increased longevity, rather than the mass production argument. Today's pumps are mostly being developed in a trial-and-error approach that involves significant test bench measurements while also using some simulations technologies such as computational fluid dynamics (CFD). Even though there are available tools to help with pump design, such as flow simulations of the ports, two-dimensional simulations of the valve dynamics, or finite element methods (FEM) simulations of the stationary parts, until recently there were not many tools to completely simulate the heart of the pump, the rotating parts. The rotating kit of an axial piston machine consists of the cylinder block, pistons, slippers, and the non-rotating parts include the valve plate and swashplate. Over the past decade, there have been tremendous advances [36, 37] in simulation tools that can predict the performance of the rotating kit, therefore, it can be said that pump development will be cheaper soon since less testing has to be performed. Nevertheless, each pump will need to be tested before going into production and the larger the pump series that is being developed, the more energy is required to power these pumps. Even if a pump is scaled up or down from an existing series, it still needs to be tested to pass qualification tests. Scaling laws do not accurately apply to large hydrostatic machines that exceed 200 cm^3/rev since they have larger masses and more importantly larger diameters, which impact the sliding velocities of the lubricating gaps and dictate their operation at lower turning speeds. Furthermore, temperature and pressure deformation effects have varying impacts on the rotating kit of larger machines. The safety factor in the wall thickness of the displacement chamber can be much larger in smaller pumps without adding much weight, which increases their lifetime; whereas fatigue failures of components, especially the brass parts, are more frequent in larger pumps. Other challenges are associated with larger pumps like their difficult assembly process due to overweight, and their increased dead volume due to hollow pistons. Therefore, it can be concluded that it is more difficult to develop a large hydrostatic machine than a smaller one, which is reflected in the low number of large units on the market, and their hefty prices. In summary, it can be argued that using multiple smaller pumps instead of one large one may be more economically feasible.

The argument that pumps are significantly slower than proportional valves, and are hence more difficult to control, is not valid. An axial piston pump, which is mostly used for DC systems, varies its displacement by changing the swashplate angle to regulate the stroke length of their pistons, as seen in Figure 5.37. The force required to change the swashplate angle is determined by the resulting moment around its swivel point. This moment is caused by the pressure difference in the displacement chamber between the first 90° of the high-pressure stroke and the second 90°. If the pressure distribution is equal, meaning that the valve plate is fully symmetric, then this moment is near zero. However, most pumps do not have a symmetric valve plate, so this moment can be in the range of a few to several thousand Nm and varies with the operating conditions. The only other additional force required to change the swashplate is its angular inertia. Even though some swash plates can be heavy, they are usually quite symmetric and have low angular inertia around their swivel point. To conclude, the forces required to change the pump displacement are highly dependent on its valve plate design, but can be as low as a few Newton for a well-designed pump.

The force to move the swashplate is exerted by a setting piston, which is powered by a pressure source, usually the control pressure line. The required setting forces

FIGURE 5.37 Axial piston pump at partial (left) and full displacement (right) using changing its swashplate angle.

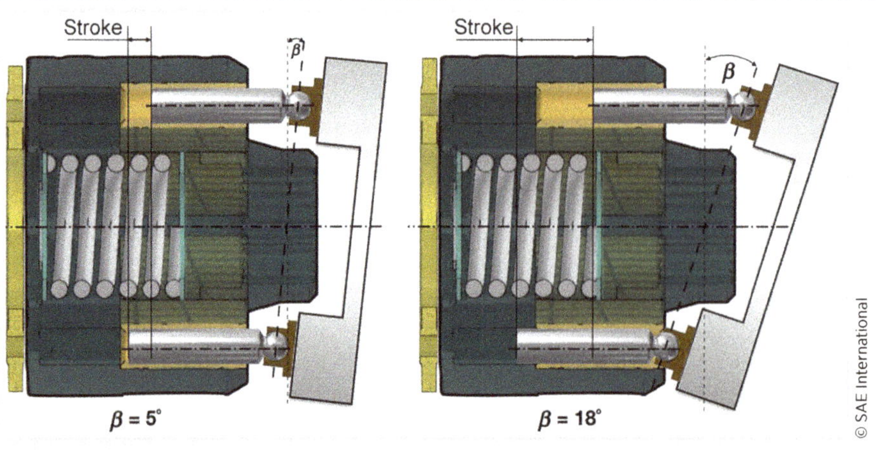

determine the size of the setting piston, which must be able to move the swashplate using the constant pressure in the control line in all operating conditions. Controlling the setting piston's motion is realized by a proportional control valve. The inertia of the setting piston and the moment of inertia of the swashplate are higher compared to the ones that act on the proportional valve. The forces and the flow to the adjustment system are the main limiters of the speed of change. Therefore, the argument that pumps have significantly slower dynamics than valves is incorrect; the pump's response can be as fast as its setting valve. A detailed publication about this topic, which shows the physically derived equations and proves the latter statement, is given in Grabbel and Ivantysynova [38]. In that work, it was also shown that a pump can be designed to require very little force to be adjusted, which means that the setting piston can be made very small to require very low volumetric flow. Lower required flow rates, in turn, reduce the required setting valve size, making it even faster while also increasing its efficiency.

Some of today's pumps are intentionally slowed down to make their systems easier to control by adding orifices in their setting line as shown in Figure 5.38. However, this does not mean that fast pumps are hard to control, it is rather the fact that hydraulic systems are complex, nonlinear, and can become unstable, if not controlled properly. By slowing down the pump, the entire system becomes less stiff and requires less effort to control. From a top-level, controlling the motion of DC actuated linear or rotary drives do not entail additional control effort nor does it necessitate the need for advanced control systems. Simple two degrees-of-freedom controllers, which combine feedforward and feedback control elements, have been properly designed and validated on multiple mobile machines such as excavators [22], wheel loaders [39], and skid steer loaders [46] to name a few. As with all technologies, more advanced control systems have been

FIGURE 5.38 Typical setting schematic of an axial piston pump.

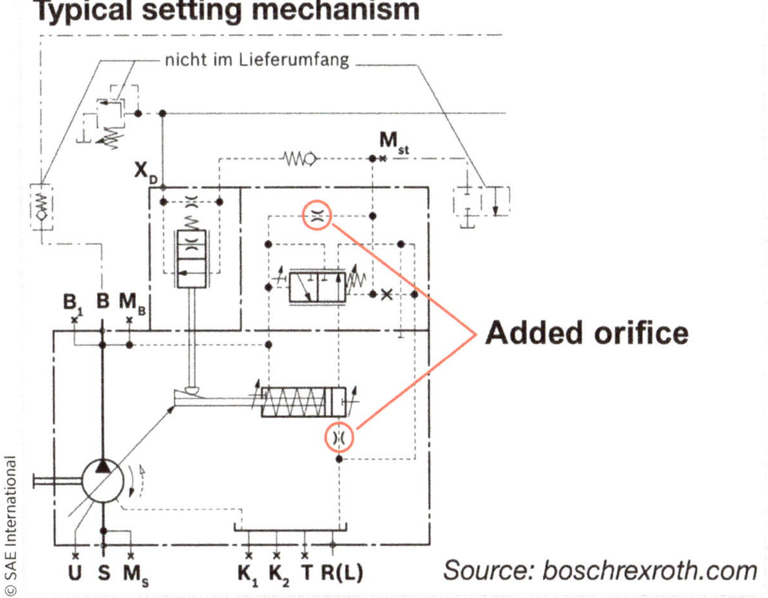

proposed for DC systems, but only for added value, targeting special performance indices, or for pure research and academic purposes in certain instances.

In conclusion, DC is a promising technology with many merits. It requires a redesign of most hydraulic machines to gain the full potential, however, the benefits can often be very high. Although most arguments show the great potential of DC adoption the industry does not implement the technology into their machines today. For a successful implementation of DC technology, OEMs need to redesign their hydraulic systems, pump manufacturers need to offer DC capable, and more importantly optimized, pumps and motors, and the next generation of fluid power engineers need to be able to understand how to properly design DC systems and how to control them.

5.3 Further Innovative DC Machine Concepts

This section shows the possibilities of combining DC systems and the technological benefits that they can offer relative to pump flow sharing, steer-by-wire, active vibration damping at the machine level, and active vibration damping at the pump level. Secondary control and hydraulic transformers are discussed at the end of this chapter.

5.3.1 Closed Circuit Flow Sharing Concept

The one-pump-per-actuator requirement in DC actuation presents a challenge for implementing the technology. A paradigm shift towards DC actuation requires compactness, cost-effectiveness, and practical drivability of a large number of hydraulic units using a single prime mover's driveshaft, on top of the promised overall system efficiency gains. How flow sharing can be implemented in an open circuit system was shown in Figure 5.26. To address the one-pump-per-actuator drawback in most DC systems, the concept of closed circuit flow-sharing was introduced in Zimmerman [21]. In this concept, a reduced number of hydraulic units is connected to a larger number of actuators through a distributing manifold populated with on/off valves. Figure 5.39 shows a simple hydraulic circuit of a displacement-controlled actuation system with flow sharing capability, where one pump is used to power two linear actuators.

The flow-sharing concept was first demonstrated on a compact excavator at the Maha Fluid Power Research Center at Purdue University in 2006. In this rudimentary setup, the operator selected, via a manual switch, the actuators to connect to the hydraulic units in two configurations: (1) a digging mode where the swing, boom, arm, and bucket were connected to a different hydraulic unit; and (2) a travel mode where the hydraulic units are connected to the tracks, offset, and blade. However, this design is limited in terms of actuator operability since only a small number of operations are possible. Moreover, the ability to use the tracks is restricted to operating in combination with the bucket, swing, blade, and offset. Hence, trench-digging or pipe-laying cycles were not possible to achieve. The challenge of automating pump switching was later tackled in Busquets's "Advanced Control Algorithms for Compact and Highly Efficient Displacement-Controlled Multi-Actuator and Hydraulic Hybrid Systems" [40],

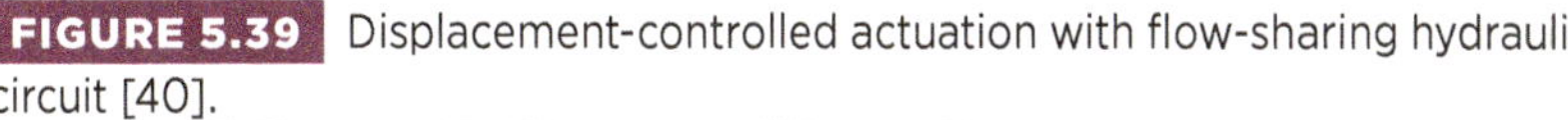

FIGURE 5.39 Displacement-controlled actuation with flow-sharing hydraulic circuit [40].

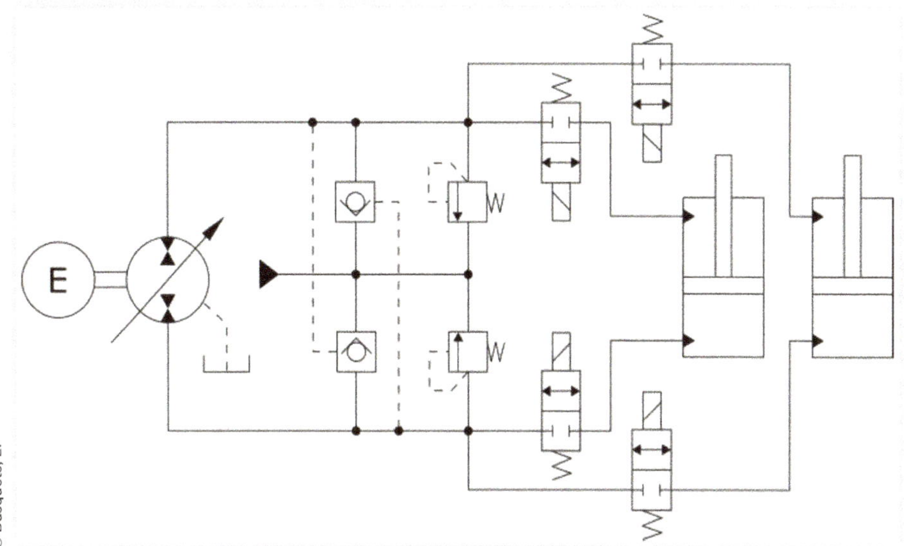

in addition to investigating multi-actuator machine design, supervisory control schemes, maximizing actuator availability, reduction in the number of hydraulic units, and compactness in the technology implementation. Another feature that was investigated is flow summing, which addresses the DC actuation limitation in actuator retraction rates, whereby combining flows from multiple hydraulic units allows certain functions (e.g., boom) to be actuated at a faster rate, thus meeting or exceeding the valve control (VC) design constraints, as previously mentioned in Section 5.2.3.

Figure 5.40 shows the hydraulic schematic of the displacement-controlled compact excavator prototype with a secondary-controlled hydraulic hybrid swing drive and flow-sharing concept, which shows the reduced number of pumps versus actuators. In this configuration, the boom can be provided with the flow from unit 2, unit 3, or unit 4 through switching valves combinations 5 and 8, 10 and 13, or 17 and 18, respectively. Due to structural limitations and the hydraulic units' sizes in this particular prototype, it is not recommended to provide flow from all three units simultaneously, but rather to actuate the boom with either one or a combination of two units.

5.3.2 DC Steer-By-Wire Technology

As shown in the previous section, it is possible to use a single hydrostatic unit for multiple actuators by simply adding on/off valves. This can lead to a holistic approach to perform every function on the mobile machine using DC-technology, without the need to add any throttling valves or to add a pump for each actuator. To demonstrate that even a critical safety function such as the steering of an on-road vehicle, such as a wheel loader, can be replaced with DC technology, a DC only steer-by-wire (SbW) system was proposed in Daher's "Novel Energy Efficient Electrohydraulic Steer-By-Wire Technology" [39]. On top of the aforementioned advantages, this can also offer a host of other advantages

FIGURE 5.40 Displacement-controlled excavator prototype with a secondary-controlled hydraulic hybrid swing drive and flow-sharing [40].

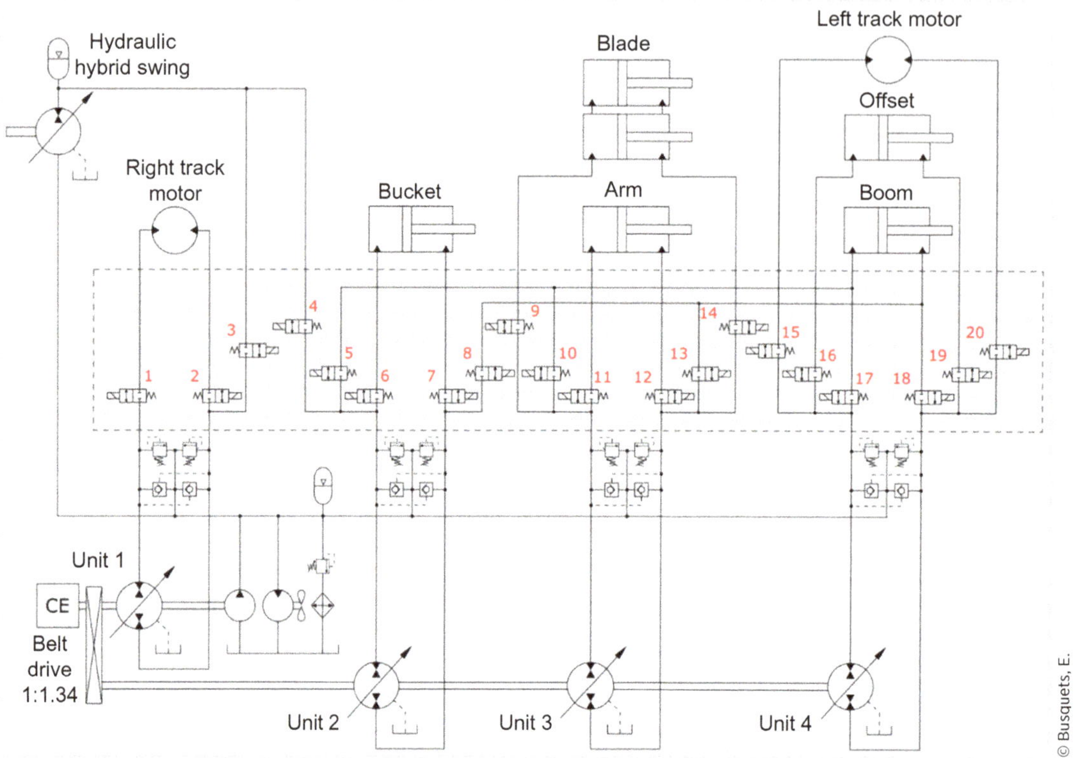

FIGURE 5.41 DC steer-by-wire system diagram [39].

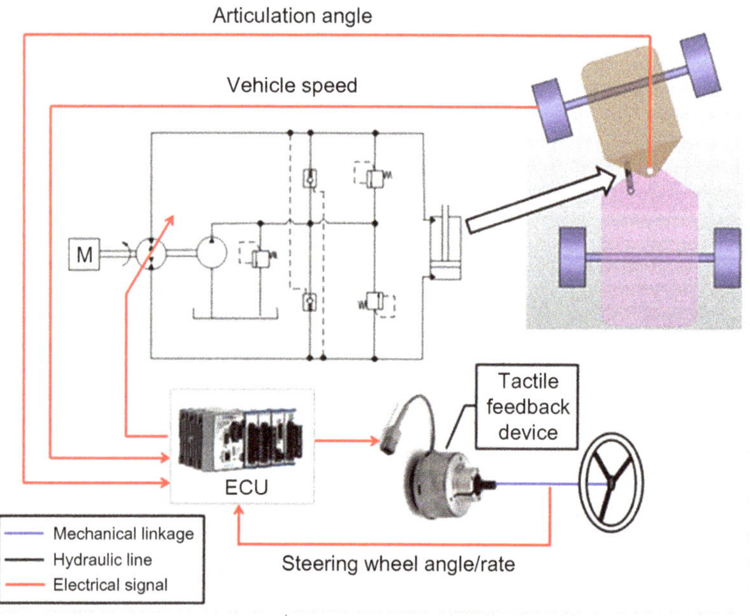

over state-of-the-art technologies that employ hydraulic control valves for motion control. The proposed DC SbW system is shown in Figure 5.41 and implemented on a compact wheel loader that featured hydrostatic steering as a baseline system. Under the same testing conditions, the DC steering system outperformed the baseline hydrostatic steering system in the areas of fuel consumption, machine productivity, and overall fuel efficiency. The new system also allows for adjustability in the steering ratio between the handwheel and the steering angle, as well as variable steering wheel feel relative to the level of torque feedback experienced by the operator. The adaptable modes result in improved operator comfort at low speeds and increased safety at high speeds. The by-wire system also opens the door for the implementation of active safety protocols that continuously monitor and correct, as deemed necessary, critical vehicle states. A yaw stability control system via active steering and a virtual yaw rate sensor was designed and validated on the prototype machine, resulting in enhanced safety and excellent correlation against a yaw rate sensor [41]. An adaptive control algorithm was designed and validated under various loads, demonstrating the machine's adaptability to varying operating and loading conditions [43]. Last but not least, the new DC steering system enabled remote and autonomous operation since it is a by-wire technology that requires no physical operator input, which is a desirable feature for all modern and future mobile machinery [44].

Analyzing the energy efficiency of the new DC SbW system versus a more conventional valve controlled counterpart was presented in Daher and Ivantysynova [45], which conveyed how effectively the two systems convert the chemical energy stored in the diesel fuel into useful mechanical energy. Experimental testing on the prototype machine showed that DC steering results in 14.5% fuel savings, 22.6% productivity gain, and a total of 43.5% fuel usage efficiency increase. Figure 5.42 shows an energy loss comparison between DC SbW and hydrostatic steering systems, with the pie charts being to scale, which explains the significant increase in fuel efficiency between the two technologies.

To add intelligence to the mobile machine, an adaptive controller in the form of an indirect self-tuning regulator (ISTR) was designed in Daher and Ivantysynova [43] to

FIGURE 5.42 Energy losses comparison between DC SbW and hydrostatic steering systems (pie charts are to scale) [45].

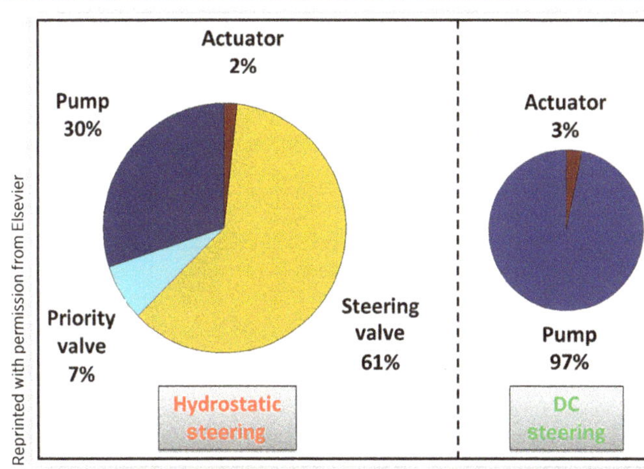

have the wheel loader cope with parametric uncertainties and uncertain nonlinearities associated with hydraulically actuated systems. The ISTR algorithm is selected given that the uncertain plant parameters are estimated in the process, which is a useful byproduct that gives insight into system properties that can be considered in future investigations related to machine's health monitoring and similar. Furthermore, a discrete adaptive control law with low computational expense was designed to suit the mobile machinery industry cost requirements. The designed adaptive scheme and estimation algorithm were validated in numerical simulations as well as experimentally on a prototype mobile machine, demonstrating the effectiveness of the proposed adaptive scheme in the face of uncertainties. The controller was able to adapt to varying load mass and inertia, which correlate to varying operating conditions that influence the system dynamics. Hence, besides offering improved fuel efficiency, the new steering technology also results in smarter machines.

Off-highway mobile machines have different security legislation regulations than their on-highway counterparts relative to safety, comfort, fuel economy, and automation. Over the past few decades, various active chassis safety control systems, architectures, and schemes have been researched and developed to improve the stability and handling of on-highway vehicles, including articulated vehicles such as tractor-trailer applications. In Daher and Ivantysynova [41] a yaw stability control system for articulated frame steering off-highway vehicles via novel steer-by-wire technology was investigated. A high-fidelity vehicle dynamics model was derived while keeping the yaw rate decoupled from the lateral acceleration, to separate the primary path-following task (driver) from the secondary disturbance-attenuation task (controller). The control algorithm was designed such that the two tasks do not hamper one another, and that the automatic controller is quickly activated for a short period to counteract instabilities, and then smoothly relinquishes control back to the human operator. Figure 5.43 shows the proposed yaw stability control system via DC SbW. Simulation and experimental testing results were obtained to validate the vehicle dynamics model, the control algorithm

FIGURE 5.43 Yaw stability control via DC SbW system [41].

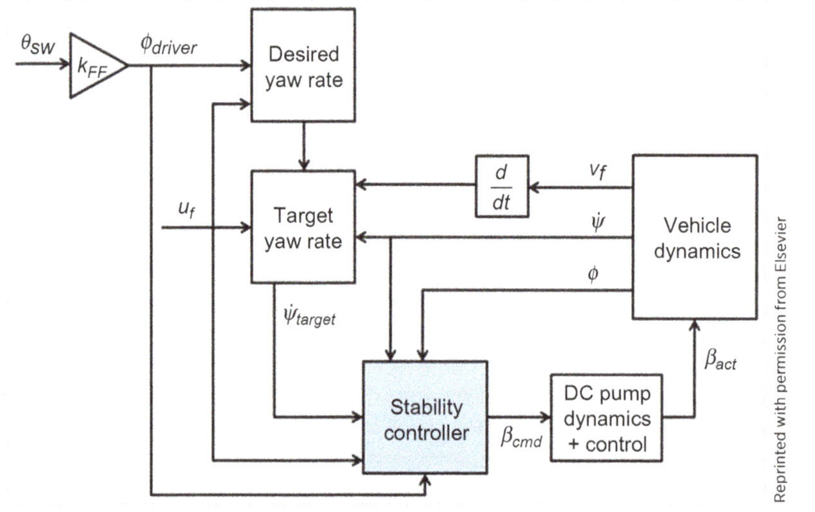

design, and the new system's efficacy in counteracting yaw instabilities on low-friction surfaces using standard vehicle dynamic maneuvers.

For added benefits, the notion of virtual sensing relative to estimating the vehicle's yaw rate by only measuring the articulation angle and vehicle speed was investigated in Daher and Ivantysynova [42]. Virtual sensing is a promising concept for yaw stability control and is an attractive option for vehicle manufactures as it reduces sensor cost, maintenance, and machine downtime. The designed yaw rate sensor was validated in simulation as well as on a prototype mobile machine by devising appropriate steering maneuvers. In summary, the SbW system based on DC actuation offers improved fuel efficiency, machine productivity and intelligence of mobile machinery.

5.3.3 Active Vibration Damping Using Swashplate Control

An additional innovative field in PCA is the capability to use the fast-moving swashplate to dampen either an actuator, or even to dampen machine vibrations of its own. Skid steer loaders are very stiff mobile machines that lack suspension systems, they are very compact and react harshly to external forces resulting in strong vibrations. Hence, active vibration damping systems can improve the operator's well-being during driving tasks. To demonstrate how dynamic and fast-reacting a pump can be, a DC powered skid steer loader was used as a demonstrator in Williamson [46]. This latter work differed from other prior work on active vibration damping in the sense that the entire machine's vibration was being damped instead of just one actuator, and without installing additional hardware such as counteracting actuators.

The hydraulic schematic of the DC skid steer loader is shown in Figure 5.44. It shows two adjustable pumps, one for each cylinder pair of the bucket and the boom. The remaining components consist of a charge pump, pressure relief valve, on-off load-holding valves, pilot-operated check valves, and accumulator for the differential cylinder volume compensation. The active vibration control was designed to compare passive damping with accumulators, damping only performed by changing the flow rate of the pump, and a combination of the two. Therefore, two accumulators were installed to compare active to passive vibration damping but were not utilized in the final solution.

FIGURE 5.44 Schematic of the DC skid steer system [46].

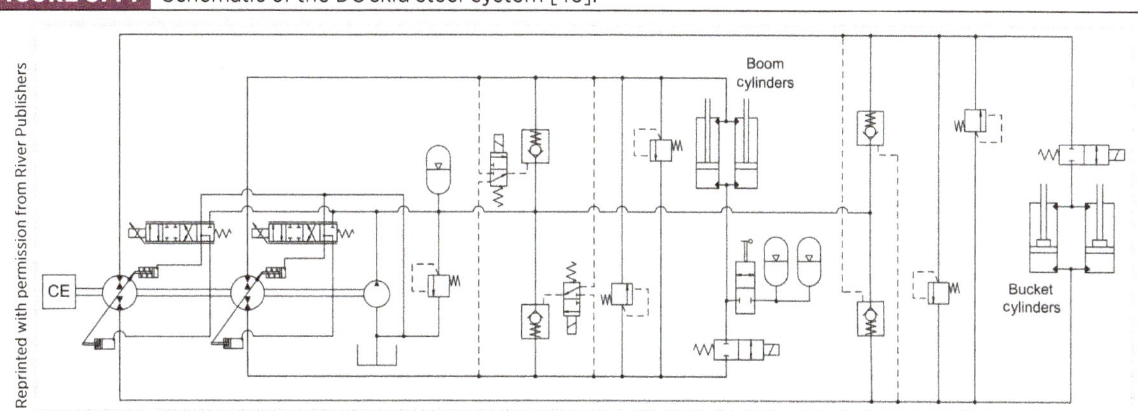

FIGURE 5.45 Control algorithm of the skid steer active vibration damping [46].

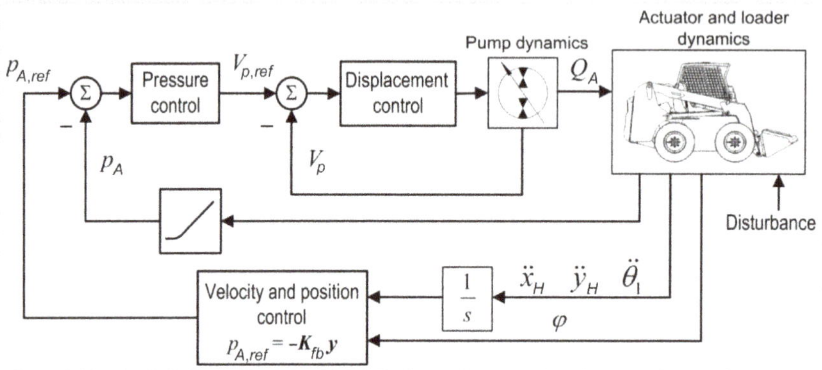

The control algorithm of the active vibration system is shown in Figure 5.45. The only required sensors are those to measure pressure, acceleration, and vehicle speed. On the DC skid-steer loader, the boom cylinders are connected to a 46 cm^3/rev axial piston pump, the swashplate is controlled with a high-speed proportional valve (40 L/min (11 gpm) at 75 bar (1,090 psi), 90 Hz bandwidth at ±10%), and an angular sensor measures the swashplate angle for position feedback. The pump closed-loop dynamics play a key part in the active vibration damping control system and are approximated by a rate-limited second-order transfer function, shown on the right of Figure 5.46. $V_{p,\, the\, ref}$ is the desired swash plate angle and V_p is the actual swash plate angle is normalized (dimensionless units). Parameter values are listed on the left side of the figure. The low-amplitude bandwidth of the pump is primarily determined by the dynamic characteristics of the control valve [38]. For larger displacements, the swashplate speed is limited by the flow rate through the control valve. Since the control pressure is rather low (20 bar [290 psi]), this flow saturation effect is significant.

Figure 5.47 shows the damped and undamped vertical acceleration of the skid steer measured at the driver's seat at 9 km/h (5.6 mph) while driving over an obstacle at 0.2 s. The un-dampened skid steer loader has its largest vertical acceleration well after it hit the bump and takes over 1 s to recover. While the damped skid steer experiences the same initial bump acceleration at 0.2 s, it can attenuate the vibrations and quickly dampen the shock in under one second, while reducing the vertical acceleration by almost 50%.

Further vehicle vibration tests were measured at low and high drive speeds (9 and 18 km/h [5.6 and 11.2 mph]) and low and high bucket masses (340 and 770 kg). The test course consisted of five 4 cm (1.6′) thick wooden boards evenly spaced on a paved concrete surface. The board's spacing was 6 m (6.6 yard) for low speed and 9 m (9.8 yard) for high speed. Four trials were repeated for each combination of speed, load, and control. Four control conditions were tested: (1) baseline condition with no extra damping, (2) passive damping with accumulators, (3) active damping by pressure regulation (virtual accumulator), and (4) active damping by multi-DOF velocity regulation (skyhook damper) with the control force proportional to vehicle velocity. Figure 5.48 compares the mean and total for each measured condition.

The table on the right side of Figure 5.48 lists the mean change in total vibration and *t*-test *P*-values compared to the baseline for each combination of load and

FIGURE 5.46 Pump dynamic parameters parameter and simplified model of pump and control valve dynamics with displacement feedback [46].

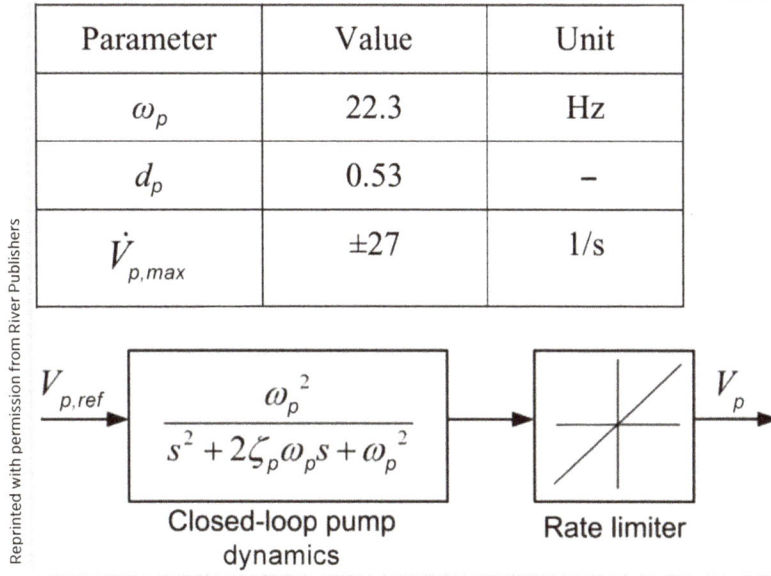

Parameter	Value	Unit
ω_p	22.3	Hz
d_p	0.53	–
$\dot{V}_{p,max}$	±27	1/s

$V_{p,ref}$ → $\dfrac{\omega_p^2}{s^2 + 2\zeta_p \omega_p s + \omega_p^2}$ → Rate limiter → V_p

Closed-loop pump dynamics

Rate limiter

FIGURE 5.47 Vertical acceleration measured at seat while driving over a single obstacle at 9 km/h with an empty bucket, with and without active "skyhook" damping [46].

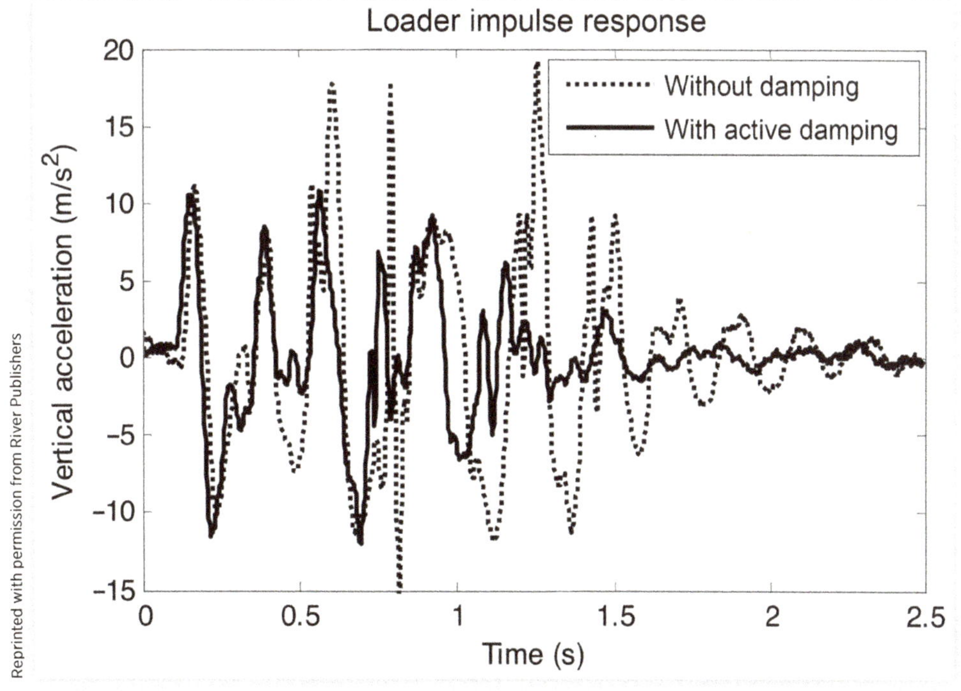

FIGURE 5.48 Vibration measurements summary (error bars denote 95% confidence intervals) and corresponding measurement data [46].

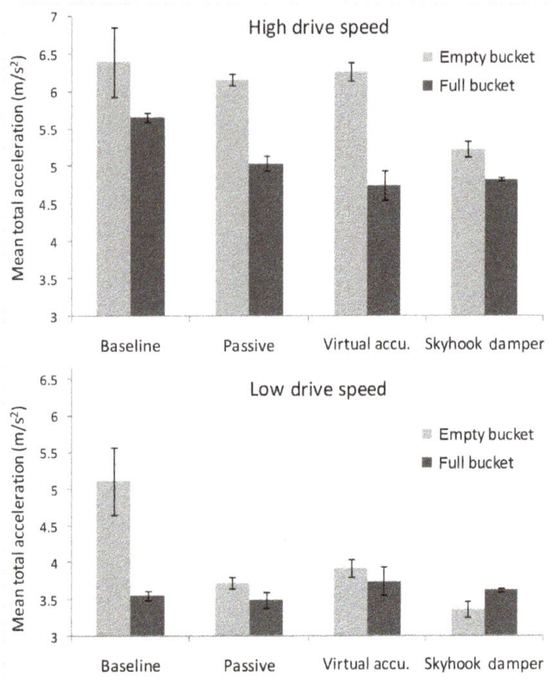

Speed/ load	Stats	Passive damping	Virtual accu.	Skyhook damper
Low low	Δ (%)	-27.1	-23.3	-34.0
	p	0.0002	0.0001	0.0000
Low high	Δ (%)	-1.6	5.6	2.2
	p	0.1794	0.0232	0.0095
High low	Δ (%)	-3.8	-2.0	-18.2
	p	0.0628	0.3547	0.0003
High high	Δ (%)	-11.0	-16.3	-14.7
	p	0.0036	0.0007	0.0010

vehicle speed. Multi-DOF velocity feedback provided the best performance overall, reducing total acceleration by up to 34%, and significantly improved ride comfort and outperformed the commercial passive accumulators. The shown example shows the ability of DC systems to accomplish active vibration damping without the need to install additional hardware. A control law that uses a multi-DOF version of the well-known "skyhook damper" principle was the most successful in damping the vibration.

5.3.4 Pump Vibration Damping Using Swashplate Control

Another innovative active vibration control concept is to dampen the pump's vibrations using the existing pump adjustment system to reduce vibrations and possibly noise emission. Such a system was investigated and experimentally validated in Kim and Ivantysynova's "Active Vibration Control of Swashplate-Type Axial Piston Machines with Two-Weight Notch Least Mean Square/Filtered-x Least Mean Square (LMS/FxLMS) Filters" [47]. A 75 cm^3/rev swashplate type axial piston pump was modified to implement the high bandwidth pump control system, which aims at canceling the swashplate vibrations. Vibration measurements using a tri-axial swashplate acceleration sensor were conducted to show the effectiveness of the proposed swash plate active vibration control system and algorithms. An illustration of the test set up is shown in Figure 5.49, here the swashplate (1) is connected to a lever arm (7), which, in turn, is controlled by a double-acting cylinder (6). The position of the cylinder is measured by a sensor (2).

This signal is fed back to the controller (3), which compares the desired swashplate angel with the actual one. The controller moves the setting valve (4), which sets the pressure in the setting chamber (5). The above setup is the same as in every variable swashplate type machine. Besides, the experimental setup also included direct swashplate acceleration measurement, case acceleration measurement, simultaneous multi-position microphone measurement in a semi-anechoic chamber, a high-speed direct drive servo valve, an electronic swash plate angle sensor, and a swashplate acceleration sensor. High-speed real-time controllers were proposed and realized using National Instrument LabVIEW Field Programmable Gate Array (FPGA).

FIGURE 5.49 Schematic of swashplate active vibration controller for a high bandwidth pump control system [47].

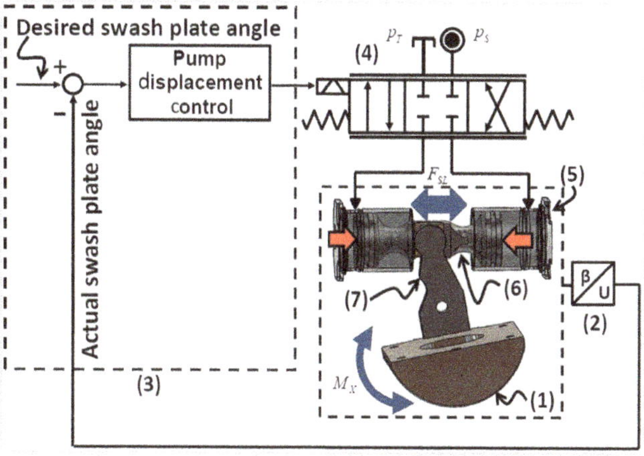

Two vibration controllers were developed for the swashplate active vibration control. One is the multi-frequency two-weight notch least means square (MTNLMS) filters with delay unit compensation, and another is the multi-frequency two-weight notch filtered-X least mean squares (MTNFxLMS) filter with offline modeling algorithm, shown in Figure 5.50, which has an estimated plant model instead of the delay unit compensation. The swashplate active vibration control procedure can be explained as follows. While the displacement controller maintains the desired swashplate angle utilizing an electronic swashplate angle sensor feedback, pump rotational speed is measured and converted to harmonic frequencies of the swashplate vibration, which is used to generate cosine and sinusoidal reference signals. The reference signals proceed to the LMS algorithm through delay units or estimated models to compensate for the pump control system dynamics and sensor delays. The LMS algorithm updates adaptive filter weights to minimize the swashplate acceleration error signal.

Multiple test results confirmed the simultaneous sound pressure level reduction with small swashplate acceleration reductions. However, the simultaneous sound pressure level reduction did not appear at all tested operating conditions. The measurement data classification demonstrated that the increased, decreased, or mixed swashplate accelerations and case accelerations could induce simultaneous sound pressure level reductions at all three microphones. At the same time, decreased swashplate accelerations and decreased case accelerations also produced the simultaneous sound pressure level increases. The experimental testing results revealed the complex relationship among swashplate accelerations, case accelerations, and pump noise generation.

Figure 5.51 shows a sample of the obtained experimental test results at 1,000 rpm rotational speed, 140 bar (2,031 psi) delivery pressure, 14.8° swashplate angle, and 25 bar (363 psi) inlet pressure setting. Only the second harmonic swashplate acceleration was reduced using the swashplate active vibration control since the second harmonic frequency is the most dominant acceleration component as shown in FFT plots. Accordingly, the second harmonic swashplate accelerations showed a 65.5% reduction. This example shows that the dynamics of the swashplate control are fast enough to even dampen out the internal pressure-dependent vibrations.

FIGURE 5.50 Multi-frequency two-weight notch filtered-X LMS filter with offline modeling for swashplate active vibration control system [47].

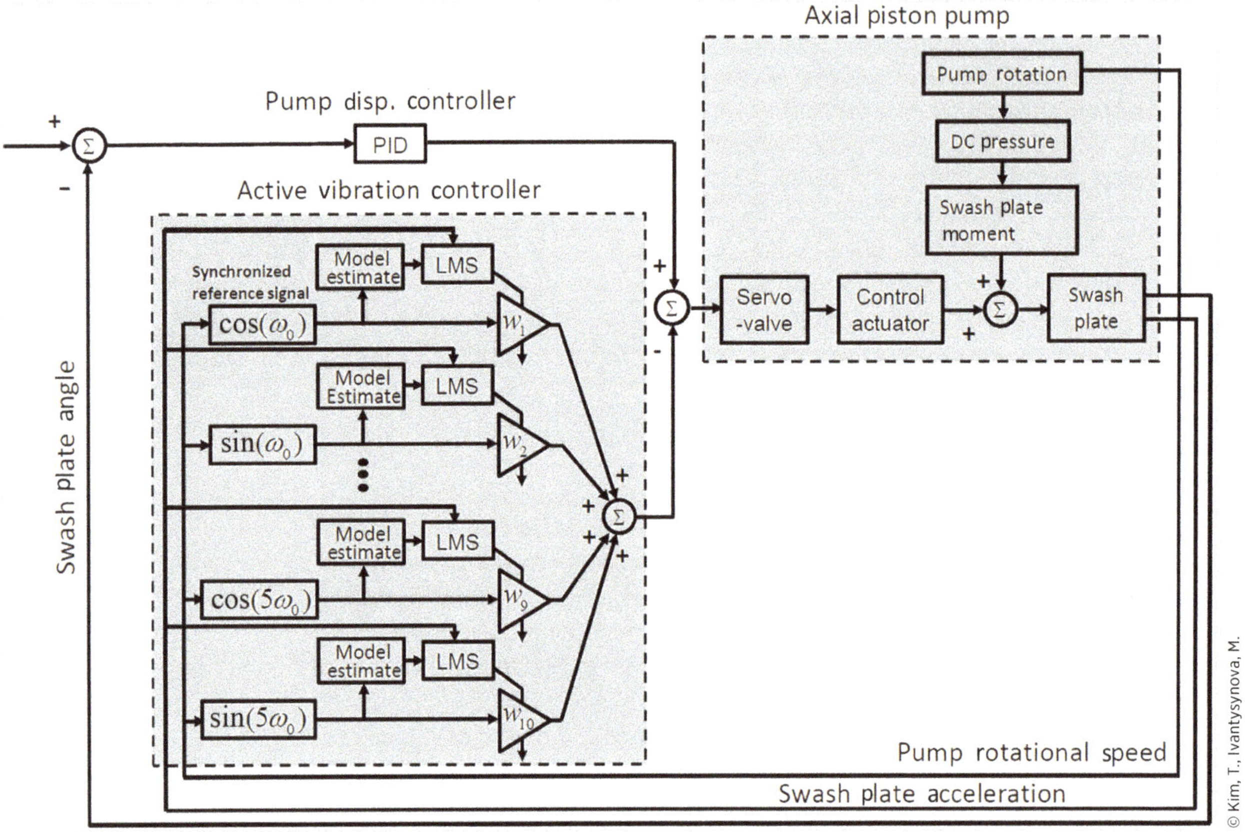

Axial piston pump
Pump rotation
DC pressure
Swash plate moment
Servo -valve
Control actuator
Swash plate
Pump disp. controller
PID
Active vibration controller
Synchronized reference signal
Model estimate
LMS
$\cos(\omega_0)$
$\sin(\omega_0)$
$\cos(5\omega_0)$
$\sin(5\omega_0)$
w_1
w_2
w_9
w_{10}
Swash plate angle
Pump rotational speed
Swash plate acceleration

FIGURE 5.51 Swashplate active vibration control experimental test results using a TNFxLMS filter with offline modeling (speed: 1000 rpm, delivery pressure: 140 bar) [47].

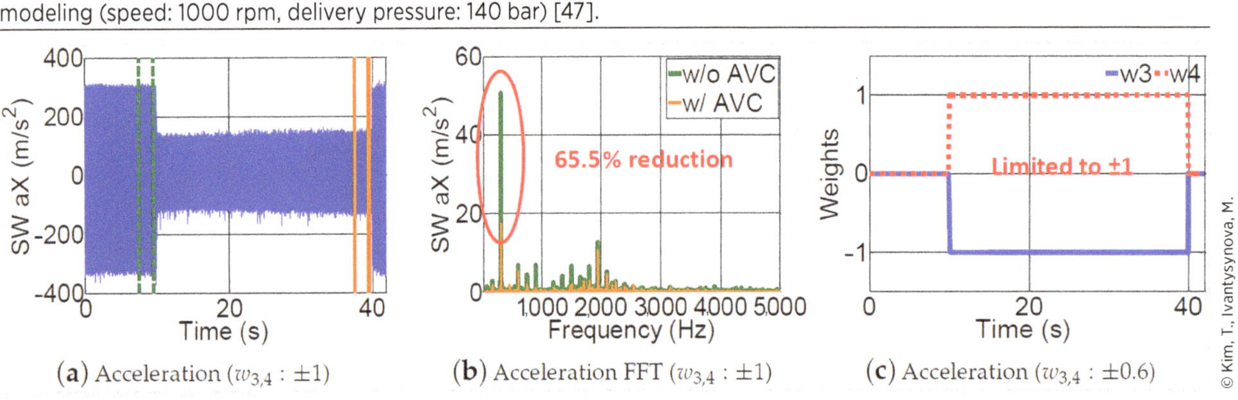

(a) Acceleration ($w_{3,4}$: ±1) (b) Acceleration FFT ($w_{3,4}$: ±1) (c) Acceleration ($w_{3,4}$: ±0.6)

In conclusion, the two examples in Sections 5.3.3 and 5.3.4 show that the swashplate can be utilized to not only dampen the vibrations of an entire machine by quickly adjusting the pump's flow rate but to also dampen the pump's very own inner kinematics. Both pumps in these examples had fast-acting valves installed on their adjustment cylinder, further showing that servo-pumps are only limited by their valve dynamics and not by their inertia.

5.3.5 Secondary Control and Hydraulic Transformers

In a closed circuit, like a PCA of Section 5.2, the actuators are flow controlled: the flow of the pump is guided to the actuator, a hydraulic motor, or a hydraulic cylinder. The pump flow, therefore, generates a rotational speed at the motor or velocity at the cylinder. Besides, the load on the systems generates the system pressure. Therefore, the speed of such a system is independent of the load and just dependent on the delivered flow.

A secondary controlled system is comparable to an electric grid system with high and low voltage lines. The systems consist of a high and low-pressure line, presenting a common pressure rail (CPR). The hydraulic pump acts as a pressure source, see Section 3.4, feeds the high-pressure line and creates, therefore, a more or less constant pressure. Typically, at least a hydraulic accumulator can be found in the high-pressure line to smooth pressure peaks and to store recuperated energy. A German patent first proposed the secondary regulation principle [48]. Examples of secondary controlled systems can be found today in airplanes and large-scale industrial applications.

For the secondary control, the actuator itself is adjustable in displacement and motion control is realized by adjusting the motor's displacement and not that of the pump [49]. A secondary controlled single actuator example is shown in **Figure 5.52**, which is from the view of the used components quite comparable to the DC system shown in Figure 5.18. Different actuators can be fed in parallel, or even series, without the need to add another pump.

The control of a secondary controlled system is completely different from a DC one. According to the basics of hydraulics, see Section 3.2, the constant pressure difference between the high and low-pressure lines generates a torque in dependence of the displacement of the hydraulic motor, therefore the actuator is torque driven. The torque will then generate the angular acceleration of the actuator. To sum up, in a flow-controlled system the speed of the actuator is controlled whereas in secondary control the acceleration of the actuator is controlled.

FIGURE 5.52 Secondary controlled motor.

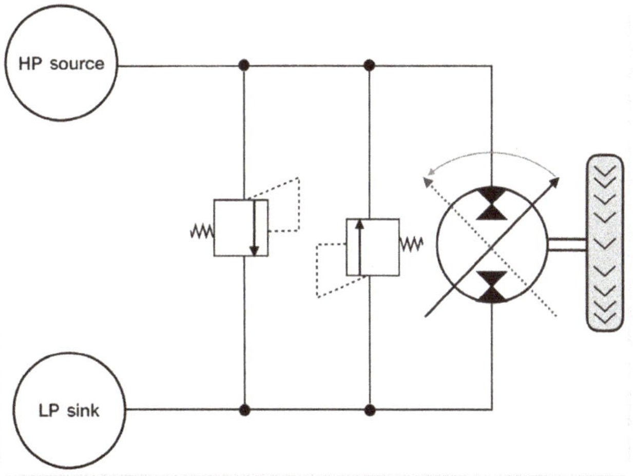

Linear drives are typically not suited for secondary control since it is quite difficult to change the area of the cylinders. However, there have been developments in that field using multiple chamber cylinders [51, 52, 53, 54, 55, 56]. This component is not in its conceptual stage anymore, as it is already implemented in a variety of machines. However, secondary control for linear drives requires a hydraulic transformer to adapt the pressure to the load.

The idea of a hydraulic transformer was presented decades ago [57, 58, 59]. During that time, energy costs became more important and therefore new efficient solutions had to be researched. The basic idea of a hydraulic transformer is to shift the input pressure and flow (port A) into the required output pressure and flow (port B) without generating losses, e.g., when throttling the pressure, see Equation (5.2) with a neglect of the losses.

$$p_A \cdot Q_A + p_B \cdot Q_B + p_T \cdot Q_T = 0 \tag{5.2}$$

The transformation can be done using a hydraulic motor and pump on the same shaft, with at least one of the units being adjustable in displacement. This element acts as a compact hydrostatic transmission that can convert one pressure and flow level into another.

In principle, there could be three different configurations of transformers:

1. A variable motor driving a constant pump,
2. A constant hydrostatic motor driving a variable pump, and lastly
3. Both units are variable giving the most flexibility.

In all cases, the motor is connected to the constant pressure system and the pump drives the actuator. Kordak [60] and Shih [61] presented the first configuration, where the required power is taken from a constant pressure system via a variable displacement motor. The motor is mechanically coupled with a fixed displacement pump and can provide the cylinder with the needed flow rate and pressure, see Figure 5.53. Shih [61] introduced the second configuration, where a constant displacement motor is mechanically connected to a variable displacement pump according to the mentioned configuration 2.

FIGURE 5.53 Principle of a hydraulic transformer to drive a cylinder on a constant pressure system [60, 61].

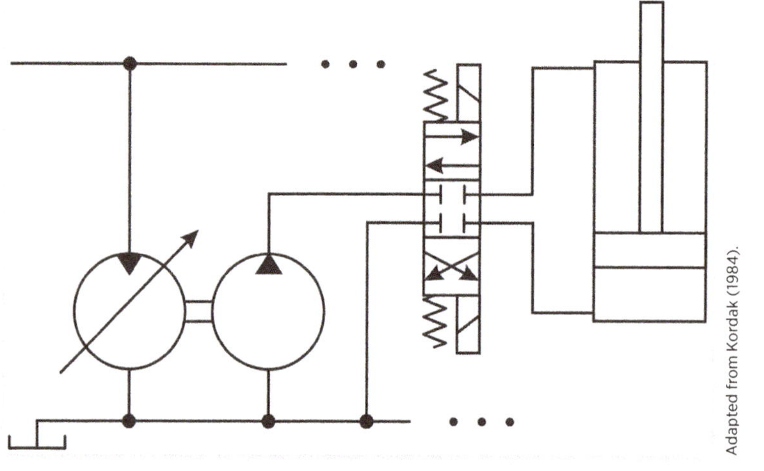

Adapted from Kordak (1984).

However, using a hydraulic transformer in a mobile machine, the efficiency of the transmission and the installation space has to be considered. In 1997, a compact three-port plate design was presented in Achten and Vael [62], containing both variable units in the space of one unit. The output pressure is set by rotating the valve plate of the unit. The idea is to move small parts and therefore get a high dynamic response of the system. The principle of rotating the valve plate is investigated in Achten and Fu [63]. Due to the compressibility of the oil, the pressure change in the valve land areas between the kidney pairs of the valve plate is dependent on the control angle of the valve. This pressure change has to match with the pressure difference between the ports of the hydraulic transformer to avoid pressure peaks and cavitation.

Shen, Karimi, and Zhao [64] give an overview of the hydraulic transformers using a three-port valve plate with a special view on the design of hydraulic transformers. Three different types according to their structure are defined there. The first one uses a rotating port plate, which can be found in Achten, Fu and Vael [62]. The second type rotates the swashplate and inherently inverts the adjustment principle. The mathematical basics and measured results of a prototype can be found in Jing's "Research on the Pressure Ratio Characteristics of a Swashplate-Rotating Hydraulic Transformer" [65]. The last type has as well a rotating port plate as a rotating swashplate.

Achten presented in "Towards Maximum Flexibility in Working Machinery, IHT Control in a Mecalac Excavator" [66] a so-called Floating Cup transformer. Nearly, 12 pistons were arranged on each side of the central rotor, see the design principle in **Figure 5.54**. The principle uses the rotor fixed rigidly on to the axle and the pistons fixed rigidly in the rotor. Unlike in slipper type and bent axis machines, the pistons in the floating cup design are press-fitted in the rotor and they do not have any degrees of freedom to move or rotate. Therefore, there is no wear and there are no friction losses between the pistons and the rotor.

The mentioned Floating Cup principle was developed and tested in an excavator [15, 66]. The system is designed with a high-pressure and a low-pressure line. The power for the actuators is taken from a high-pressure line via a hydraulic transformer. With the system, energy recovery is also possible.

The basic equations for a hydraulic transformer using a three-port valve plate can be found in Liu's "Electro-Hydraulic Servo Plate-Inclined Plunger Hydraulic Transformer" [67]. From the geometric volume of the three displacements of a hydraulic transformer, the average torque at each port can be calculated. The following transformer ratio calculation shows a highly nonlinear function. Typical pressure ratios $\lambda = \frac{p_A}{p_B}$ vary between 0 for 0° rotations angle and 3 for 100° rotating angle. A ration angle of 120° will cause theoretically an infinite pressure ratio and therefore the rotation angle is limited in real applications [67].

The speed of the transformer can be calculated by Newton's law with the three mentioned torques, the friction of the transformer, and its acceleration [67]:

$$J_{HT} \cdot \frac{d^2\varphi}{dt^2} = T_A + T_B + T_T + T_{friction} \tag{5.3}$$

FIGURE 5.54 Hydraulic transformer with variable valve plate, floating cups, and example circuit schematic [68].

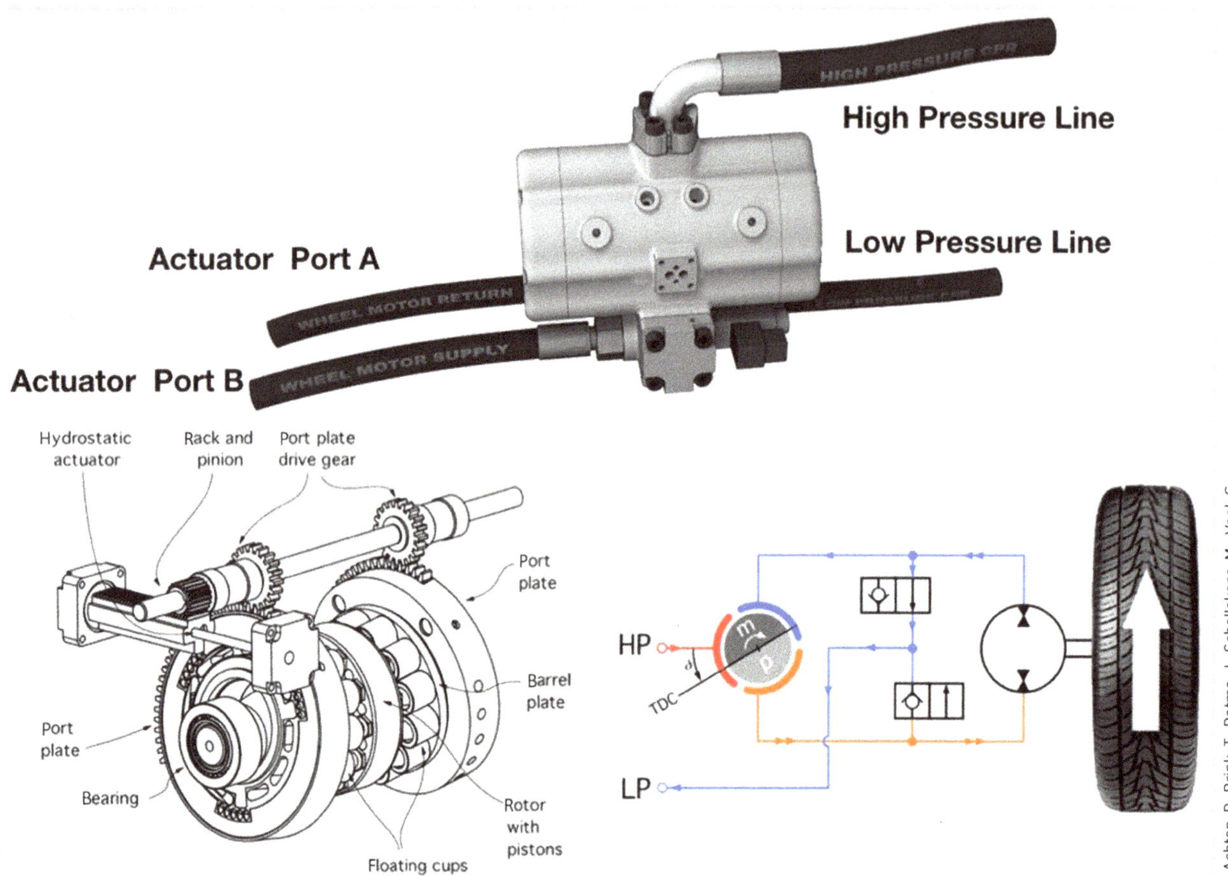

A calculation of two displacements of a three-port valve plate transformer is also possible [65]. With these displacements, a comparison and a link to the system with two separate units, as shown in Figure 5.53, are possible.

The equations for the displacements of the hydraulic transformer and the pressure ratio show that the control of the transformer is highly nonlinear, see Figure 6 in Liu [67]. A change of the angle of the port plate will change the torque in the cylinders and therefore change the acceleration and with this the speed of the transformer.

When controlling a cylinder with a hydraulic transformer, the angle of the port plate has to be changed. The speed of the transformer will result in dependence on the desired cylinder speed [64].

Due to its high nonlinearity, the control of the hydraulic transformer is a challenging task. In Liu [67], a hybrid control is discussed: for large deviations, fuzzy control is used and for small deviations a PID control. The transformer has a response time between 0.2 s for 20 bar (290 psi) load pressure and 1.2 s for 80 bar (1,160 psi). Control based on a pi-sigma fuzzy neural network (PSFNN) is suggested in Shen and Wang [69]. The control contains sum and multiplication operations in the neural network to better deal with the nonlinearities and the system modeling uncertainties. A stable and fast response is shown there.

The efficiency and the noise level of the transformer are dependent on the transformer speed and its construction. Especially at low speed, the efficiency decreases dramatically, and the speed fluctuation caused by relatively high friction forces becomes important. The Floating Cup principle overcomes this issue having more pistons, for example 24 as shown in Figure 5.54. Another possibility to additionally set the transformer speed is to swivel the swashplate on an integrated hydraulic transformer with a three-port plate as shown in "Comparative Analysis of Component Design Problems Integrated Hydraulic Transformers" [64]. Measurements are not presented there.

Tests with a hydraulic transformer on an application under real conditions are presented in different references. A control strategy, combining the switching present control angle and a fuzzy PID-controller is presented in Chao's "Control Analysis of the Hydraulic Luffing System Controlled by the New Hydraulic Transformer for Rescue Vehicles" [70] for a crane of a rescue vehicle. The lifting and descending of the crane-arm could follow the requirements and an energy-saving rate of 82% is reported.

Other applications in research today, that utilize hydraulic transformers for the crane of a forwarder, are proposed by Geiger and Zu Hohenlohe [71, 72]. In this application, energy can be directly exchanged between the inner and outer boom cylinders, which has a reduction potential of 40% of the energy by a horizontal movement of the crane.

Because of today's relatively low energy costs and due to the use of LS systems, hydraulic transformers have not been introduced into the market in a wide field of applications. Other reasons are the combined efficiency of two hydrostatic units and the need for low speed at the units when the cylinder speed is low. Lastly, the control strategy is a challenging task because of the high nonlinearity of the transformers.

5.4 Further Innovative Systems

5.4.1 Digital Hydraulics

The expression "Digital Hydraulics" or "Digital Fluid Power" is not fully defined today but it is oriented on digital electric systems. It describes, in general, a hydraulic system with several discrete states. In Linjama [51], a definition for digital hydraulic and pneumatic systems is given as follows:

> *Digital Fluid Power means hydraulic and pneumatic systems having discrete-valued component(s) actively controlling system output [51].*

Two main approaches to implement digital fluid power systems are parallel connection and switching technologies. Information about switching technologies can be found, i.e., by Scheidl [76]. Taking a valve as an example into account a parallel-connected digital hydraulic valve consists of several 2/2-switching valves, see **Figures 5.55** and **5.56**.

FIGURE 5.55 Design of a digital hydraulic valve [51].

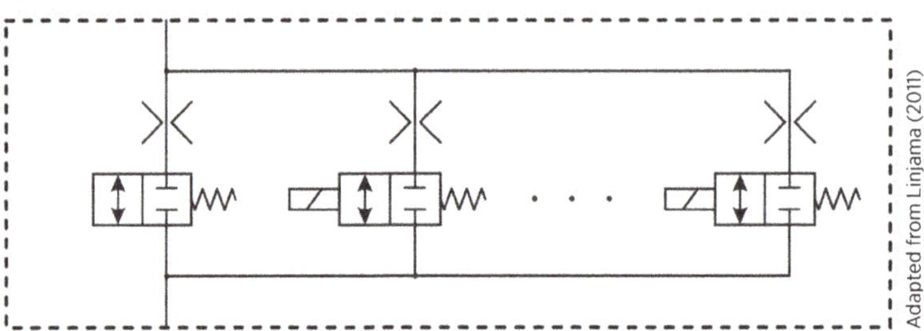

An advantageous design is to use binary coding for the ratios of the valves, like 1:2:4:8. Several N connected parallel valves will lead to 2^N opening combinations. Linjama [51] shows that 5 valves will generate 32 combinations, which will lead close to a step-less valve characteristic.

The idea of having parallel connections and using switching technologies is not new. Linjama names several examples of the past where these characteristics can be found [51]. Famous examples are an antilock braking system and an electronic fuel injection which were introduced in the 1970s and widespread in the 1980s. But taking digital fluid power systems into account, the research activities started around 2000 [51]. The newest research results are presented at conferences, like the *Workshop on Digital Fluid Power* [73].

Furthermore, digital hydraulic pumps are developed. An advantageous pump consists of several pistons and has several feeding systems, Figures 5.57 and 5.58. As discussed earlier in the chapter about hybrids a regeneration or recovery of energy is an advantageous feature in hydraulic systems for mobile working machines. A digital hydraulic pump, therefore, is typically designed as a pump-motor combination. Figure 5.57 shows an example of such a digital pump-motor combination. The different pistons can feed the high-pressure (HP) and low-pressure lines (LP) as well as the ports A and B. The LP can be connected to the tank and the HP port to a hydraulic accumulator. Energy from an actuator can be regenerated directly or stored in the accumulator. The ports A and B are connected each to a hydraulic linear actuator.

FIGURE 5.56 Suggested symbol for a digital hydraulic valve [51].

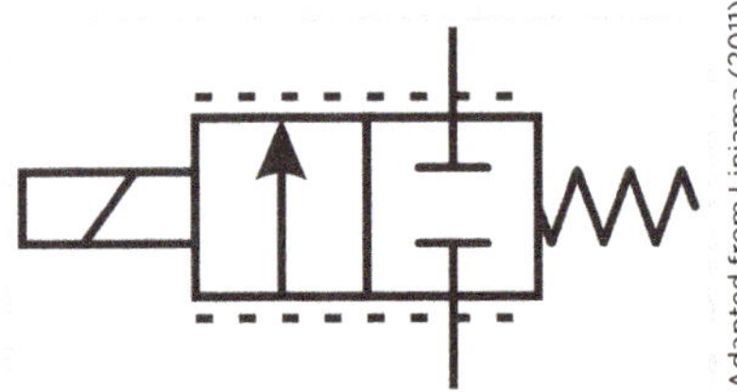

The pump-motor system shown in Figure 5.57 can also be used as a digital motor. Therefore, the system has just an A and LP line but several moving pistons. Other possibilities, like several motors with different displacement, are shown in Linjama [51].

A realized example of a digital hydraulic motor can be found in Artemis: "Technology - High Controllable, and Extremely Efficient" [74]. Every piston is controlled individually by a check valve and a solenoid valve. The switching time is specified as 30 ms, which makes a fast response possible. Compared to Figure 5.57 the pistons of the motor are the same, the inlet valve is the solenoid one and the outlet one is the check valve. During idling the solenoid valve stays open. During the pumping phase, it

FIGURE 5.57 Piston type pump-motor system (left); a hydraulic symbol of such a system (right) [75].

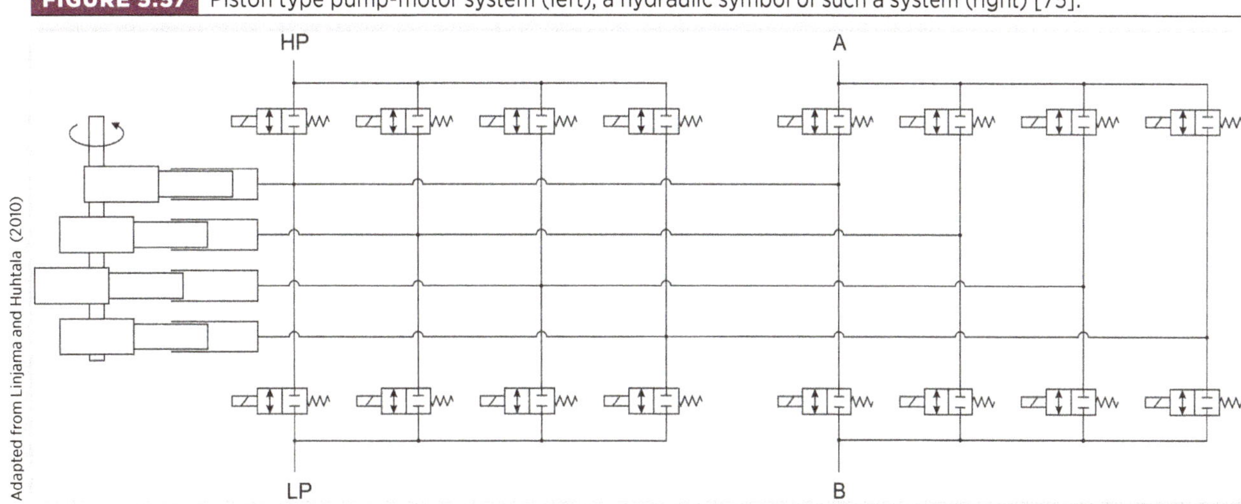

FIGURE 5.58 Hydraulic symbol of a digital hydraulic pump [75].

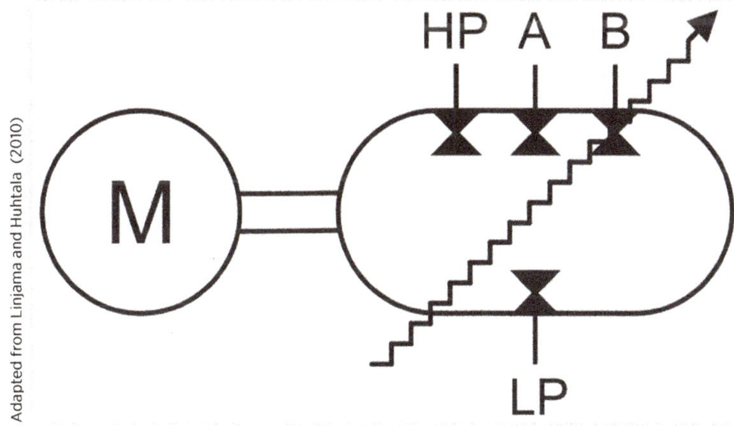

is shortly energized and therefore closed. An animation in Artemis [74] shows the working principle. The working principle allows an adaption to the required flow. The efficiency of the motor is given between 83% at 10% displacement and 94% at 100% displacement compared to another 96 cm^3 unit. Applications can be found today in off-road machines, busses, and trains.

Because flow and pressure are adaptable a double-acting cylinder is necessary as a linear actuator. In some cases, it can be of advance to use a multi-chamber cylinder. These cylinders usually consist of four chambers, two for each direction. In combination with the digital valves, the target velocity can be reached just with a small error [51]. The control of the system will then switch between different states. A four-chamber cylinder

FIGURE 5.59 Four chamber cylinder [51].

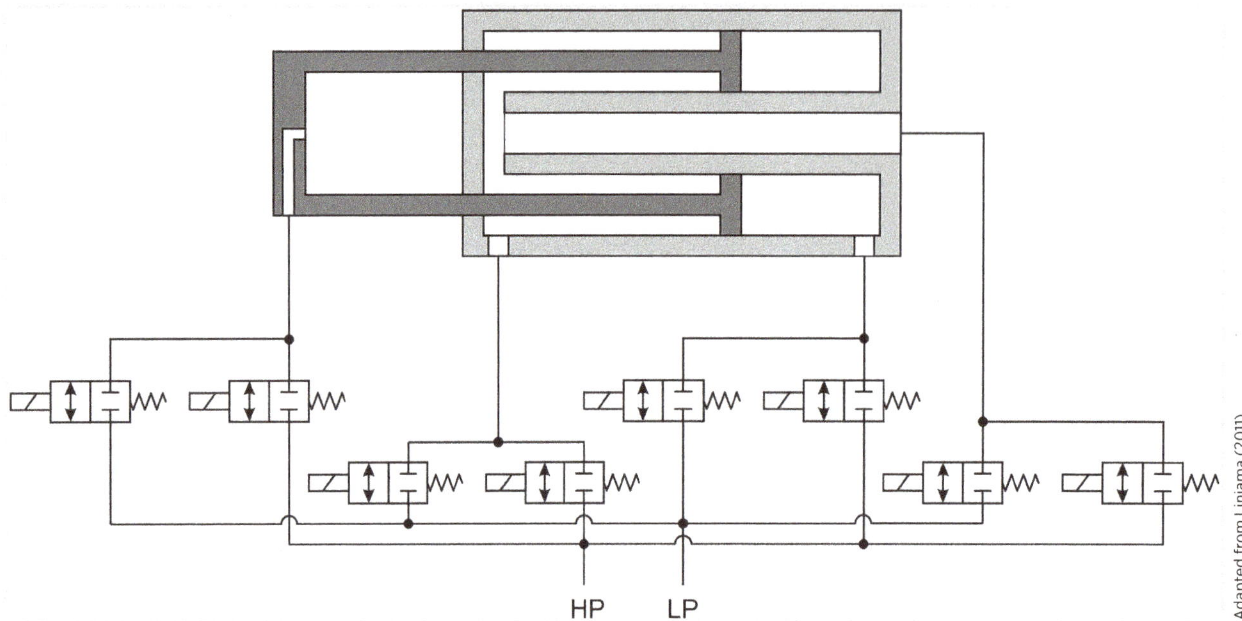

and a possible design are presented in Figure 5.59. In theory, 81 combinations can be realized with three pressure levels (high, medium, and tank pressure) and the four-chamber cylinder. Taking the useful combinations there will be 14 left for one direction of the cylinder movement [49, 50].

With these elements, a system can be built as shown in Figure 5.60. The hydraulic pump is connected via its ports A and B to the two cylinders acting as the actuators. The return line for the cylinder drives is connected through the digital valve to the LP which acts here as the tank. The pump has an HP accumulator to recover energy and gets its oil also from the LP. The 2/2-valves can be realized as switching valves because the flow is adapted by the pump and the pressure level is set in dependence of the load. Just using a common pressure line for the cylinders would require a digital valve and/or a multi-chamber cylinder.

The control of the system needs to switch a lot of valves and therefore has to make many decisions. A smooth adaption to system changes is a challenge of the control. Therefore, a lot of research can be found in the field of control for these systems. Some possibilities of power flow of a digital pump-motor unit are shown in Linjama's "Digital Fluid Power – State of the Art" [51] or the controllability of flow is discussed in Linjama and Huhtala's "Digital Hydraulic Power Management System – Towards Losses Hydraulics" [75]. An overview of controls in digital hydraulics is presented in Scheidl [76] and the possibility to use model-based control is discussed in Linjama, "Model-Based Control of a Digital Hydraulic Transformer-Based Hybrid Actuator" [77]. In the last-mentioned publication, the controller used a simplified model of the system for control. The results show, that the system can be controlled using model-based control in general

FIGURE 5.60 Digital hydraulic system with two actuators [75].

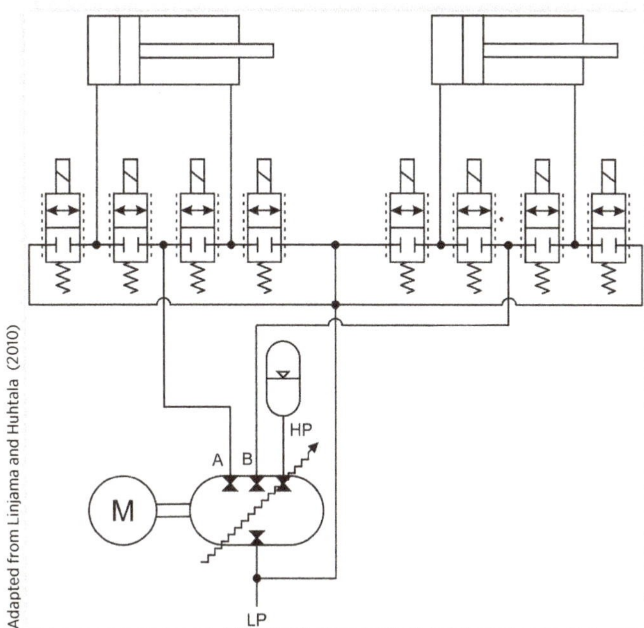

but the performance must be optimized. A great advantage of the system compared to a proportional standard one is a reduced pump size because peak flows are supported by the hydraulic accumulator.

The field of digital hydraulics is a fairly new one. The obvious downsides of the system are the addition of numerous valves and lines, which could lead to cost and reliability issues. Currently, researchers are working to reduce costs per valve and develop failsafe systems that keep working even if one or several valves fail [75, 78, 79].

5.4.2 Independent Metering

While there has been considerable improvement on the VCA by going from constant pressure levels to advanced LS systems, which utilize intelligently designed pressure compensators to ensure smooth and easy-to-control movements, the basic principle has not changed whereby flow is throttled to perform motion control. The same can be said about independent metering, which is the most efficient valve-controlled technology and works by splitting the meter-in and meter-out control edges of the valves - making them independent from each other. This results in more efficient throttling and even enables energy recovery. Figure 5.61 illustrates the different levels of individualization possible when starting to decouple the valve blocks. On the left side is the basic principle of throttling the inlet and/or outlet of an actuator, depending on its working mode. The right side of the figure illustrates what the system could look like if all metering edges have been decoupled from each other. The system topology opens up for differential modes of operation of the hydraulic consumers. These free flow paths are the most

FIGURE 5.61 Different levels of VCA in independent metering configurations [80].

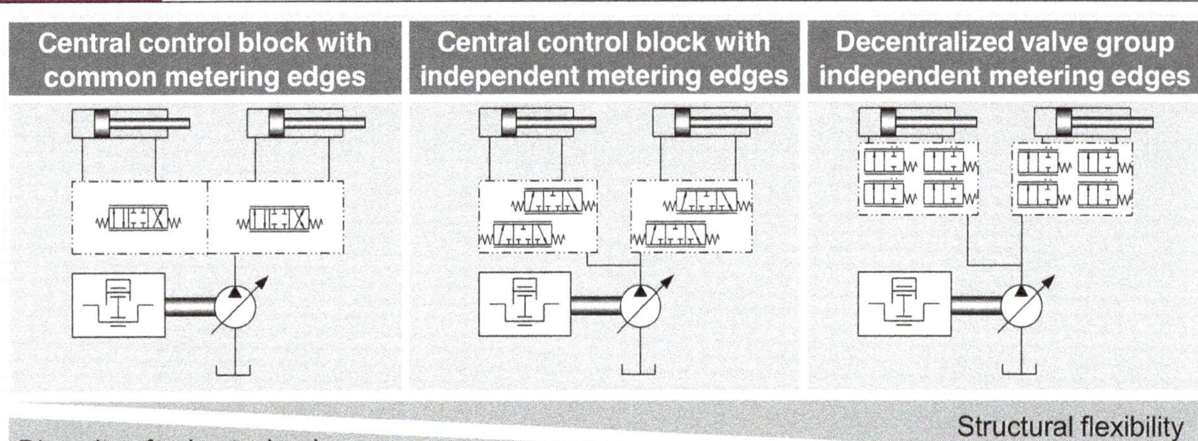

significant characteristic of independent metering systems allowing regeneration in many variants depending on the valve structure.

While this technology adds numerous components to the hydraulic system, it opens up the flexibility and the possibilities for energy recovery, simplification, and standardization of the valve technology and the possibility to add software-enabled features such as vibration damping or different user profiles. Independent metering is predicted to be the next big step in VCA since it is an evolutionary development and will gain acceptance in the market very quickly. In combination with PCA, controlling the movement of the actuator when the valves are fully open, this technology delivers very promising efficiency improvements. Once such systems have been established on the market, it might pave the way for a possible next step, which can change the valves completely to switching valves and have the pump control the actuation at all times.

5.4.3 Sustainable Energy Sources

As discussed before, mobile working machines need to be as compact and efficient as possible to meet the customer requirements in the best possible way. Battery-electric machines will be an optimal solution for machines in indoor applications and with a relatively low energy demand during the day. Although a strong development focus in electric machines can be seen, for a wide range of them the combustion engine will remain the best primary energy provision in the foreseeable future.

To reduce greenhouse gas emissions and to stop global warming, sustainable and CO_2-emission free solutions for mobile working machines have to be found. Table 5.2 shows the energy density of different energy carriers. Taking today known energy carriers into account, hydrogen, methane, and biofuels have the highest potential for mobile working machines to act as sustainable energy storage [4].

For a comparison of different energy storage for mobile working machines in Geimer [4] a machine with a diesel tank for 500 L was chosen as the reference system. Taking the energy amount and the efficiencies of the different energy transmissions into account,

TABLE 5.2 Energy density of energy carriers [4].

Elements	The energy density (MJ/kg)
Electric double-layer capacitor	0.00036–0.036
Lead battery	0.09
Lithium-ion battery	<0.54
Ethanol	26
Rape oil	37.2
Diesel	43.2
Petrol	44
Methane (biogas)	50
Hydrogen	120

a battery of more than 15 tons is needed to run an electric machine in similar conditions. For this reason, electric drives will not be dominant in a wide range of mobile machines.

Using biofuels as alternative energy carriers, the machines need not be changed in general, just the combustion engine must be adapted to the fuel. An intensive discussion is running about how to produce fuels. Using cultivated plants, an ethical discussion needs to be held about the planting of food and energy products. Using biologic waste, the quality of the raw material changes and the production gets more effort. Today, the production of synthetic fuels is being researched and in the pre-serial state.

From the energy density view, hydrogen seems to be a good alternative. Although the weight is very low, the needed volume, even of liquid hydrogen and not taking the tank size into account, is a minimum four times higher than the diesel volume, Table 5.3. Here, just the energy carrier was taken into account. The size of the tank has to be added.

Another alternative is to use hydrogen with a fuel cell. Taking an efficiency of 50% of a fuel cell into account, the needed volume of liquid hydrogen would be 1,431 L (378 gal). Again, this is also too much additional volume for use as an alternative in mobile working machines. Besides, the whole machine has to be changed for electric power as the primary energy carrier.

In contrast, methane could be a promising alternative. To replace the 500 L diesel tank, a volume of 854 L (226 gal) liquid methane would be necessary, an additional volume of 60%. This is quite a lot, but apart from bio-fuels the lowest volumes for sustainable energy carriers.

TABLE 5.3 Energy density of energy carriers [4].

Energy storage	Energy density (MJ/kg)	Volume (L)	Mass (kg)
Lithium-ion battery	<0.54	Not calculated	>4,500
Diesel	43.2	500	415
Petrol	44	543	407
Hydrogen (200 bar (2,900 psi))	120	8,788	149
Hydrogen (700 bar (10,200 psi))	120	2,511	149
Hydrogen (liquid)	120	2,047	149
Methane (200 bar (2,900 psi))	50	2,656	359
Methane (liquid, −167°C)	50	854	369

The use of liquid methane is well known today in the field of overseas shipping and trucks. The tanks of trucks, used for liquefied natural gas, which is quite similar to methane, are of the same size as needed for mobile working machines. Therefore, in Weberbeck [81] possible configurations of working machines in different applications are studied. All these configurations use technologies to the state-of-the-art.

Therefore, today it is from a technical point of view possible to use sustainable energy storage in mobile working machines which are, taking a well-to-wheel balance into account, CO_2-emission free during the lifetime. State-of-the-art technologies from other fields can be used there. Today, such technologies are not used just for economic reasons.

References

1. Schumpeter, J.A., *The Theory of Economic Development* (London/New York: Routledge, 2017).
2. Borbély, E., "J.A. Schumpeter und die Innovationsforschung (J.A. Schumpeter and the Innovation Research)," in: *6th International Conference on Management, Enterprise and Benchmarking*, Budapest, Hungary, May 30–31, 2018.
3. Sledzik, K., "Schumpeter's View on Innovation and Entrepreneurship," Management Trends in Theory and Practice, Hittmar, S. (ed.), 2013.
4. Geimer, M. and Ays, I., "Sustainable Energy Storages for Mobile Machines," in: *Symposium on Efficiency of Mobile Machines and Their Processes*, Braunschweig, Germany, March 10–11, 2015, doi:10.5445/IR/1000047733, last view April 10, 2019.
5. Ays, I. and Geimer, M., "Methane-Fuel Cell CCS-Drive: The Emission-Free Working Machine," in: *7th Conference on Hybrid and Energy Efficient Drives for Mobile Working Machines*, Karlsruhe, Germany, February 20, 2019, in: *Conference Proceedings* (KIT Scientific Publishing), 143-163, doi:10.5445/IR/1000091557, last view April 10, 2019.
6. Reick, B., "Methode zur Analyse und Bewertung von stufenlosen Traktorgetrieben mit mehreren Schnittstellen (Method for Analyzing and Evaluating Continuously Variable Tractor Transmissions with Multiple Interfaces)," in: *Karlsruher Schriftenreihe Fahrzeugsystemtechnik*, vol. 64 (Karlsruhe: KIT Scientific Publishing, 2018).
7. Himmelsbach, R. and Pohlenz, J., "eCVT for Tractors – Continuously Variable Driving and Electric Power for Implement Drives," in: *LAND.TECHNIK AgEng 2019*, Hannover, Germany, November 8–9, 2019, in: VDI Berichte, 2361 (Düsseldorf: VDI Verlag, 2019).
8. Murrenhoff, H., *Grundlagen der Fluidtechnik Teil 1: Hydraulik (Basics of Fluid Power, Part 1: Hydraulics)*, 6th edn. (Aachen: Shaker, 2011).
9. Zhang, Z., "Linear Actuators with Electric Power Supply: Electrification of Components and Efficiency Evaluation of the Overall Machine Electrification," https://www.tu-braunschweig.de/Medien-DB/ilf/forschung/handout_dbu_zhang.pdf, last visit February 2, 2019.
10. Carl, B., Ivantysynova, M., and Williams, K., "Comparison of Operational Characteristics in Power Split Continuously Variable Transmissions," SAE Technical Paper 2006-01-3468, 2006, https://doi.org/10.4271/2006-01-3468.
11. Kumar, R., "A Power Management Strategy for Hybrid Output Coupled Power-Split Transmission to Minimize Fuel Consumption," Dissertation, West Lafayette, 2007.

12. Onorti, S., Serrao, L., and Rizzoni, G., *Hybrid Electric Vehicles - Energy Management Strategies* (London: Springer, 2016), doi:10.1007/978-1-4471-6781-5.

13. Reif, K. (Ed.), *Fundamentals of Automotive and Engine Technologies* (Wiesbaden: Springer Fachmedien, 2014), doi:10.1007/978-3-658-03972-1.

14. Thiebes, Ph., "Hybridantriebe für mobile Arbeitsmaschinen (Hybrid Drives for Mobile Working Machines)," Dissertation, Karlsruhe Institute of Technology, Karlsruhe, in Karlsruher Schriftenreihe Fahrzeugsystemtechnik, vol. 10 (KIT Scientific Publishing, 2011, doi:10.5445/KSP/1000025625.

15. Achten, P. et al., "Dedicated Design of the Hydraulic Transformer," in: *3rd International Fluid Power Conference*, Aachen, Germany, March 5–6, 2002, in: *Conference Proceedings* (Shaker Verlag, 2002), 233-248.

16. Renius, K.Th., Knechtes, H., and Geimer, M., "Agricultural Tractor Development, Yearbook Agricultural Engineering," 2010.

17. Nagel, Ph., "Entwicklung einer Betriebsstrategie zur Energierückgewinnung in hybriden Mehrverbrauchersystemen (Development of an Operating Strategy for Energy Recovery in Hybrid Multi-Consumer Systems)," Dissertation, Karlsruhe Institute of Technology, Karlsruhe, in: Karlsruher Schriftenreiche Fahrzeugsystemtechnik, vol. 46 (KIT Scientific Publishing, 2016), doi:10.5445/KSP/1000051743.

18. Prinz zu Hohenlohe, F., "Phlegmatisierung als Tugend in der Mobilhydraulik - Das Energiespeichersystem des Kranvollernters HSM 405H2 (Phlegmatisation as a Virtue of Mobile Hydraulics - The Energy Storage System of a HSM 405H2 Crane Harvester)," in: *3rd Conference "Hybridantriebe für mobile Arbeitsmaschinen" (Hybrid Drives for Mobile Working Machines)*, Karlsruhe, Germany, February 17, 2011, in: Karlsruher Schriftenreihe Fahrzeugsystemtechnik, vol. 7 (KIT Scientific Publishing, 2011), 151-162.

19. Filla, R., "Optimizing the Trajectors of a Wheel Loader Working in Short Loading Cycles," in: *13th Scandinavian International Conference on Fluid Power (SICFP2013)*, Linköping, Sweden, June 3–5, 2013.

20. Geimer, M. and Synek, P.-M., *7th Conference "Hybride und energieeffiziente Antriebe für mobile Arbeitsmaschinen" (Hybrid and Energy Efficient Drives for Mobile Working Machines)*, February 20, 2019, in: Karlsruher Schriftenreihe Fahrzeugsystemtechnik, vol. 69 (KIT Scientific Publishing, 2019).

21. Zimmerman, J.D., "Toward Optimal Multi-Actuator Displacement Controlled Mobile Hydraulic Systems," Purdue University, West Lafayette, 2012.

22. Zimmerman, J. and Ivantysynova, M., "Hybrid Displacement Controlled Multi-Actuator Hydraulic Systems," in: *Proceedings of the 12th Scandinavian International Conference on Fluid Power*, Tampere University of Technology, Tampere, Finland, 2011, 217-233.

23. Hippalgaonkar, R. and Ivantysynova, M., "Optimal Power Management of Hydraulic Hybrid Mobile Machines — Part I: Theoretical Studies, Modeling and Simulation," *Journal of Dynamic Systems, Measurement, and Control* 138, no. 5 (2016): 051002-1-051002-23, doi:10.1115/1.4032742.

24. Hippalgaonkar, R. and Ivantysynova, M., "Optimal Power Management of Hydraulic Hybrid Mobile Machines - Part II: Machine Implementation and Measurements," *Journal of Dynamic Systems, Measurement, and Control* 138, no. 5 (2016): 051003, doi:10.1115/1.4032743.

25. Schneider, M. et al., "Green Wheel Loader - Development of an Energy Efficient Drive and Control System," in: *The 9th International Fluid Power Conference (IFK)*, Aachen, Germany, March 24–26, 2014.

26. Sprengel, M. and Ivantysynova, M., "Coupling Displacement Controlled Actuation with Power Split Transmissions in Hydraulic Hybrid Systems for Off-Highway Vehicles," in: *Fluid Power and Motion Control: FPMC 2012* University of Bath, Bath, UK, 2012, 505-517.

27. Karl-Erik, R., "Hydraulic Hybrids - The New Generation of Energy Efficient Drives," in: *Conference on Fluid Power Transmission and Control (ICFP)*, Hangzhou, China, 2009.

28. Sprengel, M. and Ivantysynova, M., "Investigation and Energetic Analysis of a Novel Hydraulic Hybrid Architecture for On-Road Vehicles," in: *The 13th Scandinavian International Conference on Fluid Power (SICFP2013)*, Linköping University, Linköping, Sweden, 2013, 87-98.

29. Rahmfeld, R. and Ivantysynova, M., "Energy Saving Hydraulic Actuators for Mobile Machines," in: *1st Bratislavian FLUID POWER Symposium*, Častá-Píla, Slovakia, 1998, vol. 49, 1–11.

30. Lodewyks, J., "Differentialzylinder im geschlossenen hydrostatischen Kreislauf (The Differential Cylinder in a Closed Hydrostatic Circuit)," Dissertation, RWTH Aachen, 1994.

31. Heybroek, K. et al., "Mode Switching and Energy Recuperation in Open-Circuit Pump Control," in: *Proceedings of the 10th Scandinavian International Conference on Fluid Power (SICFP'07)*, May 21–23, 2007, 197–209.

32. Ivantysyn, R. and Weber, J., "Novel Open Circuit Displacement Control Architecture in Heavy Machinery," in: *The 8th FPNI PhD Symposium on Fluid Power*, Lappeenranta, Finland, 2014, 1–8.

33. Michel, S. and Weber, J., "Electrohydraulic Compact-Drives for Low Power Applications Considering Energy-Efficiency and High Inertial Loads," in: *7th FPNI PhD Symposium on Fluid Power*, Reggio Emilia, Italy, 2012, 869-888.

34. Ivantysynova, M., "Lecture Notes on Fluid Power Components and Systems," https://engineering.purdue.edu/Maha/, 2004.

35. Shang, L., "A Path toward an Effective Scaling Approach for Axial Piston Machine," Purdue University, West Lafayette, 2019.

36. Schenk, A., "Predicting Lubrication Performance Between the Slipper and Swashplate in Axial Piston Hydraulic Machines," Purdue University, 2014.

37. Pellegri, M. and Vacca, A., "Numerical Simulation of Gerotor Units," https://www.gfpsweb.org/?q=content/numerical-simulation-gerotor-units-6, last visit May 29, 2020.

38. Grabbel, J. and Ivantysynova, M., "An Investigation of Swash Plate Control Concepts for Displacement Controlled Actuators," *International Journal of Fluid Power* 6, no. 2 (2005): 19-36.

39. Daher, N., "Novel Energy Efficient Electrohydraulic Steer-By-Wire Technology," Purdue University, West Lafayette, 2014.

40. Busquets, E., "Advanced Control Algorithms for Compact and Highly Efficient Displacement-Controlled Multi-Actuator and Hydraulic Hybrid Systems," Purdue University, West Lafayette, 2016.

41. Daher, N. and Ivantysynova, M., "Yaw Stability Control of Articulated Frame Off-Highway Vehicles via Displacement Controlled Steer-by-Wire," *Control Engineering Practice* 45 (2015): 46-53.

42. Daher, N. and Ivantysynova, M., "A Virtual Yaw Rate Sensor for Articulated Vehicles Featuring Novel Electro-Hydraulic Steer-by-Wire Technology," *Control Engineering Practice* 30 (2014): 45-54.

43. Daher, N. and Ivantysynova, M., "An Indirect Adaptive Velocity Controller for a Novel Steer-by-Wire System," *ASME Journal of Dynamic Systems, Measurement, and Control* 136, no. 5 (2014): 051012.

44. Daher, N. and Ivantysynova, M., "A Steer-by-Wire System that Enables Remote and Autonomous Operation," SAE Technical Paper 2014-01-2404, 2014, https://doi.org/10.4271/2014-01-2404.

45. Daher, N. and Ivantysynova, M., "Energy Analysis of an Original Steering Technology that Saves Fuel and Boosts Efficiency," *Energy Conversion and Management* 86 (2014): 1059-1068.

46. Christopher Williamson, S.L., "Active Vibration Damping for an Off-Road Vehicle with Displacement Controlled Actuators," *International Journal of Fluid Power* 10, no. 3 (2009): 5-16.

47. Kim, T. and Ivantysynova, M., "Active Vibration Control of Swash Plate-Type Axial Piston Machines with Two-Weight Notch Least Mean Square/Filtered-x Least Mean Square (LMS/FxLMS) Filters," *Energies* 10, no. 5 (2017): 645.

48. "Antriebssystem mit hydrostatischer Kraftübertragung (Drive System with Hydrostatic Power Transmission)," German Disclosure Document P2739968.4, September 6, 1977.

49. Dreher, T., "The Capability of Hydraulic Constant Pressure Systems with a Focus on Mobile Machines," in: *6th FPNI – PhD Symposium*, West Lafayette, 2010, 579-588.

50. Xiao-Ming, W., Quin, L., and Xin, L., "The Four-Chamber Hydraulic Cylinder," in: *IEEE/CSAA International Conference on Aircraft Systems (AUS)*, Beijing, China, October 10–12, 2016, https://ieeexplore.ieee.org/stamp/stamp.jsp?arnumber=7748139, last view March 28, 2019.

51. Linjama, M., "Digital Fluid Power – State of the Art," in: *12th Scandinavian International Conference on Fluid Power (SICFP)*, Tampere, Finland, May 18–20, 2011.

52. Heemskerk, E., Bonefeld, R., and Buschmann, H., "Control of a Semi-Binary Hydraulic Four-Chamber Cylinder 2015," in: *Proceedings of the 14th Scandinavian International Conference on Fluid Power (SICFP)*, May 20–22, 2015.

53. Heybroek, K. and Sahlman, M., "A Hydraulic Hybrid Excavator Based on Multi-Chamber Cylinders and Secondary Control – Design and Experimental Validation," *International Journal of Fluid Power* 19, no. 2 (2018): 1–15.

54. Kolks, G. and Weber, J., "Symmetric Single Rod Cylinders with Variable Piston Area? A Comprehensive Approach to the Right Solution," in: *Proceedings of the BATH/ASME 2018 Symposium on Fluid Power and Motion Control*, Bath, UK, September 12–14, 2018, 1–10.

55. Linjama, M., Vihtanen, H.-P., Sipola, A., and Vilenius, M., "Secondary Controlled Multi-Chamber Hydraulic Cylinder," in: *Proceedings of the 11th Scandinavian International Conference on Fluid Power*, Linköping, Sweden, 2009.

56. Linjama, M., Niemi-Pynttari, O., Laamanen, A., and Huhtala, K., "Parallel Pump-Controlled Multi-Chamber Cylinder," in: *Proceedings of the ASME/BATH 2014 Symposium on Fluid Power & Motion Control (FPMC2014)*, New York, 2014.

57. Konus, H., "Hydraulic transformer," U.S. Patent 3.627.451, 1970.

58. Reynolds, R.W., "Hydraulic intensifier system," U.S. Patent 4.077.746, 1974.

59. Tyler, H.P., "Fluid intensifier," U.S. Patent 3.188.963, 1962.

60. Kordak, R., "Praktische Auslegung sekundärgeregelter Antriebssysteme (Practical Design of Secondary Controlled Systems)," *O+P "Ölhydraulik und Pneumatik"* 26, no. 11 (1982): 795-800.

61. Shih, M.C., "Untersuchung einer Zylindersteuerung durch Hydro-Transformator am Konstant-Drucknetz (Investigation of a Cylinder Control by Hydro-Transformer at the Constant Pressure Network)," Dissertation, RWTH University of Aachen, 1984.

62. Achten, P., Fu, Z., and Vael, G., "Transforming Future Hydraulics: A New Design of a Hydraulic Transformer," in: *5th Scandinavian International Conference on Fluid Power (SICFP)*, Linköping, Sweden, 1997.

63. Achten, P. and Fu, Z., "Valving Land Phenomena of the Innas Hydraulic Transformer," *International Journal of Fluid Power* 1, no. 1 (2000): 39-47, doi:10.1080/14399776.2000.10781081, last visit February 8, 2019.

64. Shen, W., Karimi, H.R., and Zhao, R., "Comparative Analysis of Component Design Problems Integrated Hydraulic Transformers," *International Journal of Advanced Manufacturing Technology* 103, no. 1-2 (March 2019): 389-407, doi:10.1007/s00170-019-03543-2.

65. Jing, C. et al., "Research on the Pressure Ratio Characteristics of a Swash Plate-Rotating Hydraulic Transformer," *Energies* 11, no. 6 (June 2018): 1612, doi:10.3390/en11061612.

66. Vael, G.E.M., Orlando, E., and Stukenbrock, R., "Towards Maximum Flexibility in Working Machinery," IHT Control in a Mecalac Excavator, 2004, https://www.innas.com/assets/hydraulic-transformer.zip, last visit February 8, 2019.

67. Liu, C. et al., "Electro-Hydraulic Servo Plate-Inclined Plunger Hydraulic Transformer," *IEEE Access* 4 (2016): 8608-8616, doi:10.1109/ACCESS.2016.2628355.

68. Achten, P., van den Brink, T., Potma, J., Schellekens, M. et al., "A Four-Quadrant Hydraulic Transformer for Hybrid Vehicles," in: *Proceedings of the 11th Scandinavian International Conference on Fluid Power (SICFP)*, Linköping, Sweden, June 2-4, 2009.

69. Shen, W. and Wang, J., "Adaptive Fuzzy Sliding Mode Control Based on Pi-Sigma Fuzzy Neutral Network for Hydraulic Hybrid Control System Using New Hydraulic Transformer," *International Journal of Control, Automation and Systems* 17, no. X (2019): 1-9, doi:10.1007/s12555-018-0593-9.

70. Chao, Z. et al., "Control Analysis of the Hydraulic Luffing System Controlled by the New Hydraulic Transformer for Rescue Vehicles," *International Journal of Innovative Computing, Information and Control* 15, no. 1 (2019): 275-289, doi:10.24507/ijicic.15.01.275.

71. Geiger, C. and Geimer, M., "Efficiency Optimisation of a Forestry Crane by Implement Hydraulics with Energy Recovery," in: *Land. Technik AgEng 2017*, November 10-11, 2017, in: VDI-Berichte No. 2070 (VDI-Verlag, 2017), 175-184.

72. Zu Hohenlohe, F., Geiger, C., and Geimer, M., "Hybridantrieb für den Ladekran einer Forstmaschine (Hybrid Drive for the Crane Function of a Forestry Machine)," in: *7th Conference "Hybride und energieeffiziente Antriebe für mobile Arbeitsmaschinen" (Hybrid and Energy Efficient Drives for Mobile Working Machines)*, February 20, 2019, in: Karlsruher Schriftenreihe Fahrzeugsystemtechnik, vol. 69 (KIT Scientific Publishing, 2019).

73. N.N., in: *Workshop on Digital Fluid Power*, Linz, Austria, February 28 and March 1, 2019, Organizers: Johannes Kepler University, Tampere University of Technology, and Aalborg University.

74. Artemis, "Technology - High Controllable, and Extremely Efficient," http://www.artemisip.com/technology/, last view March 27, 2019.

75. Linjama, M. and Huhtala, K., "Digital Hydraulic Power Management System - Towards Losses Hydraulics," in: *3rd Workshop on Digital Fluid Power*, Tampere, Finland, October 13-14, 2010.

76. Scheidl, R., Manhartsgruber, B., and Winkler, B., "Hydraulic Switching Control - Principles and State of the Art," in: *Proceedings of the First Workshop on Digital Fluid Power*, Tampere, Finland, October 3, 2008, 31-49.

77. Linjama, M., "Model Based Control of a Digital Hydraulic Transformer-Based Hybrid Actuator," in: *BATH/ASME Symposium on Fluid Power and Motion Control FPMC*, Bath, UK, September 12-14, 2018.

78. Pedersen, N., Johansen, P., and Andersen, T., "Event-Driven Control of a Speed Varying Digital Displacement Machine," in: *Symposium on Fluid Power & Motion Control (FPMC)*, Sarasota, FL, 2017.

79. Rømer, D. et al., "Analysis of Dynamic Properties of a Fast Switching On-Off Valve for Digital Displacement Pumps," in: *Symposium on Fluid Power & Motion Control (FPMC)*, Bath, UK, 2012.

80. Weber, J. et al., "Novel System Architectures by Individual Drives," in: *10th International Fluid Power Conference (IFK)*, Dresden, March 8–10, 2016.

81. Weberbeck, L. et al., "Liquefied Natural Gas in Mobile Machines," *ATZ Offhighway Worldwide* 9, no. 4 (2016): 38-45, doi:10.1007/s41321-016-0532-8.

CHAPTER 6

Selected Machine Examples

Dr.-Ing. Rainer Bavendiek, Dr. Martin Kremmer, and Dr.-Ing. Herbert Pfab

6.1 Tractor (By Martin Kremmer)

6.1.1 Introduction

6.1.1.1 Formal Definition: There are various formal definitions for the term tractor and tractor attachments. The definitions presented and referred to in this chapter are those specified in ANSI/ASAE and ISO standards and in the European Directives and used by standards from the technical committee for tractors and machinery for agriculture and forestry ISO/TC23 or European type approval and market surveillance as being the most relevant ones.

An agricultural tractor is formally defined in ANSI/ASAE S390 and ISO 12934 section (3.1) as a "self-propelled agricultural vehicle having at least two axles and wheels, or endless tracks, particularly designed to pull agricultural trailers and pull, push, carry and operate implements used for agricultural work (including forestry work), which may be provided with a detachable loading platform." [3] (Copyright ASABE. All excerpts from the text of this standard have been used with permission of the ASABE.)

Under this definition, a tractor is also part of the group of "agricultural vehicles (2.3)," which can be an "agricultural tractor (3.1)," "self-propelled machinery (3.2)," or a "trailer or interchangeable towed machinery which is primarily intended to be used in agriculture including occasional road travel." [3]

A tractor is furthermore defined as being part of "agricultural field equipment" which comprises "agricultural tractors (3.1), self-propelled machines (3.2), implements, and combinations thereof designed primarily for agricultural field operations." [3]

The Tractor Mother Regulation (TMR) published under (EU) No. 167/2013 on the approval and market surveillance of agricultural and forestry vehicles valid since January 1, 2018, also provides a definition of the term "tractor" for the purpose of that regulation [42].

6.1.1.2 Relevance and Impact of Mechanized Agriculture and Global Structural Changes: The mechanization of agriculture has been enabling significant structural changes to the population in rural and urban areas and to the workforce in agriculture throughout the world. Whereas 40.9% of the world's employed males were working in Agriculture in 1992, this number decreased within the following 25 years by 13.3%–27.6% in the year 2017 according to a recent data published by the World Bank [57]. Over the same period of time, in North America the proportion of employed males in agriculture sank from 4% to 2%, respectively [57]. This is a global trend but happens on different levels in different regions and goes along with a continuous increase in the size of farm operations and reduction in the total number of farms in higher developed countries. There are estimations of around 570 million farms worldwide, three-quarters of which operate an agricultural area of less than 1 ha and commonly in the form of family farms. However, 88% of the agricultural land of 4.9 million ha worldwide is farmed by 91 million farms in farm sizes of 2 ha and beyond [16]. The countries with the world's largest proportion of their land devoted to agriculture are China (11%), Australia (9%), and the United States (8%). In Australia, the average farm size increased from 1,900 ha in 1960 to 3,200 ha per farm operation in 2010, demonstrating the trend in increasing farm operations in high-developed countries.

There are many factors contributing to the global structural changes in agricultural operations and they include in addition to the factors mentioned above advancement in biotechnology, genetics, and agricultural policies. Whereas in the year 1900 a single farmer was able to produce food for around four people, 113 years later, in 2013, he was able to feed almost 144 people [9].

Undoubtedly, the developments in agricultural machinery and tractors in particular play an important role. The continuous increase in power of agricultural tractors and their ability to pull wider and heavier implements has led to larger working widths and higher operating speeds of implements or to multiple applications, carried out in one pass in the field. These factors, of course, resulted in a reduction in human workforce per farmed area.

Since the beginning of mechanization in agriculture, the evolution of agricultural tractors has enabled a significant increase in the productivity in evolving farm operations and has also impacted the perception on agriculture in rural and urban areas throughout the world.

6.1.1.3 Historical Development: In the nineteenth century, when steam engines were established in industrial production, numerous attempts were made to use such external combustion engines also for other purposes in the form of either portable engines or traction engines. In agriculture, steam engines were increasingly being used for powering stationary threshing machines [44], mills, or saws and also for pulling plows or other soil engaging elements through a field. Upcoming machine forms were the so-called locomotives (also referred to as locomobiles or traction engines) by which threshing machines could be moved and powered using a transmission belt on a particular worksite, respectively. In the 1850s, John Fowler in the UK developed a so-called steam plow with locomotives on one or either side of the field to pull a plow by a rope through the soil (see **Figure 6.1**).

However, the large weight of locomotives limited their practical use in farming and transport operations. It was the invention and adoption of internal combustion engines that became the breakthrough for the development and establishment of tractors [46, 99].

Gasoline-fueled engines were built in 1887 by the Charter Gasoline Engine Company of Sterling, Illinois (USA), for example, which also produced a small number of traction engines by the year 1889 [49]. In the United States, in 1897 the Hart-Parr Gasoline Engine

FIGURE 6.1 Steam plow (German: Dampfpflug). Left column (Dampfpflug I.) one machine system (top) with locomotive of one machine system (middle) and balance plow (bottom) and right column (Dampfpflug II.) two machine system (top) with locomotive of two machine system (middle) and turn-cultivator (bottom). Translated from German into English [41]. (Digital image provided by Stadtarchiv Kerpen, Germany).

Dampfpflug I.
Fig. 1. Einmaschinensystem.
Fig. 2. Lokomotive des Einmaschinensystems.
Fig. 3. Balancierpflug.

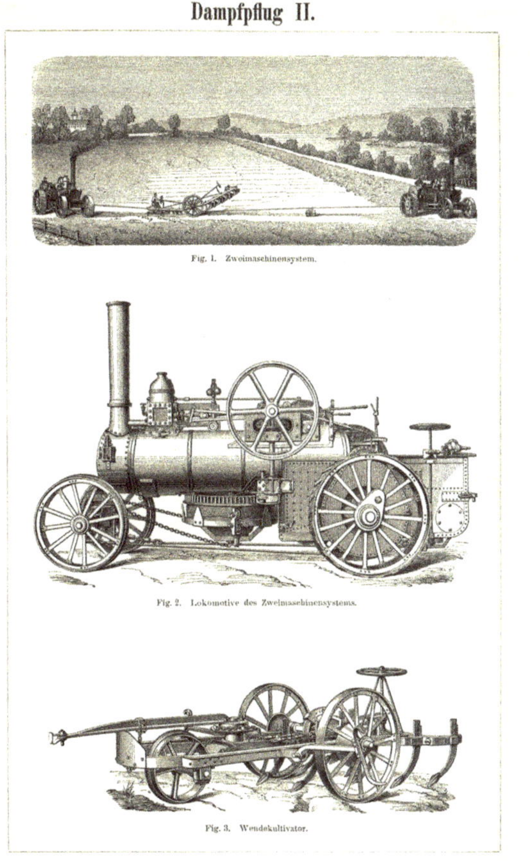

Dampfpflug II.
Fig. 1. Zweimaschinensystem.
Fig. 2. Lokomotive des Zweimaschinensystems.
Fig. 3. Wendekultivator.

Company was founded and delivered their first tractor, the Hart-Parr No. 1 by the year 1901. Between 1906 and 1918, more than 4,100 Hart-Parr 30-60 two-cylinder horizontal engine tractors were built [49]. By 1914, the Waterloo Gasoline Tractor company started production of the Waterloo Boy, a tractor running on kerosene. The company was purchased by Deere & Company in 1918 marking the beginning of John Deere tractor history.

Different agricultural machine forms were developed and presented to the public around the same time in Europe. Robert Stock in Berlin presented the first self-propelled motorplows in 1908 and founded the Stock-Motorplough Company 2 years later. The Heinrich Lanz Aktiengesellschaft in Mannheim started production of the Landbaumotor in 1912, a four-cylinder 52 kW (70 hp) gasoline tractor to which was attached a rotary tiller system, Köszegi [18]. But the dominating machine form for agricultural tractors became the so-called standard agricultural tractor (see Section 6.1.2.2). This machine form was and still is produced by many tractor manufacturers in the world. Two very popular early tractors were the Fordson starting in 1917 in North America and the Lanz Bulldog starting in 1921 in Europe.

Henry Ford & Son Inc. went into the tractor business with mass production of tractors in 1917 which were later named as Fordson tractors by applying the principles being successfully used to produce the Ford Model T automobile. An example of an early mass-produced tractor is the Fordson Model F tractor (Figure 6.2). Almost 9 years later,

FIGURE 6.2 Fordson Model F tractor (sold from 1917) with four-cylinder engine 15 kW (20 hp) with 4,118 ccm engine displacement and 3/1 transmission [19]. (Digital image provided by the University of Chicago Library, Chicago, USA).

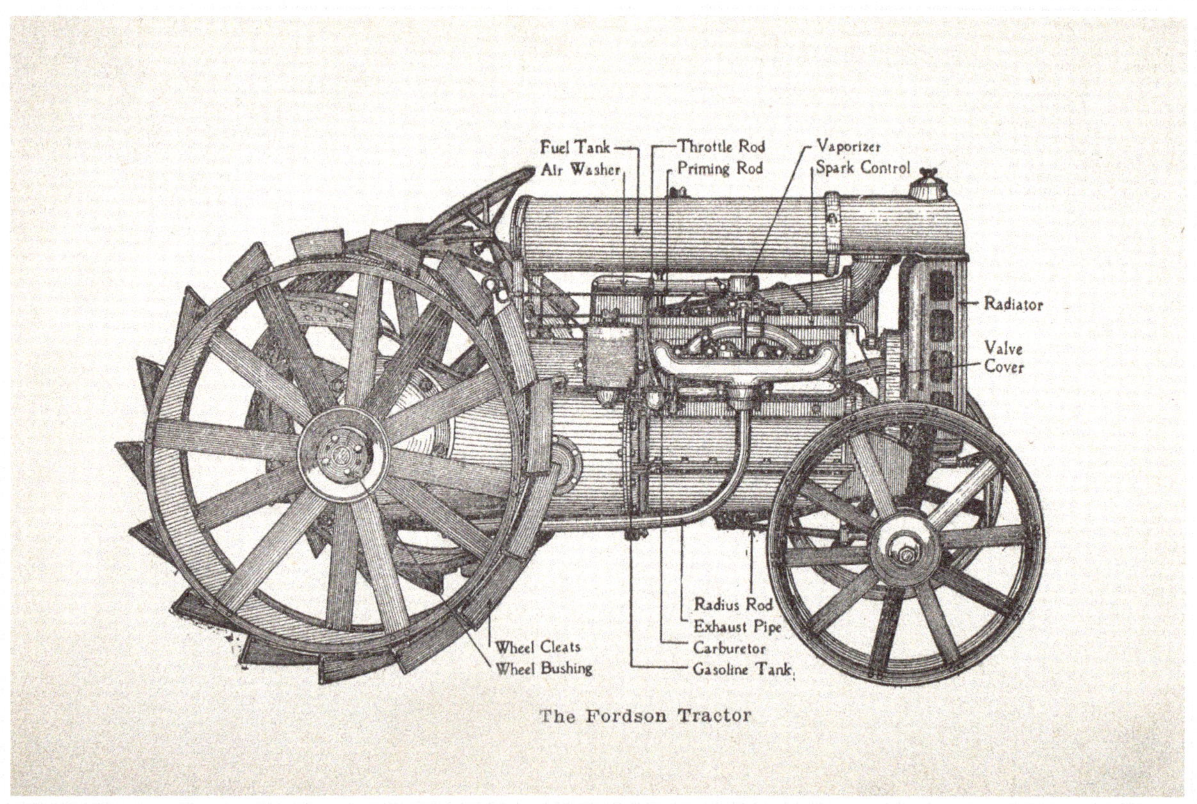

in 1926, Ford built 99,101 tractors per year reaching an overall population of almost 650,000 Fordson tractors in the following year 1927 [8, 19, 53].

In North America, row crop tractors became popular. Row crop tractors were specifically designed for operating between rows of crops without damaging the plants. Farmall tractors, such as the Farmall Regular, were early and successful examples of this machine form in the early 1920s.

After World War I, in 1921, Lanz introduced to the public a crude oil tractor, the Lanz Bulldog HL12, at the exhibition of the German Society of Agriculture (DLG) in Leipzig. These one-cylinder tractors were simple, robust, and could be operated using almost any low-octane combustion fuel. Mass production started in 1926 with the HR2 Bulldogs (16/21 kW (22/28 hp)). Nearly 620,000 Bulldogs were built and sold to farmers from 1921 until 1960 [18]. Figure 6.3 shows a cross section through a HR7/HR8 Bulldog.

Major innovations in the history of tractors include the use of a practical PTO (power take-off) by International Harvester (IH) in 1918 in the United States [40]. Eight years later, in 1926, the Equipment Manufacturers Institute (EMI) published a PTO,

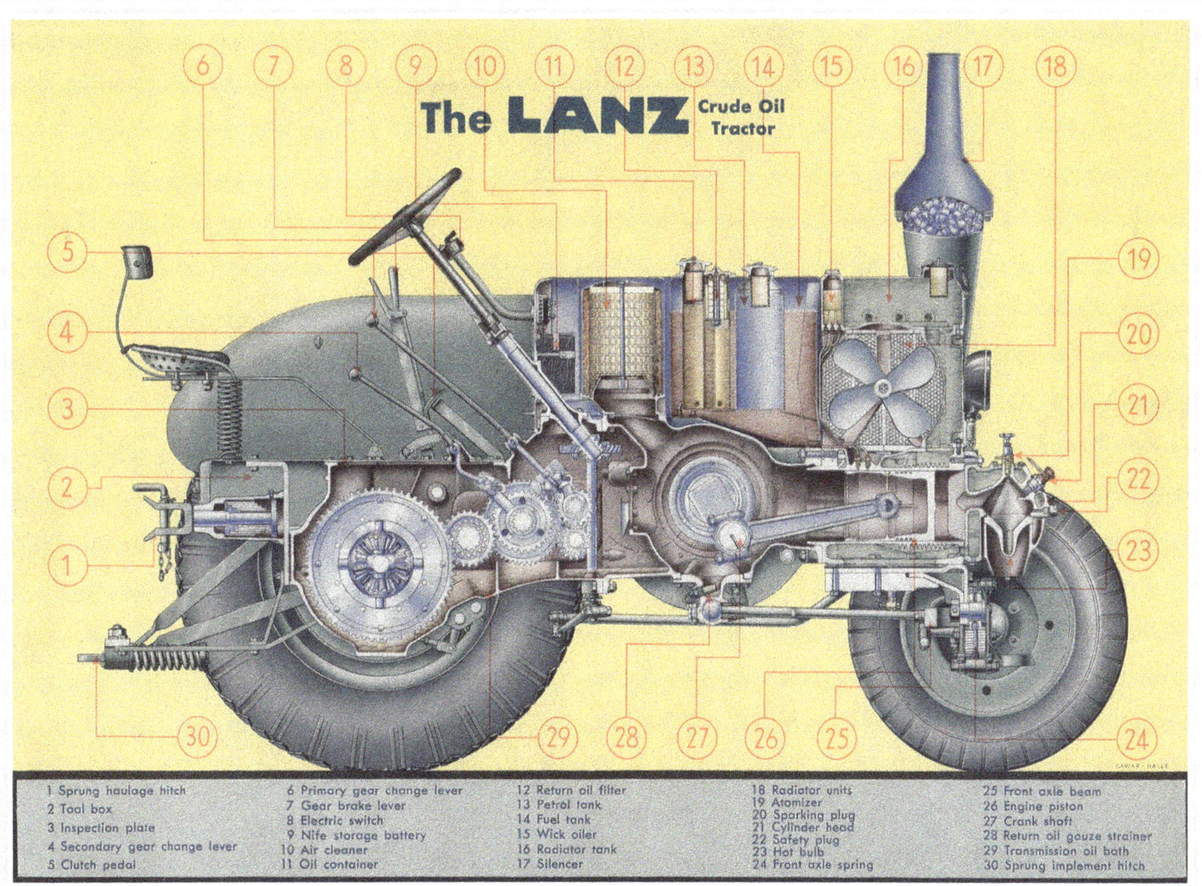

FIGURE 6.3 LANZ Bulldog HR7/HR8 (sold from 1935) with one-cylinder crude oil hot-bulb engine (10,226 ccm engine displacement, 22–41 kW (30–55 hp) model dependent) and 6/2 manual transmission.

master shield, and drawbar relationship standard jointly with the American Society of Agricultural Engineers [40].

To improve use and exchange of implements, on February 12, 1925, Harry Ferguson patented an "apparatus for coupling agricultural implements to tractors and automatically regulating the depth of work" [17].

Where traction engines and early tractors were running either on steel wheels or on wheels to which were later attached solid rubber bands, the introduction of pneumatic tires in 1932 provided a breakthrough for improved operator ride comfort and enabled higher speeds of agricultural tractors. For example, the LANZ Eilbulldog (1937) was designed for speeds up to 32 km/h (20 mph) adopting such tires [18].

In 1954, powershift transmissions were introduced on International Harvester tractors in the United States and 4 years later, in 1958, on Ford tractors (Select-O-Speed 10/2 Transmission).

Bosch developed Electronic 3-Point Hitch Control and filed in 1977 the patent DE0000027311641A [20]. The so-called Electrohydraulic Hitch Control (EHR) was adopted by Massey-Ferguson (MF) Tractors in 1978.

Hydrostatic-mechanic power-split continuously variable transmissions (CVTs) were developed by Fendt and presented to the market in 1996 in their 926 Vario series tractors.

At Agritechnica 2007, John Deere presented an integrated crankshaft-driven generator on the 7530 E-Premium tractors to produce 20 kW (27 hp) of electrical energy for both tractor internal and external usage to power implements such as electrically driven fertilizer spreaders, for example.

The introduction of US TIER I/EU STAGE I emission regulations for off-road vehicles in 1997 initiated an era that significantly affected the development efforts and expense for agricultural tractors. In recent years, technologies, established for on-road vehicles such as, exhaust gas recirculation (EGR), diesel particulate filters (DPF), oxidation catalyst and selective catalytic reduction (SCR) systems with associated diesel exhaust fluid (DEF) tanks, dosing units and infrastructure, and combinations of the above, were implemented on tractors in emission regulated markets including Europe and North America. Since 2015, TIER IV/STAGE IV emission regulations for tractors 130 kW $\leq P \leq$ 560 kW were put in place, and STAGE V (EU only) regulations are scheduled to be introduced in 2019. For further reference, see also Section 3.1.

At the present time, tractors are built with various technology specification and power levels to satisfy both legal and customer requirements in different regions all over the world. Renius, for example, distinguishes five levels of technology for tractors with two axles to describe the spectrum of specifications of worldwide tractors [45, 46, 99].

Although different machine forms appeared over time, such as the Tool Carriers (Lanz 1951), Intrac (Deutz 1972), Unimog (Böhringer 1948), or crawler tractors and tractors for specialty applications including vineyard and orchard, the standard agricultural tractor is still the most common machine form of an agricultural tractor.

There are several hundred tractor manufacturers in the world. John Deere, AGCO, CNH, Mahindra, Kubota, and YTO are among the largest manufacturers. **Figures 6.4** and **6.5** show typical examples of recently produced agricultural tractors from two different regions: Europe (AGCO Fendt in Figure 6.4) and North America (John Deere in Figure 6.5).

FIGURE 6.4 AGCO Fendt 800 Vario in a typical road transport operation, common to many European farm or contractor operations.

FIGURE 6.5 John Deere 9RX in corn planting, typical to North American farm operation.

6.1.1.4 **Markets and Industries - Production Facilities, Sales Volume, and Population:** Today, tractors are produced and sold all over the world. The wide spread of factory locations has a historical heritage but also is a result of more recent strategic management decisions to supply the demand of the worldwide customer base for agricultural tractors. Tractors still represent a significant amount of mass with associated costs for sourcing, manufacturing, and shipping (and importing or exporting) into target countries.

In Europe, there are around 24 production sites for agricultural tractors, the majority of which are in Northern Italy, Germany, France, Turkey, and England (see Figure 6.6).

As shown in Figure 6.7, in Central/North America, the main tractor factories are located in the Midwest and in the southern and southeastern states of the United States and in central Mexico.

There are only a few production sites in South America. Brazil and Argentina (Figure 6.8) are serving primarily the local markets.

In Asia (Figure 6.9), the centers of tractor production are undoubtedly India and China. In addition, there are also production sites in South Korea, Japan, and Pakistan.

For the year 2017, the Mechanical Engineering Industry Association in Germany (VDMA) estimated a total worldwide sales volume of around 2,154,000 new tractors as shown in Figure 6.10 [1]. Then, 9% of tractors were sold in Europe and 11% in North

FIGURE 6.6 Production sites for tractors in Europe (selection).

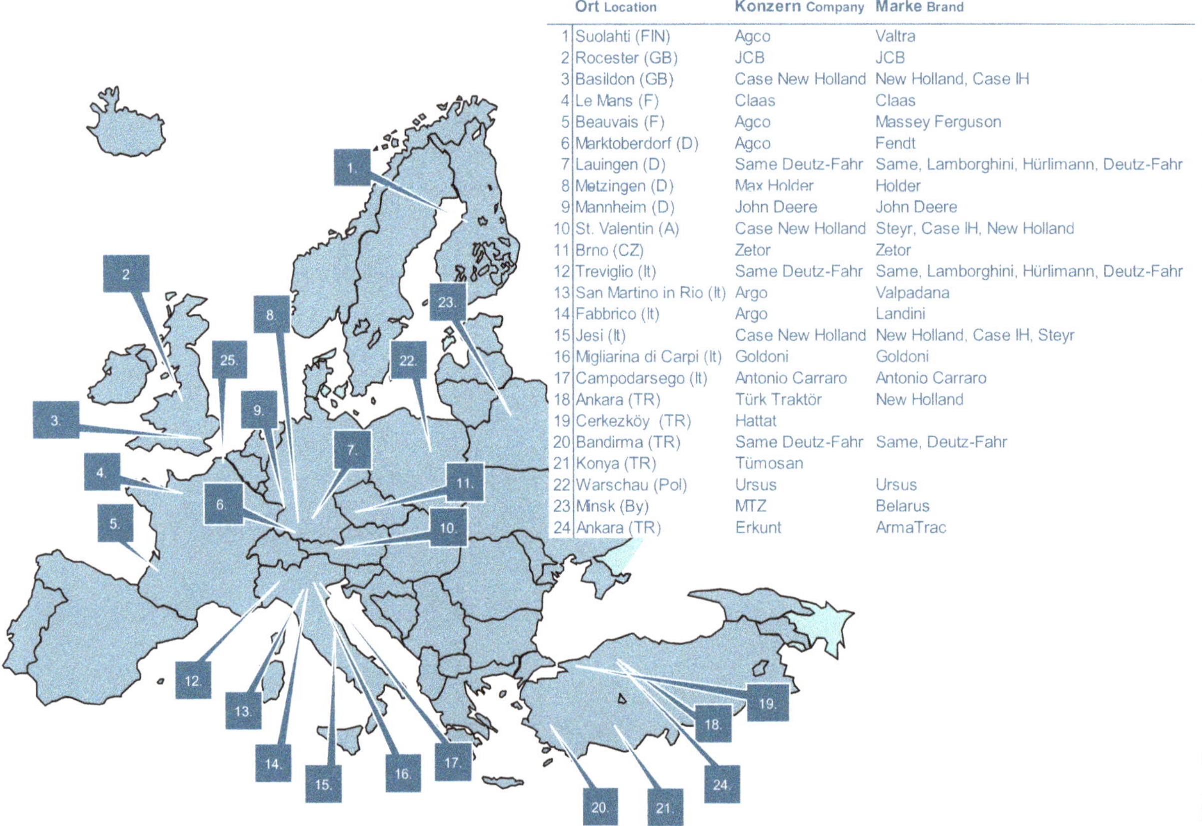

	Ort Location	Konzern Company	Marke Brand
1	Suolahti (FIN)	Agco	Valtra
2	Rocester (GB)	JCB	JCB
3	Basildon (GB)	Case New Holland	New Holland, Case IH
4	Le Mans (F)	Claas	Claas
5	Beauvais (F)	Agco	Massey Ferguson
6	Marktoberdorf (D)	Agco	Fendt
7	Lauingen (D)	Same Deutz-Fahr	Same, Lamborghini, Hürlimann, Deutz-Fahr
8	Metzingen (D)	Max Holder	Holder
9	Mannheim (D)	John Deere	John Deere
10	St. Valentin (A)	Case New Holland	Steyr, Case IH, New Holland
11	Brno (CZ)	Zetor	Zetor
12	Treviglio (It)	Same Deutz-Fahr	Same, Lamborghini, Hürlimann, Deutz-Fahr
13	San Martino in Rio (It)	Argo	Valpadana
14	Fabbrico (It)	Argo	Landini
15	Jesi (It)	Case New Holland	New Holland, Case IH, Steyr
16	Migliarina di Carpi (It)	Goldoni	Goldoni
17	Campodarsego (It)	Antonio Carraro	Antonio Carraro
18	Ankara (TR)	Türk Traktör	New Holland
19	Cerkezköy (TR)	Hattat	
20	Bandirma (TR)	Same Deutz-Fahr	Same, Deutz-Fahr
21	Konya (TR)	Tümosan	
22	Warschau (Pol)	Ursus	Ursus
23	Minsk (By)	MTZ	Belarus
24	Ankara (TR)	Erkunt	ArmaTrac

FIGURE 6.7 Production sites for tractors in North-Central America (selection).

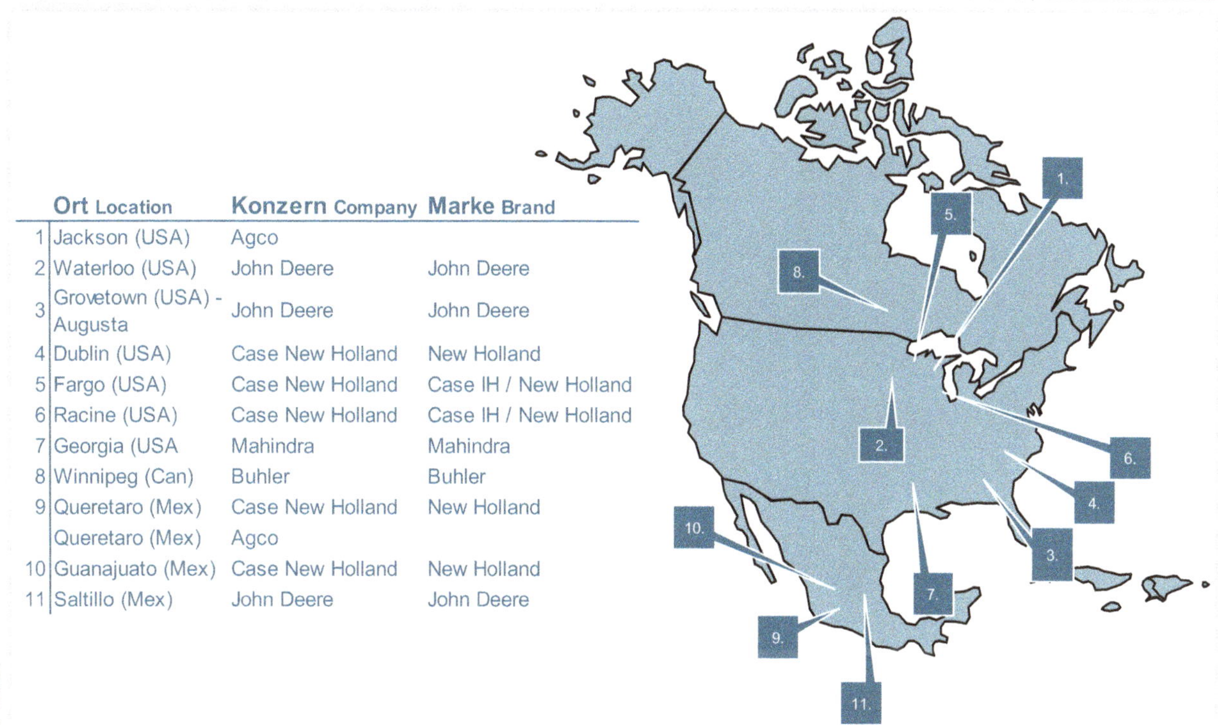

	Ort Location	Konzern Company	Marke Brand
1	Jackson (USA)	Agco	
2	Waterloo (USA)	John Deere	John Deere
3	Grovetown (USA) - Augusta	John Deere	John Deere
4	Dublin (USA)	Case New Holland	New Holland
5	Fargo (USA)	Case New Holland	Case IH / New Holland
6	Racine (USA)	Case New Holland	Case IH / New Holland
7	Georgia (USA	Mahindra	Mahindra
8	Winnipeg (Can)	Buhler	Buhler
9	Queretaro (Mex)	Case New Holland	New Holland
	Queretaro (Mex)	Agco	
10	Guanajuato (Mex)	Case New Holland	New Holland
11	Saltillo (Mex)	John Deere	John Deere

FIGURE 6.8 Production sites for tractors in South America (selection).

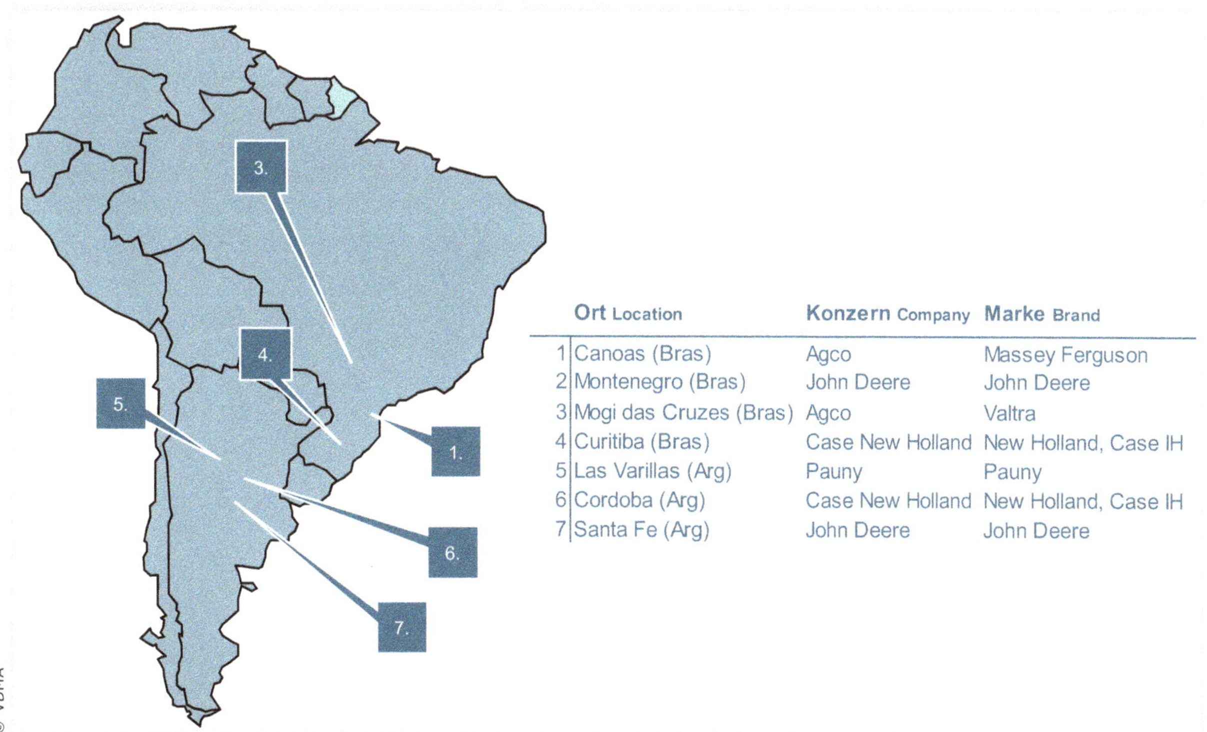

	Ort Location	Konzern Company	Marke Brand
1	Canoas (Bras)	Agco	Massey Ferguson
2	Montenegro (Bras)	John Deere	John Deere
3	Mogi das Cruzes (Bras)	Agco	Valtra
4	Curitiba (Bras)	Case New Holland	New Holland, Case IH
5	Las Varillas (Arg)	Pauny	Pauny
6	Cordoba (Arg)	Case New Holland	New Holland, Case IH
7	Santa Fe (Arg)	John Deere	John Deere

FIGURE 6.9 Production sites for tractors in Asia (selection).

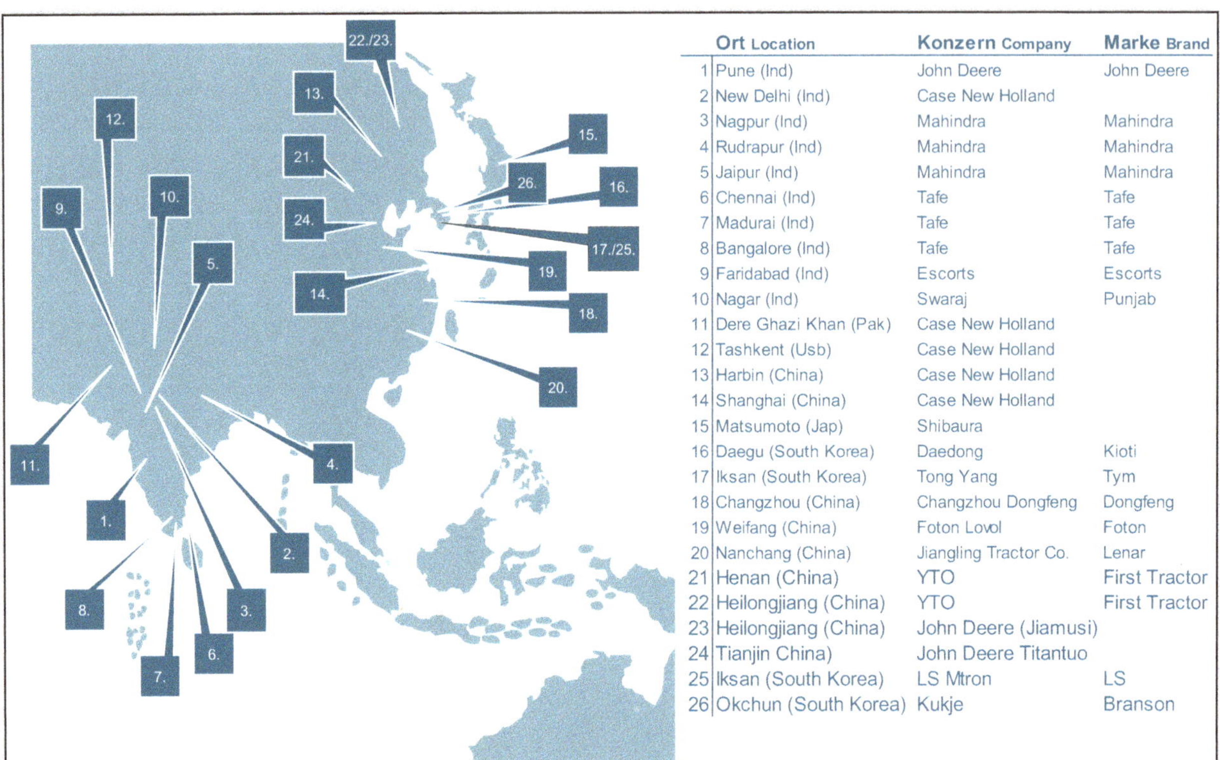

	Ort Location	Konzern Company	Marke Brand
1	Pune (Ind)	John Deere	John Deere
2	New Delhi (Ind)	Case New Holland	
3	Nagpur (Ind)	Mahindra	Mahindra
4	Rudrapur (Ind)	Mahindra	Mahindra
5	Jaipur (Ind)	Mahindra	Mahindra
6	Chennai (Ind)	Tafe	Tafe
7	Madurai (Ind)	Tafe	Tafe
8	Bangalore (Ind)	Tafe	Tafe
9	Faridabad (Ind)	Escorts	Escorts
10	Nagar (Ind)	Swaraj	Punjab
11	Dere Ghazi Khan (Pak)	Case New Holland	
12	Tashkent (Usb)	Case New Holland	
13	Harbin (China)	Case New Holland	
14	Shanghai (China)	Case New Holland	
15	Matsumoto (Jap)	Shibaura	
16	Daegu (South Korea)	Daedong	Kioti
17	Iksan (South Korea)	Tong Yang	Tym
18	Changzhou (China)	Changzhou Dongfeng	Dongfeng
19	Weifang (China)	Foton Lovol	Foton
20	Nanchang (China)	Jiangling Tractor Co.	Lenar
21	Henan (China)	YTO	First Tractor
22	Heilongjiang (China)	YTO	First Tractor
23	Heilongjiang (China)	John Deere (Jiamusi)	
24	Tianjin China)	John Deere Titantuo	
25	Iksan (South Korea)	LS Mtron	LS
26	Okchun (South Korea)	Kukje	Branson

FIGURE 6.10 Sales of new tractors worldwide (2013–2017) [1].

Sales of new tractors in the Agrievolution countries (in units)

	2013	2014	2015	2016	2017	%
Canada	27.542	28.144	24.215	22.164	25.570	15
United States	201.851	208.274	204.962	211.194	220.006	4
Brazil	65.089	55.635	37.385	35.963	36.976	3
Japan (> 30 hp)	24.721	20.944	22.203	18.393	18.173	-1
China (> 30 hp)	584.564	574.622	556.575	420.189	487.404	16
Korea (> 25 hp)	12.853	10.548	11.338	10.662	8.933	-16
India	619.159	592.942	483.769	569.066	659.303	16
Russian Federation	40.158	37.558	21.837	17.913	22.042	23
Turkey	52.285	59.458	66.788	70.178	72.352	3
Europe	184.335	185.682	171.701	167.941	189.443	13
of which: France	42.656	33.127	33.828	31.760	35.242	11
of which: Germany	36.248	34.611	32.220	28.248	33.695	19
of which: Italy	19.018	18.178	18.428	18.341	22.705	24
of which: United Kingdom	13.490	13.526	12.112	12.025	13.768	14
of which: Spain			10.628	11.513	12.025	4
World	**2.200.000**	**2.130.000**	**1.936.994**	**1.906.234**	**2.153.555**	13

China: statistic definition changes in 2013 and 2014; Europe includes: Austria, Belgium, Bosnia Herzegovina, Croatia, Czech Republic, Denmark, Estonia, Finland, France, Germany, Greece, Hungary, Iceland, Ireland, Italy, Latvia, Lithuania, Luxembourg, Moldova, Netherlands, Norway, Poland, Portugal, Serbia & Montenegro, Slovakia, Slovenia, Spain, Sweden, Switzerland and United Kingdom

Source: Agrievolution, VDMA

America (Canada plus USA), whereas the majority of tractors (54% of the entire sales volume – or 1,174,000 tractors) were registered in India, China, Japan, or Korea [1].

From a production standpoint, the largest producer of agricultural tractors in the world is India where according to the VDMA 768,000 tractors were produced in 2017 [55]. For 2019, John Deere estimates the annual world production volume of tractors to be around 1,800,000 units. But the size and average power of tractors vary between different regions and markets for which they are produced.

Selected countries provide a good track record of their tractor population. For example, in Germany, all tractors participating in road traffic have to be registered with an ordinary license plate. According to the German Federal Motor Transport Authority (Kraftfahrtbundesamt (KBA)), on January 1, 2019, there were 1,482,722 tractors for agriculture or forestry registered with a license plate with an average age of 28.6 years [38]. The overall population of tractors in use all over the world can only be estimated. The Food and Agriculture Organization of the United Nations (FAO) was collecting such data up until 2009 and since then, the database is no longer maintained. According to the VDMA [54], there were around 31,000,000 tractors in use in 2008.

6.1.1.5 **Tractor Functionalities:** The four basic functionalities of an agricultural tractor can be derived from its definition according to ANSI/ASAE S390.6 (ISO 12934:2013) [3] (Copyright ASABE. All excerpts from the text of this standard have been used with permission of the ASABE.)

- "Pulling"

  Tractors are pulling trailers or implements used for agricultural work from, to, and in the field. This includes, for example, agricultural trailers on the road and towed implements such as balers or plows operating in the field.

- "Pushing"

  Tractors are pushing implements used for agricultural work in the field. For example, front-mounted implements in the form of front three-point hitch-mounted mowers or mulchers are pushed through agricultural grassland.

- "Carrying"

  Tractors are carrying attached implements or loaders from, to, and in the field but also applied weights used for proper ballasting of the tractors.

- "Operating"

  Tractors are operating attached implements through mechanical, hydraulic, or electrical energy at specified interfaces (see Section 6.1.3.3).

6.1.2 Concept and Configurations

6.1.2.1 **Type Approval and Road Regulations:** A wide variety of regulations have been established for tractor type approval and for the use of tractors on public roads, some of which are decided upon at the international or national level, others for which the local authorities are responsible.

The European Union is one of the highest regulated tractor markets. Up until December 31, 2017, conformity has been assessed with respect to 2003/37/EC type approval of agricultural or forestry tractors which includes 24 different subjects [10]. Since January 1, 2018, this directive has been succeeded by the so-called Tractor Mother Regulation: (EU) 167/2013 which is valid for all tractors that first apply for a license plate after that date. The new regulation contains 62 requirements for the purpose of vehicle EU type approval and is valid for tractors of categories T and C as well as for trailers of category R and interchangeable towed equipment of category S [42].

For tractors sold after January 1, 2015, the conformity assessment process is covered by 2015/504 Implementing Regulation to the Administrative Requirements (RAR) [15].

There are also different laws established in different countries on operating a tractor with or without attached or trailed implements on public roads.

In Germany, for example, this is defined through the road traffic licensing regulation StVZO (Strassenverkehrszulassungsordnung) [51]. A regular general inspection is mandatory every year for tractors with transport speeds above 40 km/h (25 mph) and every 2 years for tractors below 40 km/h as regulated in §29 StVZO [51]. The §32 StVZO [51] furthermore specifies the range of permissible dimensions of tractors and attached trailers and implements operating on public roads in terms of maximum permissible width (§32, 1 StVZO), height Height (§32, 2 StVZO), and length Length (§32, 3 StVZO) of vehicles with and without attached trailers and implements. The regulation also includes measures for the maximum permissible axle loads in §34, 4 StVZO and weight (§34, 5 StVZO) [51] and other measures for agricultural tractors and combinations with trailers or implements, operated on public roads. There are exceptions though (§70 StVZO [51] and §29 StVO [52]), if the legal requirements mentioned above cannot be met. An overview of permissible regulations for agricultural tractors on public roads in Germany is provided in AID 2016 [2], for example.

6.1.2.2 **Tractor Categories:** ANSI/ASAE S390 and ISO 12934 differentiate different main tractor types (and concepts) as follows: "Standard agricultural tractor (3.1.1)," "Two-wheel drive tractor (3.1.2)," "All-wheel drive tractor (3.1.3)," "Standard track-laying tractor (3.1.4)," "Articulated tractor (3.1.5)," "Small tractor (3.1.6)," "Compact utility tractor (3.1.7)," "Extra wide tractor (3.1.8)," "High speed tractor (3.1.9)," and "Specialized tractor (3.1.10)" with further distinctions in selected categories [3] (Copyright ASABE. All excerpts from the text of this standard have been used with permission of the ASABE.).

Tractor categories are defined in the Tractor Mother Regulation (EU) No. 167/2013 on the approval and market surveillance of agricultural and forestry vehicles [42] as "Category T1" through "Category T4" for all wheeled tractors and "Category C" for all track-laying tractors. The main parameters which distinguish the subcategories T1 through T4 are the "minimum track width," the "unladen mass," and the "ground clearance" of the tractor, respectively [42].

6.1.2.3 **Configuration of Standard Agricultural Tractor:** The following chapter describes the concept and configuration of a standard agricultural tractor.

The chassis of standard agricultural tractors is configured in a block design, in a frame design, or in any configuration that uses partially elements from block or frame design to constitute the vehicle. This is also often referred to as semi-frame or partial frame (see Section 2.1).

There are tractors built in frame design that use a structural frame which connects the rear axle and front support for the front axle. The main parts of the powertrain including the engine block or engine oil pan and the significant parts of the drivetrain are nonstructural. This enables the use of standard off-the shelf powertrain options and configurations constrained elastically within the same vehicle concept without the necessity for generating individual castings for each powertrain option. In addition, front and rear axle options can be exchanged to build new configurations. Bolt patterns for the attachment of front hitches or front loaders are available on the frame itself without any modification to engine or drivetrain components. External loads induced by the wheels or by attached implements are supported by the frame which responds elastically under such loads to prevent localized stress concentrations and structural failures.

Frame design has successfully been adopted by tractor manufacturers producing different, customized configurations within the same vehicle design concept in small and large quantities. For example, John Deere adopted this design concept on the 6,000 tractor series in 1992 and since then is using it on 6 Series tractors built in numerous different modular configurations for customers all over the world (see **Figure 6.11**, for example).

In a block design configuration, the rear axle assembly is connected rigidly to the transmission (and clutch housing, if available) and to the structural engine block to which is attached the front axle (and front attachment). Often, the oil pan of the engine is a structural part to support the front axle. Hole patterns for attachments such as front loader brackets are directly designed into the castings of the powertrain. All loads externally applied have to be supported by the structural castings of the powertrain. As a result, tractors using a block design respond much stiffer to such loads. If the tractor is equipped with a front

FIGURE 6.11 Standard agricultural tractor in frame configuration connecting rear axle assembly with front support and nonstructural engine and drivetrain: John Deere 6215R Series.

hitch, the attachment points for the front hitch can be integrated directly into the structure. In addition, the front end can be optimized for improved turning radii of the front tires.

Block design has successfully been used by tractor manufacturers building tractors in large quantities (or for a long time) in the same or similar powertrain housing configuration. It is also being used by tractor manufacturers building smaller tractors or tractors of lower specification level in markets outside of North America and outside of the European Union, for example. Whereas the use of structural castings provides many degrees of freedom in the design of the tractor and particularly in its front end, changes to the castings or adoption of new or other modules necessitate additional engineering efforts and associated cost.

There is a significant population of tractor manufacturers using semi-frames or partial frames. In many cases, the engine oil pan is a structural part connecting with the frame elements to a structural rear axle assembly (see Figure 6.12). An example for a tractor configuration with a frame element that comprises an integrated oil pan is provided in Figure 6.13.

In a standard agricultural tractor, the rear axle assembly often comprises a centered differential housing to which are attached two rear axle housings on either side with final drives and axle shafts either in the form of flange axles or rack-and-pinion axles for variable rear tread settings. A part of the rear axle assembly is also the three-point hitch, the PTO, drawbar and all rear coupling devices which are either built in or are solidly connected at flanges to the differential housing. This assembly has defined interfaces for structural drivetrain housings (in block design) or for the connecting elements of a structural frame (in frame or semi-frame designs).

FIGURE 6.12 Example for rear axle assembly with attached CVT transmission (ZF Terramatic). Source: ZF.

FIGURE 6.13 Standard agricultural tractor in a configuration with frame element that comprises an integrated oil pan to which is attached the front chassis and front hitch (CLAAS Axion 870 Series). Source: CLAAS.

There are nonpowered and powered front axles with a mechanical front wheel drive commonly driven through a central shaft from the rear axle assembly. The standard front axle is a two- or three-piece structural beam, pivoting around the front support structure of the tractor through, for example, a centered shaft connected to it. Typical pivoting angles of front axles on agricultural tractors are between 9° and 15° in either direction after which the front axle makes contact with a mechanical oscillation stop.

Front axles are either nonsuspended or suspended with a variety of different suspension kinematics used by tractor manufacturers. Examples are independent link suspension, triple-link suspension, and many others. When the front wheel drive is engaged, the front axle transmission ratio is usually chosen such that the circumferential speed of the front wheels is between 1.5% and 4% larger than those of the rear wheels causing the front axle to "pull" rather than being "pushed." This front wheel lead is important for effectively using the all-wheel drive, for better steerability of the tractor and for avoidance of premature wear of both tires and transmission, in particular. As a result of this, the front wheel drive is either manually or automatically turned off when the tractor operates in transport speed.

Nonpowered front axles enable realization of smaller turning radii due to the lack of drivetrain components to the front and narrower wheels (and front fenders) being commonly used in comparison to powered axles. The use of turnable front fenders improves possible turning radii which are widely used on front axles of agricultural tractors [39].

The ability to generate traction forces is significantly higher on a four-wheel drive tractor in comparison to a two-wheel drive tractor of the same weight; it is even higher than the contribution from the additional weight on the front axle being used for traction. This is primarily a result of the so-called "multi-pass effect" [21] where the wide front wheels of a four-wheel drive tractor precompress the soft soil and increase the shear strength of the soil for the rear tires resulting in a higher traction coefficient for the rear tires running in the same track (see Section 2.3).

Certain tractor applications, particularly those operating in row crops, require particular wheel tread settings. For example, for tractors being used for cultivating in standing corn, planted at 760 mm (30 in.) row width, typical tread settings are either 1,524 mm (60 in.) or 2,286 mm (90 in.) for single wheels or 1,524 mm (60 in.) and 3,048 mm (120 in.) for dual-wheel configuration, respectively, to avoid contact of tires and plants during the operation. The wheel tread is defined as the distance between the mid-plane of the wheels on one axle. It can also be defined as

$$t = f + 2 \cdot o \tag{6.1}$$

There are various technical solutions available for adjusting the wheel tread to the desired measure: rims with different fixed offsets can be used or rims with adjustable discs that can be turned or repositioned to adjust the rim offset and subsequently the tread setting (Figure 6.14). There are also special types of front axles, such as two-wheel drive front axles (2WD), and rear axles which allow adjustment of the tread setting (Figure 6.15). In addition, there are spacers that can be mounted in between the flange and the wheel to adjust the tread setting.

Since the beginning of the development of agricultural tractors, engineering efforts have been carried out to reduce the amount of whole-body vibrations and dynamically changing tire loads induced into the vehicles.

These efforts have led to a significant increase in operator ride comfort and health making possible increased operating speeds. At the same time, the minimization of dynamic tire load changes has resulted in increasing driving stability in terms of improved steerability and braking. Reduction in dynamic tire load changes also improved the durability of the vehicle structure and of any hard surface track the vehicle is driving on. Finally, the reduction in dynamic loads improved the productivity of the tractor operation, for example, through minimization of the effect of power hopping [4].

Suspension systems being used on agricultural tractors include seat suspension, cab suspension, pitching absorber, such as on the three-point hitch, front axle suspension, and front and rear axle suspension and these systems may affect operator safety and operator comfort differently as shown in Table 6.1.

Seat, cab, pitch absorber, and front axle suspension are the most common suspension systems used by standard agricultural tractors as shown in Figure 6.16.

FIGURE 6.14 Different adjustment possibilities A-H of disc and rim to generate different wheel offsets and tread settings.

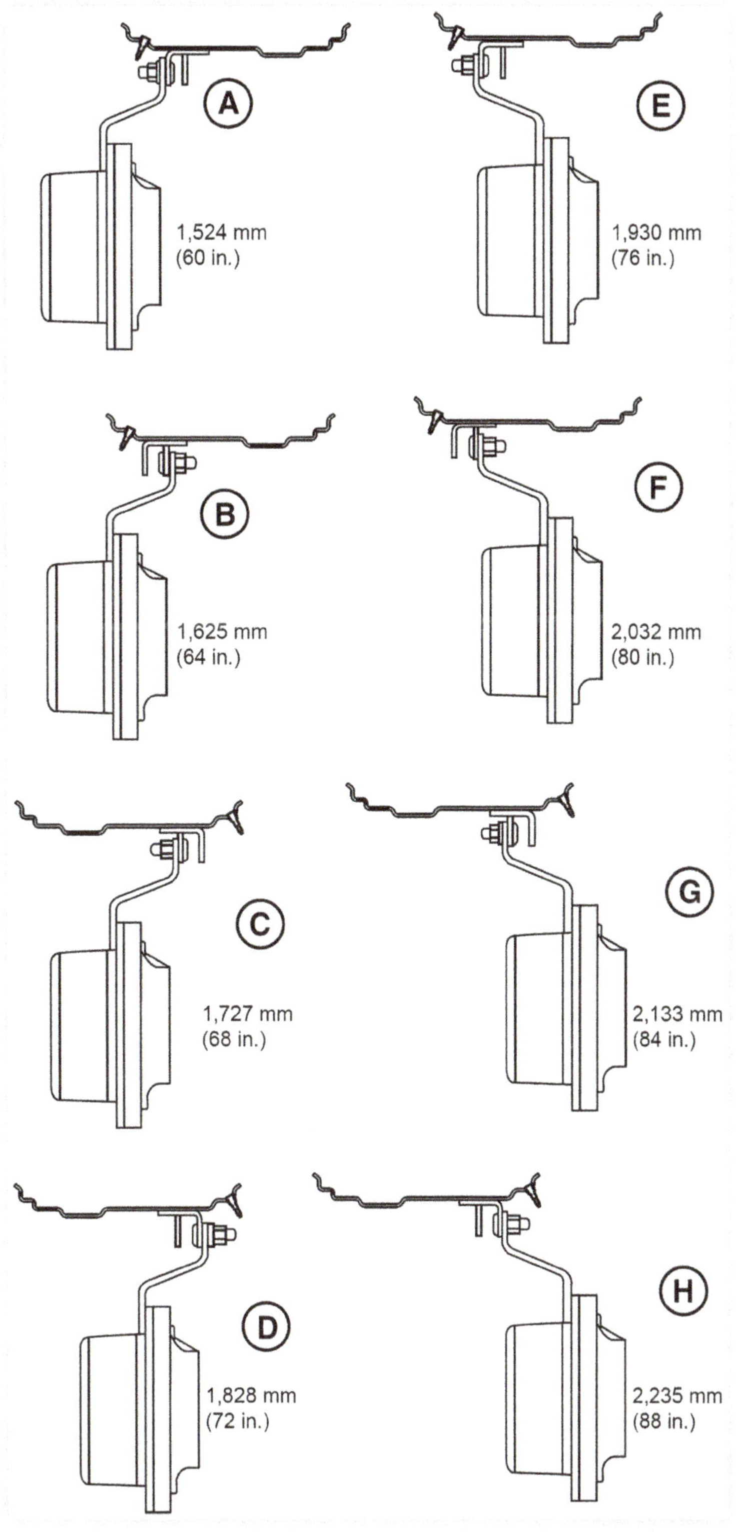

FIGURE 6.15 Left: tread-adjustable 2WD front axle; right: rack-and-pinion rear axle to set up different tread settings.

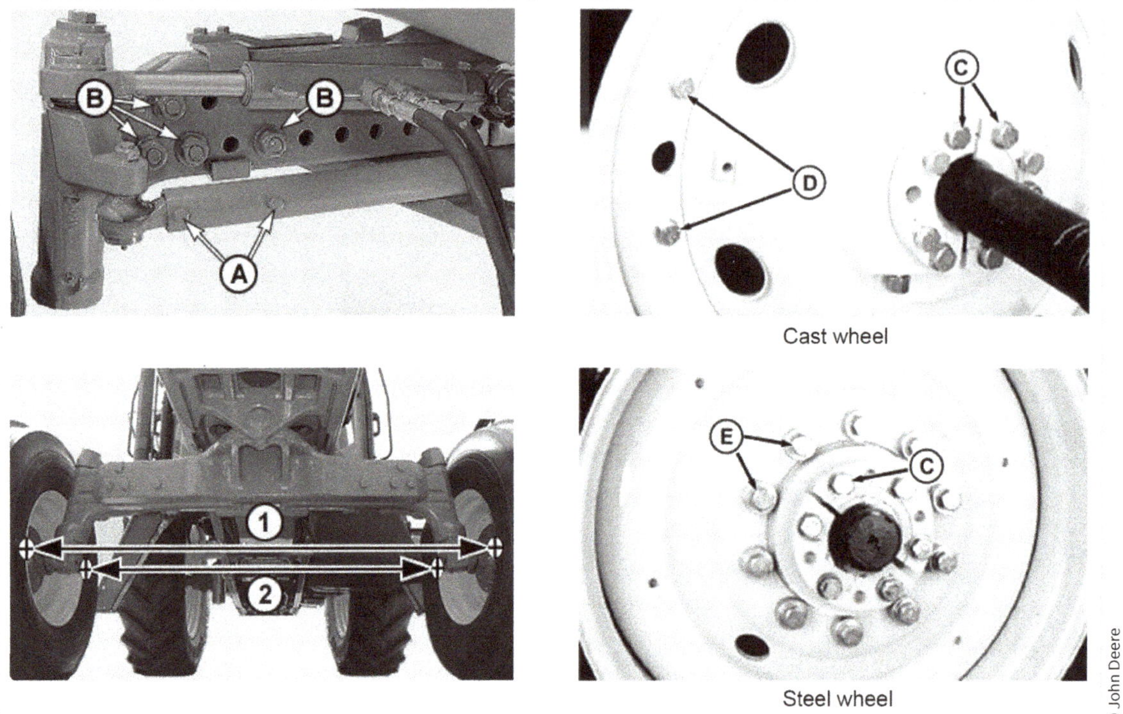

TABLE 6.1 Comparison of different suspension systems on agricultural tractors in terms of effectiveness regarding operator safety, operator comfort, and development effort. Translated from German into English with amendments from Prof. Dr.-Ing. Peter Pickel [56].

	Positive influence on		
Measure	**Operator safety**	**Operator comfort**	**Development effort**
Seat suspension	Small (indirect)	Very large	Small-ordinary
Cab suspension	Small-large	Small-large	Large
Pitching absorber	Large	Small-large	Small
Front axle suspension	Very large	Small	Large
Front and rear axle suspension	Very large	Large	Very large
Front axle and cab suspension	Very large	Large	Large

© Von Holst, Ch

6.1.3 Interfaces for Applications

6.1.3.1 **Standardization:** There are a number of tractor-implement interfaces specified and standardized. Such interfaces include mechanical coupling to agricultural trailers or implements and standards for PTO, hydraulic, electric, and braking of attachments. An overview of applicable standards is provided in Figure 6.17 and discussed below.

FIGURE 6.16 Example for cab suspension, seat suspension, and front axle suspension (CLAAS Axion 850 Series).

FIGURE 6.17 Overview of international standards for tractor implement/trailer attachments and drives.

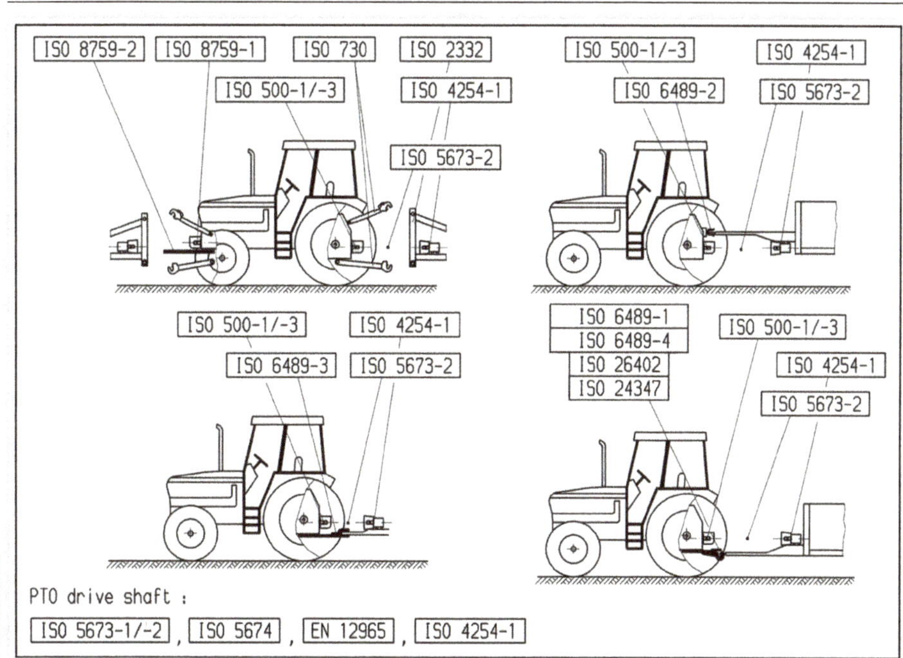

6.1.3.2 Mechanical Interfaces

6.1.3.2.1 Rear Coupling Devices. There are various forms of mechanical coupling devices between tractors and trailers and between tractors and towed implements, respectively.

In Germany, for example, rear coupling devices for trailers have to be type approved according to §22a and §43, 1 of the road traffic licensing regulation (StVZO [51]). The type approval procedure determines for each rear coupling device both vertical load S at the coupling point and the so-called D-value [13]. The D-value is a means that relates the trailer mass A to the maximum permissible mass of the tractor G to which the trailer is attached. This relation is expressed in Equation (6.2).

$$D = \frac{G \cdot A \cdot g}{G + A} \tag{6.2}$$

The rear coupling devices can be classed as either upper attachment devices or lower attachment devices where the upper attachment devices are usually pin-type coupling devices with corresponding towing eyes. Table 6.2 shows typical examples in the form of an automatic clevis-type coupling device (upper attachment), a ball-type coupling

TABLE 6.2 Examples of rear coupling devices: (a) automatic clevis-type rear coupling device, (b) ball-type coupling device, and (c) drawbar of category 2.

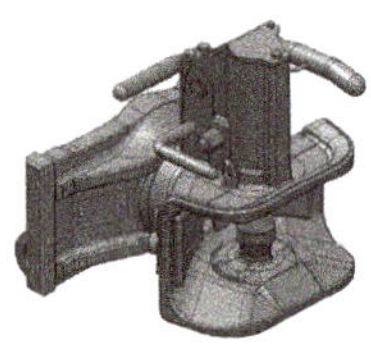	Automatic clevis-type rear coupling device (38 mm bolt diameter)
	Ball-type coupling device (80 mm ball diameter)
	Drawbar of category 2 (31.5 mm coupler bolt diameter)

device, and a drawbar (lower attachment). There are special requirements for certain countries including Italy (CUNA) and Switzerland [11, 12].

To each rear coupling device on the tractor side, there is one (or multiple) corresponding rear coupling device(s) on the implement side which can be used without compromising the function of the attachment. A chart showing the compatibility of these different tractor and implement coupling devices is provided in Figure 6.18.

In addition to the upper attachment devices, there are various forms of lower attachment devices commonly applied in different countries.

Pickup hitches with hook (either telescopic or in the form of drop-down hitches) as defined in ISO 6489-1 [23] are typically used by tractors and trailers or attached implements in the United Kingdom and in selected Scandinavian countries.

Coupling devices with a cylindrical pin, namely piton fix and specified in ISO 6489-4 [25], are typically used in France.

Ball hitches with ball-shaped coupling devices of 80-mm diameter are commonly used in Germany and many other European countries. This rear coupling device is the only one by which vertical loads of up to 4,000 kg (at speeds lower than 40 km/h (25 mph)) can be applied to an agricultural tractor.

Drawbars are commonly being used by the North American tractor population but also by a large population of tractors throughout the world for towing implements. ISO 6489-3 [24] describes six categories (sizes) of drawbars for tractors of different sizes characterized by PTO power at rated engine speed as listed in Table 6.3.

FIGURE 6.18 Overview of standards relevant to rear coupling devices [13].

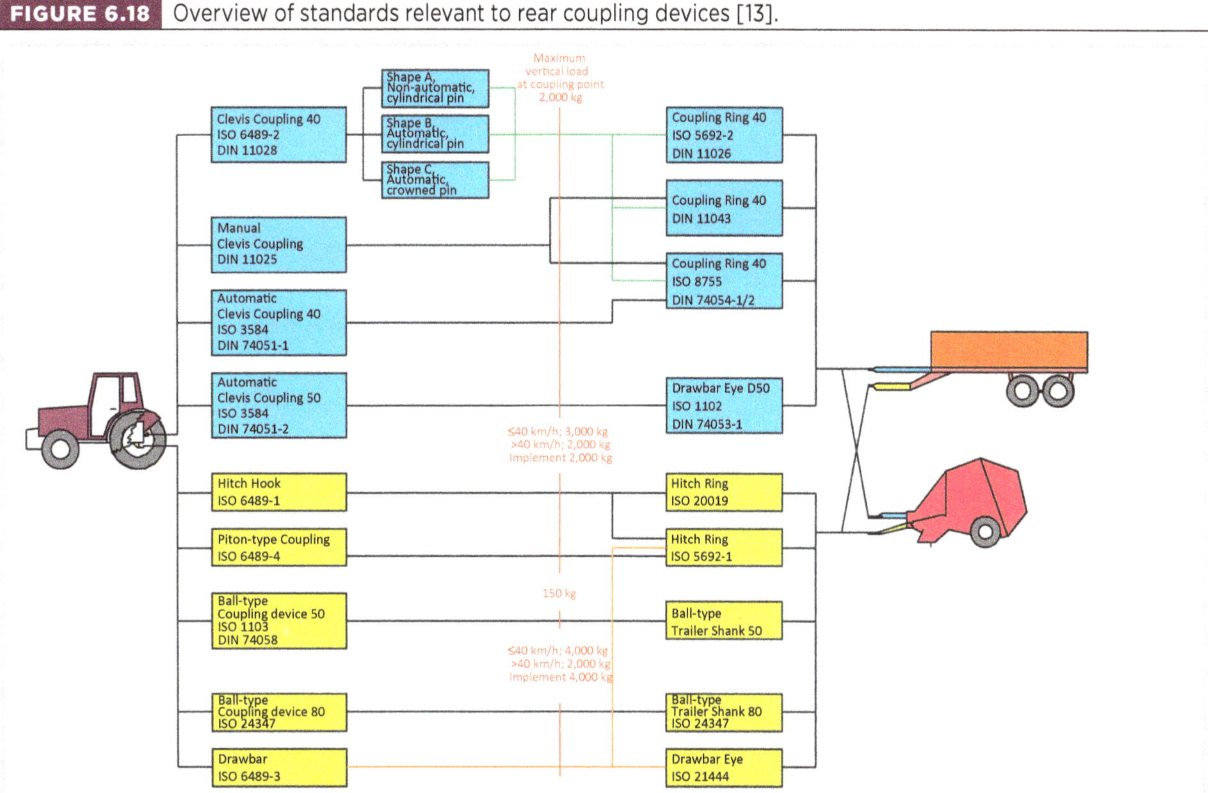

TABLE 6.3 Drawbar categories according to ISO 6489-3. ©ISO. This material is reproduced with permission of the American National Standards Institute (ANSI) on behalf of the International Organization for Standardization. All rights reserved [24].

Drawbar category	Values in kilowatts PTO power[a] at rated engine speed
0	≤28
1	≤48
2	≤115
3	≤185
4	≤300
5	≤500

[a] Determined in accordance with ISO 789-1 or OECD code 1 or 2. If PTO power is not available, use 86% of engine power as determined using ISO 14396.

ISO 6489-3 [24] furthermore specifies for each drawbar category the location of the drawbar with respect to the ground and with respect to the end of the PTO stubshaft (Figure 6.19) to maintain compatibility with PTO-driven implements.

Note: In addition to the rear coupling devices elaborated above in this chapter, there are various other types used in different countries. Whereas the type of coupling devices on tractors on the one side matters for vehicle type approval, it is primarily a result of

FIGURE 6.19 Tractor drawbar and location above ground and with respect to the tractor PTO stubshaft of the tractor according to ISO 6489-3. ©ISO. This material is reproduced with permission of the American National Standards Institute (ANSI) on behalf of the International Organization for Standardization. All rights reserved [24].

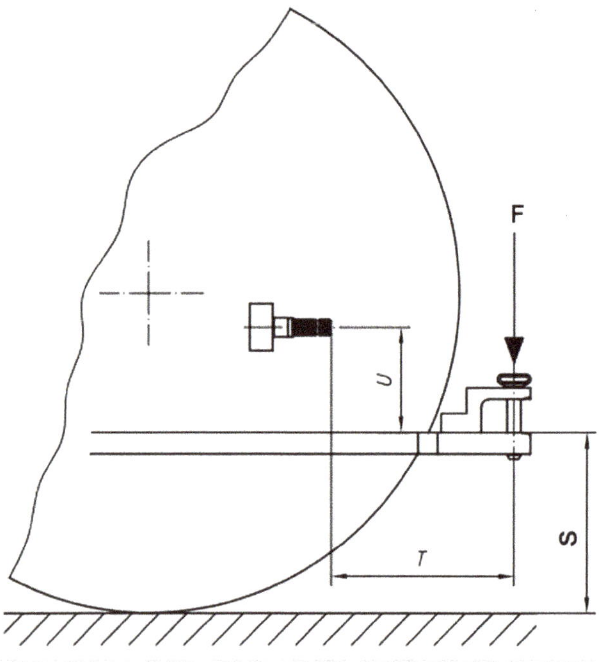

the population of implements in the particular country which are to be attached to that tractor.

6.1.3.2.2 Mechanical Interfaces - Rear and Front Three-Point Hitches. Standard agricultural tractors all over the world are equipped with rear three-point hitches according to ISO 730 [27] to connect or disconnect ballast or implements. In addition to the rear hitch, tractors (particularly in Europe) are also often equipped with front three-point hitches. Whereas in most countries implements are directly connected to the implement attachment points, in North America, U frame couplers according are frequently used in between the three-point hitch and the implements [33]. To center the draft links of the rear hitch during road transport of implements, different types of exterior stabilizers (or chains) or so-called sway blocks with wear plates are being used. The latter is more common to the North American tractor population.

To ensure compatibility to implements, the geometry of the front and rear three-point hitches is standardized in ISO 730 [27]. A three-point hitch comprises two lower links (or draft links) and an upper link (or center link). There are attachment points at each end of the links with defined diameters and dimensions. ISO 730 defines different hitch categories (Table 6.4) that include the dimensions of the links and the attachment points but also of the lower hitch point span and the implement mast height. In addition, ISO 730 provides for each category information about the kinematics in the form of the required power range (or lift range) and the location of the three-point hitch with respect to the end of the PTO-stub shaft (L-dimension) and subsequently with respect to the rear axle of the tractor (Figure 6.20).

The lift performance of the three-point hitches is assessed using ISO 789-2 [35] which provides the setup and procedure for determination of the maximum and throughout hitch lift capacity for the coupling plane and for a plane 550 mm behind the coupling plane.

The three-point hitch transfers forces and moments from attached implements to the chassis and subsequently to the drive wheels of the tractor. As a result, during pulling, the slip and resistive forces may change. There are also operations where the weight of an attached implement, for example, a mounted fertilizer spreader or sprayer, may change during operation.

TABLE 6.4 Rear three-point hitch categories 1N/1, 2N/2, 3N/3, and 4N/4 according to ISO 730. ©ISO. This material is reproduced with permission of the American National Standards Institute (ANSI) on behalf of the International Organization for Standardization. All rights reserved. Source: [27].

Category	PTO power at rated rotational frequency of engine[a] kW
1N	Up to 35
1	Up to 48
2N/2	30-92
3N/3	60-185
4N/4	110-350

[a] Determined in accordance with ISO 789-1.

FIGURE 6.20 Rear three-point hitch according to ISO 730 (2014) and reference to U frame coupler according to ISO 11001-1. ©ISO. This material is reproduced with permission of the American National Standards Institute (ANSI) on behalf of the International Organization for Standardization. All rights reserved. Source: [32]

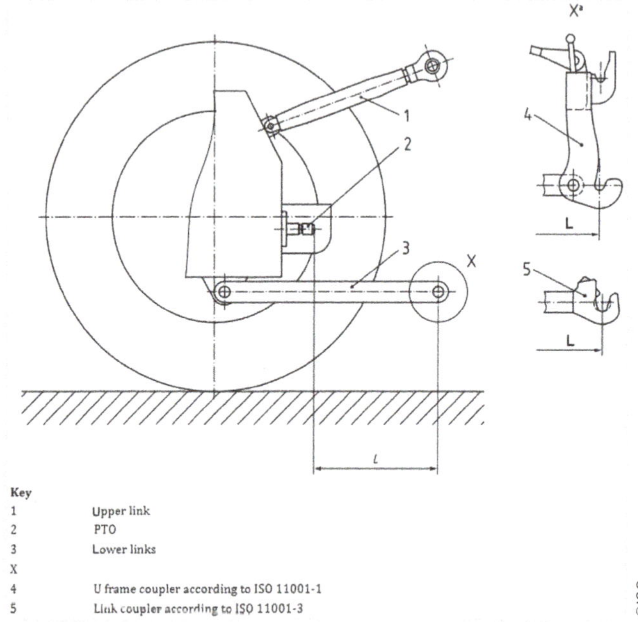

FIGURE 6.21 Electrohydraulic hitch control setup. Source: [7].

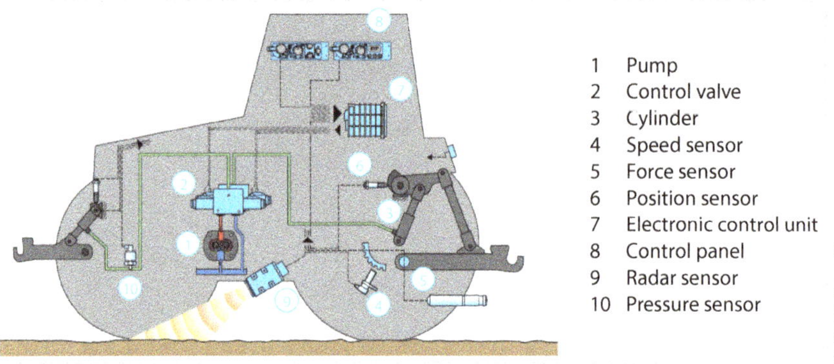

For these reasons, the height of the three-point hitch can be controlled. This may result in preventing the tractor from stalling, in optimizing the tractive power, and/or in an optimized work result by, for example, maintaining a constant working depth (or height) of the attached implement.

Modern three-point hitches are electrohydraulically controlled as shown in Figure 6.21 for a rear and front hitch control.

The hydraulic pump (1) supplies an oil flow to the hitch control valve (2) that controls the hitch cylinders (3). This cylinder acts on the lower links to lift, hold, or lower attachments. In an electrohydraulic hitch control, the command value, set by the driver in the

FIGURE 6.22 Electrohydraulic hitch control valve for load sensing circuits. Source: [6].

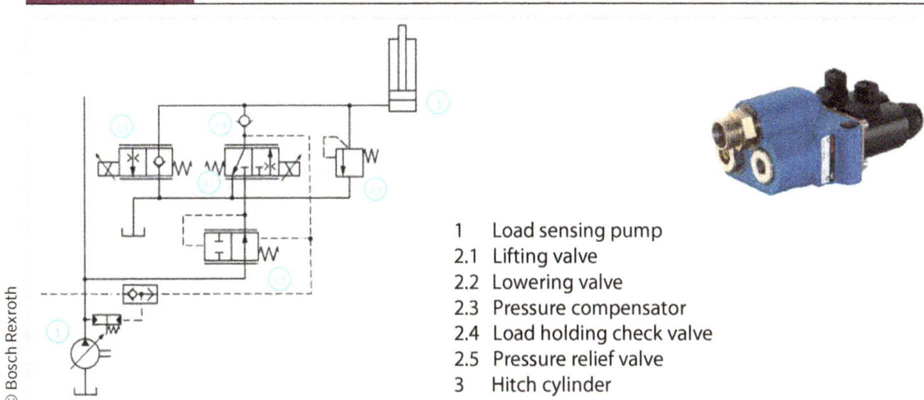

control panel (8), is compared with the actual values, measured by the force (5) and position sensor (6), and the deviation of the desired value is compensated in a closed-loop control with the hitch control hydraulics (1, 2, and 3).

Figure 6.22 shows a typical electrohydraulic hitch control valve for a load-sensing circuit. When activated, the lifting valve (2.1) opens the flow from the pump (1) to the hitch cylinder (3), while the pressure compensator (2.3) assists by keeping the pressure drop over the lifting valve (2.1) constant. The lowering valve (2.2) opens the return flow. The load holding check valve (2.4) prevents the hitch from moving when not actuated.

When the hitch is in position control, the displacement of the cylinder (3) is controlled in a closed loop with the help of a position sensor (6), connected by a radial cam on the hitch (Figure 6.21). This control mode is ideal for plowing even in terrain or soft soil to keep the implement at a constant depth. As soon as the ground becomes inhomogeneous or uneven, the position control may result in large variations in the draft force on the hitch. In these conditions using draft control (with the force sensor (5)), the ploughing operation is improved, and the hitch maintains a constant force on the implement. Very often, an adjustable ratio between position and draft control modes can be set at the control panel (8). With this mixed control, the changes in the working depth due to different ground resistances, which occur with the pure traction force control, are limited.

To reduce the front axle load changes when transporting heavy attachments and thus increase the steerability, the force (5) and position sensor (6) are used for measuring the dynamic hitch state while driving. With the help of the hitch control valve (2), an active oscillation damping of the hitch is performed.

By comparing the actual driving speed, measured with a radar sensor (9), and the wheel speed with a speed sensor (4), wheel slip can be detected and reduced by the hitch control.

An ideal compression of the farmland with packer rollers can be reached by means of closed-loop pressure control on the front hitch using an addition pressure sensor (10) in combination with the front hitch control valve.

6.1.3.3 **Interfaces for Energy Transfer:** To power or control attached implements or specific functions of these implements, agricultural tractors have interfaces for transferring mechanical, hydraulic, or electric energy.

6.1.3.3.1 Mechanical Energy Transfer (PTO). The power take-off (PTO) is a tractor interface for transferring kinetic energy in the form of torque and rotational velocity from a tractor to an attached implement. The PTO is specified in ISO 500 which comprises three parts 1–3.

ISO 500-1 [29] describes general specifications, safety requirements, dimensions for master shields, and clearance zones. Four different PTO types are specified, each of which has a distinct diameter and number of splines, an associated nominal PTO rated rotational velocity and a recommendation for the transmitted PTO power at rated engine speed (Table 6.5).

Examples for rear PTO stubshafts that are exchangeable are provided in Figure 6.23.

ISO 500-2 [30] specifies for narrow-track tractors the dimensions for master shield and clearance zone (Figure 6.24), and ISO 500-3 [31] defines the main PTO dimensions and spline dimensions.

6.1.3.3.2 Hydraulic Energy Transfer. In agricultural tractors, hydraulic systems are used to support tractor internal functions such as steering or three-point hitch control,

TABLE 6.5 Rear PTO types according to ISO-500-1. ©ISO. This material is reproduced with permission of the American National Standards Institute (ANSI) on behalf of the International Organization for Standardization. All rights reserved. Source: [29].

PTO type	Nominal diameter (mm)	Number and type of splines	Nominal PTO rated rotational frequency (min^{-1})	Recommended PTO power at rated engine speed[a] (kW)
1	35	6 straight splines	540	<65
			1,000[b]	<110
2	35	21 involute splines	1,000	<130
3	45	20 involute splines	1,000	<300
4	57.5	22 involute splines	1,300	<450

[a] Determined in accordance with ISO 789-1 or OECD code 2.

[b] This option is not available in North America.

FIGURE 6.23 Exchangeable rear PTO stubshafts of types 1, 2, and 3.

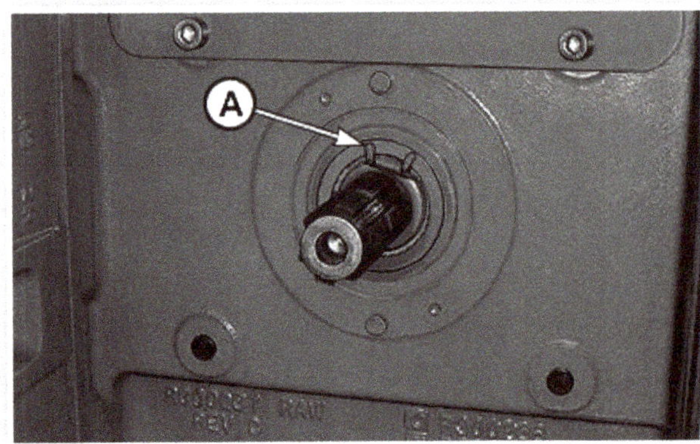

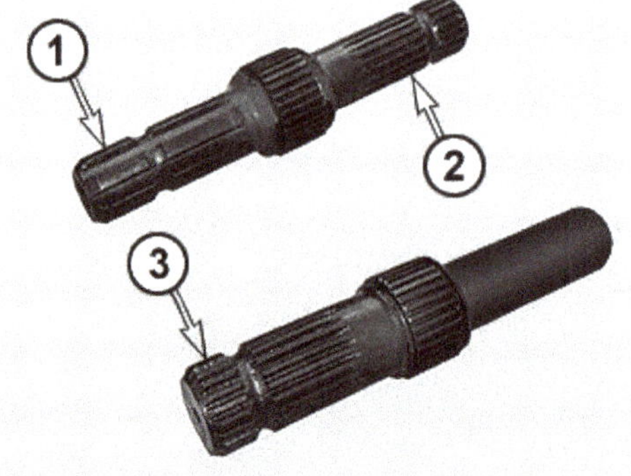

FIGURE 6.24 Rear PTO master shield and clearance zone (side view) around the PTO of a tractor according to ISO-500-1. ©ISO. This material is reproduced with permission of the American National Standards Institute (ANSI) on behalf of the International Organization for Standardization. All rights reserved. Source: [29].

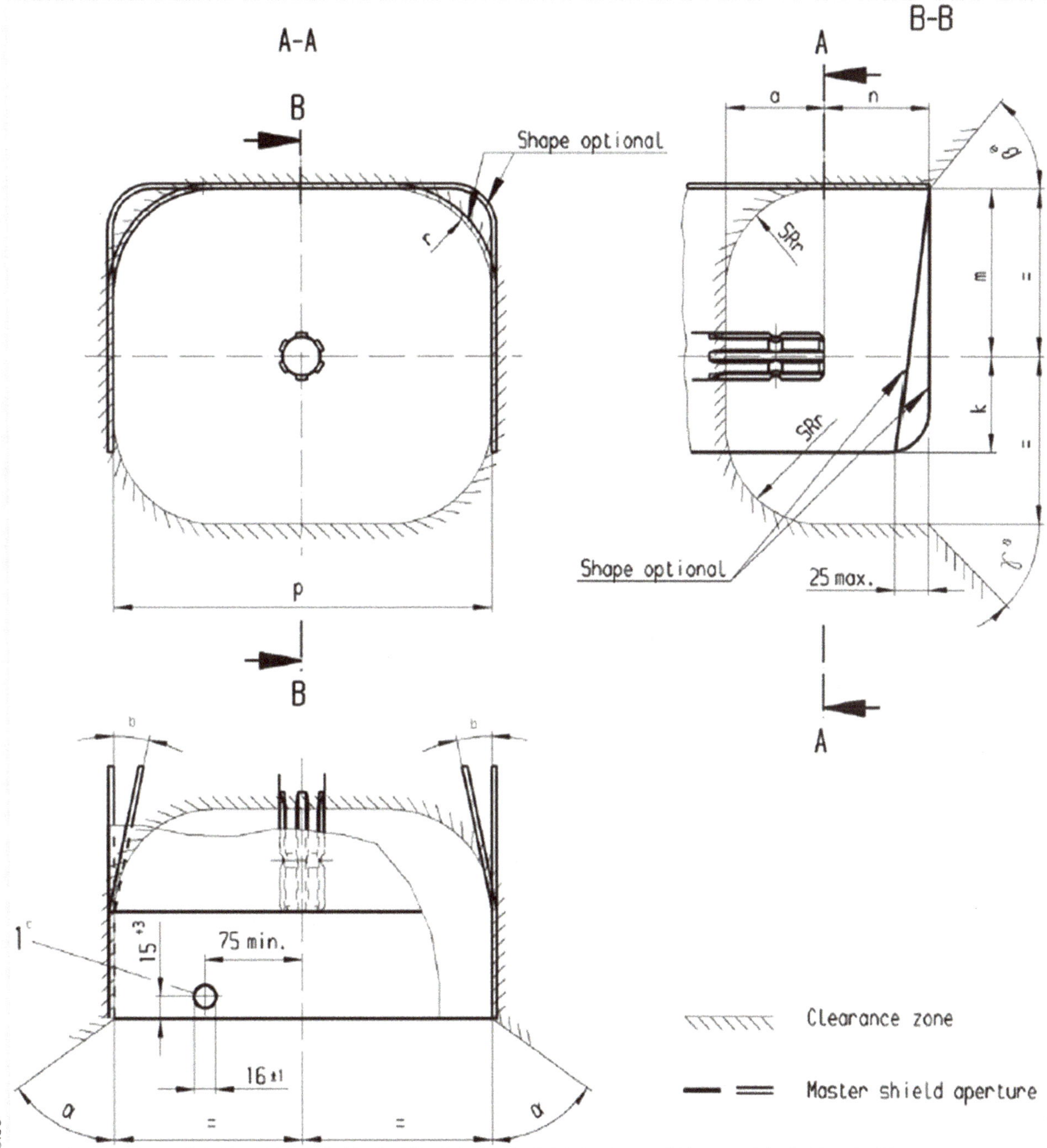

FIGURE 6.25 Hydraulic circuit of a closed center tractor hydraulic system with load-sensing and hydraulic couplers with power-beyond for attached implements. Translated from German into English. Source: [50].

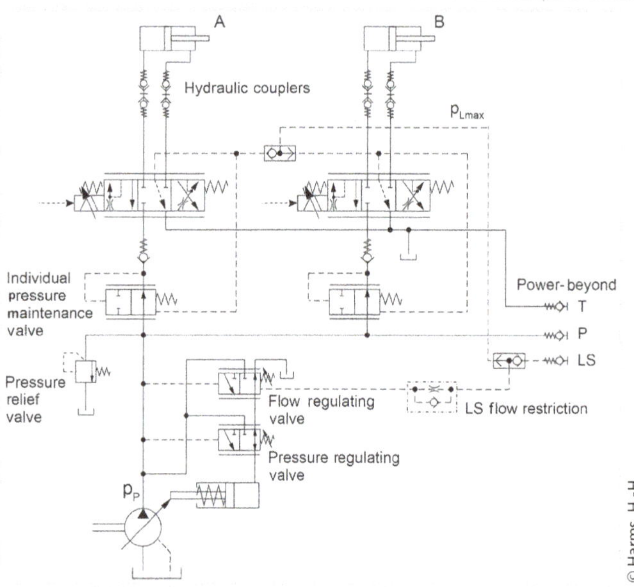

for example. In addition, these hydraulic systems have standardized interfaces to enable provision of hydraulic energy in the form of flow and pressure to attached implements or trailers. The hydraulic energy consumers can be very different ranging from intermittently actuated and/or controlled hydraulic cylinders to permanently operated hydraulic motors. There are applications where proportional characteristics are required and applications where the flow and/or hydraulic oil pressure needs to be adjustable.

In modern agricultural tractors, load-sensing (LS) hydraulic systems with variable displacement pumps are used, a hydraulic circuit of which is shown in Figure 6.25. These tractor hydraulic systems allow actuating hydraulic consumers directly from the tractor through hydraulic couplers where the hydraulic pressure and flow can be adjusted to the needs of the connected hydraulic valves. By means of a power-beyond interface, external hydraulic systems on implements can be connected to the tractor where the interface represents only a source for hydraulic energy in the form of pressure and flow.

The interfaces for hydraulic energy transfer on a typical agricultural tractor are shown in Figure 6.26. The tractor is equipped with five selective control valves (SCVs), power-beyond, and also has ports for pressure-free return oil return from the implement. Figure 6.27 shows different types of attached hydraulic consumers for the same tractor hydraulic interface.

These interfaces are also standardized to ensure compatibility between tractor and attached implements. The most important technical standards for hydraulic energy transfer from tractors to implements are as follows:

- ISO 5675 (2008) – Agricultural tractors and machinery – General purpose quick-action hydraulic couplers [28]
- ISO 10448 (1994) – Agricultural tractors – Hydraulic pressure for implements [36]

FIGURE 6.26 Hydraulic connections for implements and hydraulic motors on a John Deere 6215R tractor.

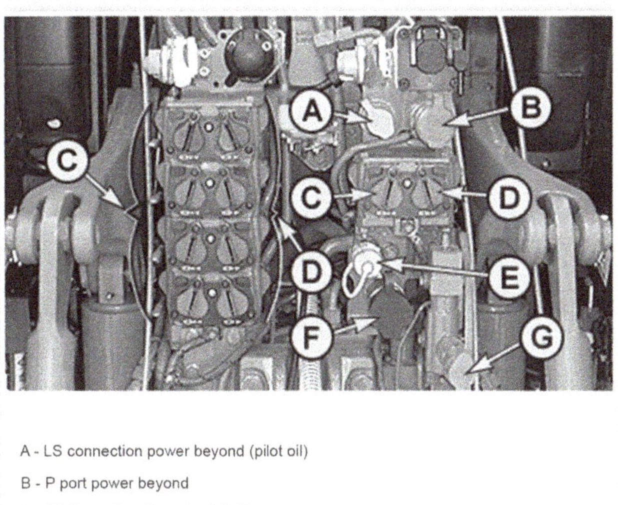

A - LS connection power beyond (pilot oil)
B - P port power beyond
C - SCV couplers for extend (I–V)
D - SCV couplers for retract (I–V)
E - R port power beyond pressure-free (leak oil)
F - R port power beyond pressure-free (great amounts of oil)
G - Coupler for hydraulic trailer brake

- ISO 11471 (1995) – Agricultural tractors and machinery – Coding of remote hydraulic power services and controls [37]
- ISO 17567 (2005) – Agricultural and forestry tractors and implements – Hydraulic power-beyond [26]

6.1.3.3.3 Electric Energy and Data Transfer. There are standard seven-pole connectors for electric power transfer and lighting, described in ISO 1724 [22] commonly being used in Europe and SAE J560 [48], usually applied in North America, and they are not compatible.

The ISOBUS as elaborated in detail in Section 4.4.1.10 is a data network for the communication between tractor and implement, standardized in ISO 11783 [34]. The standard comprises 14 parts in which part 1 lays out network description and requirements, part 2 defines the physical layer, and part 3 defines the data link layer. Whereas parts 2–5 of ISO 11783 establish rules for communication and parts 6–14 elaborate machine control functionalities including diagnostics [34, 47].

6.1.4 Tractor Applications

6.1.4.1 **Classification:** Tractors are primarily being used for agricultural purposes but also in construction and forestry, in municipal and other special applications, for example, in industry.

In agriculture, there are numerous different applications for tractors throughout the year. The use spectrum ranges from field operations to transport operations and work on the farm site itself.

FIGURE 6.27 Hydraulic connections for implements and hydraulic motors on a John Deere 6215R tractor.

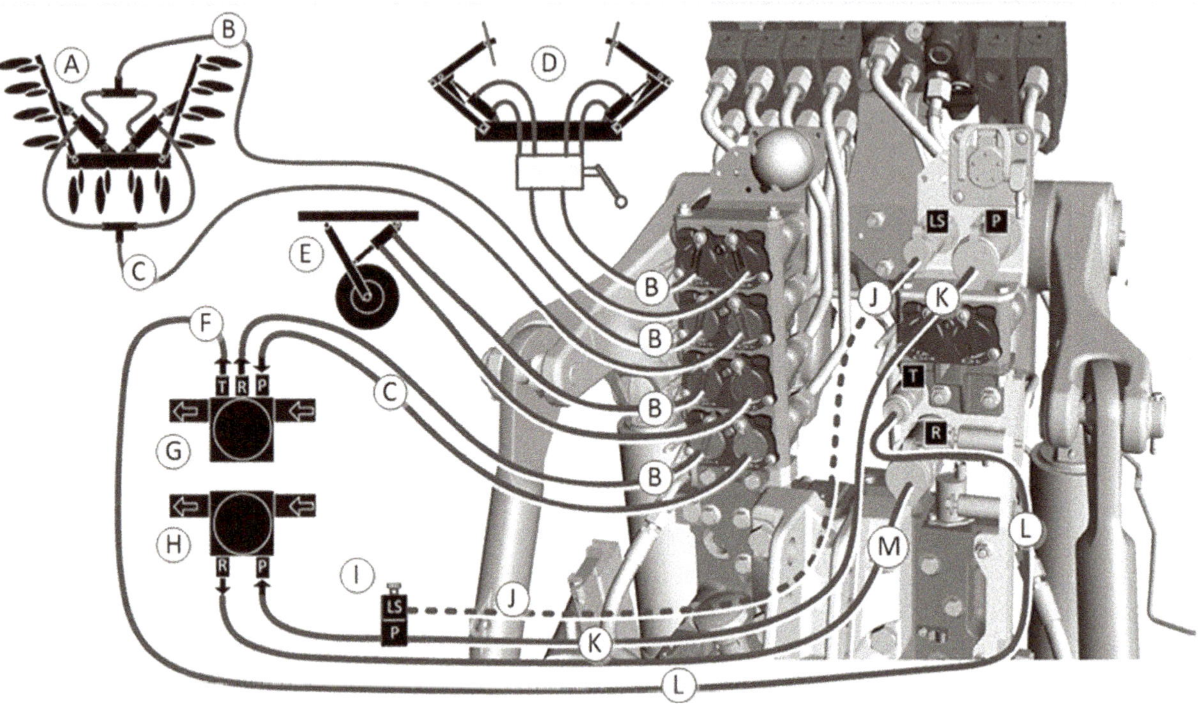

A - Folding mechanism for example (double-acting cylinder connected in parallel).	D - e.g. with track markers. Two double-acting cylinders are actuated separately by means of a commutation valve.	G - Hydraulic motor with leak-off port	J - Hose for LS pilot oil, connected to Power Beyond valve	M - Return port for large oil quantity
B - SCV port for extend.	E - Actuating a gauge wheel for example by a double-acting cylinder.	H - Hydraulic motor (e.g. for pump) without leak-off port.	K - Hose for pressure oil, connected to Power Beyond valve.	
C - To SCV port for retract.	F - Drain port of a hydraulic motor, in order to empty the motor case.	I - Control valve for integrating a hydraulic application in the PFC hydraulic system.	L - Hose for draining the hydraulic motor case (leak-off oil returned, in order to empty the case of the hydraulic motor).	

In construction and forestry, tractors are commonly being used on the construction worksites or in the forest, respectively. When used for municipal purposes, tractors carry out tasks on and along paved surfaces, roads, and on grasslands.

There are also large regional differences in the use profile and in the average annual use of tractors in different parts of the world. Whereas in Brazil, for example, it is quite common to operate a tractor for more than 2,500 hours per year in professional sugarcane production, there are tractor applications on regular farms in Germany, where tractors are used no more than a few hundred hours per year in mixed-use patterns [43].

Many factors contribute to these regional differences including the location, size and structure of the worksite, the type of crops and agricultural practices to grow them, the size and technology levels of tractors and implements, the number and education of tractor operators, and many others.

6.1.4.2 **Implements:** A definition and classification of implements is provided in ANSI/ASAE S390 and ISO 12934 which differentiates "mounted implements (3.3)," "semi-mounted implements (3.4)," and "balanced towed implements (3.5)" [3] (Copyright ASABE. All excerpts from the text of this standard have been used with permission of the ASABE.).

"Mounted implements (3.3)" are defined as a "device or machine that performs a specific operation and which is normally attached to a tractor or a self-propelled machine and fully carried by the tractor or self-propelled machine (3.2)." According to this definition, "implements can be mounted on the front, the rear, the load platform (if any) and/or between the axles of the agricultural tractor (3.1) or self-propelled machine (3.2)" [3].

In contrast, semi-mounted implements are formally identified as "interchangeable towed machinery (2.4) with one axle or axle group of land wheels and a drawbar towing device, which cannot move relative to the agricultural vehicle (2.3) allowing the transmission of vertical forces from the towed agricultural vehicle through the drawbar to the towing agricultural vehicle" [3].

Whereas the latter, "balanced towed implements," are defined as "interchangeable towed machinery (2.4) which is designed to be towed by an agricultural vehicle (2.3) that changes or adds to its functions but does not transmit any vertical force(s) through the hitch to the towing agricultural vehicle." [3]

6.1.4.3 **Trailers:** Supplemental to the definition of implements, ANSI/ASAE S390 and ISO 12934 define "Trailers (3.6)" as "trailed agricultural vehicles (2.3)" and differentiate these vehicles in the form of "balanced load carrying trailers (3.6.1)" and "semi-mounted load carrying trailers (3.6.2)" [3] (Copyright ASABE. All excerpts from the text of this standard have been used with permission of the ASABE.).

Under this definition, trailers are equipped "with wheels or endless tracks, intended mainly to carry loads and designed to be towed by an agricultural tractor (3.1) or self-propelled machine (3.2)" [3].

6.1.4.4 **Application Profile:** The requirements and tractor use of customers are very different from region to region and from farm structure to farm structure in the world. Farm operations often use multiple vehicles with different power and specification levels for different applications. Different crop types, agricultural practices, and level and skill of available labor also drive tractor specifications significantly.

For example, a tractor being used by a Brazilian sugarcane grower for cane haulage is significantly different in its use pattern than a tractor that is operating in a rice field in India or one that is being used by a large arable farm in North America for tillage and planting or by a Western European contractor for mixed operations with road transport.

Furthermore, although the interfaces between tractor and implement are standardized, there are worldwide a large number of different types (and sizes) of implements on the farms and not every tractor is compatible with each implement.

Knowledge about representative customer application profiles for agricultural tractors is essential for tractor manufacturers in the design of agricultural tractors. Such profiles are a prerequisite for the determination of valid duty load cycles that can be used in design and verification of tractors and tractor component durability. Such user profiles are thus considered proprietary information by the tractor manufacturers.

A recent literature review on representative use pattern of agricultural tractors revealed that recent publications on such data are lacking and that previously published data from the 1980s are only partially applicable to today's farms due to the structural changes in agriculture that have taken place since [14]. This work includes data from studies of more than 500 agricultural farms [5] which are categorized in (1) soil and tillage, (2) seeding, nurture, and protection, (3) harvest, (4) farmsite, (5) transport, and (6) miscellaneous work. Use pattern and average annual use times for different sizes of farm operations have also been reported [43]. The average annual use of a tractor is dominated by the size of the farm operation, the total number, and size of the tractors with respect to all other available tractors. The use spectrum presented for tractors, ranges from (1) plowing, (2) other tillage work, (3) organic fertilization, (4) inorganic fertilization, (5) seeding, (6) row crop harvesting, (7) mowing, swathing, and tedding of grasslands, (8) harvesting of grass, (9) daily forage collection, (10) transport operations within the farmsite, (11) transport operations outside of the farmsite, (12) frontloader work, and (13) others.

In Eckstein [14], representative duty load cycles for durability analyses of agricultural tractors are developed and a use pattern model with the percent use of implements for three different types of farm operations is elaborated for (1) livestock farms, (2) arable farms, and (3) mixed farm operations in Germany (see Tables 6.6 and 6.7). Depending on the total number and engine power of tractors and on the size of implements available on that farm types, the percent use or use pattern for an individual tractor on said farm can be derived as a subset of this.

The application profile will deviate from this for tractors operating in different regions and different farm structures or farm sizes in the world. It will also deviate for tractors being used in nonagricultural applications including construction and forestry, municipal, or other applications, such as industrial applications.

6.1.4.5 **Ballasting:** The ability of a tractor to be reasonably ballasted on all axles is an important matter. On one hand, tractors have to be ballasted to ensure driving stability, ride comfort, and compliance to regulations, particularly when driving on public roads. When implements or trailers are attached to tractors, the vehicle weight,

TABLE 6.6 Use pattern of agricultural tractors for different types of farm operations in Germany: (1) livestock farm, (2) arable farm, and (3) mixed farm. Distribution summary for different work processes. Translated from German into English. Source [14].

	Livestock farm (%)	Arable farm (%)	Mixed farm (%)
Primary tillage	14	30	22
Secondary tillage	9	19	14
Seeding, fertilizing, and maintenance	14	26	20
Pasture and feeding	21	0	12
Transport	24	16	20
Loader	18	9	12

TABLE 6.7 Use pattern of agricultural tractors for different types of farm operations in Germany: (1) livestock farm, (2) arable farm, and (3) mixed farm - detailed distribution for different work processes and implements. Translated from German into English. Source: [14].

Work process	Implement	Livestock farm (%)	Arable farm (%)	Mixed farm (%)
Primary tillage	Subsoiler	1.0	4.0	2.0
	Cultivator	5.0	13.0	9.0
	Plow	5.0	14.0	9.0
	Disc harrow	3.0	8.0	4.0
Secondary tillage	Harrow	2.0	5.0	3.0
	Rotary harrow	3.0	5.0	3.0
	Seedbed combination	5.0	12.0	8.0
Seeding, fertilizing, maintenance	Seeder	3.0	12.0	7.0
	Slurry tanker	6.0	3.0	4.0
	Manure spreader	3.0	0.0	1.0
	Sprayer	1.0	5.0	3.0
	Fertilizer spreader	5.0	0.0	3.0
Pasture, feeding	Mower/combination	8.0	0.0	5.0
	Tedder	3.5	0.0	2.0
	Swather	4.5	0.0	3.0
	Baler	4.0	0.0	3.0
	Forage wagon	3.0	0.0	2.5
	Mulcher	1.0	0.0	0.5
	Fodder mixing wagon	8.0	0.0	5.0
Transport	Trailer	3.0	8.0	6.0
	Empty tractor	3.0	2.0	3.0
Front loader	Earthworks	2.0	1.0	1.0
	Material handling	6.0	5.0	4.0
	Farm work	12.0	3.0	9.0

axle loads, and subsequently load under each tire may change. It is therefore important to comply with the following:

a. Minimum axle loads (e.g., 20% of the empty mass on the front axle (StVZO §34 [51]))
b. Sufficient axle loads (to meet braking requirements) [42]
c. Not exceeding maximum permissible axle loads as specified by the manufacturer
d. Not exceeding maximum permissible axle loads to operate vehicle on public roads (StVZO §34 [51])
e. Not exceeding the maximum permissible tractor weight as specified by the manufacturer
f. Not exceeding the maximum permissible tractor weight to operate vehicle on public roads (StVZO §34 [51])
g. Not exceeding maximum permissible tire load capacities (as specified by the manufacturer)

On the other hand, the ability of a tire or track to transfer forces in soft soils in the field is directly proportional to the applied vertical load on said tire, as elaborated in detail in Section 2.3. There are various means to apply loads to tires or tracks of agricultural tractors. These include the following:

a. Standard chassis-mounted weights, such as front-mounted weights
b. Rear three-point hitch or front three-point hitch-mounted weights
c. Wheel or rim weights
d. Water filling of tires
e. Weights with variable center of gravity, such as CLAAS Flex-Weight
f. Weight applied through load transfer from three-point hitch for traction or stability
g. Load transfer from trailers attached to rear coupling devices

High vertical loads may result in higher traction capability but may also damage the structure of the soil and its ability to provide adequate crop growth. Increasing wheel slip improves the traction capability only up to the point of maximum traction efficiency. But from an agronomic perspective, larger wheel slip is unfavorable. This is true particularly for operating on grassland.

In summary, there are competing requirements and objectives for proper ballasting of tractors coming from a regulatory framework, from the desire to optimize traction, from use of the available engine power to maximize machine productivity, and from an agronomic point of view.

6.2 Wheel Loader (By Herbert Pfab)

According to ISO 6165 (Figure 6.28), the loader is a basic type of earth-moving machine. The standard categories' loaders are further divided into swing loaders and skid steer loaders. In any case, loaders are mobile working machines, whose main task is to load loose material or piece goods, whereby the working equipment (bucket) is moved in the working cycle both by the traction drive and by moving work functions (lift arms). The material's location is changed from the loading site to the unloading site by moving the vehicle itself. This is necessary for systematic differentiation from excavators. The longitudinal movement of the bucket is primarily performed by the traction drive and the vertical movement by the lifting device. The tilting kinematics ensure that the bucket is filled when it is tilted in and that the material is discharged when it is tilted out. When handling piece goods, the tilting kinematics have no function. These goods are set down on an individual load carrier, such as a euro pallet. This basic function can be compared with a forklift truck.

The wheel loader is operated intermittently during the loading process. The machine's controls are never stationary within a loading cycle. The cycle can be divided into individual phases, during which the machine drives are controlled very differently.

Loaders are designed with wheels or crawler tracks (Figure 6.29). Both machines differ considerably in terms of their machine concept. This chapter is essentially limited to loaders with wheeled chassis.

FIGURE 6.28 Earth-moving machinery. Source: [58].

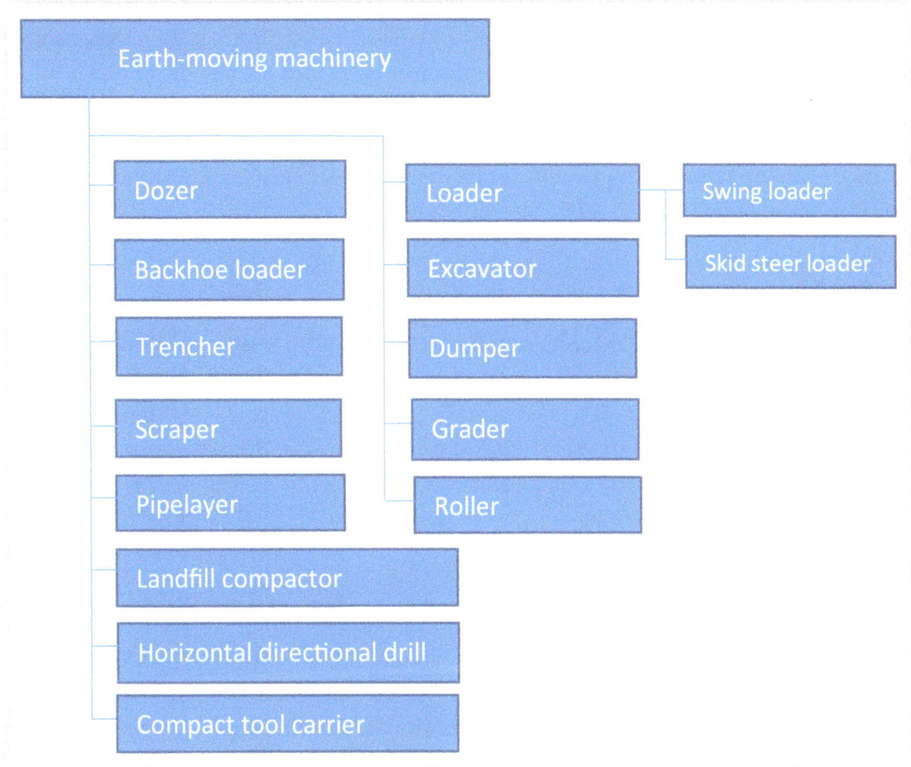

FIGURE 6.29 Wheeled and tracked loader.

Wheel loaders are defined as earth-moving machines. However, their applications are diverse and very often outside of the construction industry, as well.

The wheel loader market with operating weight between 5 and 35 tons has a volume of currently approx. 100,000 units per year. China accounts for a good 30,000 of these machines, and other low-regulated markets (LRC) cover further approx. 25,000 wheel loaders. China, the United States, and Germany are the world's most important markets in this segment. Japan is very important for small wheel loaders. Quantities are subject to large annual fluctuations. These figures apply to 2017.

6.2.1 Application

The loading process of the machine is a recurring cycle. This consists of the following phases: bucket filling, reversing, forward travel to the unloading site, unloading, reversing, and forward travel to the loading position (**Figure 6.30**). The material to be loaded is stored in the form of a pile. At the unloading site, the material is discharged into a mobile container (truck), into a stationary facility (unloading bunker for further processing in a system), or stationary storage (landfill, stockpiling). Normally, the loading and unloading sites are positioned in such a way that the distances between the two are as short as possible.

Material pick-up (bucket filling): Building materials are usually extracted in a quarry. If the wheel loader is used at the excavation site, it is referred to as an extraction or mining operation. If the material is loaded again later for further use, this is a rehandling operation. In the extraction operation, the material must be removed from its natural composition or after blasting. The task in material pick-up consists of breaking the material loose and filling the bucket with the loosened material. In the rehandling operation, the breaking loose process is mainly omitted. The extraction operation of the wheel loader is limited to low soil classes. In the case of more solid material such as larger

FIGURE 6.30 Short Y-cycle, typical for wheel loaders, dynamic behavior of driveline and hydraulic [59].

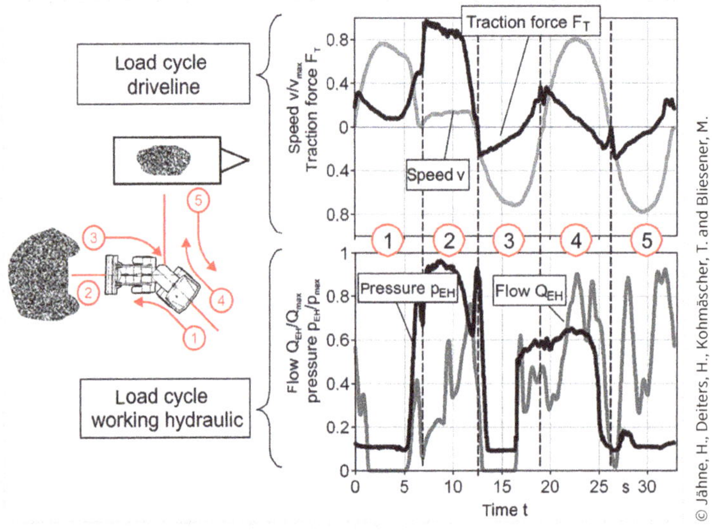

FIGURE 6.31 Wheel loader handling logs with a log grapple and required kinematics.

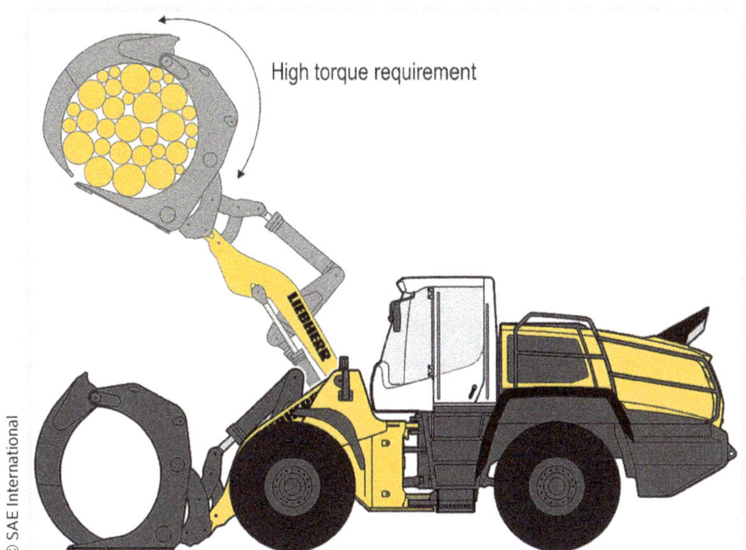

moraine gravel, excavation with the hydraulic excavator or a previous separate loosening with the crawler or the ripper tooth on the hydraulic excavator is necessary.

Discharging process: For discharging into a truck, the wheel loader must lift the bucket with its pivot point above the side wall height. Overloading heights of 4 m are required for specifically light materials such as wood chips. Optional equipment (high dump bucket) or long-reach lift arms are used to increase the unloading height. During stockpiling, the machine drives onto the pile and thus expands the discharging option. Only the front axle is ascending on the pile. In special cases, the loading material must be discharged smoothly. For example, logs are lifted over the truck's stanchions with the log grapple, and the fully loaded grapple is tilted out and then opened slowly (Figure 6.31). In this case, the tilting kinematics must be able to move the full payload at any tilting angle. The same applies to buckets with hold-down devices.

Transport: If possible, the loading operation is organized in such a way that no actual transport route results. The movement is in the form of a Y (Y-cycle). The transport distance is determined by the maneuvering distances of the wheel loader and the size of the discharging container like an articulated truck. In the case of stationary unloading sites, a variable loading location results in a variable transport route. This is typical for extraction operations and feeding into a bunker or crusher. This is called the load-and-carry operation. The sole purpose of road travel with a wheel loader is to reach another site of operation. This can be done on its own axles or by truck transport.

Wheel loaders are occasionally used for purposes other than loading. For example:

- Snow clearance in winter service with plough or hydrostatically driven rotor plough or blower also in combination with a rear-mounted spreader device.
- Use of a hydrostatically driven sweeping brush.
- Fields of application are sectors of industry-related customer groups such as building construction, civil engineering, road construction, mining, tunneling, industry, demolition and debris processing, recycling and waste processing,

forestry and timber industry, gardening and landscaping, municipal use, agriculture, and rental.

- In the construction industry, the machine is often used in many different ways, with different equipment. In industry or mining, on the other hand, a single, consistent cycle prevails. The machine is equipped specifically for the field of application. This affects mainly the bucket (or general attachments) and the tires.
- Loading material: Loading materials are typical for the field of application.
- Mining: Raw materials (limestone, hard rocks, coal, ores, salt, clay, various rocks and minerals), overburdon, blasted or mined on the face, mining operations, or rehandling crushed or cleaned and sifted fragments.
- Construction industry: Supply to mixing plants, handling sand, gravel, and soil, handling building materials and construction elements, and paving stones and slabs.
- Waste recycling: Supply and disposal of sorting or incineration plants.
- Industry: various materials, steel scrap, aluminum scrap, wood chips, broken glass, slag.
- Agriculture: Cereals, fertilizers, bales, silage.
- Forestry and timber industry: Logs, bark mulch.
- Gardening and landscaping: Soil, organic matter, plants.
- Municipal use: Snow, supply to municipal facilities (heating power stations, etc.). In Scandinavia, wheel loaders are also used for clearing snow from country roads. Those wheel loaders are designed for maximum speed up to 50 km/h (31 mph).

Due to the different properties of the materials mentioned above, there are different specifications for the equipment to be configured (bucket size, bucket geometry, wear protection, blade shape, tooth system, hold-down device, etc.). The basic machine is equally equipped for all fields of application and therefore for all materials. The user's operation-related configuration essentially refers to the loading equipment and the tires. Sometimes, certain protective devices are typical for user areas.

Another characteristic of a field of application is the intensity of machine use. The wheel loader is a key machine in the production chain in mining. This means that the availability requirements are high and the machines are also used in multiple shifts. About 3,000 operating hours per year are typical, occasionally 6,000 operating hours per year. The same applies to use in industry and in waste and recycling management.

In the construction industry, the loaders are used for approx. 1,500 operating hours per year with seasonal fluctuations. The average annual usage also increases with machine size.

6.2.2 Machine Concepts

Due to the size and position of the forces introduced by the equipment, wheel loaders are generally designed with a welded steel chassis (frame). All components are attached to the vehicle frame. Compare the structural component design of agricultural tractors. The following frame-mounted main components should be mentioned (**Figure 6.32**): Diesel engine, cooling system, driving transmission, front and rear axle, control block for working equipment, lift cylinder, tilt cylinder, steering cylinder, lift arm, driver's cab, rear ballast, fuel tank, hydraulic tank, hoods and fenders, and access ladder.

FIGURE 6.32 Basis design of a wheel loader structure with articulated steering.

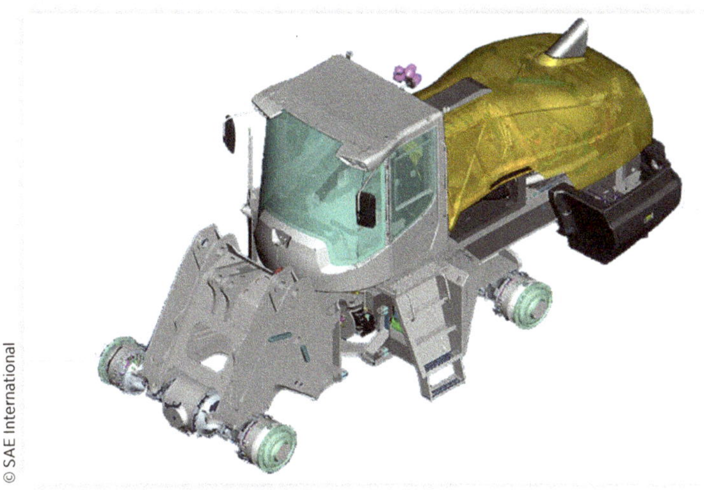

The external forces are transmitted to the frame through the lift arms. The two pivot points of the bucket arm, the bottom side of the two lift cylinders and the bottom side of the tilt cylinder, are the points of introduction. The frame is supported by the bolted-on front axle and the mostly oscillating rear axle on the chassis. The chassis suspension is rigid so there is no spring-loaded and damped connection between the wheels and the frame. The frames are designed as torsional stiff boxes.

Wheel loaders weighing 7 tons or more are equipped with a frame steering system (articulated steering, Figure 6.33). On smaller machines, three different steering modes are used: articulated steering, king-pin steering, and drive steering (skid steer). The articulated steering has the advantages of robust rigid axles, high maneuverability, lateral positioning of the loading material when standstill, and equally spinning axle drives, even when driving in curves. The disadvantages are in the design effort for the articulated joint and the poorer stability against tilting in the fully steered position. In contrast, the two-axle (all-wheel steering) king-pin steering design is flexible in the control system,

FIGURE 6.33 Turning characteristics of relevant steering mechanism.

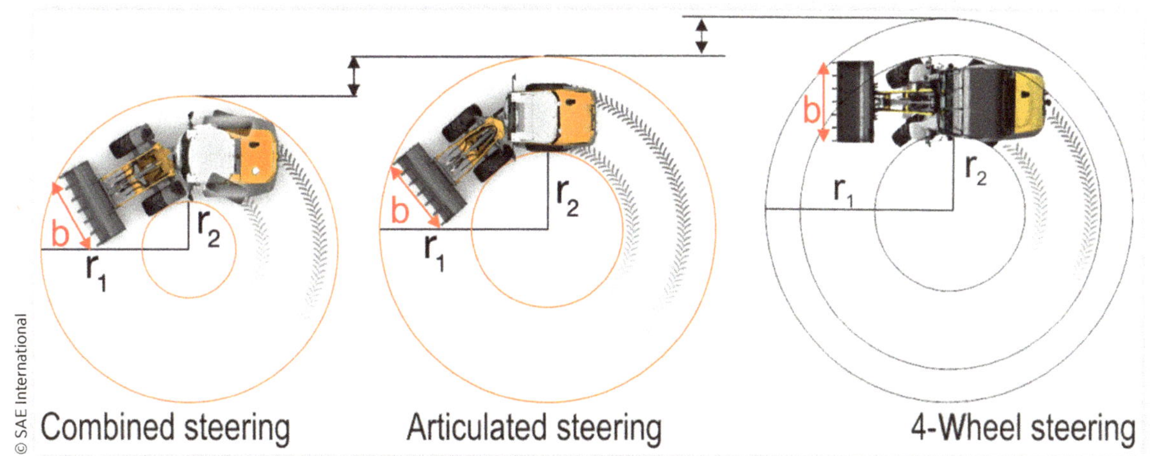

simpler in the frame construction, and possesses safety against tilting, independent of the steering angle. Skid steering is extremely maneuverable (turn on the spot). However, the short wheelbase does not allow good driving stability at higher speeds.

6.2.2.1 **Classification:** The primary customer benefit of the wheel loader is that a quantity of material is loaded per cycle. This load indicates the size of the machine. If the machine is loaded by the so-called tipping load (**Figure 6.34**), the wheel loader tips over around the front axle. In normal bucket operation, the size of the standard bucket is defined so that half of the tipping load is reached. Therefore, the safety factor against tipping is 2. The conversion of bucket capacity to half the tipping load is based on a material weight of 1.8 tons/m^3, an average value for common building materials. The capacity of the standard bucket and the tipping load are equivalent nominal values for the description or classification of a wheel loader. The operating weight of the vehicle (vehicle mass without payload) is more descriptive than the tipping load. In case of the typical center of gravity of the machine components, the tipping load is approximately 75% of the operating weight (with the articulated steering in a straight position). The tipping load depends on the lever of the bucket to the front axle (reach). Standard lift arms have a low variation for a specific machine size. The wheel loader size is selected in such a way that an integer cycle value results during standard operation. The handling capacity of the industrial plant or construction project can be adjusted by the number or size of the loading vehicles. In addition, care must be taken to ensure that the transport vehicles do not have long standing times during loading (no more than three or four loading cycles) and that there is no queue for the transport vehicles (daily fluctuation in handling capacity). Wheel loaders featuring a 3- to 4.5-m^3 standard bucket have established themselves in the sand and gravel business.

FIGURE 6.34 Tipping load is the main purpose of the machine.

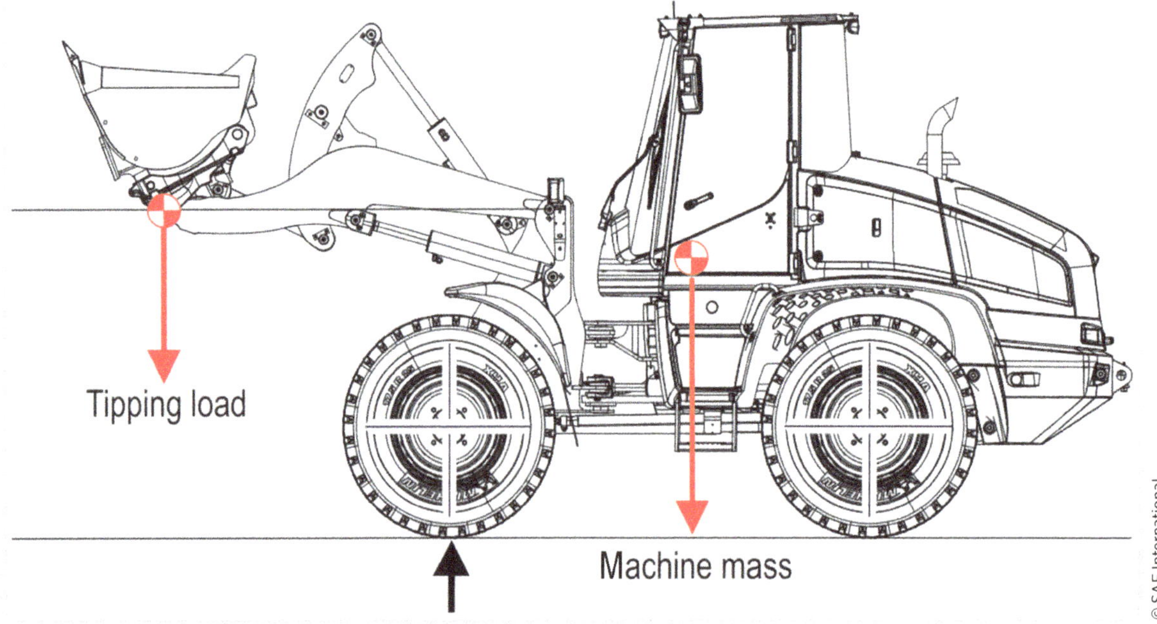

Wheel loaders are manufactured in series up to an operating weight of 250 tons. Machines with an operating weight of less than 5 tons are called the compact wheel loaders. These are designed for simple operations or rental companies. Small and medium wheel loaders up to 15 tons operating weight are used in diverse operations. Machines weighing more than 15 tons are used for mass handling. Sand and gravel handling and other rehandling operations are included. Wheel loaders weighing 30 tons or more are used in mining for the extraction of raw materials. When building new roads, loaders weighing between 24 and 100 tons are used. The range of industrial applications is wide and extends from 5 tons (in case of recycling, up to 100 tons (steelworks)). Increasingly, agriculture also relies on the wheel loader as a dedicated machine for handling material. Small loaders are used in the indoor or farmyard operations, and larger machines of up to 20 tons operating weight are used in field operations, especially in silage.

6.2.2.2 **Model Series and Platform Strategy:** As described in the section on classification, various sizes are required for wheel loaders due to a wide variety of operations. Liebherr, as an example, currently produces 15 models in the range from 5 to 33 tons (standard bucket 0.8 to 6.0 m^3) for the highly regulated countries (HRCs) in its product range. Two of these models are redundant in size (compact loader series). The 15 types are split into four model ranges. The variety of all components is referred to as a modular system. The described range can only be mastered economically and technically if a modular system is carefully planned. In addition to the physical components of the modular system (parts, components), functional systems must be defined in the modular system. These refer to mechatronic, often cross-assembly units. The software structure of the electronic control system forms a separate modular system.

The different basic types offered by a construction machinery manufacturer are referred to by the manufacturer as product lines or product groups. These include wheel loaders, mobile excavators, crawler excavators, bulldozers, crawler loaders, telehandlers, graders, ADTs, and so on. The task of product architecture is to form the modular system for all product groups of a manufacturer (or a division of the manufacturer). Uniform modules in the modular system result from this viewpoint in the area of the engine, the electronic control system, and the man-machine interfaces, to name just a few. The organization of engineering responsibilities across all product groups has a decisive influence on the width and depth of the entire modular system. A large portion of the parts and components, and therefore the elements of the construction kit, are purchased. In particular, this includes the mechanical and hydraulic drive components. Normally, these are not developed for a single OEM, but the vendor supplies a large number of OEMs, often for different product groups. The task is therefore to match suppliers' modular systems sensibly with the proprietary modular system.

Starting with diesel emission regulation according to stage IIIB and engine power above 129 kW, a complex exhaust gas purification system is required. In addition to the associated considerable costs, there are also higher requirements with regard to the sulfur content in the fuel and special requirements for the engine oil that is used. For these reasons, wheel loader manufacturers have begun to develop a parallel vehicle series for LRC markets. In addition to the difference from the diesel engine mentioned above, this series also differs in other assemblies, where local requirements can be met with

FIGURE 6.35 Rear view at the wheel loader with diesel engine positioned transversely.

less complex and more cost-effective solutions. For example, this includes the drive system, cabin equipment, and electronic control unit. In addition to the exhaust emission regulations, additional different and formal requirements must also be taken into account.

The main components relevant to the layout are the following: driver's cab, cooling system, diesel engine, travel drive transmission, axles, steering and work pump, control block, lift arms with lifting and tilt cylinder, fuel tank, hydraulic tank, and rear ballast.

The vehicle frame supports and connects all these main components. In contrast to most other construction machines, the energy source (diesel engine) of the vehicle is not mounted at the front, but rather at the rear. The reasons for this are the weight distribution, visibility requirements, and the risk of damage in the area of the working equipment. The installation position of the diesel engine results in the mechanical traction drive. The travel drive transmission is centrally mounted on the vehicle in the rear chassis, and the axle drive is supplied by the travel drive transmission from the center of the vehicle through cardan shafts. Small wheel loaders up to 15 tons operating weight have a hydrostatic traction drive. The diesel engine can then also be installed transversely at the rear of the vehicle (Figure 6.35), the traction drive transmission is small and is usually blocked with the differential housing of the rear axle, and the front axle is driven through the rear axle by means of a cardan shaft. The hydrostatic drive transmission consists of a pump and hydraulic motor, where the pump is flanged to the diesel engine and the hydraulic motor is flanged to the reduction dropbox of the rear axle. The hydraulic pump for operating the lifting device is arranged either on the diesel engine or on an auxiliary PTO of a mechanical traction drive transmission. The same applies to a separate steering pump. Thanks to the hydraulic energy transfer, the control block for distributing and metering the oil flow to the lift and tilt cylinders can be integrated adapted to the special design conditions on the vehicle. Usually, the control block is mounted centrally, inside the front section, and supplies the hydraulic cylinders of the lifting device from a short distance.

The cabin is mounted in the center of the vehicle. In articulated vehicles, it is elastically suspended on the rear section of the vehicle. This solution is the best compromise to enable good control over the entire vehicle in all phases of the working cycle, even during reverse travel. In case of large vehicles, the direct view to the rear loses importance, because reversing cameras are used for this purpose. From the driver's point of view, a higher cab position is preferred. This also improves control over the work process. The limits to the overall height lie in the transport of the machine by means of a low-bed truck road traffic requirement and height limits in halls and buildings.

The key factors influencing the choice of installation location for the cooling system include the accessibility of the fluid lines, the conditions for guiding the cooling air, the accessibility for cleaning the heat exchangers, the degree of contamination in the ambient air, the weight distribution on the machine, and the thermal load to the driver's cab. The common positions are the rear of the vehicle behind the diesel engine, between the diesel engine and the driver's cab, or, in case of transverse installation, to the side and next to the diesel engine. When arranging the hydraulic tank, particular attention must be paid to the placement of the suction hose for the working hydraulics pump. There are diverse solutions and philosophies regarding the position of the fuel tank. Familiar positions include the front section, next to or under the cabin, and at the rear of the vehicle. The tank contents last for 20-30 operating hours. When defining the components' set up, ensure that all maintenance points can be reached safely. It is advantageous if maintenance points can be reached from the ground and there is therefore no danger of falling. Access to the cabin must be designed so that it is safe, comfortable, and free of dirt.

6.2.3 Process Connection

The wheel loader is a self-propelled working machine, so the process technology (working equipment) is part of the machine itself, and the machine and all its drives are tailored to the task. The basic task is loading bulk materials, especially in the construction industry. The bucket is fixed permanently to the vehicle (direct mounting). This configuration is ideal for the wheel loader. All other operations are more or less unsuitable, and special machines are available that can fulfill other tasks better. Nevertheless, customers tend to use the wheel loader in various operations. Several tasks have already been mentioned in a previous chapter. For example, a rear-mounted attachment, such as a spreader or a trailer hitch with drawbar load, is systematically incorrect, because the loader is designed for load attached to the front. This statement has technical relevance. The interests of the sales department, which is always on the lookout for additional customer benefits, speak against this. Essentially, the basic machine suffers both technically and commercially when operating in its basic task if adapted to additional operations.

In contrast to the tractor, which basically performs different tasks throughout the seasons, wheel loader operations are organized so that a single task often results during the entire operating period. This situation leads to the optimization of the customer machine to suit a single, specific operation. If possible, the manufacturer only desires to fulfil the configuration for the customer's purpose by adapting the bucket (or other optional equipment), without changing the basic machine and with standard kinematics.

When loading loose material, the machine pushes the bucket resting horizontally on the ground into the pile. After a certain filling level, the bucket is lifted and tilted with low tractive force, so that the bucket is filled completely (**Figure 6.36**). Lifting, tilting in, and repositioning with the travel drive are often iterative processes to enable

FIGURE 6.36 Filling strategies of the bucket. Source [60].

complete and rapid bucket filling. It should be noted that the lift cylinders are loaded by the wheel tractive force and, with simultaneous control of the lifting hydraulics, the force of the lift cylinders must first overcome the effect of the tractive force, and only the excess lifting force causes the lifting device moving upward. In extreme cases, the tractive force can raise the bottom-side pressure of the lift cylinders to the level of the secondary relieve valve pressure, causing the lift cylinders to be pushed in. The most common loader kinematics is called the z-bar kinematics (Figure 6.37). The tilt cylinder, linkage, and connecting link to the bucket form the shape of a "Z" when viewed from the side. The geometry of these kinematics causes the bucket to tilt in when lifted from the loading position, and then to tilt out over the lift in the upper position. The kinematics therefore support the requirements of tilting bulk materials in and out over the stroke by means of the coupling kinematics. The driver supports this tilting movement by actively controlling the tilt cylinder. The lever ratios on the z-bar kinematics generate a maximum brakeout force in the loading position which decreases severely when tilting

FIGURE 6.37 Basic linkage geometries on wheel loaders.

in or tilting out. This property is suitable for loading loose material. It should be emphasized that the loading process requires the dynamic coordination of tractive force and lifting and tilting movement. The properties of the material to be loaded are so different that no automatic bucket filling control system has been developed so far. In other words, activation of the control levers for tractive force and for lifting and tilt cylinders is the job of the driver, and the loading result is a consequence of his qualifications (driver's influence). Automatic loading is provided by the manufacturers as an option; the logic of the software used does not meet the requirements of professional drivers.

In case of loading materials other than homogeneous bulk materials, the requirements for the kinematics are different. Forklift operation is popular and often in the standard scope of delivery, especially in the case of small wheel loaders. Instead of the bucket, a fork carrier is connected to the lift arm pins, where laterally adjustable fork prongs are used. In forklift operation, parallel guidance along the lifting movement is practical. This requires either a parallel-acting coupling gear or active parallel controls by the machine operator. For this reason, the z-bar kinematics of small wheel loaders is optimized for approximate parallel guidance.

As described above, the brakeout force is reduced when tilting out or in at the z-bar kinematics. When working with heavy equipment (high dump bucket, bucket with retaining device) and when tipping with material (e.g., logs with closed grapple), the lever ratios on the z-bar kinematics are unfavorable, and the payload in such an operation is limited. For this reason, optimized kinematics, also known as industrial kinematics, are preferred for such operations. The lever ratios on the tilting kinematics are designed so that a high tilting moment results, especially when tilting out.

6.2.3.1 **Design of Lift Arms:** Lift arms are designed similarly to a ladder frame, two side arms are connected by a cross tube. The arms are connected to the pivot at the front at the bucket pins. The lift arms are bolted at the rear in the front chassis. The frame is relatively stiff in terms of torsion in the longitudinal direction of the vehicle, thanks to the solid cast steel or steel cross tube. Besides this predominant design, designs as central tubes are also familiar. The central tube may be welded or made of cast iron. All joint bearings are equipped with cast, turned, or rolled bearing bushings. Standard z-bar lift arms feature 11 lubrication points. These must be greased at regular intervals. This is done manually or with the aid of an automatic central lubrication system, which can be adjusted to any lubrication intervals and lubrication durations, depending on the ambient conditions. The lubrication points are sealed axially so that grease is retained, and the ingress of dust is prevented. In the end stops, the movement of the links of the coupling gear is limited by defined positive stops. Stops are provided between the bucket and the bucket arm, as well as between the linkage and the cross tube. Hydraulic cylinders stop internally. All pins are secured against twisting and secured axially.

6.2.3.2 **Other Lift Linkage Geometries:** In addition to the lift arms based on coupled linkages, the telescoping lift arm is increasingly popular on wheel loaders. In place of the bucket arm, a telescope with two sections is attached, and the tilting mechanism for bucket operation is mounted at its front. This machine may not be mixed up with the telescoping loader (telehandler). In case of the telescoping loader, the telescoping boom is attached to the rear of the vehicle frame, approximately at the position of the

FIGURE 6.38 Experimental hexapod type linkage. Source: [61].

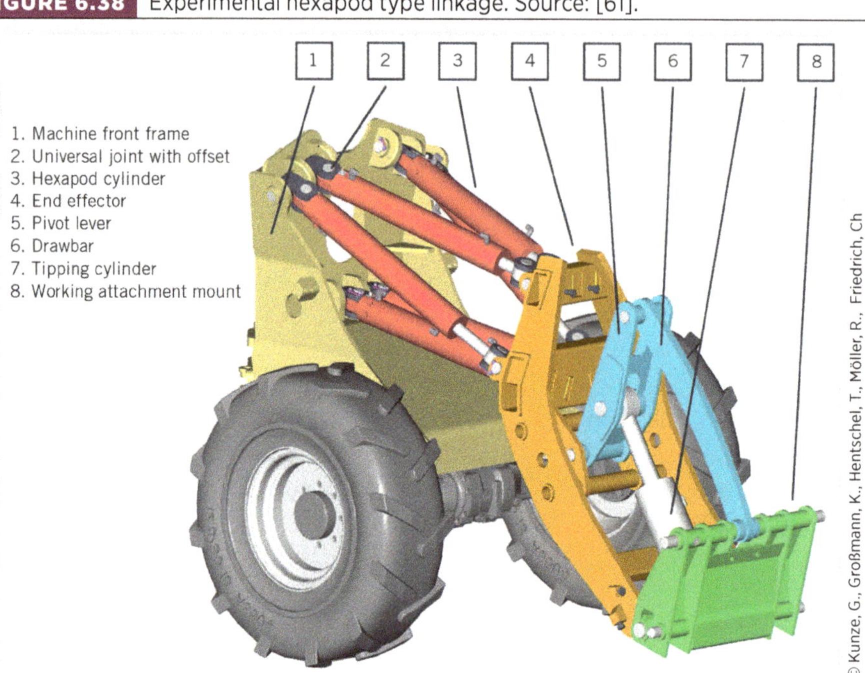

rear axle. The cabin is located to the side of the telescope. The telescoping wheel loader is primarily used for stacking piece goods. The loading cycle is slow for loading loose material. Telescoping lift arms are used on small and medium wheel loaders. In the slewing loader, the lift arms are mounted on a slewing ring on the front section. This creates an additional degree of freedom, because the bucket may be rotated around the vertical axis. For any kinematic movement of the working equipment, a wheel loader was developed at Dresden Technical University, whose lift arms were designed as a hexapod and equipped with six linear actuators (Figure 6.38). Plans for industrialization are not yet known.

Wheel loaders are all loading vehicles, although some of them have a single loading purpose, as mentioned above. The loading tasks of smaller machines tend to be more varied than with large machines. Small wheel loaders under 10 tons operating weight are therefore equipped with a hydraulic quick coupler device (Figure 6.39). This allows the bucket to be changed from the driver's seat and replaced by a second loading device, such as a fork carrier. A quick coupler device connects this device to the bucket through four contact points. The two upper contact points on the bucket are designed as hooks. The quick coupler is inserted into the open hooks from below. The two lower contact points are designed as a bolt connection. The pins arranged on both sides are actuated by a hydraulic cylinder and connect the quick coupler with the boreholes in the bucket. The bucket and quick coupler device are connected to form a kinematic rigid body. The axis of the tilting function is the connecting of lift arms and the quick coupler. Some equipment includes its own hydraulically actuated functions. These may include high dump buckets, buckets with clamping devices, and sweeping brooms. The hydraulic supply and control system are provided by the wheel loader. The connection to the wheel

FIGURE 6.39 Hydraulically operated quick coupler including coupling of hydraulic power lines.

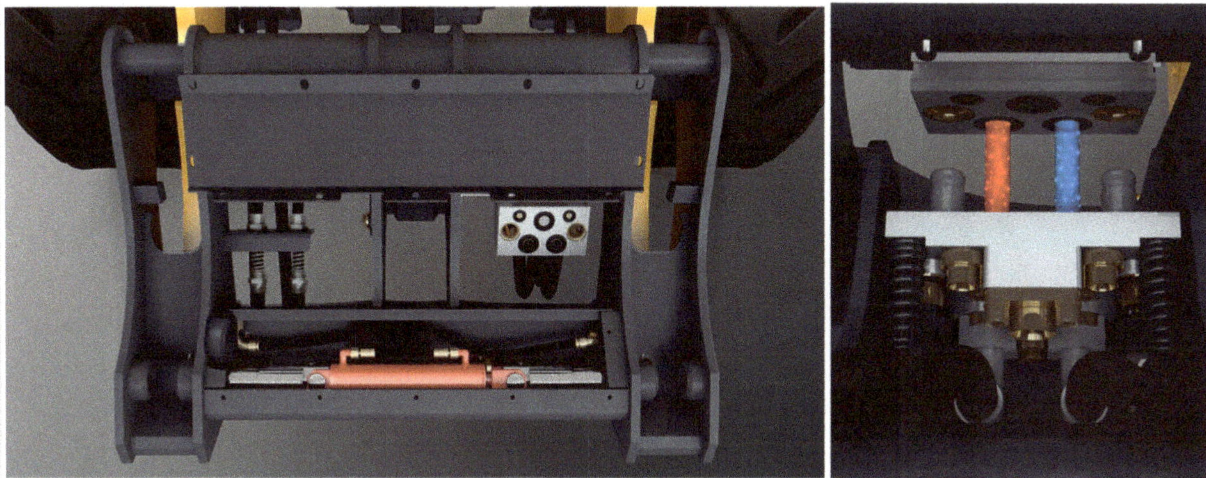

loader therefore consists of a mechanical and a hydraulic coupling. Automatic quick couplings are also available for the additional hydraulic coupling. However, manual connection of the hydraulic lines is common.

Quick couplers are optionally available for larger wheel loaders. Direct attachment is nevertheless more prevalent. This emphasizes the fact that large wheel loaders are used for a single task. In case of wheel loaders, there is no standardized interface between the attachment tool (bucket) and the quick coupler. The connection dimensions are agreed manufacturer-specifically and disclosed for use by third-party suppliers (bucket manufacturers).

6.2.3.3 **Work Tools:** Despite the uniform basic task of loading, the variety of attachment tools is very wide. In case of standard buckets, we differentiate between rehandling buckets (loose material) and earth-moving buckets. The buckets differ in their nominal dimensions in terms of capacity and outer width. The basic body of the bucket features a defined long bucket bottom and a rolling radius at the back. These dimensions are matched on one hand to the design of the tilt cylinder (brakeout force) and to the flow behavior of the loading material and therefore the filling behavior of the bucket on the other. The base body is made of fine-grained steel or wear-resistant welding steel. To increase the torsional resistance, two closed boxes are usually added as welded constructions at the back of the bucket. The bucket is optionally equipped with accessories, for example, an overspill deflector to repel overflowing material from the bearing points on the lift arm and the connecting link (Figure 6.40). In case of mining operations, deflectors welded on both outer sides of the bucket are available as an option. These should prevent the tires from being damaged by spoiled rocks.

When penetrating into the loading material, this is done by the cutting edges at the bottom and the side walls of the bucket. Depending on the operation, these cutting edges are designed differently in terms of material, geometry, and type of connection to the base body. The bottom cutting edge is either welded into or connected to the base body by a detachable bolted connection. In case of bolted on bottom cutting edges, this is

FIGURE 6.40 Basic elements of a standard wheel loader bucket.

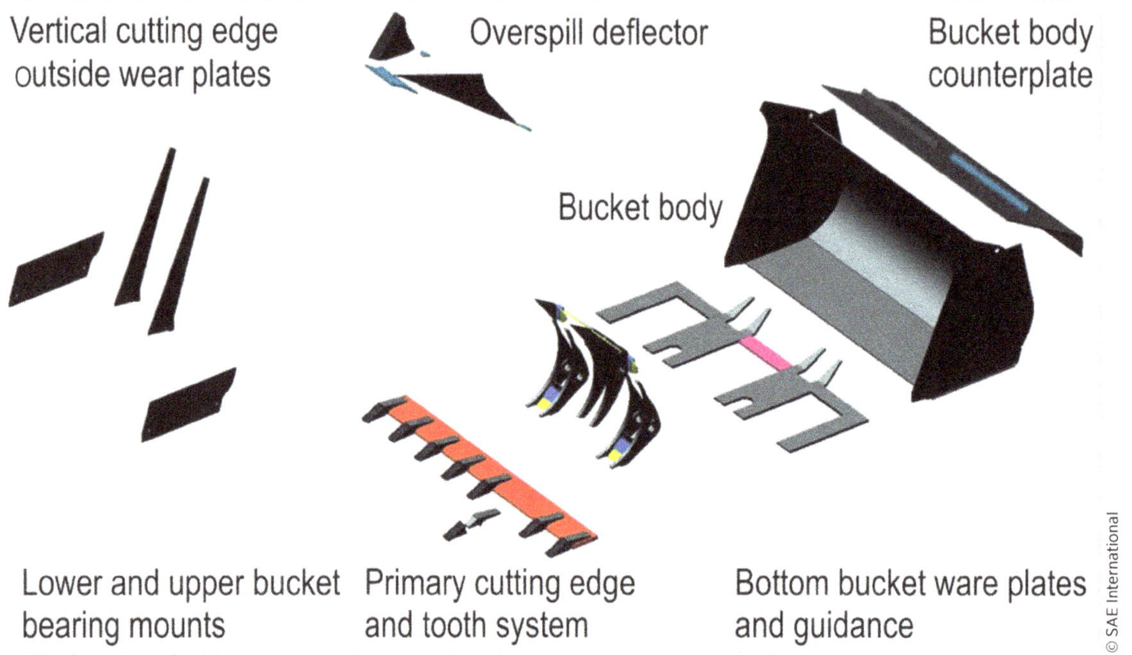

provided with a cutting edge at the front and rear, whereby the cutting edge may be turned over after one-sided wear. The bottom cutting edge is straight or trapezoidal when loading boulders; the trapezoidal shape facilitates a targeted approach into the pile and increases the specific load when penetrating the material.

To further improve the penetration behavior and to limit wear, the buckets are equipped with teeth in case of tight materials or mining operations. The teeth are firmly welded to the buckets of small wheel loaders. Otherwise, the tooth consists of the tooth holder, which is firmly welded to the bucket, and the tooth inserted on the holder. The connection is made through a plug-in system. The tooth and tooth holder are produced by forging. There are different tooth shapes for each manufacturer. The manufacturers of wheel loaders have their own tooth system; the tooth systems of third-party suppliers are often preferred by customers. The teeth may be replaced individually after reaching the wear limit. Buckets used in mining require an increased wear protection. In addition to the precautions already mentioned, replaceable wear protection devices are used between the teeth, on the bucket bottom, on the side walls, and on the skids of the bucket. Standard buckets are offered for every loader size and different specific material weight. This results in a standard bucket series defined for the entire wheel loader family. The modular bucket system can be continued in depth by modularizing the main (welding) assemblies of a bucket.

Special buckets include light material buckets, high dump buckets, multipurpose buckets, and buckets with clamp-down devices (Figure 6.41). Light material shovels are used for loading material with low specific material weight between 0.5 and 1.0 m^3. Compared with the standard bucket, the loading volume is 2–3 times greater; wood

FIGURE 6.41 Special attachements for wheeled loader application.

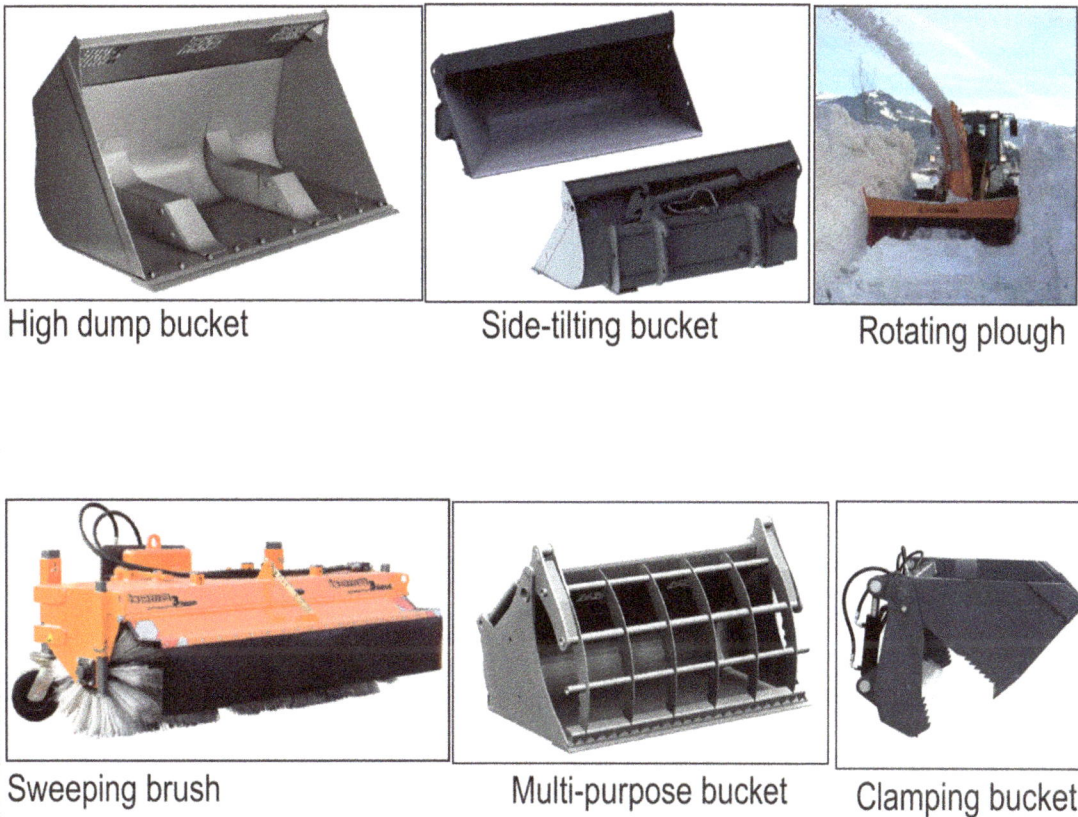

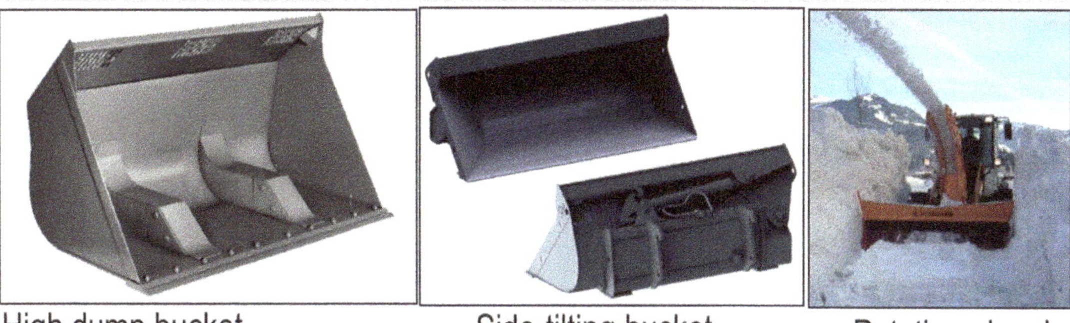

Sweeping brush

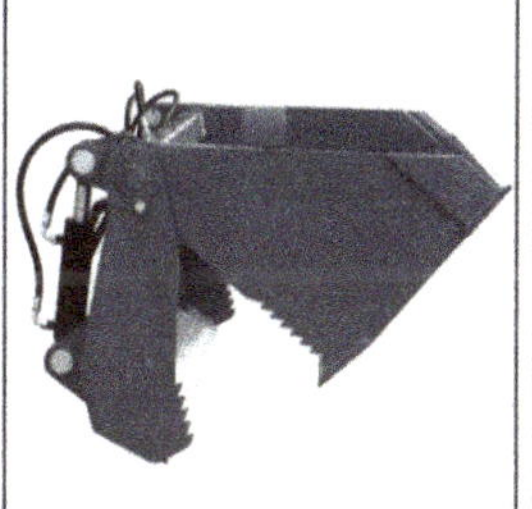

chips, compost, and waste fragments are considered light goods. In a high dump bucket, a separate dumping mechanism is integrated in the bottom of the bucket. The bucket pivot point shifts from the pivot point of the bucket to the front near the front blade. Depending on the size of the vehicle, the dumping height may therefore be increased by more than one meter. The efforts in design are high; due to the bucket mass, the payload is reduced, and the interior of the bucket is fissured. The multipurpose bucket is also called 4-in-1 bucket. This bucket is characterized by the fact that the bucket body consists of two parts, which enables a gripper function to be performed by the bucket. The pivot point of the folding function is the upper tip of the approximately triangular base body. The opening of the flap is between the bucket bottom and the back of the bucket. Using this bucket, piece goods may be gripped and lifted, the bucket filled by a gripping movement, material unloaded in a controlled manner by opening the gripper gap, and a plane created with the gripper edge. These buckets are popular with small wheel loaders.

Buckets with clamp-down devices are used for heterogeneous loading materials. The clamp-down device is mounted on the top of the bucket and is operated hydraulically. The loading material is therefore pressed into the bucket and held during transport. This bucket is often used in waste and recycling management and is attached to small- and medium-sized wheel loaders.

Side tilting buckets are used to fill channels. The bucket can tilt to the side by means of an integrated tilting device. The vehicle is driven along a channel and empties the bucket while it is driving. This covers the pipes laid in the channel with sand and therefore protects them.

6.2.4 Traction Drive

To evaluate the drive concept, it is necessary to analyze the working cycle. The loading cycle is described in Section 6.2.1. It is characterized by the dynamics of operating variables over the cycle. These include travel speed, tractive force, pressure, and flow rate on the three linear drives for lifting, tipping, and steering movement. This also leads to transient, dynamic behavior of the internal component operating parameters, such as the operating speed and operating torque on the diesel engine or the speed on the exhaust gas turbo charger. The cyclical operation and the heavily fluctuating power requirements of the drives lead to a strong partial-load operation of many main components. On the other hand, the specifications also include transport on ramps lasting several minutes with maximum travel drive power, so that downsizing with the aid of a parallel hybrid drive structure is not a fully equivalent and market-oriented drive solution. Deceleration phases in travel movement and lowering the lift arms result in a potential for energy recovery. However, the typical potential is relatively low and is accompanied with high design costs and additional drag loads.

The wheel loader consumers may be assigned to three main functions: traction drive, work function (lifting and tipping), and steering. Depending on the operation, this results in a certain share of energy turnover. This is by far the highest in terms of drive share (approximately 50%). The drive concepts of the three main drives are described below.

In addition to the dynamic operating mode, the following operating conditions must be mentioned:

- High tractive force at low travel speed close to standstill while penetrating into the pile.
- Good dosing of travel speed and tractive force in combination with actuation of the operating hydraulics.
- Frequent reversing with high tractive force requirements.
- High time share in reverse gear (close to 50% during short loading cycle).
- Permanent change in travel speed between 0 and approx. 15 km/h (10 mph) in the short loading cycle.
- Permanent change in the steering angle because of the Y-shape of the driving cycle.

The suitability of transmission technologies may be derived from this. In principle, a continuously variable transmission is particularly advantageous in this application. This is available in the form of a hydrostatic transmission, a hydraulic-mechanical power split, and an electric transmission. The hydrostatic drive also has the advantage that in case of a swash plate-design axial piston variable displacement pump, reverse travel can be easily realized by the negative actuating angle of the swash plate. No separating clutch or friction elements are required for reversing. In the case of power-split transmissions, the manufacturing costs for reversing depend on the internal structure of the transmission. In case of output-coupled transmissions, a friction clutch must be integrated; for input-coupled systems, it is possible to drive in reverse by swiveling the hydrostat (geared neutral). In case of diesel-electric drives, no further component is required for reversing, because the power rectifier reverses the traction motor.

In case of wheel loaders with an operating weight of less than 15 tons, the hydrostatic travel drive has proven to be popular. This fulfils the functional requirements in an ideal way. No additional mechanical transmission is required for machines up to 30 km/h (19 mph) end speed. The components of the drive include the travel pump, hydrostatic traction motor, reduction gear, rear axle with bevel gear drive, differential and planetary final drive, and the front axle with the same structure. The hydraulic motor is flanged with the input stage to the differential housing of the rear axle. The front axle is driven through a cardan shaft with bevel pinion speed through the rear axle input. The pump size must primarily be adapted to the performance of the diesel engine. The size of the oil motor is based on the range of the tractive force hyperbolic and therefore on the choice of the final speed in turn with maximum rim pull requirements. The available series of hydrostatic travel drive components are assigned to pressure ranges. The useful pressure drop in this operation is between 400 and 450 bar (5,802 and 6,527 psi). In case of higher demands on the transmission ratio caused by the tractive force or travel speed, the range of the drive must be increased by increasing the useful angle range of the oil motor or with a shiftable mechanical transmission. The installation of two oil motors is typical, whereby these are adjusted by speed or pressure. A hydrostat in combination with power-shift transmissions is only used with large wheel loaders. The disadvantages of a hydrostatic transmission are poor efficiency and subjectively perceptible noise development due to the operating principle. One major advantage is the good hydraulic or electronic controllability. This advantage can be applied to the previously mentioned coordination of travel and lifting movements when filling the bucket. For reasons of comfort and safety, hydrostatic traction drives are increasingly controlled electronically. This can also resolve the strict relationship between diesel speed and hydrostatic transmission ratio. If possible, the operating speed of the diesel engine should be kept low. The speed of the diesel engine is only increased if the power requirement or flow demand is high.

6.2.4.1 **Example of Liebherr 2plus2® Travel Drive:** The most advanced hydrostatic traction drive for large wheel loaders is the so-called 2plus2® drive (**Figure 6.42**). A hydraulic drive pump, two variable oil motors, and a multispeed transmission are used. The drive is used as a modular system for wheel loaders from 17 to 33 tons operating weight. The mechanical transmission is identical for all applications. In case of six wheel loader models, two pump sizes and three oil motor sizes from the supplier's series are used. All three main components are controlled electronically. The vehicle manufacturer develops the control software that runs on the OEM's master computer. The system is the result of a development cooperation of Bosch-Rexroth, Dana, and Liebherr. In the first travel range up to approx. 7.5 km/h (4.7 mph), both oil motors are connected to the output through couplings K1 and K2. The speed is increased by continuously reducing the bent-axis angle of the motors. In the second travel range, only oil motor 2 provides torque (up to approx. 17 km/h (11 mph)). In the third travel range up to 40 km/h (25 mph), motor 1 with coupling 3

FIGURE 6.42 Driveline structure of the Liebherr 2plus2® drive.

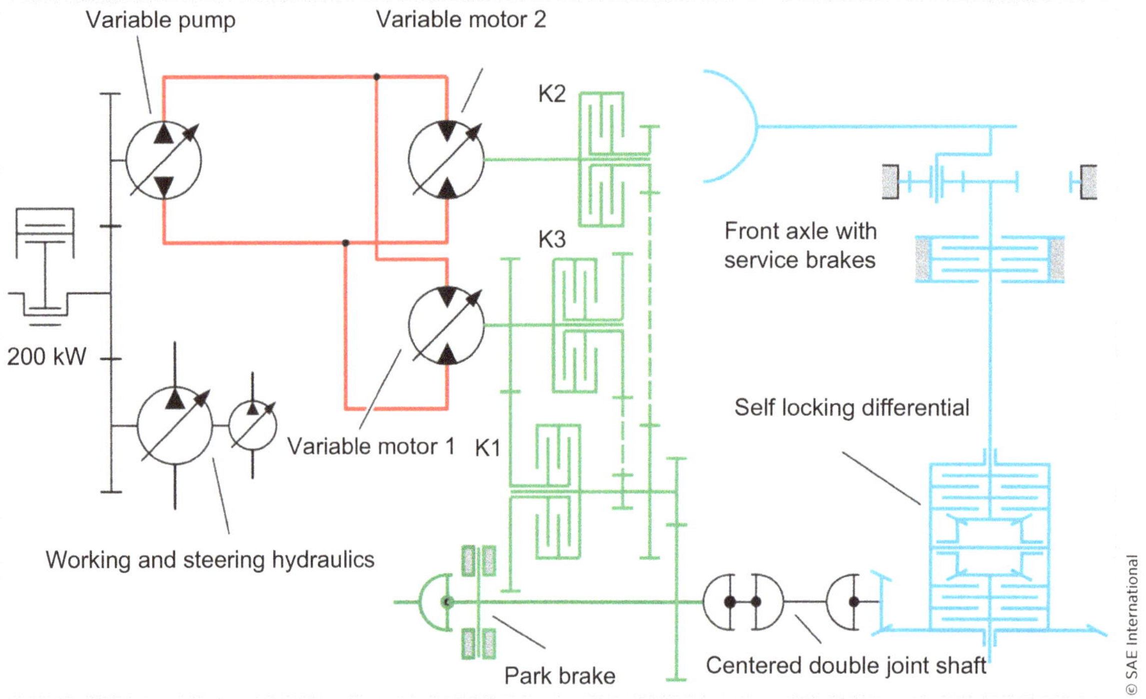

and an adequate transmission ratio is used. In the first section of travel ranges 1 and 3, the speed is increased by swivelling out the pump (at constant motor angle). Due to the system, the speed and torque range forward and backward is the same, that is, the same rimpull and maximum speed in both directions of travel.

6.2.4.2 **Power-Shift Gearbox and Torque Converter:** The power-shift gearbox dominates the market for large machines. Figure 6.43 shows the diagram of this type of traction drive. In this case, the transmission has six forward gears in two groups and three reverse gears in one group. The gears are shifted by wet multidisc clutches. The control is electrohydraulic and automated. The gears can be changed under full engine load, whereby the hydrodynamic torque converter placed between the engine and transmission absorbs differences in speed and also temporarily allows the transmission to transmit high tractive force at zero wheel speed. The torque converter transmits the force from the impeller to the turbine wheel through the kinetic energy of the fluid. Even in stationary operation, there is a loss due to slippage and hydraulic friction. Modern transmissions are therefore equipped with a lock-up clutch in the converter. This means that at higher speeds, the pump and turbine are connected rigidly and loss due to slip is avoided.

6.2.4.3 **Power-Split Transmission:** Currently, hydraulic-mechanical power-split transmissions in tractors have replaced the conventional multiple-stepped and power-shifting gear transmissions. This transmission concept is now also used sporadically for wheel loaders and, with a suitable design, can exceed the hydrostatic drive in terms of

FIGURE 6.43 Driveline structure of a power shift transmission (ZF Ergopower®).

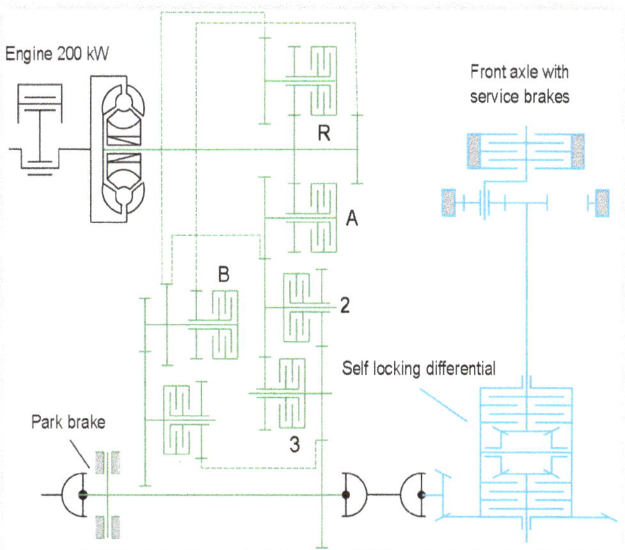

efficiency with the comfort of being continuously variable. Liebherr equips the whole range of big wheel loaders with a power-split drivetrain called Xpower® (**Figure 6.44**). The transmission platform consists of three sizes. Two of them are two-range and one is a three-range gear box. The hydrostatic units are available in two sizes. Figure 6.44 shows the transmission schematics for the two-range version. The input shaft is connected to a reversing unit with forward and reverse clutches. This drives the planetary carrier

FIGURE 6.44 Driveline structure of a power-split transmission (Liebherr Xpower®).

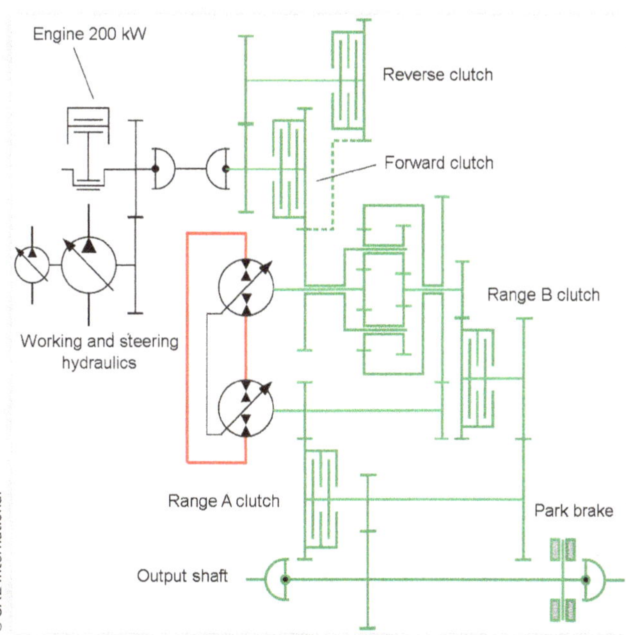

of the following planetary gear set. The power is split at the planets in the sun-driven hydrostatic power and the ring-gear-driven mechanical branch. The hydrostatic motor on the left side and the gear on the right hand side drive the common shaft which is connected to the output by clutch A. In the second range (clutch B) pump and motor function vice versa. The pump unit is now driven by the ring gear and the hydrostatic motor drives the sun gear of the first stage of the planetary gear set. The torque of the mechanical and hydrostatic path is summed at the planets of the second stage of the planetary drive. The sun gear of the second stage drives the output shaft through coupling B. Range 3 of the three-range version is analog to range 1 with adjusted output gear ratio. The transmission ratio is controlled by the angle of the hydrostatic units. At zero ground speed the pump angle is set to zero, so no torque is transmitted at high rotational speed of the pump. Only input shafts are rotating. For accelerating the vehicle, the pump increases its displacement and the motor's volume is reduced simultaneously. The transmission ratio is controlled by a transmission ECU. The ratio is commanded by the vehicle controller.

6.2.4.4 Axles: The axles in wheel loaders have the following functions:

- Distribution of the drive torque to four wheels.
- Increase in the drive torque by a factor of 15–35 times.
- Control of the drive torque distribution under different tractive conditions in the contact area of the tires (differential lock).
- Support of vertical loads by means of the trumpet arm to the wheels.
- Integration of the single wheel brakes.
- If necessary, adapt the king-pin steering including hydraulic actuation.

The axles in wheel loaders are designed as rigid axles, so both wheels are mounted in a rigid body (common housing). In case of steered wheel loader axles, both steering knuckles are mounted rotating in the trumpet arm (steerable rigid axle). The term "rigid axle" is used with regard to the function of the wheel suspension and therefore stands in contrast to an independent wheel suspension. Wheel loaders are always equipped with four wheel drive, whereby there is no differential between the front and rear axle drives. In case of articulated steered vehicles and a central articulated joint, this speed balance is not necessary because the path at front and rear wheels is equal. The axle input-side bevel gear drive converts the torque by the factor of 3–4. The differential housing contains a differential gearset with two or four bevel gears. The compensation bevel gears drive output bevel gears. On the output side, a final drive featuring planetary design is installed which increases the torque by a factor of 6. The planetary gear features a standard design to provide drive through the sun and output through the planetary carrier. The planetary carrier is connected to the wheel flange. The planetary final drive can be mounted on the differential or wheel side. Wet disc brakes, dry partial disc brakes, or dry drum brakes are used as braking devices. Dry drum brakes are mounted at the axle input. The braking torque acts on both wheels of the axle. Wet disc brakes or dry partial disc brakes (Asia) are normally preferred as wheel service brakes. The brakes are mainly pressurized hydraulically with auxiliary power. The brake pressure is up to 100 bar (1,450 psi) and is applied to a ring-shaped cylinder in wet disc brake design. Wet

FIGURE 6.45 Cross section of axle differential featuring a self-lock.

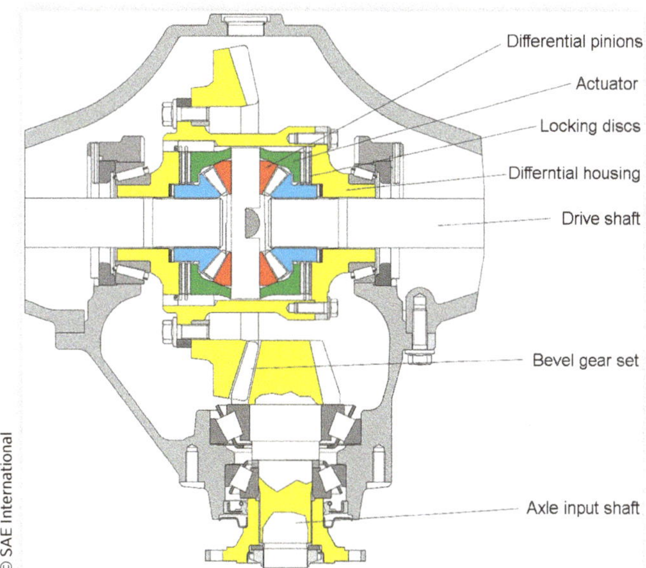

disc brakes are dimensioned so that long life is achieved and the braking requirements according to standards are fulfilled. In the case of a hydrostatic driveline and pushing operation, the diesel engine also acts as a brake and protects the service brake from excessive wear. Hence, the lining on the brake discs lasts for machine life without replacement. The following designs are used for the differential lock:

- Open differential in case of standard operations on the rear axle. During loading, the torque is mainly produced on the front axle. When the rear axle is open, a self-locking differential or a switchable 100% lock is applied on the front axle.
- Self-locking differential (Figure 6.45). The locking effect of this differential is automatically generated depending on the drive torque. The lock functions without any action of the driver and requires no sensors. The lock is designed as a wet disc lockout. The axial pressure results from the drive torque. The blocking effect is limited to from 30% to 50%.
- Switchable 100% lock: A wet multiplate disc clutch connects one output side to the differential housing so that there is equal speed on the output shaft and the differential housing. This type of lock is activated manually or by monitoring the wheel torque.

Operations with abrupt changes in the tractive force coefficient are critical. A wheel can spin freely or build up torque in a jerky manner. Typical operations are the paper industry, for example, where the front axle is driven onto the loading material. In this case, a careful selection of the lock is required, which is optionally available. Occasionally, axle disconnect is also offered. While driving on the road, the driveline of one axle may be disconnected. Wet multiplate disc clutches are also common in this case. Driveline tension can also occur in articulated steered vehicles with rigid all-wheel drive due to different tire deflection.

Depending on the steering type and geometry of the cardan shaft in relation to the articulated joint, single or double cardan joints are used between the transmission and the front axle. A special design is the centered double universal joint drive shaft, which enables balanced motion in the shaft tube due to the W-layout in the double universal joint.

6.2.4.5 **Wheels and Tires:** Even if the wheel loader is referred to as an earth-moving machine, this machine moves mainly on the solid surface with high traction capability. This does not apply to machinery used in the gardening, landscaping, and agriculture sectors. For these customers, the machines are equipped with tires equipped with lugs as on agricultural tractors if required. Tires are preferably selected according to the following ground conditions: sand, gravel, rock, clay, soil, asphalt, concrete, scrap, and recycling.

The following requirements apply to the tires: ride comfort, stability against roll, traction on required surface, self-cleaning, protection against damage, mileage, and multipurpose ability.

Ride comfort is very important because the tire is the essential suspension and damping element on the vehicle. Despite the high axle loads and the associated high internal tire pressures, the tire must have good suspension and damping properties. High-quality tires are always manufactured in radial design. These offer not only greater ride comfort but also longer life, better traction, lower rolling resistance and therefore lower fuel consumption, do not overheat easily, and offer greater protection against damage. Tires vary in profile depth (L2–L5) between 100% and 250%, whereby large profile depths mainly improve the cutting resistance. In some cases, low-profile tires are available. These are used where high payloads are driven at high speed (timber work). In case of operations with high damage protection requirements, solid rubber tires are used or the tires are filled with foam. Both measures increase puncture protection, lead to a considerably higher vehicle weight, and require a speed limit of 20 km/h (12 mph). In any case, the vehicle manufacturer must approve these tire types or compensate for them by reducing the rear ballast, for example. In case of large wheel loaders, the rim may be split for better assembly. The outer rim ring is held by a snap ring. This ring is removed for assembly. The two rim parts are secured in the circumferential direction by an antirotation device. The practical lifetime of the tires depends strongly on the conditions of use. Typical values are between 5,000 and 8,000 operating hours.

6.2.4.6 **Working Hydraulics:** The work function controls both movements of the lift linkage, lifting, and tipping the bucket. Both movements are performed by dual-acting hydraulic cylinders. Normally, both circuits are supplied by a single hydraulic pump. In larger vehicles, the consumers are controlled by a load-sensing system (LS) where the oil flow depends on the demand. The operating pressures are around 200 bar (2,900 psi) for small machines and around 400 bar (5,800 psi) for large machines. The maximum flow rate depends on the vehicle size and is dimensioned so that the loading cycle (lifting, tilting out, lowering) can be completed in approximately 12 s. For small wheel loaders, gear pumps with steering function prioritization are used as the main pump; for larger loaders, axial piston variable displacement pumps featuring a swash plate design are usually used. The swivel angle of the pump is determined by a pressure-flow regulator. In addition to the LS governor, pressure controllers and power regulators are used. Depending on the operating pressure, the control valves are designed in sectional and single-block design. Pistons are available for the lifting/lowering and tilting out/in functions and optionally for a third and fourth control circuit. The control circuits feature primary and secondary relieve valves. In case of small loaders, the control pistons are mechanically actuated; while in case of medium and large wheel loaders, the pistons are always piloted hydraulically. The pilot valves are controlled either manually or electrically. Electrohydraulic pilot controls enable a number of additional convenient functions. These include parallel movement, lift height limitation, damping at end stops, automatic lowering, automatic return to dig, and automatic weighing systems. There are two concepts for the control levers of the working hydraulics: Cross-control lever and multi-lever control. In case of multi-lever control, the lifting and tipping functions are operated by two single levers, which are controlled in the direction of the driver's forearm. Additional control circuits are controlled by additional control levers or small proportional switches integrated in the main lever.

6.2.4.7 **Steering System:** The steering system is operated hydraulically. The driver activates the rotary spool of the steering valve (also referred to as a steering servostat or orbitrol steering) through the steering wheel. The flow rate through the steering valve is determined by the rotational speed of the steering wheel. The maximum steering speed is determined by the size of the steering valve and the size of the steering pump. The steering valve is dimensioned so that the steering wheel can be turned 4–6 turns before from left to right stops. The steering system operates at a maximum operating pressure of approx. 200 bar (2,900 psi). Due to the considerable portion of the steering in the total energy consumption of the vehicle, axial piston variable displacement pumps are also used on the wheel loader as a single pump in the steering system in a load-sensing control system. This heavily reduces losses in the steering system due to flow control and the high time share involved with the steering operation. The steering movement of the articulated steering is generated by one or two steering cylinders. Small machines often only have one steering cylinder. This results in a different steering speed (differential cylinder) for left- and right-hand turns. Power-assisted steering systems must remain steerable even if the power supply fails. In case of small wheel loaders, this can be achieved using the steering valve as a steering pump. Larger vehicles require a secondary energy source. This may be a driveline-driven pump or an electrically driven auxiliary pump.

In case of large wheel loaders, which are mainly used for short loading cycles, there is an increasing demand for steering systems controlled electrically by joystick. The easy and short lever movement of the joystick on the left armrest considerably increases the steering comfort. Two steering principles are possible. On one hand, the steering speed can be controlled by the angle of the steering joystick. A second option is to use the angle of the lever to set the steering angle (articulation angle). The steering speed results from the mismatch of the lever position from the current steering angle. In this kind of steering system, the current steering angle must be measured. One has to get used to operate a machine with joystick steering. After a learning phase, this kind of system is generally well-accepted. In accordance with the rules of functional safety, steering systems have a high safety integrity level. Electrohydraulic steering systems must be designed and dimensioned accordingly. Electric and piloted hydraulic elements in the control chain are therefore designed redundantly and independently. Electrohydraulic steering systems can also be approved for use on public roads, if the system structure is suitable.

6.2.4.8 **Brake System:** The classic braking system from medium wheel loaders upward is the hydraulic-pump-accumulator braking system. This has been in use for decades. Air-pressure-operated brake systems are not common for wheel loaders, because there is usually no operation with trailers on public roads. In Asia, air pressure brake systems are also used in wheel loaders with dry partial disc brakes. In the hydraulic brake system, hydraulic oil from hydraulic accumulators is pushed to the brake pistons in the axles through the foot-operated brake valve. The hydraulic accumulators are charged by a gear pump through an accumulator charge valve. In case of wheel loaders with a maximum speed of 40 km/h (25 mph), the brake system is designed with two circuits. One circuit each operates the brakes for one axle. The status of the accumulator is monitored by pressure sensors. The accumulators are dimensioned so that eight repeated braking operations are possible without reloading the accumulators. In addition to the service brake, an electrically actuated, negatively acting parking brake is used. This acts on a dry disk brake on a drive shaft to one axle, for example.

6.2.5 Drivers Cab

Several aspects must be taken into account when developing the cab.

- The available installation space between the front section of the vehicle and the engine compartment or cooling system must be used optimally.
- The basic cabin structure must be simple and cost-effective, consisting four columns and with strength and energy absorption capacity in accordance with the standard (ROPS).
- Direct and safe access in combination with good maintenance accessibility to the diesel engine and the cooling system must be ensured as required.
- Optimum visibility to the front, to the side into the articulation area, and to the rear on the ballast edges must be possible. Even though a reversing camera has become standard on large machines, many drivers also drive backward direct view. An essential aspect of visibility is suitable illumination of the working areas with work lights.
- An optimal cab climate can be achieved by air conditioning in two or three levels, with high air volumes but low air speeds. The air conditioning system must allow a rapid cooling curve and a low steady-state temperature.

The display concept must be ergonomically positioned and user-friendly (**Figure 6.46**). The challenge lies in the fact that the properties of the machine control system can be optimized to suite the driver and the operating conditions by means of a large number of parameters. These settings are accessible through a settings menu. Drivers often struggle when trying to make all the settings on a multilayer control monitor.

On one hand, the control monitor should be large and give a clear overview; on the other hand, it must not restrict the view. Additional monitors are optionally available for a weighing device or for displaying surround view.

The central control elements include the accelerator and brake pedals, the control lever for the working hydraulics on the right hand, and the operation of the steering

FIGURE 6.46 View from the driver's seat and instrumentations.

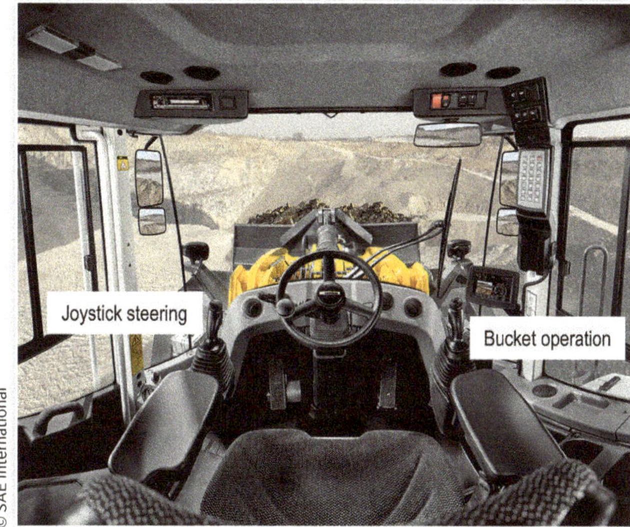

with the left hand (steering wheel or steering joystick). The steering and the control lever for the working hydraulics must be optimally and individually adjustable to the driver's body metrics.

Assistance systems include systems in the vehicle that support or relieve the driver but are not necessary to control the basic functions. These systems provide easy operation, safety, easy service and maintenance, increased efficiency and handling performance, and relieve the driver of routine tasks. Systems of this type are also generally familiar from on-road vehicles. In any case, the focus of working machines is not only on support during travel but also and especially on improvements in the execution of the working process. These systems are partly based on technologies and sensors that have been developed for automotive application and may also be used for work functions at a reasonable cost.

Assistance systems to improve visibility are the most important. These primarily improve safety. These include the surround view (360° panoramic view from a top perspective), a front camera for large equipment, and the popular reversing camera. In case of these systems, the driver must continue to look at the image and recognize dangers for himself. Images may also be interpreted automatically using an object recognition algorithm. Initial application in construction machines makes it possible to recognize people in different poses. The driver is actively warned, and the recognized person is indicated. Depending on the operation, wheel loaders are often driven at the limits of a collision. A general warning in the event of an imminent collision would often draw the driver's attention away for no reason. Front cameras are used with large-view-blocking equipment and monitor the close range during forward travel. The lens is mounted on a tripod above the cab for a better angle of view. Alternatively, these cameras may be mounted at the front of the equipment in a protected area. Other assistance systems influence the movement of the lift arms. The controlled return of the bucket tilt angle to the ground-parallel alignment after dumping is common. This "bucket return-to-dig" function may be extended with automatic lift arm lowering. All the driver has to do is check the area of emergency, and the movements are carried out at the push of a button. Any stops of the movement of lift and tilt cylinders are also possible. For example, the lifting height may be limited. This may be used in buildings to avoid collisions or to save energy. Other kinematic aids include parallel guidance and lift movement damping before hard stops. The automatic loading system, which carries out the bucket filling process without any control by the driver, is particularly important. Such systems have been developed and are offered as an option. However, the function is still limited, and the experienced driver prefers to control this action manually. In special operations, wheel loaders are remote-controlled or move autonomously in sections of routes. These may include subsurface mining operations, contaminated environments, or very consistent operating conditions that are easy to automate. Another area of the assistance systems involves the acquisition and analysis of process data. For wheel loaders, the most important process variable is the weight of the loaded material. The current payload may be recorded on the lift arms by means of an optional measuring system and is generally available for further use inside and outside of the machine.

6.2.5.1 **Telematics Systems:** Machine and process data should be analyzed primarily to optimize machine utilization. To increase machine availability, data describing the machine status should be evaluated by the service department (condition monitoring). Faulty or not allowed machine states are recorded electronically and stored in the control system as service codes. Knowing how the wheel loader operates, where it is located, and how often it is used is essential for effective fleet management. Finally, the engineering department should collect data on the operation profiles and the component load spectrum. This is the only way to achieve real data for dimensioning of the main components (Figure 6.47).

The data for these user groups are recorded in an electronic control unit and transmitted at regular intervals to a stationary server through a radio module. Users can view the relevant data based on defined access permissions. Evaluation is performed offline by the user groups by accessing the server data. This technology is part of the basis for digitizing construction site processes (BIM).

FIGURE 6.47 Telematic system and data processing for engineering purpose.

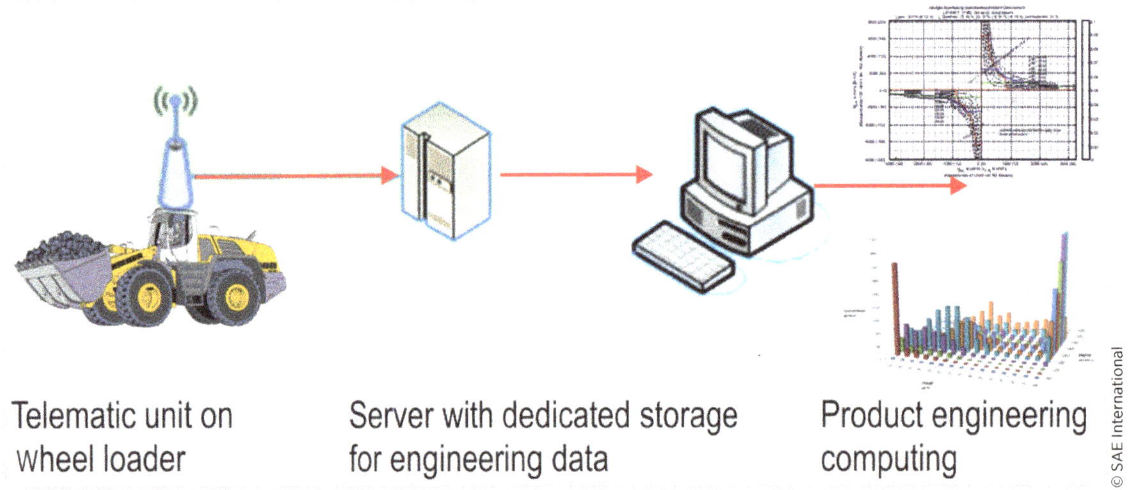

6.3 Forklift Truck (By Rainer Bavendiek)

The forklift truck has been around for a century, but today it is found in every warehouse operation around the world. The first forklifts were developed as a result of the manpower shortages caused by the First World War. During the early 1920s, the design of the forklift evolved from a tractor with an attachment to a dedicated machine with a vertical lifting mast. In 1924 the "CLARK Tructractor Company" introduced the first forklift truck with an IC engine inside as shown in Figure 6.48, and in 1928 the first forklift truck equipped with a mast with hydraulics instead of chains and cables was offered in the United States [66]. Development of the forklift advanced with the advent of the Second World War with the forklift playing a part in the handling of materials for armies throughout the world. After the Second World War, the development of the forklift gained momentum and battery-driven forklifts made an appearance in the 1950s as well as more specialized forklifts such as the narrow aisle reach truck. In 1953, a torque converter replaced the dry friction clutch and the first serious safety issues as an overhead guard came up in 1964 [66]. In the 1960s and 1970s, improvements in the electronic controls made forklifts more versatile as companies began to look at warehouse efficiency. Today, the forklift truck can be powered by a number of fuel options including gasoline, diesel, electrical battery, compressed natural gas (CNG), and liquid propane gas (LPG).

The first 4.0–5.0 tons engine hybrid forklift was developed and sold by Mitsubishi Heavy Industries, Ltd. that run on diesel and a lithium-ion battery [91]. This powertrain is equipped additional to the conventional construction with two electrical motors, one for traveling and one for lifting as shown in Figure 6.49. While driving with low speed only the electrical motor powers the fork lift truck (blue arrows on the left side). At higher speed, the IC engine is used too (red arrows on the left side). Braking is realized by regeneration through the electrical machine which recharges the lithium-ion battery (yellow arrows on the left side). An idling stop function is integrated. For lifting with heavy loads, both drives work (blue and red arrows on the right side).

FIGURE 6.48 Historical forklift truck (Clark Material Handling Company) [66].

FIGURE 6.49 Hybrid forklift truck: Energy flow (Mitsubishi Heavy Industries, Ltd.) [91].

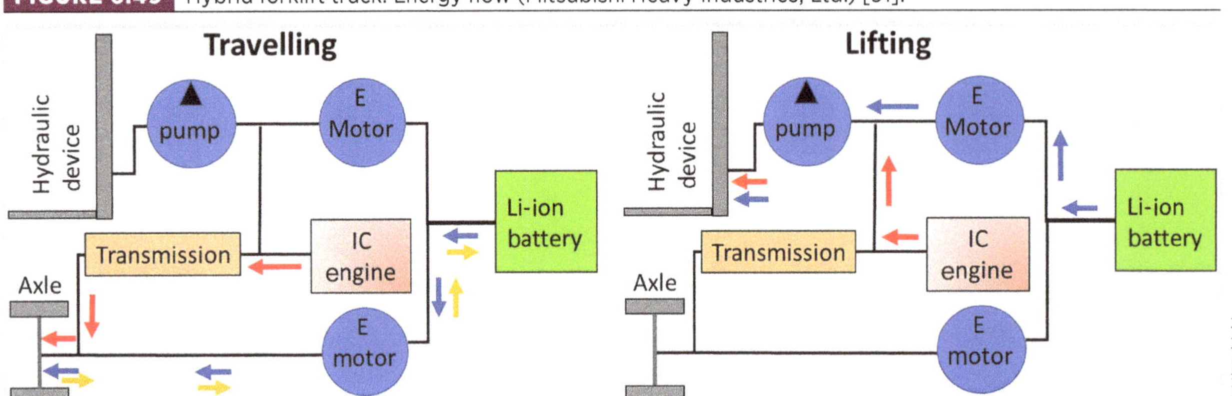

The world of material handling changes dramatically by standardization of the load pick-up. In 1956, the first tanker was loaded with 58 containers [67]. Containers have been standardized in ISO 668 and have been available in 10-, 20-, and 40-ft length [87]. Nowadays, many different containers for special usage as dry bulking or cooling are standardized, too. In 1961, European railroad companies reached an agreement for a standardized wooden pallet [73], the so-called "Euro-pallet," for transportation goods. Since 1968, box pallets have been standardized in EN 13626 [72], which are used very often for the in-plant transportation. Most of the goods to be transported could be stored on a pallet or in such a container. Unfortunately, these standards for the pallets are only used in Europe and not worldwide, such as in India or in China for the transportation out of the plants. However, the load handling attachments are standardized international too, see Section 6.3.8.

6.3.1 Classification and Standards

6.3.1.1 Definition Forklift Truck: A **forklift truck** is a powered industrial truck used to lift and move materials over short distances. A forklift truck can handle auxiliary devices in and outside the warehouse. In the logistics, an auxiliary device bundles singular goods up to bigger units. Such auxiliary devices could be differentiated as given below:

- Noncarrying auxiliary devices as bags
- Carrying auxiliary devices as pallets or box pallets
- Carrying and enclosing auxiliary devices as shelf boards

6.3.1.2 Forklift Classifications: The operation of a forklift truck is very different regarding the tasks and the loads, the load handling devices and the environmental conditions as already described in Chapter 2. Therefore, in the United States, there are defined seven classes of forklifts to consider the different fuel options and the usage as shown in Table 6.8. This table gives an overview about the main different kinds of forklift trucks. The first column contains a rough description of the main characteristics, the second column describes the ideal usage of the truck, and the third column the main benefits. In column 4, there is only one example shown. Additional examples could be found at the

TABLE 6.8 Classes of forklift trucks [92].

Class	Description	Ideal Uses	Benefits	Example
I	Electric motor rider trucks, cushion or pneumatic tires	Loading/unloading tractor-trailer; handling pallets	Electric means no emissions, minimal noise	
II	Electric motor narrow aisle trucks	Operating in tight spaces, handling pallets, picking/storing inventory	Can be used to gain more storage space in same warehouse footprint	
III	Electric motor hand trucks or hand/rider trucks with a tiller	Unloading deliveries from tractor-trailers; short runs in smaller area	Rider and walk-behind options	
IV	Internal combustion engine trucks (solid/cushion tires)	Moving pallets from the loading dock to storage, vice versa	Cushion tires great for low-clearance situations	
V	Internal combustion engine trucks (pneumatic tires)	Versatile; trucks can handle single pallets to loaded 40-foot containers	Mostly for outdoor use, but also indoors in large warehouses	
VI	Electric and internal combustion engine tractors	Commonly used for hauling of pulling loads rather than lifting; versatile	Example: airport "tugger" towing luggage carts	
VII	Rough terrain forklift trucks	Great for lumberyards/construction sites where crews need to lift building materials to high elevations	Some are equipped with telescoping mast to provide far greater reach	

United States Department of Labor, Occupational Safety & Health Administration [92]. There are many companies worldwide which exist which act as a manufacturer with a complete product range and provide the market with all those different trucks.

6.3.1.3 **Standards:** As described above in the 50th and 60th of the last century, the standardization of the transport equipment came up. Therefore, it was a need to standardize load handling equipment worldwide, too. In the beginning it started

TABLE 6.9 Most important standards for forklift trucks

Standard	Headline	Information
ISO 3691	Industrial trucks	Safety requirements and verification
ISO 3691-1	Self-propelled industrial trucks, other than driverless trucks, variable reach trucks and burden-carrier trucks	
ISO 3691-2	Self-propelled variable-reach trucks	
ISO 3691-3	Additional requirements for trucks with elevating operator position and trucks specifically designed to travel with elevated load	
ISO 3691-4	Driverless industrial trucks and their systems	
ISO 3691-5	Pedestrian propelled trucks	
ISO 3691-6	Burden and personnel carriers	
ISO 3691-7	Regional requirements for countries within the European Community	
ISO 3691-8	Regional requirements for countries outside the European Community	
ISO 2330	Fork-lift trucks—Fork arms	Technical characteristics and testing
ISO 5053	Powered industrial trucks	Terminology
ISO 6292	Powered industrial trucks and tractors	Brake performance and component strength
ISO 6055	Industrial trucks. Overhead guards.	Specification and testing
DIN EN 16796	Energy efficiency of industrial trucks	Test methods
ISO 13564	Powered industrial trucks	Test methods for verification of visibility
ISO 22915	Industrial trucks	Verification of stability
ISO 20898	Industrial trucks	Electrical requirements
VDI 2198	Type sheets for industrial trucks	

national-wise. For example, in the United States, it started regarding safety for powered industrial trucks in 1969 with ASME B56-1 [63]. In Europe, some organizations such as BS in the United Kingdom or DIN or VDI in Germany worked out national standards. At end of the 80th of the last century, the standardization was transformed into the European standards (Machinery Directive 89/392/EEC [90]). Since the 90th many of the national and international standards consolidated in ISO standards. The overall international worldwide standard is ISO 3691 [85]. Meanwhile, most of the safety issues regarding forklift trucks are covered by standards, mainly worldwide. The most important standards (international and national) could be found in Table 6.9.

6.3.2 Requirements and Main Functions

6.3.2.1 **Requirements:** As already mentioned above, the requirements are very different regarding, for example, the transport goods, the environment conditions (indoor or outdoor), the quality of the floor surface, the constricted room, and the duty cycle. Of

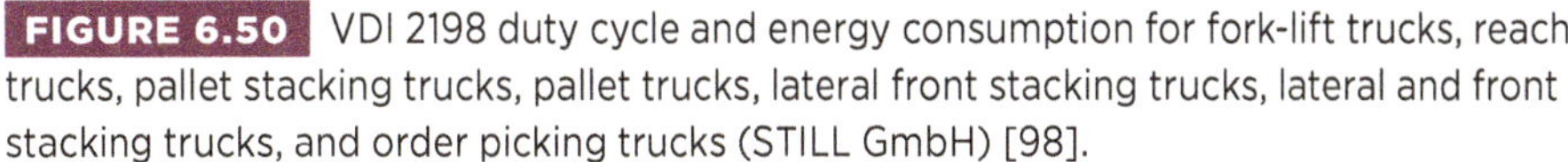
FIGURE 6.50 VDI 2198 duty cycle and energy consumption for fork-lift trucks, reach trucks, pallet stacking trucks, pallet trucks, lateral front stacking trucks, lateral and front stacking trucks, and order picking trucks (STILL GmbH) [98].

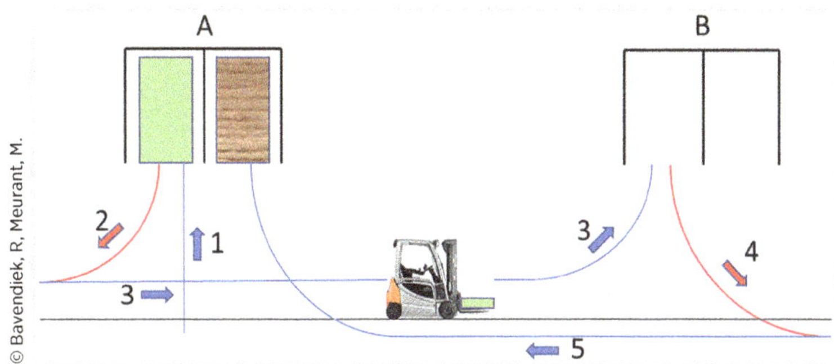

course, duty cycle is the most different factor in the usage of a forklift truck. Sometimes, it is far away from any standard. Sometimes, it is very hard for the customer to choose the most sufficient forklift truck for his/her special use case and the right supplier for it. The German standard VDI 2198 could be helpful to compare the performance data of the different manufactures [98]. Depending on the kind of the forklift truck, there are two different duty cycles described in VDI 2198. For trucks equipped with forks, Figure 6.50 shows the duty cycle. On position A, the load is colored green, the pallet wooden.

Typical for a duty cycle, a load has to be transported from one place A to a place B. Very often, loading a truck in a truck company, the forklift truck has to return without load to place A. Therefore, the duty cycle contains several actions:

- Driving to place A where the pallet (or the load) has to be picked up (1).
- Positioning the truck in front of the load by maneuvering and steering.
- Pick up the load with the load-handling device – normally by lifting and tilting the mast and using attachments such as a side shifter and clamps (see Section 6.3.8).
- Driving backward (2).
- Transport the load to destination B (3).
- Positioning the truck to deposit the pallet (or the load).
- Deposit the pallet (or the load) at destination B.
- Driving backward (4).
- Driving to place A back (5).
- The same procedure with load and so on.

The distances between place A and B (20 m or 30 m (22 or 33 yards)) depend on the kind of the truck. For checking the energy consumption, this duty cycle has to be repeated 60 times in 1 h, but the pick-and-load cycles should be simulated.

Coming out of the different use cases of a forklift truck, there are many different and special requirements to be fulfilled which lead mainly to many different functions. They can be clustered into several main functions and each with several subfunctions.

6.3.2.2 **Main Functions:** To fulfill this duty cycle, a forklift truck has in general 11 main functions as shown in Figure 6.51. In the following chapters, the main

FIGURE 6.51 Main functions of a forklift truck [95].

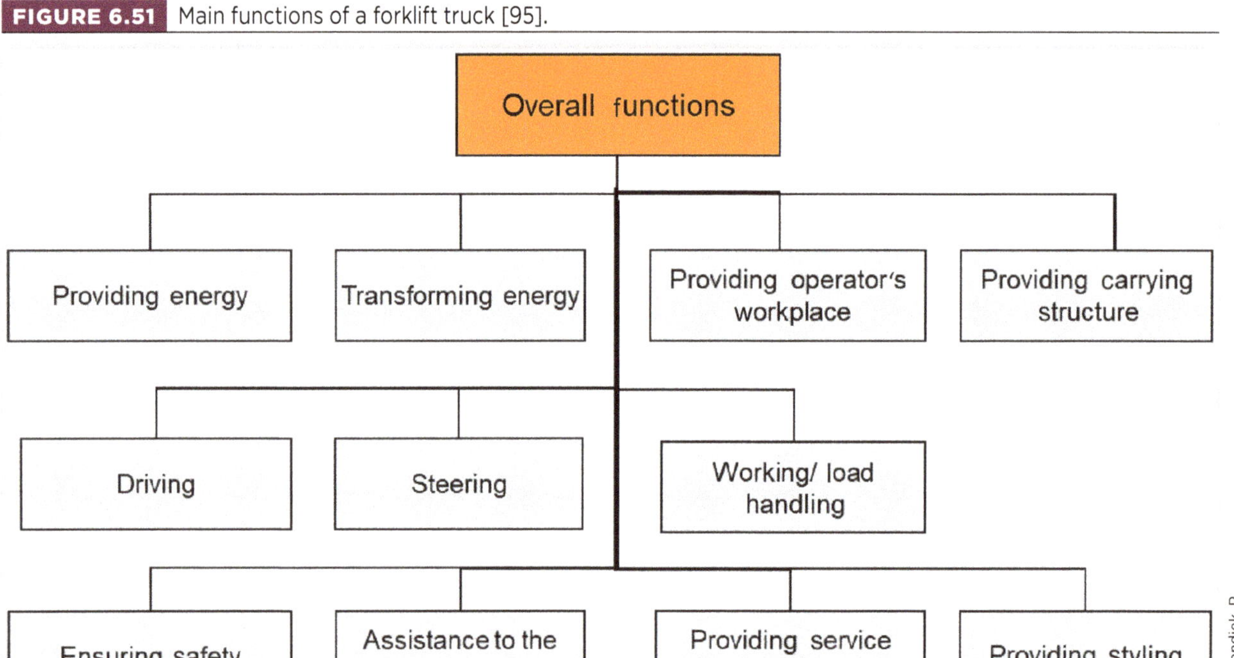

components of a forklift truck are described to fulfill these functions. At first, the truck has to provide the powertrain with energy which could be electrical through a battery or through different kinds of fuel such as gas, diesel, LPG, CNG, or hydrogen. The powertrain transforms the energy into mechanical energy for driving, steering, and load handling. A chassis structure is required to carry all components of the forklift truck, often designed as a frame as described in Chapter 2 and Section 6.3.3. On the frame, there is often mounted the operator's workplace with all the controls for driving, steering, and working/load handling, such as a steering wheel, accelerator and braking pedal, or levers for the hydraulics (see Section 6.3.6). Additionally, there are often overall functions integrated to ensure safety and to give the operator assistance. To minimize the total cost of ownership (TCO), easy entrance to all repair and maintenance positions should be ensured. Finally, the styling of the forklift trucks has shaped up well in between the last two decades.

Each of these main functions could be detailed in up to five levels of subfunctions. An example for providing electrical energy on an electrical AC-powered tuck is shown in Figure 6.52. There are three additional levels of subfunctions added. In the first subfunction level, the energy (e.g., on 80 V) is provided through a converter into the powertrains for driving, steering, and working as shown in Figure 6.52. Through a transformer, the electrical energy supplies all additional electrical components as the controls at the operator's workplace, the display panels, devices as heating for the cabin and the seat, window washers, searchlights, reading lamp and an optional radio, as well as for the electronical controls and the transport safety devices on different voltage levels.

FIGURE 6.52 Subfunction providing electrical energy [95].

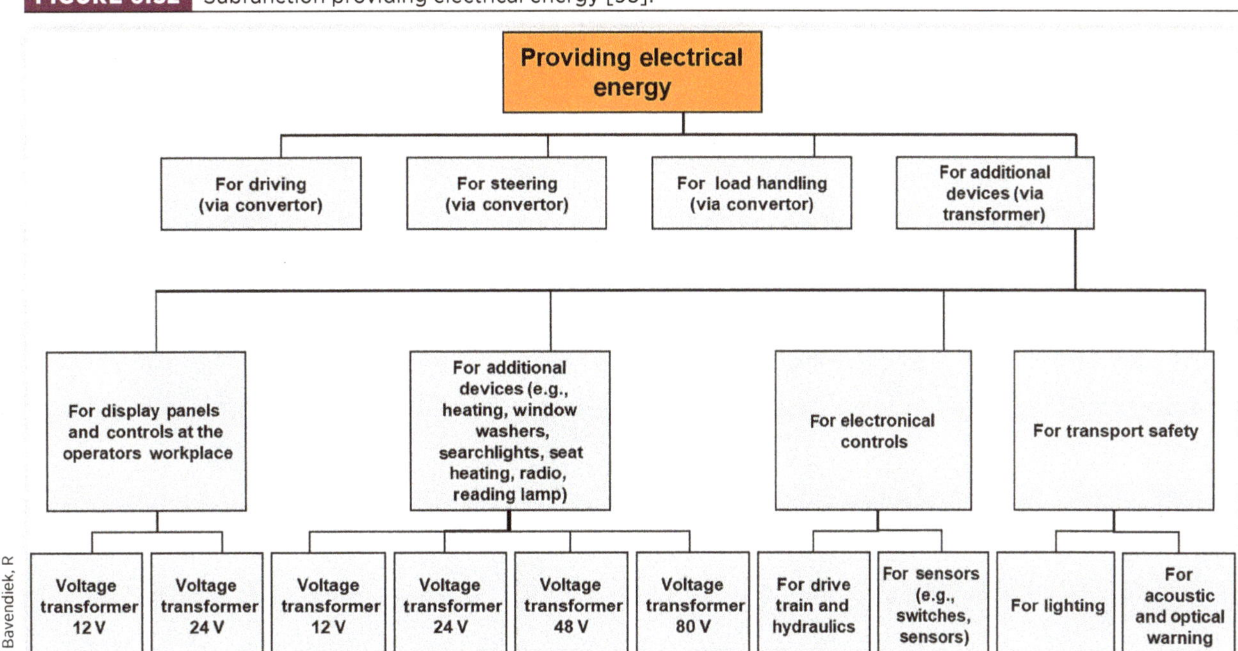

As accounted in Chapter 2, it is very important to fulfill as much functions as possible by creating modules like seating the operator, the cabin, and axles, but some could only by designing a system such as steering.

There are some very important requirements which cannot be fulfilled by only one component/module or system. Some important examples are the stability of the truck, visibility, functional safety, noise, and vibrations.

6.3.2.3 Overall Functions:

6.3.2.3.1 Stability. The stability of a forklift truck is in principle defined by the type of construction (three or four wheels) and the kind of rear axle (center pivot plate or pendulum axle, see Figure 6.53 (shown without the steering function)): The axle is mounted at the frame/counterweight through the red bearings. In case of the pendulum rear axle, it can turn around the connection of the two bearings, that is, around the forward traveling direction of the truck. The center pivot axle can be designed with one or two wheels and the wheels can only rotate around the vertical axis of the bearing.

There are several traveling and maneuvering situations which could be critical for the stability of a forklift truck. A forklift truck can tilt over lateral or to the front, if the truck gets while operation into an instable (driving) situation, such as traveling very slowly (or fast) longitudinally or laterally on a sloping level (failure of the static balance) or traveling fast into a curve or over traveling an obstacle (failure of the dynamic balance). Tilting to the front can only happen by traveling with lifted load because with lowered load the fork arm will prevent the tilting. Traveling with lifted load would be misuse or not in accordance with the regulations, but is necessary by transferring the load into/out

FIGURE 6.53 Different kinds of rear axles.

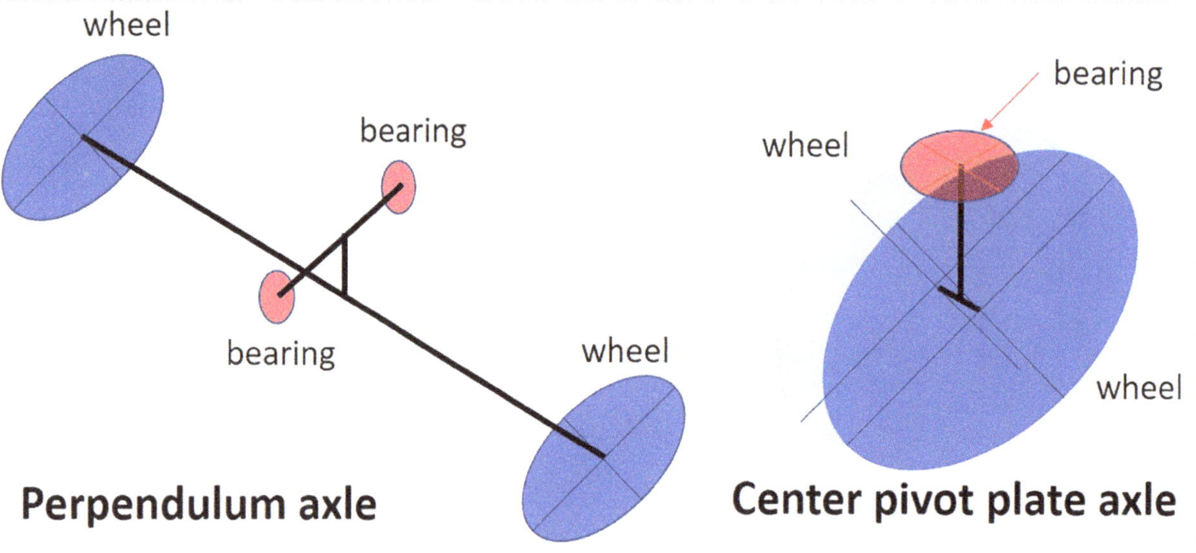

of the rack or by loading/unloading a motor truck. A short jerk by braking or accelerating the forklift truck or a floor roughness initiates the tilting. Very often the operator steers additional because of lack of space in a narrow aisle so the forklift truck will tilt lateral.

This tilting behavior depends, in principle, on the design of the rear axle. Most important is the position of the tipping line. These tipping lines are defined by both outside of the turn wheels in case of a three-wheel truck and are located therefore on the floor level as shown in Figure 6.54 left-hand side with the red line. Running with high speed in a curve, the inside of the turn wheel will raise and the counterweight will hit the ground so that the new tipping line will be defined by the chassis/frame as shown in Figure 6.54 left-hand side with the blue line [65].

Four-wheel trucks are normally equipped with a pendulum axle to realize a stable driving. There are three cases of tilting: The chassis, the pendulum axle, and the whole truck. In case of a lateral tilting, it starts with the tilting of the chassis: The tilting line runs from the outside of the driven wheel A to the pendulum hinge C and is located

FIGURE 6.54 Tilting lines of a three-wheel forklift truck [65].

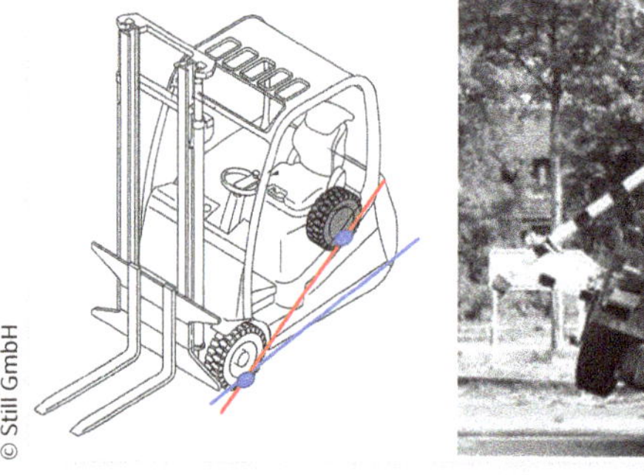

FIGURE 6.55 Tilting lines of a four-wheel forklift truck.

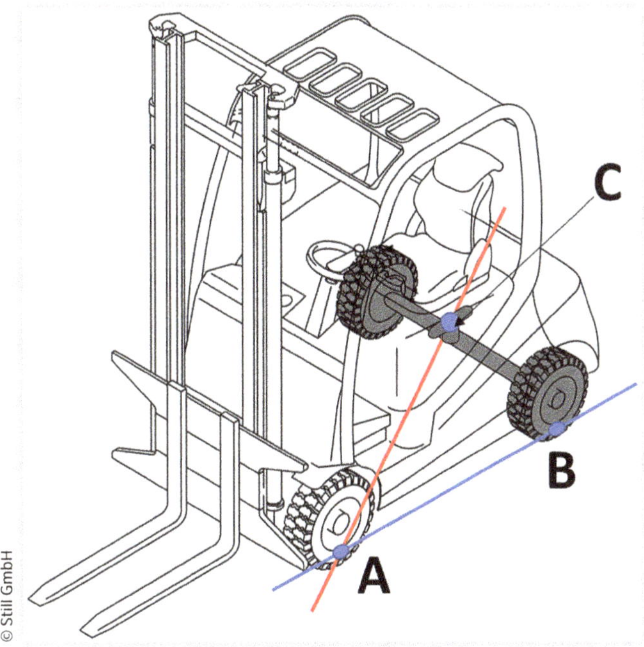

therefore oblique in the room and not flat on the ground as shown through the red line in Figure 6.55. Around this tilting line, the forklift truck moves up until the block of the pendulum axle is reached. The inside of the driven wheel will raise from the ground. If the tilting torque raises, the forklift truck will tilt around the second tilting line by both outside wheels A and B (blue line). In case of tilting the whole truck, the pendulum axle and the chassis act as one unit. Regarding the former tilting of the chassis, the center of gravity was moved a little bit so that the distance between the center of gravity and the tilting line is getting smaller which reduces the stability. Therefore, it is very helpful to design the pendulum hinge as high as possible to optimize the maximum possible turning speed [65].

The requirements and the test procedures regarding the stability of a forklift truck are defined in ISO 22915 [82]. The dynamic behavior in the main critical operations as described above is simulated by static tests on a tilting platform. For class I, IV, and V trucks, there are four main tests for longitudinal and lateral stability defined in ISO 22915-2. Two tests are conducted with lifted load while maneuvering in a curve and straight on, one test traveling forward with load and mast tilted backward and one test without load traveling around a curve. For this case, there exists an additional standard EN 16203 in Europe for an additional dynamic test [75]. As the result of those tests, the manufacturer has to define a load diagram which should be placed on the truck so the operator can check it while operation. An example is shown in Figure 6.56.

The horizontal line is the maximum load; the curve in the middle is defined by test No. 1 and the third part by test No. 3.

To generate a fine forward stability, it is very helpful to minimize the load arm (distance from the center of gravity from the load and the center of the front axle [c + x])

FIGURE 6.56 Stability diagram of an electrically powered 3.5-ton forklift truck for different load centers [93].

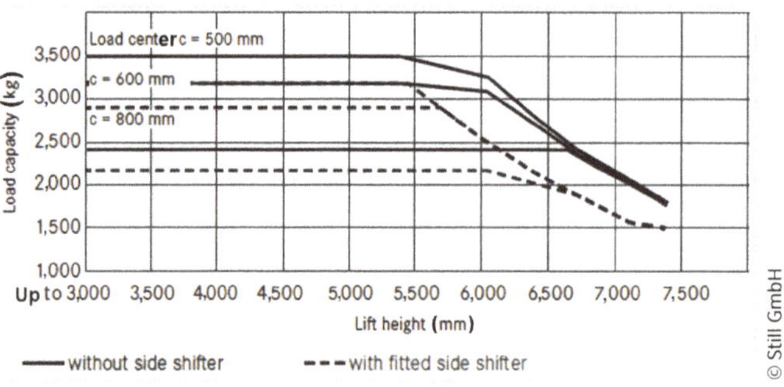

FIGURE 6.57 Load arm on a four-wheel forklift truck.

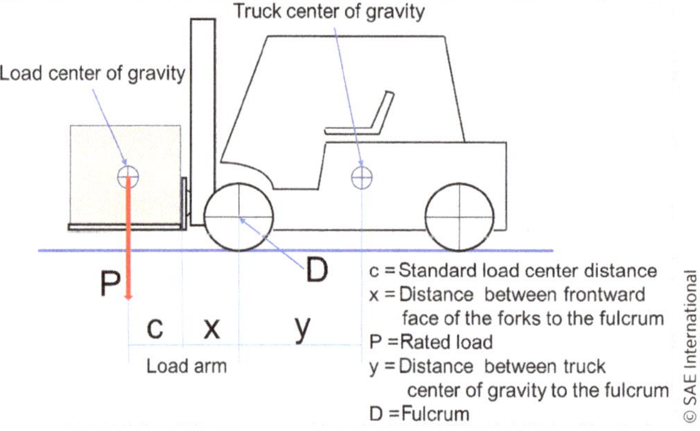

as shown in Figure 6.57. The dimension c is defined in the above standards, and x could be influenced by design. On the other hand, the center of gravity from the whole forklift truck should be placed as much as possible to the back, but it is a compromise regarding the lateral stability.

Regarding lateral stability aspects, a four-wheel forklift truck has advantages for higher velocity, but in narrow aisles a three-wheel truck has a smaller turning radius. The steering axle is arranged for better maneuvering in the back and the steered wheel could be rotated up to 90° in case of a three-wheel truck.

6.3.2.3.2 Visibility. The all-around visibility is very important for a safe operation of every mobile machine as already mentioned in Chapter 2. Because of the very different types of forklift truck as defined in ISO 3691, the all-around visibility is very special for every truck [85]. The requirements are defined in ISO 13564 and the operator has to recognize a squatted or standing person while traveling or maneuvering in the working area in front and to all sides around the truck [80]. Particularly with regard to the design of the mast, this has to be considered while working out the mast concept. For bigger trucks, such as container handling trucks, it would be very helpful to use auxiliary equipment like camera systems.

6.3.2.3.3 Human Vibrations. The primary purpose of ISO 2631 is to define methods of quantifying whole-body vibration in relation to the following:

- Human health and comfort
- The probability of vibration perception
- The incidence of motion sickness [84].

It is concerned with whole-body vibration and excludes hazardous effects of vibration transmitted directly to the limbs. In forklift trucks, there are two important areas where those vibrations could influence the operator. The most important vibration will impact the operator through the seat. There are measurement systems available in the market which contain a three-axis acceleration sensor in a very flat seat cushion made from a kind of rubber material to place it directly on the operator's seat. In this way, it is possible to measure the vibrations on a whole shift based on DIN EN 14253 by the operation company to fulfill the requirements of the employment protection laws, for example, 2002/44/EG [62, 74]. The vibration indicator can be calculated as described in ISO 2631-1 [84]. This measurement system can also be used by the manufacturer to optimize the design of the operator's compartment and the seat to minimize the exploration for the operator. In case of uneven floors, the operation company can equip the fork lift truck with an air-suspension seat.

6.3.2.3.4 Functional Safety. As already mentioned in Section 4.5, the aspects of the functional safety for mobile machines are based on EN ISO 13849 [81]. This has to be fulfilled for every forklift truck by the manufacturer.

6.3.3 Frame and Counterbalance

As described in Section 2.1, the chassis of a forklift truck has to fulfill plenty of requirements and has to carry at least all components of the truck. As already mentioned above, there exist very different trucks regarding the different duty cycles on the market. Thus, the chassis is very different, too. The chassis of a common forklift truck (classes I, IV, V, VI, and partly VII) includes the frame itself and the counterweight, and sometimes the overhead guard (Linde H25 (type 392) and E20 (type 381)).

6.3.3.1 Counterweight

As the name already says the main function of this component is the counterweight to the load. If the moment equilibrium around the axis of rotation in the center of the front axle (D) is made, the load (P) through the load arm (c + x) and the counterweight through the gravity center distance (y) have to be in balance, see Figure 6.57. Regarding ISO 22915, the mass should be so high that the rear axle weight should be around 1.4 higher than the rated capacity [82]. The material is normally gray cast iron. Nowadays, the outer shape of the counterweight has to fit into the styling of the whole forklift truck.

There are generally three different types in use:

- Only the function of a weight.
- Additional to be a carrying part of the frame concept in the rear with bending-resistant connection to the left and the right parts of the frame.

FIGURE 6.58 Frame and counterweight of an electrically powered four-wheel truck [95].

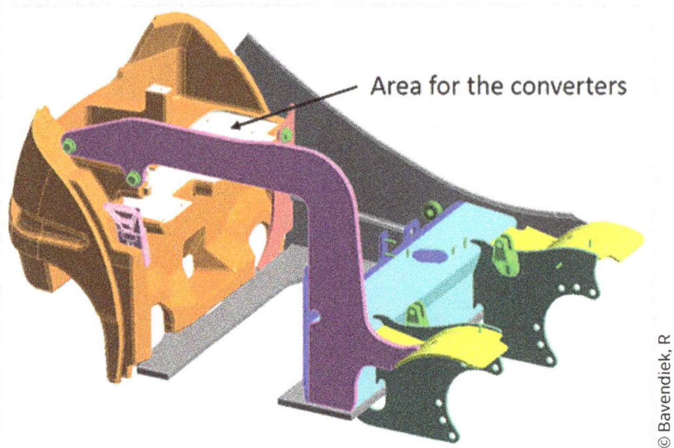

- With places to mount some components such as the rear axle, converters, cooler, or bearing of an IC engine. These places normally have to be machined. Particularly with regard to converters, the assembly level has to be machined very fine that the waste heat of the converters can be transferred through thermal conductance paste into the counterweight as shown in Figure 6.58.

6.3.3.2 **Frame:** Generally, the frame should be designed to be bending-resistant and with low torsion. The frame will carry most of the components of the forklift truck directly or indirectly, such as the front axle, the IC engine, the overhead guard, the powertrain, the tilting cylinders, or tanks. The material is weldable and formable steel and is build up as a package of different semi-finished metal sheets. Because of the complex structure, it is advisable to use the finite element method (FEA) for the design of the frame to optimize the material inventory and to generate a sufficient durableness and lifetime.

The concept of a frame of a common forklift truck (classes I, IV, V, VI, and partly VII) depends on the type of components to be used:

- A right and a left singular part, very often reversed left to right. In the rear connected to the counterweight, in front with the front axle to build up a closed structure. A challenge are the tolerances coming out of the welding processes, in particular at the connection to the front axle. The tanks for the fuel and for the hydraulic oil could be part of the left/right part of the frame, but they have to be purged very well after the welding process to protect the fuel/oil system against the slag.
- A complete closed package. All important components as both axles, the IC engine, the mast, and the counterweight are fitted to this kind of a frame. A big advantage is an easier assembly and a very good serviceability. However, this concept needs more material.
- A mixture of the above concepts.

A specific characteristic is the frame of an electrically powered forklift truck. The frame has to be designed around the battery (standardized in many countries) which takes a lot of space in the truck. The trucks have only very narrow space between the battery and the outer part of the frame on the left- and right-hand side because the truck shall be designed as compact as possible. For multishift operation, it is necessary to take the discharged battery out the truck and to put the charged battery in it again. A very common concept is to take the battery with a crane to the top through a slot in the overhead guard. A big advantage is a symmetric frame concept, but the swinging battery in a crane with narrow space to each side includes a high risk for the operator and the truck regarding damage. Therefore, at the beginning of this century there was developed an asymmetric frame concept to take out the battery to the right side as shown in Figure 6.58. A challenge was the creation of a sufficient stiffness of the right-hand side of the frame.

6.3.4 Powertrain

As already described in Chapter 3, the powertrain contains all components regarding the power flow from the energy storage (fuel tank or battery) to the wheels and to the area of operation (mast/fork arms). The energy origins could be fuel such as gasoline, diesel, compressed natural gas (CNG), and liquid propane gas (LPG) as well as lead-acid or lithium-ion batteries and at least some hybrids. In the inner area, electrically powered trucks become more and more commonly accepted. An overview is given in Figure 6.59.

For better maneuvering, forklift trucks normally steer with the rear axle and the front axles contain the drive components. In special cases, like reach trucks in rough terrain, a three-wheel drive is used.

6.3.4.1 **IC Engine Trucks:** Worldwide, counterbalanced forklift trucks are mainly equipped with IC engines. The crankshaft of the IC engine powers the drive train and normally the hydraulics with mechanical energy. Sometimes, the hydraulic pump is

FIGURE 6.59 Drive train in forklift trucks.

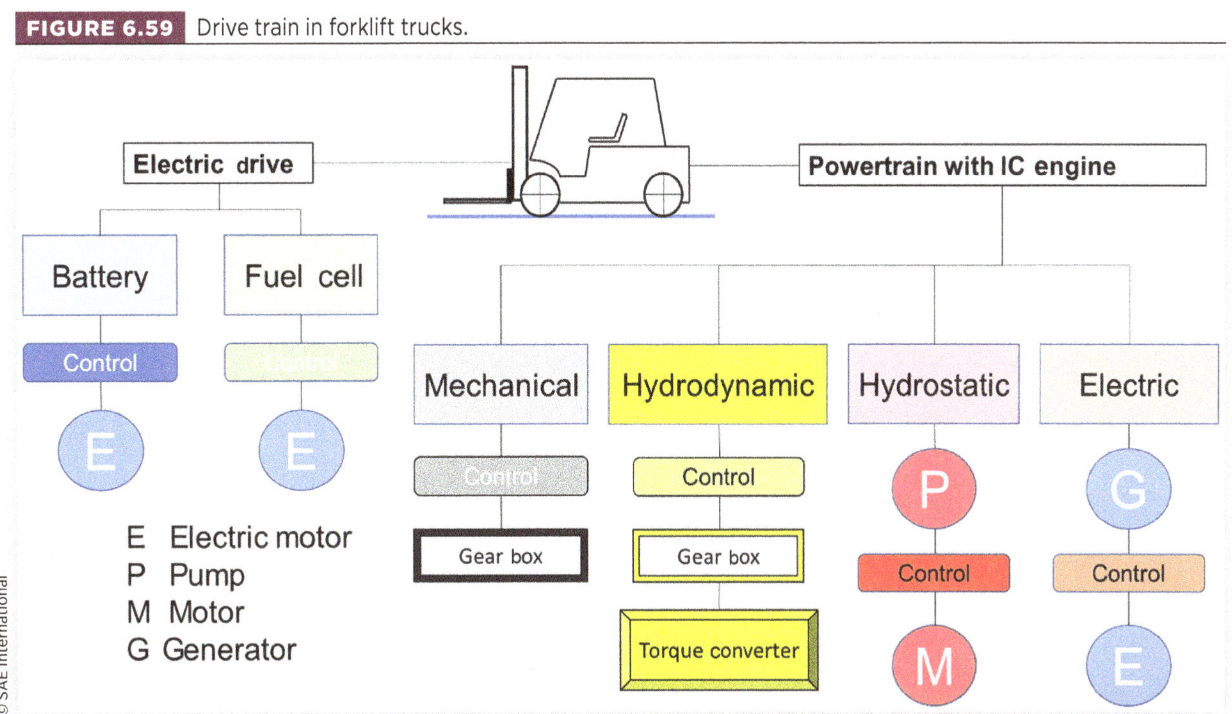

FIGURE 6.60 Hydrodynamic transmission for a forklift truck [77].

mounted at a power take-off (PTO) at the IC engine. In general, there are four different types of traction drives on the market:

- **Mechanical gearbox**

  As described in Section 3.3.1, this type is very similar to a car drive train with a manual shift. There is a clutch pedal to operate the clutch to connect the IC engine with the mechanical gear box with normally two gears to the front and one for the rear. Through a differential, the wheels of the front axle will be rotated. This is the easiest and cheapest drive train and only economic in countries with very low wages. The handling capacity and the operator's comfort are the lowest in comparison to all the others.

- **Hydrodynamic transmission, Figure 6.60**

  This is worldwide the most common drive train in fork lift trucks and is shown generally in Section 3.3.2. The moment of torque is transferred from the IC engine through the torque converter and the differential to both front wheels. The advantages appear while lifting and transporting on longer distances, smoothly starting and reversing, continuous acceleration, manageable complexity, and easy maintenance and service. Through an inch-pedal, it is possible to use the maximum engine power for lifting while maneuvering slowly.

- **Hydrostatic transmission**

  This traction drive concept was – introduced already in Section 6.3.3 – used in forklift trucks (Linde) nearly 60 years and has several advantages as a very high starting torque for running in short distances and numerous and fast reversing, sensitive driving and using the hydraulics independently, and a very high handling of goods with a very good efficiency. This traction drive needs low service. But the manufacturing costs are higher in comparison to a hydrodynamic transmission. In Europe, the concept is getting more and more common. As an example, Figure 6.61 shows a modern hydrostatic powertrain which is driven by electronical controllers and CAN buses. The front axle contains two hydrostatic motors. The pressure level is up to 450 bars (6,500 psi).

FIGURE 6.61 Modern hydrostatic transmission in a forklift truck [77].

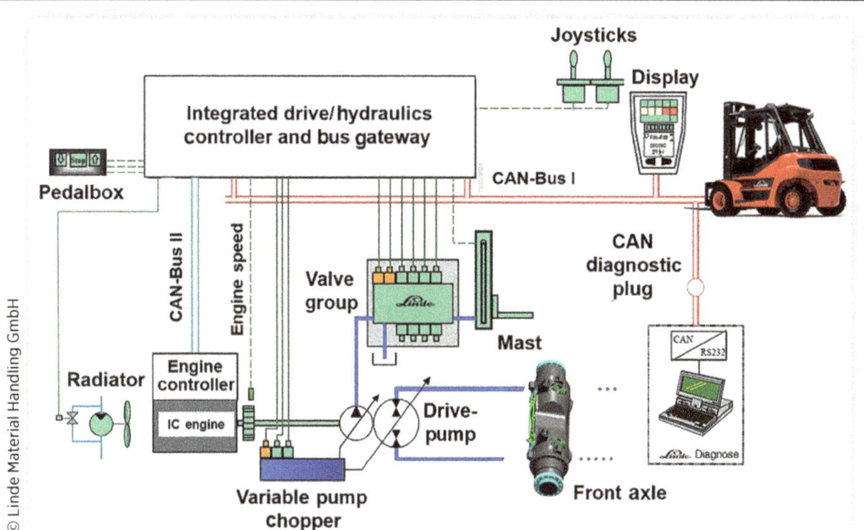

- **Diesel-electric transmission**

This type of transmission is often used in locomotives and since more than 40 years in forklift trucks (Still) through DC-drives. Nowadays, AC technology is used as shown in Figure 6.62.

On the crankshaft of the IC engine, the synchronous alternator and the hydraulic pump are mounted. The converter transforms the AC into DC current in a range between 150 and 750 V. Through the brake resistor, the forklift truck brakes without any wearing. In a modern variant, there is in parallel to the brake resistor an energy

FIGURE 6.62 Diesel-electric transmission [77].

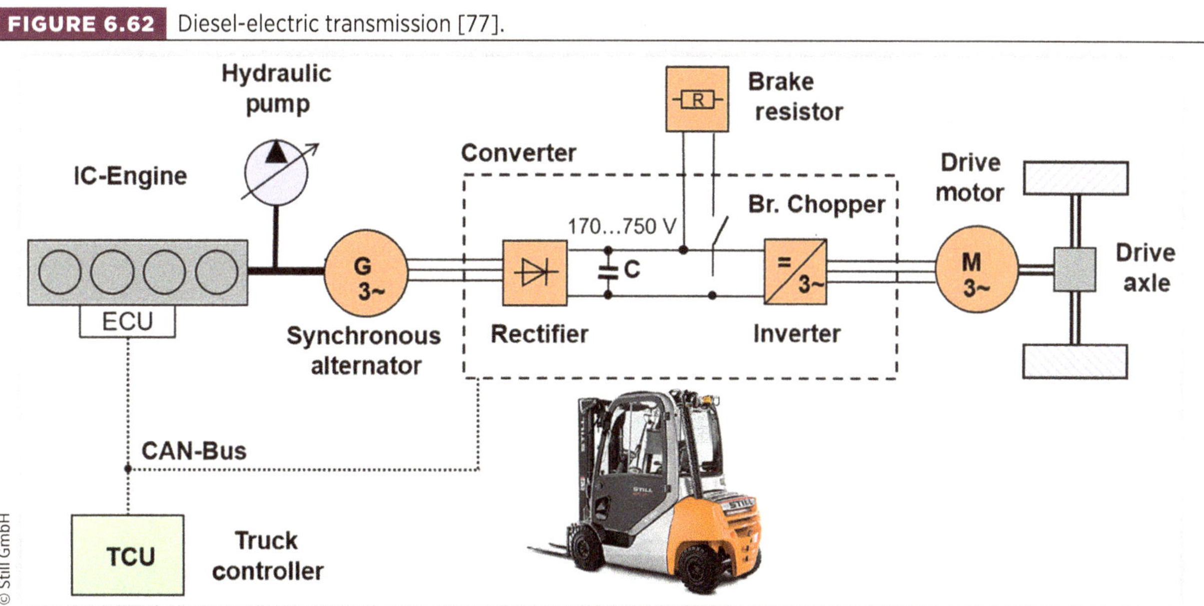

storage like ultra-caps or a Li-ion battery to reuse the brake energy for acceleration of the forklift truck through a DC/DC convertor. The front axle contains one rotating field-controlled AC motor and a differential or a two-AC motors or two AC motors.

6.3.4.2 **Electrically Powered Trucks:** In the past, electrically powered forklift trucks had less power than IC trucks with the same rated capacity. Up to 1990, there have been series-characteristic DC motors common. Those trucks had to brake through drum brakes or shoe brakes. Smaller electrical trucks up to two tons rated capacity are available as a three-wheel (Figure 6.63) truck for narrow aisles. Those trucks are generally equipped with a two-motor front axle. The rear axle has one or two very narrow wheels to generate a very small turning radius. Therefore, the rear wheel could be turned up to 90°. In a four-wheel truck, one- or two-motor drives are used.

Since 1990, more and more shunt-wound motors got common on the front axle. Therefore, it was possible to regenerate the braking energy into the traction battery by a four-quadrant chopper which saves up to 15% of the energy consumption. But those motors still use carbon brushes which are stressed highly while braking by a high current (high wear). Ten years later, AC drives came up as shown in Figure 6.63.

There are two different autonomously power trains, one for the driving and the other for the hydraulics. The drive motor runs in a four-quadrant mode, and the pump motor only in a one-quadrant mode. The temperature of the convertors and the motors is checked to avoid overheating, and the drive motor needs a rotation sensor for the control. The operator can run the truck through CAN bus using the truck control unit TCU.

6.3.4.3 **Wheels:** There are three different types of tires available on forklift trucks. On rough terrain in outside areas, there are pneumatic tires useful because the axles do not have any suspension. The comfort for the operator is the best one, but the risk of damage is very high and the stability conditions are the worth. On a dead level floor – necessary in narrow aisles such as for man-up trucks with lift heights up to 14 m (15 yards) – cushion tires (bandages) are recommendable to maximize the stability behavior. A compromise for forklift trucks regarding classes I, IV, V, VI, and partly VII are superelastic tires which are very common nowadays because the risk of damage is small, the rolling resistance is low, and the stability conditions are quite well. In principle, they are filled inside completely with a damping rubber material.

FIGURE 6.63 Power train of an AC electric fork lift truck [77].

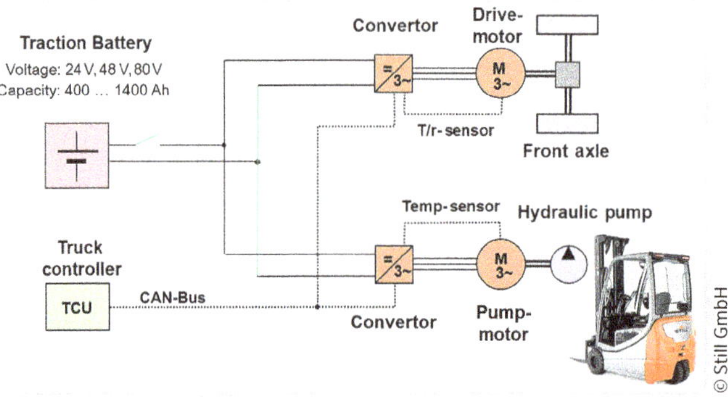

6.3.5 Hydraulics

The hydraulic system has two (in case of a hydrostatic drive three: traction drive) main tasks: Working (moving the forks) and steering. The main components are shown in Figure 6.64. Generally, the hydraulic system of an IC truck and an electrically powered truck is the same. As explained in Section 6.3.4, the pump of an IC truck is directly connected to the crankshaft of the IC engine, and the pump of an electrical truck has a separate motor. For cost reasons, a fixed displacement gear pump is used normally. A variable displacement pump could save energy on an IC truck. An internal gear pump can reduce the noise level on electrically powered trucks. The pressure control valve could be mounted directly on the pump or on the priority valve or be integrated into the control valve and allows a working pressure up to 250 bars (3,626 psi).

The material of the tank is made from plastics or steel. The big advantage of a plastic tank is the flexibility in the outer shape in narrow rooms. Steel has a better thermal conductivity and could be integrated in the frame and helps for its stiffness as shown in Figure 6.58. Very often, the filter is integrated into the tank. There are different concepts in use. A coarse filter in the suction pipe joint at the bottom of the tank in combination with a return flow filter could be designed into the tank at its top in combination with the air bleeding filter. This concept is used very often in the hydraulic systems with a mechanical-operated directional control valve. If a proportional control valve is integrated into the hydraulic system, a higher quality of cleanness of the oil is required, and therefore a high-pressure filter is usually installed between the pump and the priority or control valve. The main advantages of a proportional control valve are low forces, sensitive control of the load, and the valve must not be located beneath the operator.

Through the control valve, the operator can handle the load. If a forklift truck (classes I, IV, V, VI, and partly VII) is not equipped with an attachment, the control valve has generally two different sections in one block: One for lifting and lowering the load and

FIGURE 6.64 Hydraulic system of a fork lift truck with an electrical powertrain.

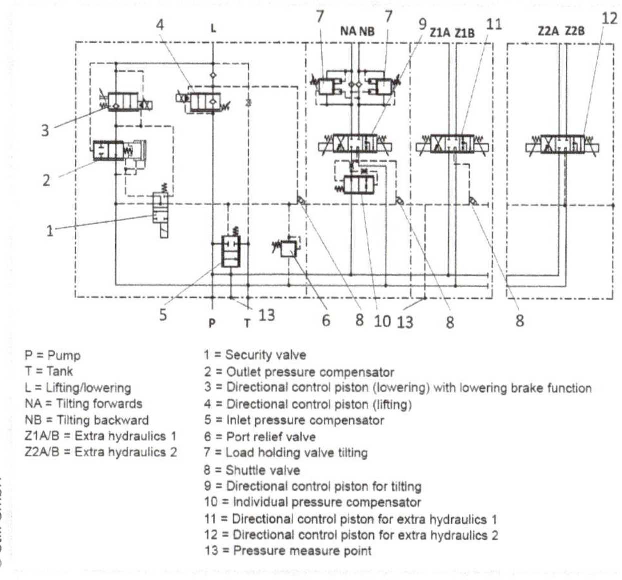

the basic functions and the other one for tilting the mast. If an attachment like a side shifter or a fork adjusting device is mounted on the fork carriage, there are up to two additional sections (11 and 12, see Figure 6.58) adjoined to the block. In the section of the basic functions and lifting/lowering are integrated as follows:

- Port P for the pump, T for the tank, and L for the connection to the lifting cylinder.
- The security valve (1) ensures the operator to be seated.
- The directional control piston (3 and 4) for lowering with an integrated lowering brake function and for lifting.
- The (2 and 5) inlet and outlet pressure compensators to realize the load-sensing function.
- The port relief valve (6) as a control valve to protect the system against overload pressure.

The tilting section contains the following:

- Ports NA and NB to connect both ports of the tilting cylinder(s).
- Load holding valves (7).
- Directional control piston (9) for tilting.
- Individual pressure compensator (10) to avoid forward tilting regarding the impact of the load at high lifting height.

In case of an electrically powered truck, this control valve (9) could be designed in combination with an appropriate control of the speed of the pump to spend as less as possible energy as shown in **Figure 6.65**.

The speed of the pump is controlled to deliver exactly the flow rate which is required by the working equipment instead of running it with maximum speed. Through the shuttle valves (8) in every section of the control valve block, the highest working pressure is impressed into the hydraulic circuit and the load-sensing system (5) realizes a work pressure with an offset of only a couple of bars generated by the feather of the load-sensing valve (5) (pump pressure).

FIGURE 6.65 Energy consumption of a fork lift truck with an electrical powertrain.

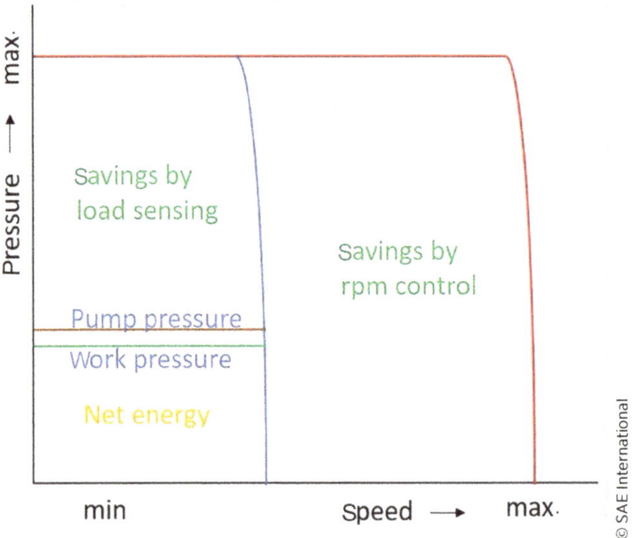

To save manufacturing costs, there is often only one pump installed for working and steering. For safety reasons, the steering function is more important and so through a priority valve the oil flow will provide the steering always first. An alternative concept uses two pumps on one shaft with different aligned displacements.

From the tank, the pump generates the oil flowing through pipes and flexible hoses (O-ring seals for long-term density if the fittings will be unscrewed often) through priority and control valve to the cylinders of the mast or to the steering device which could be a cylinder or a hydraulic motor.

If high power is required and the oil volume is very limited, an oil cooler should be integrated into the hydraulic system located in the backflow between the control valve and the tank. The tank itself should be equipped with wash plates to minimize the danger of cavitation.

Figure 6.66 shows an example of a steering system with an own pump and a tank. More information about steering systems are given in Section 2.4.

FIGURE 6.66 Steering system of a forklift truck (Danfoss Power Solutions) [68].

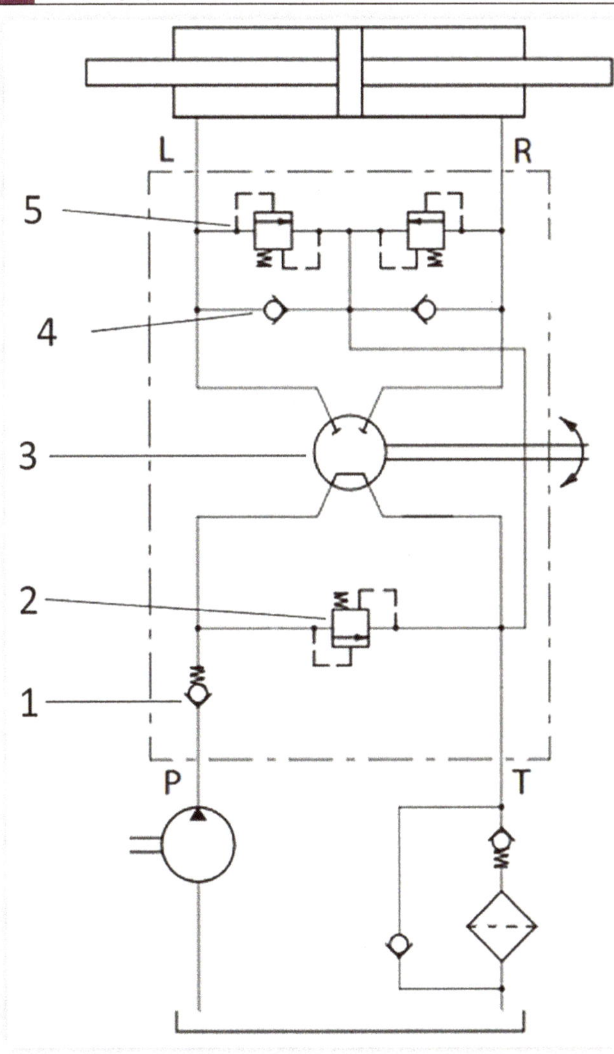

The steering unit, normally an Orbitrol in mobile machines, is inside the line-dotted area. A check valve (1) protects the pump against pressure peaks from the steering cylinder. A port relief valve (2) protects the Orbitrol (3), because an Orbitrol has a lower maximum pressure than the working hydraulics. The two pressure relief valves (5), namely the shock valve, protect the steering system against pressure peaks coming from the steering cylinder (like if the wheel runs against a step) from both sides. The check valves (4) ensure the connection to the other side of the cylinder.

6.3.6 Operator Working Place

The overhead guard with all components has a big effect on the styling of a forklift truck. Therefore, the shaping, the optical impression, and the haptics of all components have to underline the styling concept of the whole forklift truck.

6.3.6.1 Overhead Guard/Cabin: The operator working place has been developed dramatically in between the past decades. The first forklift trucks had no overhead guard until 1967 when Clark bolted an overhead guard [66]. There are different concepts available in the market:

- Made from steel and bolted with three or four rods on the chassis as shown in Table 6.8
- A self-contained steel design mounted through damping elements on the chassis as shown in Figures 6.62 and 6.63
- A part of the chassis made from steel to fix the tilting cylinders at the top of the overhead guard as shown in Figure 6.61 (Linde).

For safety reasons, the overhead guard must fulfill the requirements of ISO 6055, and static and dynamic tests must be conducted with weights and a wooden cube to fall 10 times on the top of the overhead guard around the operator's head without exceeding a maximum deformation. These requirements are in contradiction to optimize the visibility trough the top of the overhead guard especially while maneuvering with lifted loads [86]. To optimize the all-around visibility, the vertical rods and those on the top of the overhead guard should be as small as possible (less as the distance between the two eyes of the operator) [80]. The instrument panel in front of the overhead guard and the rear valance should be as deep as possible to generate a very good visibility to the forks or to the back.

Depending on the operating conditions, the customer can choose different variants as given below:

- An overhead guard without any doors and shields for operation inside of buildings
- A cabin with shields at the front and the back and plastic doors which protect the operator from weather conditions such as rain and wind. Those doors could be rolled up for easy entry during good weather conditions. The overhead guard is closed on the top by a shield made from glass or plastics like Makrolon©
- A closed cabin with doors and shields for operating outdoor. Both doors and the back shield could be opened. These cabins are normally equipped with a heating or/and an air condition, wipers at the front and back shield, sometimes on the glass on top of the overhead guard, a radio, or a sun visor.

The lightning devices are very often mounted on the overhead guard.

To reduce the noise inside the cabin, every flat level of the cabin as the bottom side of the shield could be glued with acoustical boards. A movable sun visor should be installed for outdoor use. A (panorama) mirror inside and convex mirrors outside could help control the surrounding of the operating truck.

6.3.6.2 **Ascent/Descent:** Particular smaller forklift trucks used for picking should have an easy ascent/decent. The maximum step height is limited by ISO 3691 to 500 mm (20 in.) [85]. Therefore, many trucks have additional steps on both sides underneath the floor plate (Figure 6.62). The surface of the floor plate and the steps should be nonskid, for example, rubber material or corrugated sheet metal. The design in that area should have roundness parts as the hood of the battery or the engine compartment.

A grip at the rod of the overhead guard helps the operator for ascending the truck.

6.3.6.3 **Seat:** For forklift trucks and for every mobile machine, the seat is very important for the operator because he/she sits on it the whole shift. The seat shall be located in the truck to satisfy 5 percentiles up to 95 percentiles persons and should be as comfortable as possible. Therefore, the seat has many important functions:

- To accommodate the operator
- To generate sufficient suspension
  - Damping material between the frame of the seat and its cover
  - A mechanical spring mounted between the frame of the seat and the basis of the seat cushion
  - An air-suspended seat
- Adjustments such as for the seat in longitudinal and vertical direction, backrest, weight of the operator
- Duo-sensitive two-point seat belt
- Heating
- Backrest extension
- Armrest together with control and display elements
- Lumbar support
- Storage box
- Fore/aft isolator
- Lateral isolator

On forklift trucks (classes I, IV, V, VI, and partly VII), the seat is located on the hood for the battery/engine a little bit (ca. 80 mm (3.1 in.) for a 2 tons truck) asymmetrical to the left to optimize the visibility onto the right fork arm as shown in Figure 6.67. To avoid damages at the spinal column of the operator, some seats are turnable up to 20° to the left for forklift trucks which must travel long distances backward. The disadvantage is the poor visibility to the right side. For this purpose, camera systems are coming up. The seat cover could be synthetic leather (outdoor) or cloth material (indoor or closed cabin). Some seats are equipped with a restrain system working with turnable brackets on each side if the usage of the seat belt is too laborious. Such restrain systems could be mounted on the cover of the battery or IC engine, too.

FIGURE 6.67 Working place of a forklift truck (Still GmbH) [93].

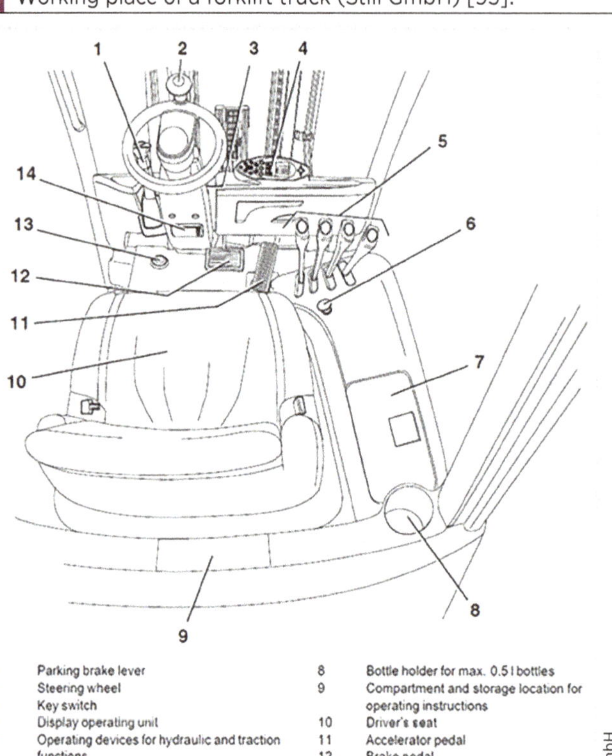

6.3.6.4 **Restrain System:** Because of plenty of accidents tilting over lateral in the past as explained in Section 6.3.2, there is a device necessary which protects the operator not to be hurt by parts of the overhead guard touching the ground at the end of the tilting phase. This could be fulfilled by several different systems, such as by a safety belt or bars which are mounted on the seat, cabin doors, and safety bars from the front to the rear bar of the overhead guard which could be used like cabin doors.

6.3.6.5 **Control and Actuating Devices:** Around the operator, there are many control and actuating devices positioned as shown in Figure 6.67:

- Parking brake (1)

  The activation of the parking brake depends on the kind of parking brake:

  - Manual activated brake through a lever located on the instrument panel or on the hood over the battery or IC engine or through the left foot on the floor plate. In some cases, a switch is mounted on the instrumental panel if the control device needs a force intensification by hydraulics or by an electric motor. For cost reasons, the force transmission is very often realized through a rope with a length as short as possible.
  - Spring-loaded brake is activated by a switch normally on the instrumental panel because a force intensification is needed anyhow. A risk analysis has to ensure that the parking brake does not brake inadvertent.

- Steering wheel (2)

  It is located in trucks (classes I, IV, V, VI, and partly VII) in front of the operator and could normally be turned toward the operator through a lever (14) and sometimes adjustable in the height. The diameter depends on the volume of the steering valve. The steering valve itself could be mounted directly at a very short steering column or with a longer column under the floor plate and should be mounted elastic to avoid incommodious noise. To realize fast steering, a knob (2) is connected normally on the steering wheel as shown in Figure 6.67. The operator uses this knob with his left hand.
- Key switch (3)

  There could be different places in trucks (classes I, IV, V, VI, and partly VII) for the key switch. In Figure 6.67, it is mounted on the right side of the steering column. This switch has several functions similar to cars: To fulfill the requirement of ISO 3691 to authorize the operator, to switch on the electrical power, and to start the IC engine [85].
- Display operation unit (4)

  There are plenty of solutions available on different trucks. Some have only one device for display and switches, some separate them. Some use toggle switches similar to cars as shown in Figure 6.67 upper right-hand side. They are located under hoods at the top of the overhead guard or in the instrumental panel or in the armrest. If they are integrated into the display membrane switches are normally used for cost reasons and they need less room.
- Operating devices for hydraulic functions (5)

  As already shown in Section 6.3.5, there are two different hydraulic control valves with up to four levers or a multifunctional grip in the market:
  - Mechanical operated directional control valve (5).

    The levers are located on the right-hand side of the operator. The form depends on the position of the valve. One position is above the floor plate, so the levers are very short and turned longitudinal to the front as shown in Figure 6.67. The other position is under the instrumental panel and the levers look like a neck of a swan to ensure access to the right side of the forklift truck. These levers move up and down.
  - Proportional control valve.

    This operating device is very often integrated into the right-hand armrest of the seat (10) mounted in trucks (classes I, IV, V, VI, and partly VII). Same examples are shown in **Figure 6.68.** In the upper part left- and right-hand side, there are four levers in different sizes which work in the same way as those from the mechanical operated controls. In the middle of the figure a classical joystick works. In the lower part left-hand side, both levers could be moved crossways. An overlay in every direction is possible to move the load by lifting/lowering and tilting with only one lever. The attachment could be moved by the second lever. A trained operator can control the load very fast with this device but smooth. The right-hand solution is typical for reach trucks and therefore used in mixed fleets. For safety reasons, an emergency off switch (6) is necessary for this kind of control.

FIGURE 6.68 Control devices for proportional valve.

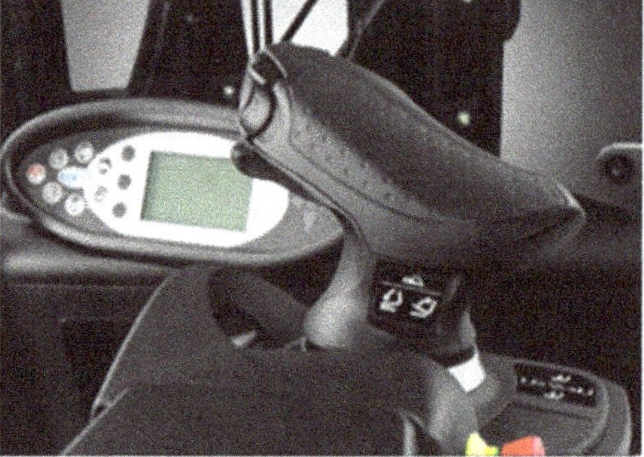

- Driving direction switch.

  Forklift trucks with one accelerator pedal need a driving direction switch. For maneuvering often, it should be easy to control it normally with the right hand. Therefore, it is very often integrated into the front end of the armrest beneath the hydraulic levers as shown in Figure 6.68.

- Storage opportunities (7, 8, 9)

  As shown in Figure 6.67 especially in smaller trucks up to 2 tons capacity (classes I, IV, V, VI), space is very poor in the operator's compartment. Therefore, some place is necessary for the manual, a bottle and some personal things such as a mobile phone or pencil which could be stored under the lid of the compartment (7) which has a clamp for some papers.

- Accelerator pedal (11)

  It is placed in the floor plate on the right side similar to a car. There are many different technical solutions available from different manufacturers. It depends on the kind of direction indicator switch. If this function is done through the accelerator pedal, there are different solutions in the market:

  - Two pedals (Linde), one left-hand side of the steering column for the rearward's direction and one right-hand side for the forward direction.
  - Two pedals (Hyster) on the right-hand side of the steering column directly side by side, the left one for the rearward's direction and the right one for the forward direction.
  - One pedal (Hyster) on the right-hand side of the steering column with a knob on it to differentiate the driving direction.

FIGURE 6.69 Warehouse management system on a forklift truck.

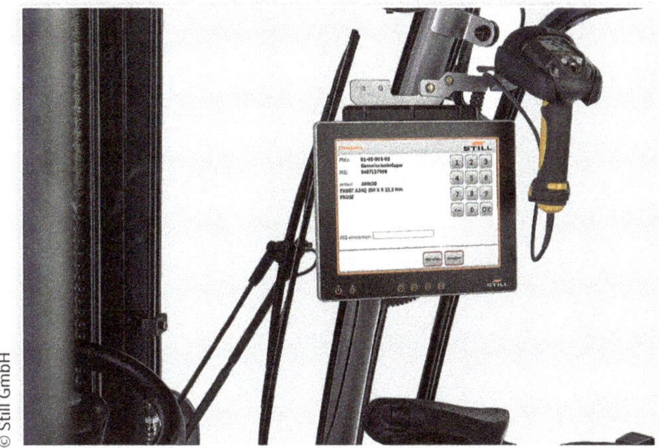

© Still GmbH

- One pedal on the right-hand side of the steering column and a switch at the steering column or at the hydraulic control lever.

- Service brake (12)

  Normally, a forklift truck has a service brake pedal similar to a car on the right-hand side of the steering column left beneath the accelerator pedal. The force transmission could be realized mechanical through a rope or levers or similar to a car by hydraulics.

- Alarm horn (13)

  The horn switch could be integrated into the control device for the hydraulics as shown in Figure 6.68 or be placed in the floor plate as shown in Figure 6.67 or in the steering wheel.

- Systems for information and communication

  Those systems getting more and more important in the future (Section 6.3.7). A radio with Bluetooth and USB could be installed in a hood at the top of the overhead guard.

  A socket plug is usually used for the diagnostics and could be located under a hood or in the display operation unit.

  A fleet management system is more and more equipped in combination with an authentication device through RFID card or a key pad.

  On trucks (classes I, II, II, and VI) warehouse management systems could be mounted as shown in Figure 6.69 to integrate those tucks into the logistic chain.

6.3.7 Electric Controls and Software

As shown in the chapters above, all main components have nowadays electric and electronic parts integrated as the powertrain, especially an electrical one, and the operators' working place. Therefore, the electrical system has several main components by itself and has to connect them all with the different control units through the wiring system.

The main components are as follows:

- Battery
- Lighting system
- Sensors
- Actuators as shown in Section 6.3.6
- Displays as shown in Section 6.3.6
- Wiring
- Fuse box
- Control units
- Software for the different control units.

The electrical system is similar for IC- and electrically powered trucks.

- Battery

  IC trucks use 12- or 24-V systems and use car technology and those components as starter batteries.

  Electrically powered trucks are designed around the traction battery as shown in Section 6.3.3. Regarding to the maximum power of the powertrain, there are different voltages common in different markets:

  - In the United States and Japan, 24, 36, and 72 V are well-known
  - In Europe, 24, 48, and 80 V in lead technology are common which are standardized in IEC DIN EN 60254 and DIN 43535 (24 V), DIN 43531 (48 V), DIN 43536 (80 V), or British Standards [69, 70, 71, 79].

  24 V is used for rated capacity up to 1.5 tons, 48 V up to 2 tons, and 80 V up to 8 tons.

  Li-ion batteries are coming up in warehouse trucks and on counterbalanced trucks.

- Lighting system

  Forklift trucks could be equipped with adjustable working headlights in the front and the back which could be mounted on the overhead guard or on the mast to optimize the visibility in the working area. A yellow flash light on the top of the overhead guard could help recognize the operating truck.

  If the truck is running on public roads, a lighting system is necessary as used in a car. Inside the cabin, an operator compartment lighting is very helpful.

  As a warning device, a blue light system as shown in **Figure 6.70** could be mounted on the truck [88]. It consists of two bright, antiglare lights that project a blue spot onto the ground. They can be set to either shine constantly or flash. The light spot is positioned several meters ahead of the forklift truck, whether it is traveling forward or backward.

  Nowadays, more and more lamps are equipped with LED lights.

- Sensors

  There are plenty of sensors installed in a forklift truck today including the below:

  - Speed sensors for control the traveling or lifting speed as already shown in Section 6.3.4
  - Temperature sensors to protect the components against overheating

FIGURE 6.70 Blue light warning device for safety. [88].

  - Pressure sensors to control the load
  - Accelerator sensors to detect an accident
  - Current and voltage sensors to control the powertrain and the battery
  - Angle sensor to control the mast
  - Resistant strain gauge to control the load
  - Camera beneath the forks or for forward traveling in case of poor visibility

  These sensors are integrated into different devices and connected to the control units through cables or CAN bus.

- Wiring

  The wiring system has different kinds of cable, those for power and those for information. It is helpful to use sealed plugs in areas of worth environment conditions such as salted water. The cables could be protected by a slotted and corrugated plastic pipe or by a protective tape wrapped around the cables.

  A CAN bus system is used more and more to connect all devices which needs information. A power port, controlled through CAN bus, can be used to split the electrical power locally especially for the lighting system to avoid too much cables in the truck.

- Fuse box

  Similar to cars, a fuse box is installed with relays and fuses.

- Control units

There exist several control units in a modern forklift truck for different functions. The central truck control unit is the main controller and connected through CAN bus with all the other control units.

For more complex main functions, additional control units are necessary, like for the IC engine (stage 5 of Regulation (EU) 2016/1628)), invertors, power convertors (could be mounted directly on the frontaxle), and the proportional valve [76].

- Software

  Every control unit needs its own software to generate the desired functions. The necessary information from different sensors and control units are provided through CAN bus. This system allows additional features for operator assistance, comfort, and safety issues including the following:

  - Curve speed control – Helps prevent the tilting of the truck with high speed in a curve.
  - Mast positioning control system – Is an optically or electronically operating assistant that supports the operator in determining the precise vertical position of the forks.
  - Dynamic mast control – Reduces the swinging of higher masts to handle the lifted load. At the top of the mast accelerator sensors measure the movement, and through additional sensors the lift height and the mass of the load are captured. The control unit regulates the electrical reach powertrain to compensate the dynamic oscillation and the deflection of the mast [89].

- Fleet management system

  Provides the operating company with relevant field data to optimize the fleet, generates an operator access control, and reduces travel speed in dangerous zones etc.

6.3.8 Mast and Attachments

6.3.8.1 Main Requirements:

- As shown in Figure 6.57, a minimized distance (x) from the front of the fork carriage and the pivot point of the front axle is very important for the stability of the truck.
- A favorable relationship of the lifting height to the height of the mounted lowered mast.
- A very stiff design of the mast to avoid torsion and bending.
- Optimized visibility through and around the mast.

In the world market, there are two different styles of mast connection typical:

- *European style*: The mast bearing/pivot point is at the lower front of the front axle and the tilting cylinders are located on the level of the floor plate or at the top of the overhead guard (Linde).
- *US/Japanese style*: The mast bearing is located around the front axle and the tilting cylinders are located on the level of the floor plate. This type allows a very small distance (x) and a very good relationship of the lifting height to the height of the mast.

FIGURE 6.71 Types of mast for forklift trucks.

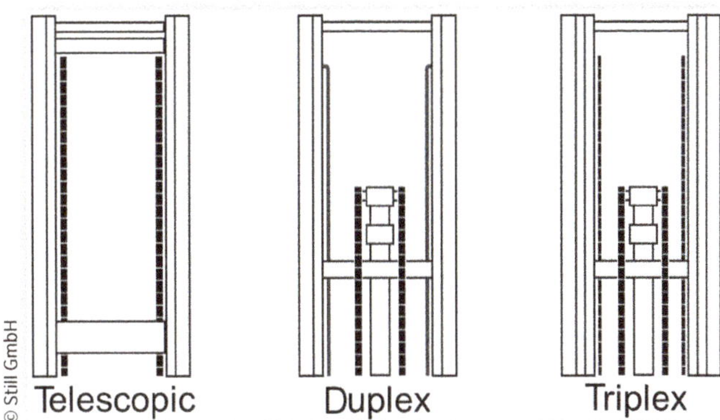

Depending on the needed lift height and the usage, there are several different types of mast available as shown in Figure 6.71:

- *Telescoping mast*: Cheapest type with good visibility through the mast with lifting heights up to 5 m (5.5 yards).
- *Duplex mast*: Lifting heights up to 5 m (5.5 yards) and the load will be at the top of the mast caused by the cylinder in the middle.
- *Triplex mast*: Lifting heights up to 8.5 m (9.3 yards) and the load will be at the top of the mast caused by the cylinder in the middle, too.
- *Quadruplex mast*: Lifting heights up to 12 m (13.1 yards) with poor visibility.

6.3.8.2 Main Components:

- Outer mast with the bearings to the front axle at the bottom and those for the tilting cylinders and cross members at the bottom and the top
- Inner mast with track rollers and cross members at the bottom and the top
- Fork carriage with track rollers and a guidance bar standardized in ISO 2330 [83]
- Minimum two hydraulic cylinders with pipe break protection valves
- Chains, tubes, pipes, and guide rails.

6.3.8.3 Main Attachments: Plenty of attachments for the very different tasks of transportations and loads have been developed in the past. For most of those attachments as shown in Figure 6.72, the interface is the fork carriage regarding ISO 2330 [83]:

In Europe, a side shifter is used very often to move the forks sideways to an optimized position to pick up the load or drop it down onto the right position. A fork-adjusting device with up to 16 forks could rise the handling capacity like in the beverage industry.

In any case, the stability of the forklift truck has to be adapted using an attachment (see Section 6.3.2).

6.3.9 Trends and Future Developments

Electrically powered trucks will eliminate more and more IC trucks in a range up to 8 tons rated capacity. IC trucks will become more and more extensive and expensive to fulfill the requirements of the exhaust regulations. Plenty of operating companies avoid

FIGURE 6.72 Different attachments for forklift trucks for special use [97].

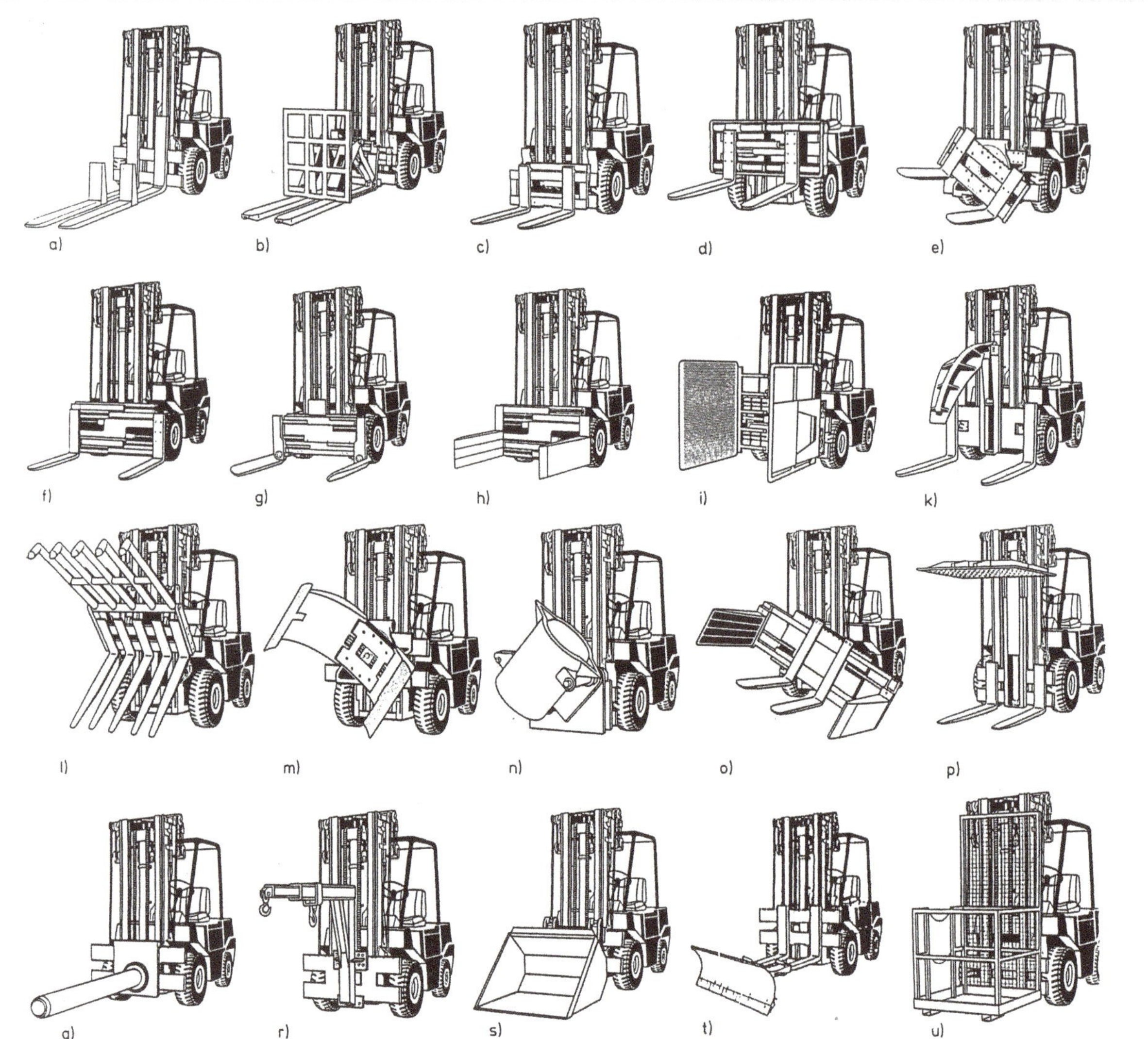

a) teleskopic forks	f) fork clam	l) scap grappels	q) poles (for carpet or coils)
b) pusher fork	g) fork clamp with turnable forks	m) roll clamp	r) crane jibs
c) side shifter	h) Bale clamps	n) charging device	s) bulk buckets
d) fork adjusting device	i) Big area clamp	o) pallet inverter	t) snow plows
e) rotator	k) wood grappels	p) load stabilizer	u) work basket

the exhausts inside of the buildings (see TRGS 554 [96]). The electrically powered trucks get more powerful using AC technology as shown in Section 6.3.4.

The handing capacity will rise because of more powerful powertrains for more speed; optimized ergonomics and assistance systems will increase the fatigue-proof operation as shown in Section 6.3.7.

The increase in sensors accompanied with sensor fusion and more powerful controls helps automatize driving and load pick up/drop down by raising the security with respect to ISO 13849 [81]. Therefore, automated guided vehicle (AGV) systems have come up again in the past years which are caused by such high sophisticated sensors. They get more flexible and are higher automated with and more and more without operator. For example, in 2016 Still GmbH presented an order picker which followed the operator autonomously [94].

The digitalization will influence the forklift truck use dramatically. For example, telematic systems in combination with fleet management systems will increase the operational availability of the trucks by predictive maintenance and remote diagnostics. The reliability of the goods to be transported will increase clearly by labeling them through a bar code or an RFID chip (and scanning while good pick-up) to ensure the exact position at any time of every good inside the warehouse to optimize the inventory control and registration.

A big trend regarding modern assembly concepts in the intralogistics to just in sequence generates very small batch sizes and transport units and will increase the operation of tugger trains more and more like in the automotive industry.

References

1. AGRIEVOLUTION, VDMA, "Tractor Market Report Calendar Year 2017," Agrievolution Economic Committee, Mechanical Engineering Industry Association in Germany, Agricultural Machinery, Frankfurt a.M., Germany, 2018, 5pp.
2. AID, *Landwirtschaftliche Fahrzeuge im Strassenverkehr (Agricultural Vehicles in Road Traffic)*, 23rd edn. (Bonn: AID Infodienst Ernährung, Landwirtschaft, Verbraucherschutz e.V., 2016).
3. ANSI/ASAE S390.6 (ISO 12934:2013) (DEC2016), "Tractors and Machinery for Agriculture and Forestry – Basic Types – Vocabulary." (Copyright ASABE. All excerpts from the text of this standard have been used with permission of the ASABE.)
4. Wiley, J.C. and Turner, R.J., "Power Hop Instability of Tractors, ASABE Distinguished Lecture Series No. 32," *Agricultural Equipment Technology Conference*, Louisville, KY, February 10-13, 2008, ASABE Publication No. 913C0108.
5. Auernhammer, H., "Einsatzdaten größerer Ackerschlepper – Ergebnisse einer Erhebung (Operational Data of Larger Agricultural Tractors - Results of a Survey)," *Landtechnik* 38, no. 11 (1983): 458-463.
6. Bosch Rexroth, "Bosch Rexroth EHR5 Product Documentation RE 66125," July 2013.
7. Bosch Rexroth, "Bosch Rexroth HER Product Poster lt761805e_1999-03," 1999.
8. Bryan, F.R., *Rouge: Pictured in Its Prime* (Detroit, MI: Wayne State University Press, 2003b), ISBN:0-9727843-0-6.

9. Deutscher Bauernverband DBV, "Situationsbericht (Situation Report) 2014/15," Trends und Fakten zur Landwirtschaft (Trends and Facts of Agriculture), Berlin, S. 15f.

10. Directive 2003/37/EC of the European Parliament and of the Council of 26 May 2003 on Type-Approval of Agricultural or Forestry Tractors, Their Trailers and Interchangeable Towed Machinery, Together with Their Systems, Components and Separate Technical Units and Repealing Directive 74/150/EEC (Text with EEA Relevance).

11. DIN 11026:1989-04, "Agricultural Tractors and Machinery; Trailers Coupling Ring 40 with Reinforced Shaft with Socket; Dimensions," dated April 1989.

12. DIN 74054-1:1989-01, "Mechanical Connections between Towing Vehicles and Trailers; 40 mm Drawbar-Eye with Sleeve," dated January 1989.

13. DLG Merkblatt 387, *Anhängevorrichtungen an Traktoren (Trailer Hitches for Tractors)*, 2nd edn. (2013). Translated from German into English. All information without guarantee and liability.

14. Eckstein, C., "Ermittlung repräsentativer Lastkollektive zur Betriebsfestigkeit von Ackerschleppern (Determination of Representative Load Collectives for the Fatigue Calculation of Agricultural Tractors)," Dissertation, Technical University Kaiserslautern, Maschinenelemente und Getriebetechnik Report vol. 26, 2017, ISBN:978-3-95974-064-7.

15. Commission Implementing Regulation (EU) 2015/504 of 11 March 2015, L85/1, ELI, http://data.europa.eu/eli/reg_impl/2015/504/oj, last visit June 19, 2019.

16. The Food and Agriculture Organization of the United Nations FAO, "What Do We Really Know about the Number and Distribution of Farms and Family Farms in the World?," Background Paper for The State of Food and Agriculture, 2014, http://www.fao.org/docrep/019/i3729e/i3729e.pdf.

17. Ferguson, H., "Apparatus for coupling agricultural implements to tractors and automatically regulating the depth of work," British Patent No. GB253566A, 1926.

18. Häfner, M., *LANZ Bulldog Erfolgsgeschichte eines Klassikers von 1921 bis 1945 (LANZ Bulldog Success Story of a Classic Machine from 1921 to 1945)* (Bath, UK: Paragon Books, 2011), 14, ISBN:978-1-4454-4121-4.

19. Manly, H.P., *The Ford Motor Car and Truck; Fordson Tractor: Their Construction, Care and Operation* (Chicago, IL: Frederick J. Drake & Co, 1919), 261.

20. Heiser, J. and Kobald, W., "Einrichtung zur Hubwerksregelung (Device for controlling the hitch)," Robert Bosch GmbH, German Patent No. DE0000002731164A1, 1979.

21. Holm, I.C., "Multi-Pass Behaviour of Pneumatic Tyres," *Journal of Terramechanics* 6, no. 3 (1969): 47–71.

22. ISO 1724:2003, "Road Vehicles - Connectors for the Electrical Connection of Towing and Towed Vehicles - 7-Pole Connector Type 12 N (Normal) for Vehicles with 12 V Nominal Supply Voltage," dated November 2003.

23. ISO 6489-1:2001-01, "Agricultural Vehicles - Mechanical Connections between Towed and Towing Vehicles - Part 3: Dimensions of Hitch-Hooks," dated January 2001.

24. ISO 6489-3:2004-06, "Agricultural Vehicles - Mechanical Connections between Towed and Towing Vehicles - Part 3: Tractor Drawbar," dated June 2004.

25. ISO 6489-4:2004-02, "Agricultural Vehicles - Mechanical Connections between Towed and Towing Vehicles - Part 4: Dimensions of Piton-Type Coupling," dated February 2004.

26. ISO 17567:2005, "Agricultural and Forestry Tractors and Implements - Hydraulic Power Beyond," dated May 2005.

27. ISO 730:2009-01, "Agricultural Wheeled Tractors - Rear-Mounted Three-Point Linkage - Categories 1N, 1, 2N, 2, 3N, 3, 4N and 4," dated January 2009.

28. ISO 5675:2008, "Agricultural Tractors and Machinery - General Purpose Quick-Action Hydraulic Couplers," dated August 2008.

29. ISO 500-1:2014-04, "Agricultural Tractors - Rear-Mounted Power Take-Off Types 1, 2, 3 and 4 - Part 1: General Specifications, Safety Requirements, Dimensions for Master Shield and Clearance Zone," dated April 2014.

30. ISO 500-2:2004-09, "Agricultural Tractors - Rear-Mounted Power Take-Off Types 1, 2, 3 and 4 - Part 2: Narrow-Track Tractors, Dimensions for Master Shield and Clearance Zone, dated September 2004.

31. ISO 500-3:2014-04, "Agricultural Tractors - Rear-Mounted Power Take-Off Types 1, 2, 3 and 4 - Part 3: Main PTO Dimensions and Spline Dimensions, Location of PTO," dated April 2014.

32. ISO 730:2014 (2009/Amd1: 2014), "Agricultural Wheeled Tractors - Rear-Mounted Three-Point Linkage - Categories 1N, 1, 2N, 2, 3N, 3, 4N and 4.

33. ISO 11001-1:2016-02, "Agricultural Wheeled Tractors - Three-Point Hitch Couplers - Part 1: U-Frame Coupler," dated February 2016.

34. ISO 11783:2017, "Tractors and Machinery for Agriculture and Forestry - Serial Control and Communications Data Network Part 1-14," dated December 2017.

35. ISO 789-2:2018-022018, "Agricultural Tractors - Test Procedures - Part 2: Rear Three-Point Linkage Lifting Capacity," dated February 2018.

36. ISO 10448 (1994), "Agricultural Tractors - Hydraulic Pressure for Implements," dated July 1994.

37. ISO 11471:1997, "Agricultural Tractors and Machinery - Coding of Remote Hydraulic Power Services and Controls," dated May 1997.

38. Kraftfahrtbundesamt, Fahrzeugzulassungen (FZ) Bestand an Kraftfahrzeugen und Kraftfahrzeuganhängern nach Fahrzeugalter 1. Januar 2019 (Vehicle Registrations Quantities of Motor Vehicles and Their Trailers by Age of Vehicle January 1st 2019).

39. Kremmer, M., "Kotfügel für die gelenkten Räder eines Fahrzeuges (Fender for the steered wheels of a vehicle)," Xaver Fendt & Co. 98562 Marktoberdorf, German Patent No. DE3607000A1, 1987.

40. Mayhew, R.D., "Agricultural Tractor/Implement Drivelines," *Winter Meeting of the American Society of Agricultural Engineers*, Atlanta, GA, December 14, 1994, ASAE Distinguished Lecture No. 19, 1-28, 913C1994, https://elibrary.asabe.org/abstract.asp?aid=20438, last visit June 19, 2019.

41. *Meyers Konversations-Lexikon (Meyers Dictionary of Conversations)*, 5th edn. (Leipzig/Wien: Verlag des Bibliographischen Instituts, 1893-1897.

42. Regulation (EU) No. 167/2013 of the European Parliament and of the Council of 5 February 2013 on the Approval and Market Surveillance of Agricultural and Forestry Vehicles (Mother Regulation).

43. Olfe, G. and Schön, H., "Einsatzzeiten von Schleppern bei unterschiedlichen betrieblichen Verhältnissen (Operating Times of Tractors under Different Operating Conditions)," *Grundlagen der Landtechnik* 34, no. 6 (1984): 236-243.

44. Pitts, J. and Pitts, H., "Machine for Threshing and Cleaning Grain," U.S. Patent 542, 1837, https://www.uspto.gov/patents-application-process/search-patents.

45. Renius, K.Th., "Global Tractor Development: Product Families and Technology Levels," *Actual Tasks on Agric. Engineering*, Opatiya, March 12-15, 2002, 87-95.

46. Renius, K.Th., *Traktoren - Technik und Ihre Anwendung (Tractors - Technology and Their Application)*, 2. Durchgesehene Auflage (Frankfurt/Main, BLV Buchverlag, 1987), ISBN:3-405-13146-4.

47. Rothmund, M. and Wodok, M., "ISOBUS - eine systematische Betrachtung der Norm ISO 11783 (ISOBUS - A Systematic Consideration of ISO 11783 Standard)," GIL Jahrestagung, University of Hohenheim, 2010, 163-166.

48. SAE J560, "Primary and Auxiliary Seven Conductor Electrical Connector for Truck-Trailer Jumper Cable," 2016.

49. Sanders, R.W., *Vintage Farm Tractors*, 4th Printing edn. (Voyageur Press, 1996), ISBN:08965820804.

50. Schumacher, A. and Harms, H.-H., "Potenzial von Traktormanagementsystemen mitleistungsverzweigten Getrieben (Potential of Tractor Management Systems with Power-Split Transmissions)," *"Hybridantriebe für mobile Arbeitsmaschinen" (Hybrid Drives for Mobile Working Machines)* (Karlsruhe: WVMA e.V. Wissenschaftlicher Verein für Mobile Arbeitsmaschinen), 2007, 17-29.

51. StVZO, "Straßenverkehrs-Zulassungs-Ordnung (StVZO) vom 26. April 2012 (BGBl. I S. 679), die zuletzt durch Artikel 1 der Verordnung vom 20. Oktober 2017 (BGBl. I S. 3723) geändert worden ist (StVZO dated April 26, 2012 (BGBl. I p. 679), Which Was Last Amended by Article 1 of the Ordinance of October 20, 2017 (BGBl. I p. 3723))," 2012.

52. StVO, "Straßenverkehrs-Ordnung (StVO) vom 6. März 2013 (BGBI. I S. 367), zuletzt geändert durch Artikel 2 der Verordnung vom 15. September 2015 (BGBL. I S. 1573) (StVZO dated March 6, 2013 (BGBl. I p. 367), which was last amended by Article 2 of the Ordinance of September 15, 2015 (BGBl. I p. 1573))", 2013.

53. http://www.tractordata.com/farm-tractors/000/2/1/217-fordson-fordson-f-engine.html.

54. VDMA, "Tractor Report 2013," Mechanical Engineering Industry Association in Germany, Agricultural Machinery, Frankfurt a.M., Germany.

55. VDMA, "Tractor Report 2018," Mechanical Engineering Industry Association in Germany, Agricultural Machinery, Frankfurt a.M., Germany, 37pp.

56. Von Holst, Ch., "Vergleich von Reifenmodellen zur Simulation der Fahrdynamik von Traktoren (Comparison of Tire Models for Simulation of the Driving Dynamics of Tractors)," Dissertation, TU Berlin; Forschungsbericht VDI Reihe 14 Nr 102, Duesseldorf, VDI Verlag, 2001.

57. The World Bank, "World Development Indicators," Producer and Distributor, Washington, DC, 2018, https://data.worldbank.org/indicator/.

58. DIN EN ISO 6165:2013-02, "Earth-Moving Machinery - Basic Types - Identification and Terms and Definitions [ISO 6165:2012]," dated February 2013.

59. Jähne, H., Deiters, H., Kohmäscher, T., and Bliesener, M., "Antriebsstrangkonzepte mobiler Arbeitsmaschinen (Powertrain Concepts for Mobile Working Machines)," *6th Conference Construction Machinery Technology (Fachtagung Baumaschinentechnik)*, Dresden, October 5-6, 2006.

60. Filla, R. and Frank, B., "Towards Finding the Optimal Bucket Filling Strategy through Simulation," *15th Scandinavian International Conference on Fluid Power, SICFP'17*, Linköping, Sweden, June 7-9, 2017.

61. Kunze, G., Großmann, K., Hentschel, T., Möller, R. et al., "Research Projekt 'Hexapod-Mobima,'" BMBF-Funding Code 03FO3182, Final Report, Technical University Dresden, June 2013.

62. Directive 2002/44/EG of the European Parliament and of the Council of 25 June 2002 on the Minimum Health and Safety Requirements Regarding the Exposure of Workers to the Risks Arising from Physical Agents (Vibration), L 177/13, dated July 6, 2002.

63. American Society of Mechanical Engineers (ASME), "B56.1: Safety Standard for Powered Industrial Trucks," 1969.

64. Bruns, R., "Technische Logistik 1/Materialflusstechnik (Technical Logistic 1/Material Flow Technology)," Helmut Schmidt University, Lecture, 2017.

65. Bruns, R. and Höppner, O., "Dynamische Standsicherheit von Flurförderzeugen (Dynamic Stability of Industrial Trucks)," Helmut Schmidt University Hamburg, University Research, 2006.

66. CLARK, "Company History," https://www.clarkmhc.com/company/history, last view June 20, 2019.

67. CONTAINERBASIS, "Geschichte der Container (History of the Container)," June 12, 2014, https://www.containerbasis.de/blog/allgemein/geschichte-der-container/, last visit June 20, 2019.

68. Danfoss, "Technische Information – Lenkung – Allgemeines, Lenkungskomponenten (Technical Information – Steering – General, Steering Components)," Danfoss Catalogue: BC00000096de-DE0401, September 2014.

69. DIN 43531:2012-06, "Lead-Acid Batteries – Traction Batteries 48 V with Cells of Dimensions Series L in Accordance with DIN EN 60254-2 for Industrial Trucks – Dimensions, Weight, Design," dated June 2012.

70. DIN 43535:2012-06, "Lead-Acid Batteries – Traction Batteries 24 V with Cells of Dimensions Series L in Accordance with DIN EN 60254-2 for Industrial Trucks – Dimensions, Weight, Design," dated June 2012.

71. DIN 43536:2012-06, "Lead-Acid Batteries – Traction Batteries 80 V with Cells of Dimensions Series L in Accordance with DIN EN 60254-2 for Industrial Trucks – Dimensions, Weight, Design," dated June 2012.

72. EN 13626:2003-08, "Packaging – Box Pallets – General Requirements and Test Methods," dated August 2003.

73. DIN EN 13698-1:2004-01, "Pallet Production Specification – Part 1: Construction Specification for 800 mm × 1200 mm Flat Wooden Pallets," dated January 2004.

74. EN 14253: 2008-02, "Mechanical Vibration – Measurement and Calculation of Occupational Exposure to Whole-Body Vibration with Reference to Health – Practical Guidance," dated February 2008.

75. EN16203:2014-11, "Safety of Industrial Trucks – Dynamic Tests for Verification of Lateral Stability – Counterbalanced Trucks," dated November 2014.

76. Commission Implementing Regulation (EU) 2016/1628 of 14th September 2016, OJL 252, http://data.europa.eu/eli/reg/2016/1628/oj, last visit June 29, 2019.

77. Frerichs, L. and Neuf, O., "Drive Train Concepts for Forklift Trucks – A Spectrum of Innovative Developments," *VDI-Congress "Getriebe in Fahrzeugen"*, Friedrichshafen, 2011.

78. Frerichs, L., "Mobile Arbeitsmaschinen und Nutzfahrzeuge (Mobile Working Machines and Commercial Vehicles)," *5th Conference "Techniken der Intralogistik" (Techniques of Intralogistics)*, Technical University Braunschweig, Lecture, 2017.

79. DIN EN 60254-2:1989-04, "Lead-Acid Batteries; Lead-Acid Traction Batteries; Dimensions of Cells and Terminals and Marking of Polarity on Cells," dated April 1989.

80. ISO 13564:1:2012-09, "Powered Industrial Trucks – Test Methods for Verification of Visibility, Part 1: Sit-On and Stand-On Operator Trucks and Variable-Reach Trucks up to and Including 10 t Capacity," dated September 2012.

81. ISO 13849-1:2016-06, "Safety of Machinery – Safety-Related Parts of Control Systems – Part 1: General Principles for Design," dated June 2015.

82. ISO 22915-1:2017-02, "Industrial Trucks - Verification of Stability - Part 1: General," dated February 2017.

83. ISO 2330:2002-05, "Fork-Lift Trucks - Fork Arms - Technical Characteristics and Testing," dated May 2002.

84. ISO 2631-1:1997-05, "Mechanical Vibration and Shock - Evaluation of Human Exposure to Whole-Body Vibration - Part 1: General Requirements," dated May 1997.

85. ISO 3691-1:2017-05, "Industrial Trucks - Safety Requirements and Verification - Part 1: Self-Propelled Industrial Trucks, Other Than Driverless Trucks, Variable-Reach Trucks and Burden-Carrier Trucks," dated May 2017.

86. ISO 6055:2013-10, "Industrial Trucks - Overhead Guards - Specification and Testing," dated October 2013.

87. ISO 668:2013-08, "Series 1 Freight Containers - Classification, Dimensions and Ratings," dated August 2013.

88. Linde, "Blue Light for Greater Safety [LINDE BLUESPOT®]," https://www.linde-mh.com/en/About-us/Innovations-from-Linde/Linde-BlueSpot.html, last visit June 29, 2019.

89. Linde, "Safe Handling at the Highest Level [DYNAMIC MAST CONTROL®]," https://www.linde-mh.com/en/About-us/Innovations-from-Linde/Dynamic-Mast-Control.html, last visit June 29, 2019.

90. Council Directive of 14 June 1989 on the Approximation of the Laws of the Member States Relating to Machinery (89/392/EEC), *Official Journal of the European Communities*, No. L 183/9, dated June 29, 1989, ELI: http://data.europa.eu/eli/dir/1989/392/oj, last visit June 29, 2019.

91. Ogawa, I., Futahashi, K., Teshima, T., and Akahane, F., "Development of the World's First Engine/Battery Hybrid Forklift Truck," *Mitsubishi Heavy Industries Technical Review* 47, no. 1 (March 2010), https://www.mhi.co.jp/technology/review/pdf/e471/e471046.pdf, last visit June 20, 2019 46–50.

92. United States Department of Labor, "Occupational Safety & Health Administration: Powered Industrial Trucks (Forklift) - Forklift Classifications," https://www.osha.gov/SLTC/etools/pit/forklift/types/classes.html, last visit June 29, 2019.

93. Still, "Elektrostapler (Electric forklift truck) RX 60 2.5 - 3.5 t," https://www.still.de/fahrzeuge/gabelstapler-und-lagertechnik/elektro-stapler/rx-60-25-35-t.html, last visit October 7, 2018.

94. Still, "Order Picker iGo Neo CX 20 - The Smart Way," https://www.still.de/en-DE/trucks/new-trucks/order-pickers/igo-neo-cx-20.html, last visit October 7, 2018.

95. Bavendiek, R., "Erfolgreiche Produktklinik an einem Gabelstapler (Successful Product Clinic on a Forklift Truck)," *Speech at TCW-Conference "Produktklinik & Produktordnungssysteme"*, Munich, Germany, September 4, 2008.

96. BAUA, Germany, "Technical Rules for Hazardous Substances - TRGS 554: 'Exhaust Emission of Diesel Engines,'" 2009.

97. Scheffler, M., Feyrer, K., and Matthias, K., Fördermaschinen - Hebezeuge, Aufzüge, Flurförderzeuge (Hoisting machines - hoists, elevators, industrial trucks) (Braunschweig/Wiesbaden: Friedr. Vieweg & Sohn Verlagsgesellschaft mbH, 1998), ISBN:3-528-06626-1.

98. VDI 2198:2018-02, "Type Sheets for Industrial Trucks," dated February 2018.

99. Renius, K.Th., *Fundamentals of Tractor Design* (Cham: Springer Nature Switzerland, 2019).

CHAPTER 7

Summary

Prof. Dr.-Ing. Marcus Geimer

The variety of tasks mobile working machines have to do is very wide and therefore many different machines and variants are developed. The work carried out should be as productive, efficient, and of high quality as possible. Examples of these machines can be found in the fields of agriculture, forestry, construction, logistics, municipal sector, and in other special applications. Although the tasks are very different, many of the technologies used there are common to these machines. The first important topic addressed in this book is how a mobile working machine is defined. Because there is no standardized definition of the machines, a mobile working machine is defined here as a machine that is mobile and characterized through a working task. This is also the reason why the task of the machine and the process the machine is implemented in, defines the construction of the machine.

Typical designs of the machines are described in Chapter 2 including their axles and their steering systems. Frame concepts of mobile working machines are block, frame, and self-supporting constructions. These concepts are compared with each other and it is shown how the locomotion is realized. Therefore, different steering systems, like Ackermann, articulated and skid steering, as well as the used types of axles are shown. In contrast to cars there is no need and often no possibility for a direct mechanical connections between steering wheel and tires.

The soil-tire contact enables the machine to transmit forces of the work functions on the ground. Especially in the field of agriculture, many research activities can be found in this field. Traction forces and traction efficiency are highly relevant for the quality of the task as well as for the efficiency and productivity of the machine. The origin of the forces as well as their calculation possibilities are explained.

The energy flow and the forces are guided through the structure of the machine. We see similar drive structures in mobile working machines and Chapter 3 is dealing with these drives. The power train has to feed the traction drive, the function drives, and the auxiliaries of the machines. The power demand of these drives can vary in a wide range, which is shown by the following examples: tractors are designed to provide the traction drive and the PTO with the maximum power of the combustion engine and excavators can transfer the whole power to the function drives.

Independently of its specific energy demand, today's commonly used primary energy provision for the machines is a combustion engine with a fuel tank as the energy storage. But battery electric vehicles are used as well for decades, e.g., in forklift trucks. This makes the machines of great interest, because mechanic, hydraulic, and electric energy conversions are already state of the art. Battery-driven machines with small energy demand, for indoor use or noise-sensitive applications, are coming up in all sectors of industry. Fuel cell energy provision may follow.

Commonly used technologies for the traction drive in combustion-driven mobile working machines are purely hydrostatic, mechanic, or hydro-mechanic power-split transmissions. Today, a clear trend from hydrostatic or hydro-mechanic drives to hydraulic-mechanic power-split drives can be seen. Electro-mechanic power-split drives are new on the market and offer the possibility of an electric PTO. Tractors are using the hydro-mechanic power-split transmissions since decades, first wheel loaders with these drives can be seen in the market. In comparison to hydrostatic and hydro-mechanic drives, the power-split versions have a high efficiency and a variable transmission ratio.

As already mentioned above, electric-driven machines are used since decades. Research activities to electrify mobile working machines can be also found often. In many cases, electric motors only replace hydrostatic rotary drives. However, this may not be the preferable way to find the best solution. An entirely new construction taking advantage of the specific benefits of each drive technology would create the best possible solution.

The function drives of mobile working machines are today mostly hydraulically. Different hydraulic systems are presented and compared in the book. The components for such systems and different function drives are as well presented. Beginning from simple constant flow to complex load-sensing systems, the different behavior of the systems is shown.

Because of its great importance the possible energy flows through a mobile working machine, beginning at the prime mover and ending on its drives with the contact to the surrounding, are presented. All possible transmissions, mechanic, hydraulic, electric and a combination of them, are described in detail.

It can be seen in today's development and solutions of mobile working machines that the significance of electronic controls is continuously increasing. Only few machines in the low-cost segment are known which are not electronically controlled. Since the newest emission regulations, such as Tier 4 final or Stage V, combustion engines need an electronic control and this gives the chance also to electrify the machines. The control architectures in the machines are as well discussed in Chapter 4 as the CAN-BUS system, which is commonly used in the machines. Today's research activities are focusing on machine-learning technologies and adaptive control structures. The aims of the control focus on autonomous systems, on the support of the operator, and on predictive maintenance.

Considering today's established technologies, where will this lead to in the future? Chapter 5 gives some answers to this question and shows promising technologies that have been successfully tested and validated in research. The replacement of the hydraulic LS-signal to an electronically pump control is under research and can be already found in some applications today. Displacement controlled systems are discussed as the systems with the highest overall efficiency. Another possibility to reduce systemic losses is the implementation of hybrid systems. An accumulator can store and therefore recover energy of the system. These systems are also of great importance for the control: The new freedoms on hybrid systems need an optimized control to realize the expected goals.

Taking these technology trends into account a picture of the future can be painted as it is seen today. From the references of this chapter the most important conferences for mobile working machines can be picked up. Following these conferences, surely future technology trends in the mobile working machines can be seen.

The book concludes at its end with the Chapter 6 that shows typical examples of mobile working machines. The selected examples are:

- Tractors as a popular example in the field of agriculture and as a typical example of an universal machine. Many implements can be attached to these machines.
- Wheel loaders as an example in the field of construction. In contrast to tractors this machine is developed for a special task. Although there is the possibility to attach implements, the machine is most often used for carrying goods over short and medium distances.
- Forklift trucks are typical machines for carrying goods. For indoor applications battery electric machines are preferred. There, a special need for the cassis is based on the possibility to change the battery.

These examples are chosen to show exemplary the wide range of mobile working machines and to give examples for the combined use of technologies.

In summary, this book gives a wide overview about the used technologies in mobile working machines and shows a lot of references where more detailed information can be found. In the references, also the most relevant conferences and journals are presented where research and development of new technologies for the machines are shown. Future developments can be followed there and therefore the book gives a good introduction into this field as a reference work. This book addresses as well new engineers to get involved in this topic as engineers who want to widen their knowledge.

about the authors

Dr.-Ing. Rainer Bavendiek (born 1957) studied mechanical engineering/vehicle technology in Braunschweig. He worked as a research assistant at the Helmut Schmidt University of Hamburg, where he wrote his Ph.D. thesis on the efficiency of hydrostatic axial piston pumps.

From 1987, he worked for the STILL GmbH (forklift trucks) in Hamburg, Germany, in several managing functions such as testing, advanced engineering, senior project manager, and head of development for electrically powered forklift trucks, and from 2008 as a module manager development for the KION Group.

In 2010, he became director Product Development of the Hako Group (municipal und cleaning technology) in Bad Oldesloe, Germany.

Since 2018, he is working as a freelancer in the field of mobile working machines/ vehicle technology and lectures at the Nordakademie Hamburg/Elmshorn in Germany.

Swen Bosch was born in 1985 in Heidenheim/Brenz, Germany. He received his B. Eng. degree in Electronics and the M.Sc. degree in Computer Controlled Systems from Aalen University of Applied Sciences, Aalen, in 2011 and 2013, respectively.

Since 2011, he has been a research assistant at the Laboratory of Power Electronics and Electrical Drives at Aalen University of Applied Sciences. Currently, he is working toward the Ph.D. degree in cooperation with the Technical University of Dresden, Germany. His main research interests include control of electrical drives as well as grid feed and compensation of reactive power and distortion power.

Prof. Dr. Naseem Daher (born 1982) received the BSME and MSME degrees from Lawrence Technological University in Michigan, USA, in 2006 and 2008, respectively. Naseem worked at TRW Automotive in Michigan, USA, between 2005 and 2011, where he was involved in the research and development of electrohydraulic braking control systems. Naseem earned his Ph.D. degree from Purdue University in Indiana, USA, with a specialty in dynamic systems, measurement, and control in 2014. He is an assistant professor at the American University of Beirut in the Electrical and Computer Engineering Department. Naseem's research interests include vehicle dynamics and control, fluid power systems, automotive active safety systems, and robotics.

Prof. Danilo Engelmann was born in Karlsruhe, Germany, in 1987. He received the Dipl.-Ing. degree in mechanical engineering from the Karlsruhe Institute of Technology in 2013. From 2013 to 2014, he worked at the Forschungs- und Entwicklungsgesellschaft für Motorentechnik, Optik und Thermodynamik mbH (MOT), as a project engineer for optical measurement in engines. From 2014 to 2018, he was a research assistant and the chair of mobile machines of the Karlsruhe Institute of Technology researching on power-trains and alternative fuels for mobile machinery. Since 2018, he is a professor at the

Bern University of Applied Sciences, in Biel. He is the head of the federally recognized laboratory for IC engines, emissions, and powertrains. He gives lectures on thermodynamic applications and powertrains.

Prof. Dr. Ludger Frerichs (born 1959) is the director of the Institute and University Professor for Mobile Machines and Commercial Vehicles (IMN) at the Technische Universität Braunschweig in Germany since 2012. After studying mechanical engineering in Osnabrück and Braunschweig, he worked as a research assistant at the Institute of Agricultural Engineering of the University of Stuttgart-Hohenheim, where he wrote his dissertation on the fundamentals of electronic tractor plow management. From 1990, he worked for CLAAS in Harsewinkel, Germany, most recently as the head of Advanced Engineering. In 2008, he became the director Product Development of STILL in Hamburg. In addition, Prof. Frerichs was engaged in a variety of professional and voluntary work. In 2000, he received the teaching assignment for the lecture "Agricultural Engineering" at the RWTH Aachen, and from 2004 to 2008 he was the chair of the Max-Eyth-Society of Agricultural Engineering in the VDI, the German Association of Engineers. While at STILL, he served as the chairman of the Scientific Advisory Council of the Intralogistics Research Association IFL. Since 2005, he is the Honorary Professor of St. István University Gödöllö in Hungary. For his commitment, he was awarded by the European Association of Agricultural Engineers EurAgEng and the World Association CIGR.

Dr.-Ing. Gerhard Ruben Geerling, born in Hennef/Sieg, Germany, in 1964, is the director/head of the Innovation, Project und Product Data Management Office Mobile Electronics and Systems. His business unit Mobile Hydraulics, situated in Neu-Ulm/Elchingen in southern Germany, is part of the Bosch Rexroth AG, the Drive & Control Company. He is a mechanical and aeronautical engineer graduated for a Master degree (Dipl.-Ing.) at the Rheinisch Westphaelisch Technical University of Aachen (RWTH Aachen) in Germany and he did his Ph.D. (Dr.-Ing.) at the Technical University of Hamburg (TUHH) in aircraft systems design in collaboration with the Technology Transfer Centre in Hamburg-Finkenwerder (THF) of Airbus Industries in Germany and France. His further professional career at industry started with Brueinghaus Hydromatik GmbH, at this time part of the holding Mannesmann Rexroth AG, and nowadays integrated in the business unit Mobile Hydraulics of the Bosch Rexroth AG. Since 1998, he works for Mannesmann and later Bosch Rexroth in various positions at engineering, innovation, product and project management, as well as marketing, technical sales, and project procurement. Gerhard Geerling is an external lecturer at the Karlsruhe Institute of Technology KIT, Karlsruhe, Germany, for planning, projection, and design of hydraulic drive systems.

Detlef Hawlitzek, born in 1959 is a Certified Electrical Engineering Technician. He has been working for ifm electronic gmbh in Essen since 1990. He started as a Project Engineer for sales, programming, and commissioning of programmable logic controllers and customer-specific control solutions. For the past 15 years, he has been responsible for global Key Account Management in the field of control systems for mobile machines with focus on sales and marketing.

Edwin Heemskerk, born in 1975 in Utrecht (Netherlands), is the head of the application support innovative concepts for mobile applications at the Bosch Rexroth AG in Germany. He studied at the Clausthal University of Technology in Germany and received his Dipl.-Ing. degree in mechanical engineering. Since 2000, he works for the Bosch Rexroth AG in the area of system engineering and innovation for hydraulics in mobile machinery.

Prof. Kalevi Huhtala (born 1957) is leading the fluid power research in Tampere University. He graduated from the Tampere University of Technology in 1996. The title of his doctor thesis is "Modelling of Hydrostatic Transmission – Steady-State, Linear and Non-Linear Models." His nomination of professorship took place in 2003. The research field of the professorship is hydraulics in mobile machines.

His main research interests are autonomous mobile machines. He has supervised more than 10 doctoral theses and more than 100 master theses. He has several scientific publications. He has been a member of scientific program committees of conferences in Aachen, Dresden, Linz, Hangzhou, Linköping, and Tampere.

Roman Ivantysyn was born in Ilava, Slovakia, in 1986. He grew up near Hamburg and migrated to the United States in 2004. He received his Bachelor and Master's degree in mechanical engineering at Purdue University, USA. Currently, he is working on his doctoral degree at the Technical University Dresden at the Institute for Mechatronical Engineering (IMD former IFD).

From 2008 to 2011, he was a research assistant at the MAHA Fluid Power Laboratory at Purdue University; here, he worked on virtual prototyping and optimization of axial piston pumps, which was also his Master's thesis topic. Afterward, he went back to Germany to work at the TU Dresden to work as a research assistant. At first, his focus was on energy-efficient mining systems with emphasis on displacement-controlled actuation. Now his research focus lies on the virtual development of inexpensive, efficient, robust, and fast actuating axial piston pumps, which will also be his Ph.D. topic.

Dr.-Ing. Torsten Kohmäscher was born in Osnabrück, Germany, in 1977. He received the Dipl.-Ing. degree in mechanical engineering and the Dr.-Ing. degree with honors from RWTH Aachen University in 2003 and 2008, respectively.

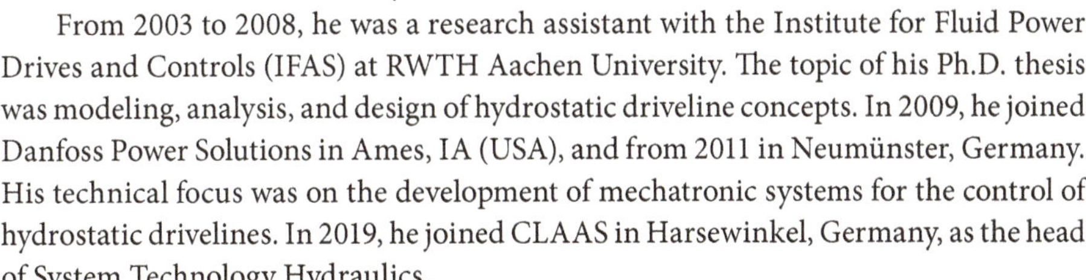

From 2003 to 2008, he was a research assistant with the Institute for Fluid Power Drives and Controls (IFAS) at RWTH Aachen University. The topic of his Ph.D. thesis was modeling, analysis, and design of hydrostatic driveline concepts. In 2009, he joined Danfoss Power Solutions in Ames, IA (USA), and from 2011 in Neumünster, Germany. His technical focus was on the development of mechatronic systems for the control of hydrostatic drivelines. In 2019, he joined CLAAS in Harsewinkel, Germany, as the head of System Technology Hydraulics.

Dr. Martin Kremmer, born 1972, is an agricultural engineer, working in industry for Deere & Company, one of the major manufacturers of tractors and agricultural machinery since 2002 in different functions in Tractor Design Analyses, Vehicle Concept Design, Implement Integration Design, Automation and Integrated Solutions Technology Strategies. Dr. Kremmer is a John Deere Fellow and the Director European Technology Innovation Center (ETIC) in Kaiserslautern, Germany. He is a member of the Executive and Board of the European Society of Agricultural Engineers (EurAgEng) and member of the Advisory Council of the Max Eyth Society of Agricultural Engineering of the Society of German Engineers (VDI-MEG). Dr. Kremmer is a named inventor of numerous patent applications and publications in the field of Agricultural Engineering. He has been teaching at the Karlsruhe Institute of Technology (KIT) since 2006 where he was promoted to an Honorary Professor in February of 2019.

Markus de la Motte, born in 1977 in Wismar (Germany), is the project leader for electrohydraulic steering systems at the Hydraulik Nord Fluidtechnik GmbH in Germany. He studied at the University of Rostock in Germany and received his Dipl.-Ing. degree in mechanical engineering. Since 2013, he works for the Hydraulik Nord Fluidtechnik GmbH in the area of system engineering and innovation for steering systems in mobile machinery.

Fabrizio Panizzolo, born in 1955 in Venice (Italy), completed in 1980 his Master's degree in Mechanical Engineering on Turbomachinery at Padova University in Italy. From 1982 to 1994, he has been involved in design and has lead Application's Engineering group on many Off-Highway drivetrain system projects for construction machinery as well as either diesel- or electrically driven material handling vehicles. From 1994 to 1997, he was in charge of Lead Research and Development Engineering, which mainly focused on axles and power-shift transmission design to cover the needs of small- and medium-sized industrial fork lift trucks. From 1997, he assumed responsibility for European Off-Highway Engineering at Dana leading both Engineering Centers, respectively, for axles and hydrostatically driven mechanical transmission and, respectively, for power-shift transmissions.

From 2010, he is leading the Research and Development Engineering teams at Dana Rexroth Transmission System, (DRTS), to produce advanced and innovative Off-Highway CVT (power-split) transmission systems. He has more than 20 granted EU and US patents on axles and transmissions.

Dr.-Ing. Herbert Pfab (born 1959) is managing director of Liebherr-Werk Bischofshofen in Austria. His company is part of Liebherr Earthmoving Technology and is responsible for the worldwide range of wheeled loaders. He is a mechanical engineer and did his Ph.D. at Technical University Munich in tractor hydraulics and graduated with a Master's degree in agricultural machinery at Silsoe College, UK. His professional career started with John Deere Werke Mannheim in tractor design. Since 1995, he works for Liebherr in various positions. Herbert Pfab is an external lecturer at the TU Munich for construction machinery.

Thomas Pippes was born in Pforzheim, Germany, in 1980. He received his Master's degree in mechanical engineering from the Technische Universität München in Germany. Since 1996, he is working in the hydraulic field in different functions with focus on predevelopment of hydraulic components, starting with developing internal gear pumps at Eckerle Hydraulic Division. He joined Hydraulik Nord Fluidtechnik GmbH, a leading company for hydrostatic steering systems, as the head of development in 2017 and is currently the CEO of Hydraulik Nord Technologies GmbH.

Dr.-Ing. Christian Pohlandt (born 1983) graduated from the Karlsruhe Institute of Technology, KIT. His research activities and Ph.D. thesis focused on machine learning and electric powertrain systems. In 2016, he joined the Daimler AG and is currently working as the Strategic Project Manager for the Serial Development of Battery Systems.

Prof. Dr.-Ing. Heinrich Steinhart was born in Freiburg/Breisgau, Germany, in 1961. He received the Dipl.-Ing. degree in electrical engineering and the Dr.-Ing. degree from the University of Karlsruhe in 1990 and 1996, respectively.

From 1990 to 1996, he was a research assistant with the Institute of Electrical Engineering, University of Karlsruhe. The topic of his Ph.D. thesis was a fast flicker compensator with a superconducting energy storage. In 1996, he joined the Daimler-Benz AG, where he was focused on the development of mechatronic systems. Since 1999, he is a professor at Aalen University of Applied Sciences, Aalen, where he is teaching power electronics and electrical drives.

Dr. Steinhart is a member of the Baden-Württemberg Center of Applied Research (BW-CAR) and the ZIM cooperation network Power Quality (NetPQ).-

mobile working machines

Marcus Geimer et al.

[PQ:]"So, what are the common characteristics of mobile working machines? It is their mobility and their work task!"

In *Mobile Working Machines*, Marcus Geimer et al. present a wide overview of the design and technology used in mobile working machines—machines that are mobile and characterized through a working task. Mobile working machines vary in complexity from highly complex process technologies to relatively simple lift function.

Mobile working machines are used in the fields of agriculture, forestry, construction, logistics, municipal sector, and other special applications. Although the tasks performed are very different, many of the technologies used are common to these machines. The machine's work function typically determines its name and its technique principle and construction. Mobile working machines include excavators, combine harvesters, forklift trucks, and waste-collecting vehicles. These mobile working machines all combine work functions, drives, and a built-in or external traction drive, a chassis, and an operator working place.

The machine and module-specific requirements of mobile working machines are considered in detail by the authors. The structure of the book follows the structure of mobile working machines, from typical machine design and construction, disciplines and technologies of traction and function drives, structure of controls and technologies used to innovative evolutionary and revolutionary machine concepts, promising technologies that have been successfully tested and validated in research, and extensive examples of mobile working machines.

Future worksites will be characterized by different levels of automation of mobile working machines: These machines can be controlled manually, remote operated, or operate autonomously. How these machines work individually and cooperatively in a fleet will be determined by future technological developments.

index